CHEMISTRY RESEARCH AND APPLICATIONS

A COMPREHENSIVE GUIDE TO CHEMILUMINESCENCE

CHEMISTRY RESEARCH AND APPLICATIONS

Additional books and e-books in this series can be found
on Nova's website under the Series tab.

CHEMISTRY RESEARCH AND APPLICATIONS

A COMPREHENSIVE GUIDE TO CHEMILUMINESCENCE

LUÍS PINTO DA SILVA

EDITOR

nova

science publishers

New York

NOTICE TO THE READER

Library of Congress Cataloging-in-Publication Data

ISBN: 978-1-53616-170-0

Published by Nova Science Publishers, Inc. † New York

CONTENTS

PREFACE

Chemiluminescence is a visibly impressive phenomenon that consists on the generation of light resulting from a chemical reaction. The general population usually associates it with the striking light-emission observed in nature from fireflies or by different organisms in seawater. For researchers, however, the main interest regarding different chemiluminescent systems is their broad analytical potential.

Chemiluminescent systems have been associated with high quantum yields, relative nontoxicity of luciferins and high signal-to-noise ratio. Arguably more important, these systems do not require photoexcitation, which diminishes the possibility of autofluorescence arising from the background signal. The absence of photoexcitation also eliminates problems associated with light penetration into biologic tissue (except in emission). These properties have made chemiluminescence a powerful tool that allows the real-time, highly sensitive, specific and noninvasive imaging and sensing of target molecules and processes, both *in vivo* and *in vitro*. It is no surprise that chemiluminescence is already widely used in the fields of medicine, pharmaceutical, molecular and cellular biology and analytical chemistry.

Both the remarkable beauty of this phenomenon and its numerous potential applications have stimulated in the last several decades significant effort in fundamental research toward elucidation of chemiluminescence. In a true multi-field approach, researchers have focused: on the identification and classification of chemi-/bioluminescent systems and organisms; identification, purification and synthesis of chemiluminescent substrates, products, intermediates and catalysts; unraveling the basic photophysics and photochemistry of the light-emitters; explaining the complex chemiexcitation pathways; among others. The importance of this work was recognized in 2008 by awarding the Nobel Prize in Chemistry to Osamu Shimomura, Roger Y. Tsien and Martin Chalfie for the discovery and development of the green fluorescent protein (GFP).

Despite this, trying to review the chemical/biochemical/biological features of chemiluminescence is a daunting task, both for specialized researchers and for students/researchers wanting to start on this field. This result from several factors. One of them is the sheer number of chemiluminescent systems known to date. Chemi-/bioluminescent systems have been found in thousands of species represented by more than 700 *genera*, being present on bacteria, insects, worms, dinoflagellates, squids, fishes and fungi (among others). Chemiluminescent systems not present on living organisms are also numerous, being continuously identified and designed new ones. Another factor is that research on chemiluminescence is ongoing for a long time, which means that some basic properties of chemiluminescent systems were described some decades ago, which can difficult finding needed information.

These considerations led to this book: an effort to produce a comprehensive guide to various aspects of chemiluminescence, both from a point of view from fundamental and applied research. My aim is to provide a reference manual focusing on different chemiluminescent reactions and organisms, by referencing and explaining both well-described and recent findings on mechanistic features and their different applications.

This was achieved by the production of different chapters, each authored by international specialists on different aspects of chemiluminescence. Chapters One, Two, Five and Seven review the fundamentals of well-known chemiluminescent systems, as are that of marine imidazopyrazinones (Chapter One), firefly bioluminescence (Chapter Two), acridinium esters (Chapter Five) and squid bioluminescence (Chapter Seven). Chapters Three and Four describe in detail the mechanisms and potential of lesser-known systems, such as fullerene and derivatives (Chapter Three) and lanthanides ions (Chapter Four). Chapters Six, Eight and Ten explain different approaches for the enhancement of chemiluminescent systems, such as the use of phenothiazine derivatives as enhancers of peroxidase-catalyzed chemiluminescence (Chapter Six), the use of metal-enhanced chemiluminescence (Chapter Eight), and the thermostabilization of luciferase (Chapter Ten). Chapter Nine explains how small model systems can help in the characterization of complex bioluminescent steps, and how they open the door for solid-state chemiluminescence in macroscopic molecular crystals. Chapters Five, Eleven and Twelve help us to understand how theoretical chemistry can revolutionize our understanding of chemiluminescent reactions. Finally, Chapter Thirteen describe several easy to perform experiments for chemiluminescent demonstrations for educational purposes. Several of these chapters refer and explain existent practical applications for these different chemiluminescent reactions.

I hope that this book will be a valuable tool regarding chemiluminescence for researchers and students, both already studying it or using it as a tool, and for those that wish to enter on this field of the first time.

Luís Pinto da Silva,
Porto, Portugal.

In: A Comprehensive Guide to Chemiluminescence ISBN: 978-1-53616-170-0
Editor: Luís Pinto da Silva © 2019 Nova Science Publishers, Inc.

Chapter 1

CHEMILUMINESCENT IMIDAZOPYRAZINONES: A STEP TO FUTURE DIAGNOSIS TOOLS

*Ara Núñez-Montenegro, Paulo J. O. Ferreira and Luís Pinto da Silva**

Chemistry Research Unit (CIQUP),
Faculty of Sciences of University of Porto, Porto, Portugal

ABSTRACT

This chapter aims to present in a clear manner the progress and evolution of the chemiluminescent studies in imidazopyrazinone-based compounds, with Coelenterazine (CLZ) derivatives as the most representative molecules. Besides that, it will cover historical developments and the fundamental principles of the chemiluminescence phenomena applied to these compounds. We will highlight the progress in applications based on the reported research.

CLZ is a substrate widely found in marine luminescent organisms and can emit light via chemi- and bioluminescent reactions. CLZ-dependent chemiluminescence has found attention in diverse fields of application since Teranishi and Shimomura proposed the mechanism of imidazopyrazinone chemiluminescence in detail, showing two light emitters, the amide anion and the phenolate anion, as the typical species involved. Among these fields, the development of (bio)analytical detection methods has become an important source for transferring technology into commercial kits. One of the most studied application is the use of CLZ analogues for the analysis of reactive oxygen species (ROS) as selective probes for monitoring superoxide. On the other hand, the current state of the art in chemiluminescent detection methods and in optical technologies provides a very promising potential use in photodynamic therapy (PDT) of cancer.

---

* Corresponding Author's Email: luis.silva@fc.up.pt.

1. INTRODUCTION

Over the years, scientists have been impressed by the ability of some organisms to produce light. This fascination has led to the willingness to isolate the molecules responsible for this phenomenon. The most famous light-producing molecule is Coelenterazine (CLZ), a substrate found in a vast number of marine luminescent organisms (including for example jellyfishes, ostracods, deep sea fishes, squids and shrimps) [1, 2]. Other well-known light-emitting compound is *Cypridina* luciferin, presented in ostracods [1, 2]. Common to these luciferins is the imidazopyrazinone scaffold; hence, molecules containing this core can be referred as imidazopyrazinone derivatives (Figure 1).

Since the 1980s, many groups have synthesized large amounts of these compounds, and evaluated their luminescent properties [3, 4]. CLZ as a model of imidazopyrazinone derivatives emits light by chemi- and bioluminescent reactions, by decomposing into Coelenteramide and CO_2. These reactions are characterized by the emission of radiation (light), being bioluminescence (BL) a type of chemiluminescence (CL) in which the chemical reaction is catalyzed by an enzyme in living organisms [5]. Most of the studies about CLZ analogues founded in the literature have been focused on the bioluminescent properties [6, 7, 8]. However, in our experience, is not easy to find comparative chemiluminescent studies and there is a lack of reviews exploring this topic. The fact that chemiluminescent reactions for CLZ have a lower efficiency or quantum yield (Φ) than bioluminescent reactions (Φ_{CL} are 0.02 while Φ_{BL} are 0.2) could be responsible for making this latter field more attractive. This difference is attributed to both the conformational stability of the emitter in the protein and the hydrophobic environment surrounding the Coelenteramide anion [9]. Given this, a comprehensive review of the findings of published chemiluminescent studies of imidazopyrazinone derivatives is intended here. It is also an attempt to summarize the existing practical applications that have been developed in the fields of bioanalysis, biotechnology and biomedicine [8, 10]. Several examples will be further explained in detail.

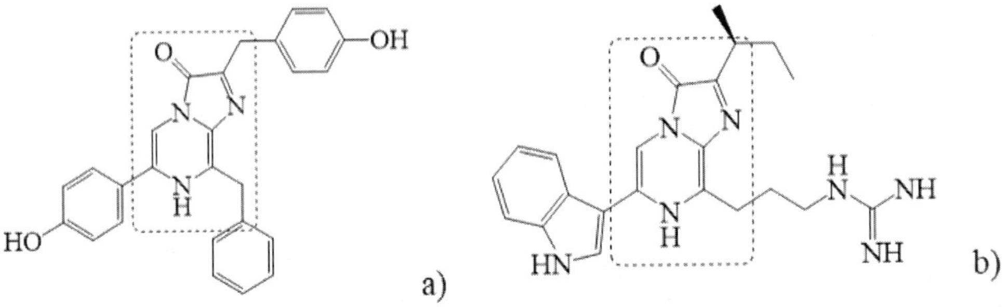

Figure 1. Imidazopyrazinone core in CLZ (a), and *Cypridina* Luciferin (b).

This chapter is structured in four sections. First, a brief description of the chemiluminescent mechanism is given for imidazopyrazinone-based compounds, following with a detailed timeline of the chemiluminescent studies, the most popular applications and finally we will present a conclusion with the prospects in this field.

2. THEORETICAL INSIGHTS OF CHEMILUMINESCENCE IN IMIDAZOPYRAZINONE COMPOUNDS

CL consists on the emission of light as a result of an optically-active excited state, formed in a chemical reaction, returning to the ground state. Understanding the role of the species involved in this process is crucial for the development of tools in real-time applications. Furthermore, the characterization of the chemiexcitation step responsible for light emission is essential for further applications in bioimaging and biomedicine.

The chemiluminescent reaction of imidazopyrazinones has been described as follows: the imidazopyrazinone scaffold reacts with molecular oxygen, which rapidly forms a peroxide that is converted into a four-membered ring, a dioxetanone intermediate. Upon decomposition of this latter species into chemiexcited oxyluciferin (called Coelenteramide in the specific case of CLZ), visible light is emitted due to radiative decay of the chemiluminophore to the ground state (Scheme 1).

However, despite decades of research, the CL mechanism is still not fully understood. Since Teranishi and Shimomura proposed two light emitters, the amide and the phenolate anions, as the typical species involved [11], some advances were reported based on theoretical calculations [12]. Different authors have tried to explain the phenomena by either the chemically induced electron-exchange luminescence mechanism (CIEEL) or the charge transfer-initiated luminescence mechanism (CTIL), but they have failed to explain some theoretical and experimental results [13-19]. More specifically, the most recent theoretical approach hypothesizes that efficient chemiexcitation results not from electron/charge transfer but is instead based on the degree of interaction between the keto and CO_2 moieties of the dioxetanone intermediate, which controls the access to a region of degeneracy between the ground and excited states [20]. So far, the experimental studies on the chemiluminescent properties of imidazopyrazinone-based compounds were based on a fluorescent spectroscopic approach, used to determine energy and kinetic chemiluminescent profiles, or by irradiating the spent solution after CL and measuring the spectrum of the radiation emitted.

Chemiluminescent reactions of these molecules have been extensively studied in aprotic solvents such as DMSO or diglyme (DGM), whether a trace catalytic amount of base is added (NaOH, KOH, acetate buffer, etc...) [21-24]. Conversely, in protic solvents (for instance, water or mixtures of MeOH/DMF, water/DMF), only a few studies are found because of the significant decrease of the CL efficiency; in fact, an increase in the percentage of water affects negatively the chemiluminescence quantum yield (Φ_{CL}) [25, 26]. Interesting, it was found that water by itself does not affect negatively the Φ_{CL}, being the inhibition attributed to the lifetime of the O_2^- radical.

Recently, combined experimental and theoretical approaches carried out by our group showed that imidazopyrazinone compounds CL have a pH-dependent behavior in DMSO (a model solution) [18, 19].

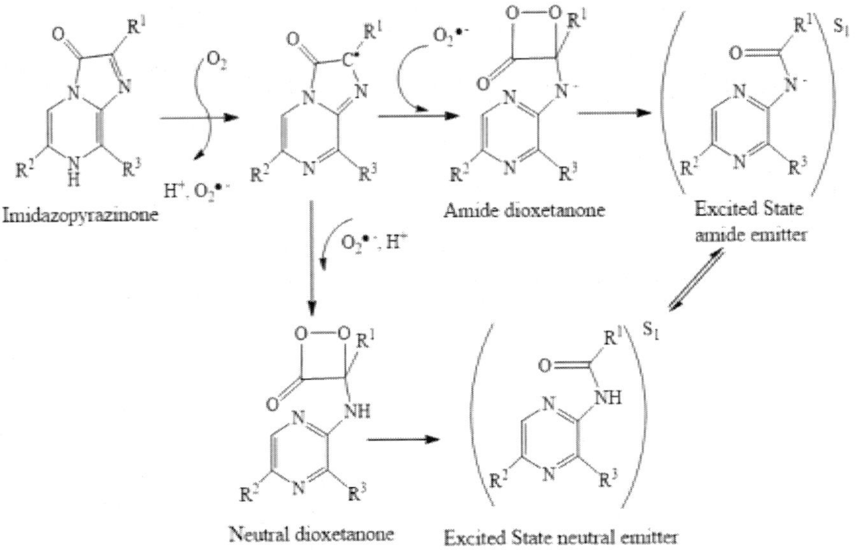

Scheme 1. Schematic representation of the chemiluminescent reaction of imidazopyrazinone-based compounds.

3. HISTORICAL NOTES: DISCOVERING SELF-ILLUMINATING MOLECULES

In the early 1960s, it was generally thought that different types of bioluminescent organisms contained chemically different luciferins. The discovery of the chemical identity of *Cypridina* luciferin [27, 28] and the further isolation of trace amounts of a new class of luciferin from *Renilla reniformis* and *Aequorea victoria* [29, 30] showed a structural similarity: all these molecules contained an imidazopyrazinone skeleton, which appeared to have a key role in the light-emitting systems. This novel BL substrate would be defined as CLZ in 1975, and its chemical synthesis performed [31, 32]. Two years later, in 1977, Goto and coworkers would elucidate its detailed structure [33]. Since then, many imidazopyrazinone derivatives have consistently appeared, and different synthetic approaches were developed. At the beginning, these molecules were prepared via a classical strategy consisting of the formation of a pyrazine ring, being the synthesis of the native CLZ one example [32]. The advances of organometallic chemistry allowed developing novel paths with metal catalyzed reactions like Suzuki coupling to avoid a long number of steps. The introduction of appropriate substituents on a pyrazine ring started from commercially available 2-chloropyrazine or 2-aminopyrazine molecules [34, 35].

The main strategy for obtaining CLZ analogues has been the modification of the substituents at the C-2, C-6 or C-8 positions of the imidazopyrazinone core, especially for an optimization of the CL efficiency (Figure 2). There are other possible modifications, for example, in the C=O or in the 3-carbonyl position, the critical active site of the bioluminescent reactions, but since BL is beyond the scope of this chapter, we will not focus on that. Herein, we pretend to summarize imidazopyrazinones involved only in chemiluminescent studies.

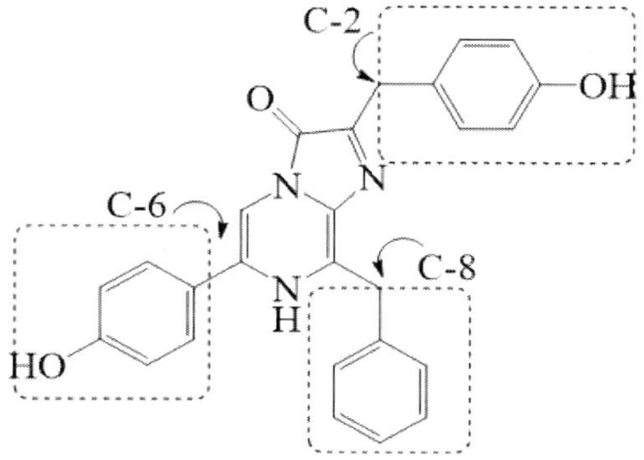

Figure 2. CLZ structure (as model molecule) and its sites of modification.

The first examples of chemiluminescent studies were developed in the 1980s to understand the chemical mechanism of *Cypridina* luciferin in BL. Different molecules were synthetized by Goto and his collaborators, and two compounds showed the most promised results: a *Cypridina* luciferin analog CLA [2-methyl-6-phenylimidazo[1,2-α]pyrazin-3(7H)-one] and its methoxy derivative MCLA [2-methyl-6-(4methoxyphenyl) imidazo[1,2-α]pyrazin-3(7H)-one] [36, 37]. CLA presented a methyl in C-2 position, a phenyl at C-6 and a hydrogen as C-8 substituent, and MCLA added a methoxy group in *para-* in the phenyl at C-6 position (Figure 3, Table 1). Both molecules showed CL in aqueous media and reacted not only with oxygen, but also with ROS (singlet oxygen, $^1O_2$, and superoxide anion, O_2^-, among others) that resulted in light emission (Table 3). This fact led to a new era of CL experiments, to develop methods for detecting O_2^-. These compounds become very useful as chemiluminescent probes in biological systems, for example in the estimation of the ability of human granulocytes and monocytes to generate O_2^- [37, 38]. MCLA became more popular than CLA, because its emission intensity was 4.6 times higher, with a λ_{max}= 465 nm (blue light emission) (Table 2). A MCLA derivative containing a fluorescein moiety, 6-[4-[2-[*N'*-(5-fluoresceinyl)thioureido]-ethoxy]phenyl]-2-methylimidazo[1,2-*a*]pyrazin-3(7*H*)-one, FCLA, was reported in an effort to achieve sensitive molecules with longer wavelengths, emitting in green (532 nm) but with a reaction rate slower than the others analogues (Figure 3) [39].

Table 1. Structures of reported imidazopyrazinone compounds

Compounds	C-2	C-6	C-8	Ref.
CLZ	CH_2PhOH	PhOH	CH_2Ph	41, 45, 46
CLA	CH_3	Ph	H	21, 24, 59, 60
MCLA	CH_3	$PhOCH_3$	H	21, 24, 59, 60
FMCLA	CH_3	$PhOCF_3$	H	25
ICLA	CH_3	indolyl	H	42
NCLA	CH_3	naphthyl	H	42
1	CH_2PhOH	$PhOH$--(CH_2)	CH_2Ph	40
2	CH_2PhOH	$PhOH$--$(CH_2)_2$	CH_2Ph	
3	CH_2PhOH	$PhOH$--$(CH_2)_3$	CH_2Ph	
4	CH_2PhOH	$PhOCH_3$	CH_2Ph	
5	CH_2PhOH	$PhN(CH_3)_2$	CH_2Ph	
6	CH_2PhOH	$Ph(OH)_2$--$(CH_2)_2$	CH_2Ph	
7	CH_2PhOH	Ph	CHOPh	
8	CH_3	PhOH	CH_2Ph	21, 41
9	Ph	PhOH	CH_2Ph	
10	CH_2Ph	PhOH	CH_2Ph	21, 62
11	$(CH_2)_2Ph$	PhOH	CH_2Ph	
12	$(CH_2)_3Ph$	PhOH	CH_2Ph	
13	CH_3	H	H	21
14	CH_3	PhOH	H	
15	CH_2Ph	H	H	
16	CH_2Ph	Ph	H	
17	CH_2Ph	PhOH	H	
18	CH_2Ph	$PhOCH_3$	H	
19	$CH_2PhCF_3N_2$	PhOH	CH_2Ph	43, 44
20	CH_2adamantyl	PhOH	Ph	45, 46
21	CH_2PhOH	PhOH	CH_2adamantyl	
22	CH_2PhOH	3-F,4-OHPh	CH_2Ph	48
23	CH_2PhOH	2,3-F_2,4-OHPh	CH_2Ph	
24	CH_3	$PhCF_3$	CH_2Ph	49
25	CH_3	PhF	CH_2Ph	
26	CH_3	Ph	CH_2Ph	
27	CH_3	$PhOCH_3$	CH_2Ph	
28	CH_3	$PhN(CH_3)_2$	CH_2Ph	
29	CH_3	$C\equiv CPh$	chlorostyrylPh	50
30	CH_3	$C\equiv CPh$	styrylPh	
31	CH_2PhN_3	PhOH	CH_2Ph	53
32	CH_2Ph	PhOH	$(CH)_2Ph$	55
33	CH_2Ph	PhOH	$CHC(CH_3)_2$	
34	CH_2Ph	PhOH	PhOH	
35	CH_2Ph	Ph	Ph	

Compounds	C-2	C-6	C-8	Ref.
36	CH₂Ph	Ph	thienyl	
37	CH₂Ph	thienyl	thienyl	
38	CH₃	PhN(CH₃)₂	H	59, 60, 23
39	CH₃	ClPh	H	
40	CH₃	NCPh	H	
41	CH₃	3-indolyl	H	
42	CH₃	3-methylindolyl	H	
43	CH₃	3-benzofuranyl	H	
44	CH₃	3-benzothienyl	H	
45	CH₃	PhCF₃	H	22
46	CH₂Ph	Ph	CH₂Ph	62, 67, 68
47	CH₂Ph	Ph	S-Ph	
48	CH₂Ph	Ph	CH₂PhCl	
49	CH₂Ph	Ph	SPhCl	
50	CH₂Ph	PhOH	SPhCl	
51	Ph	Ph	H	63
52	Ph	PhOCH₃	H	
53	Ph	N(CH₃)₂Ph	H	
54	PhOCH₃	N(CH₃)₂Ph	H	
55	3,4,5-(CH₃O)₃Ph	N(CH₃)₂Ph	H	
56	2,4,5-(CH₃O)₃Ph	N(CH₃)₂Ph	H	
57	CH₃	Ph	PhCF₃	64
58	CH₃	Ph	Ph	
59	CH₃	Ph	PHOCH₃	
60	CH₂PhF	3-OCH₃,4-OHPh	CH₂Ph	65
61	CH₂PhF	PhOH	styryl	
62	CH₂PhF	PhOH	naphthyl	
63	CH₂PhOH	3-F,4-(OH)Ph	CH₂Ph	
64	CH₂PhOH	3,4,5-(OH)₃Ph	CH₂Ph	
65	CH₂PhOH	styrylPhOH	CH₂Ph	66
66	CH₂PhOH	styrylPh	CH₂Ph	
67	CH₂PhOH	styrylPhPh	CH₂Ph	
68	CH₂Ph	Ph	OPh	67
69	CH₂Ph	Ph	SPh	
70	CH₂Ph	PhF	CH₂Ph	68
71	CH₂Ph	4-amino-3-FPh	CH₂Ph	
72	CH₂Ph	PhCH₂O	CH₂Ph	
73	CH₂Ph	4-nitrilePh	CH₂Ph	
74	CH₂Ph	5-methylfuryl-2	CH₂Ph	
75	CH₂Ph	Furyl-3	CH₂Ph	
76	CH₂Ph	3-ethoxycarbonylpPh	CH₂Ph	
77	CH₂Ph	Benzofyranyl-2	CH₂Ph	
78	CH₂Ph	Furyl-2	CH₂Ph	
79	CH₂Ph	PhOCH₃	CH₂Ph	
80	CH₂Ph	naphthyl	CH₂Ph	
81	CH₂Ph	Thienyl-2	CH₂Ph	

In 1990, Teranishi and Goto reported novel CLZ analogues with modifications at the C-6 position of the imidazopyrazinone core. Compounds containing a "bridge" (methylene, dimethylene or trimethylene) between the phenyl group and pyrazine ring were prepared to investigate the effects of conformational rigidity in comparison with the native CLZ (1-3, Table 1). CL efficiency was measured in DMSO, and the results indicated that the compound with a dimethylene bridge (2) had the higher efficiency, with a light yield 2.3 times greater than the efficiency founded in CLZ (1.0 relative light yield, RLY) (Table 4). They also prepared other analogues having an electron-donating group, OCH_3 or $N(CH_3)_2$, instead of a hydroxyl group at the 4-position of the phenyl group (4 and 5, Table 1). The results showed that CL efficiencies in DMSO did not increase (0.92 and 0.93, respectively), suggesting that electron donation from the phenyl group is not necessary for the high CL efficiency of CLZ. The last group of studied compounds were designed with a hydrogen group available to make hydrogen bonding with one of the nitrogen atoms in the emitter (6 and 7, Table 1). They showed a decrease in the light emitting efficiency [40].

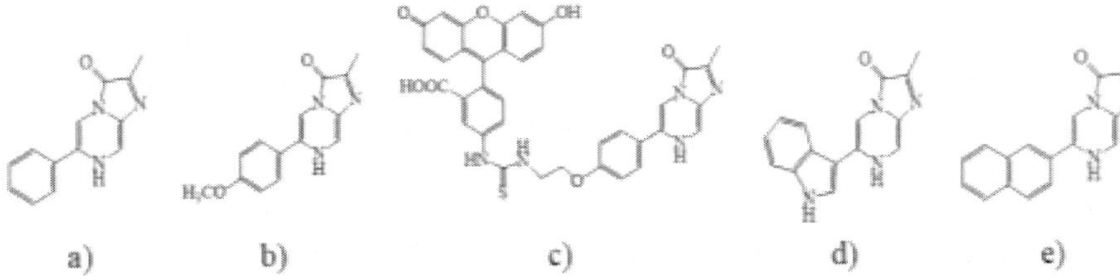

Figure 3. a) CLA, b) MCLA, c) FCLA, d) ICLA and d) NCLA structures, respectively.

Between 1991 and 1992, Tsuji and coworkers prepared CLZ analogues with different substituents at the C-2 position. They evaluated the influence of CH_3, Ph, CH_2Ph, $(CH_2)_2Ph$ and $(CH_2)_3Ph$ in the CL in various media (DMSO, DMSO-NaOH and DGM-acetate) (8-12, Table 1). They concluded that the wavelength maxima for CL and fluorescence was apparently dependent on the base concentration (Table 2). The results demonstrated that the light yields for almost all the synthetic analogues were larger than that of CLZ, except for C-2 substituted analogue with R= Ph (9), which showed a negligible light emission [41].

At the same time, Ohashi et al. studied the luminescence mechanism with imidazopyrazinone compounds substituted at the C-2 (CH_3, CH_2Ph), C-6 (H, C_6H_5, C_6H_4OH, $C_6H_4OCH_3$) and C-8 (H, CH_2Ph) position to characterize the emitter of CL (13-18, Table 1) [21]. Examining the CL spectra of these compounds and the fluorescence spectra of the corresponding Coelenteramide analogues emitting species, they found that the structure of the emitter depended on the concentration of proton under the CL conditions (DGM-acetate and DMSO-NaOH), the same trend predicted before by Tsuji [11].

Table 2. Chemiluminescent properties (emission maxima) of imidazopyrazinone derivatives in aprotic solvents

Compounds	λmax (nm)				Ref.
	DMSO[a]	DGM-acetate[b]	DMSO-base[c]	CH₃CN-TMG	
CLZ	474/475/465	422, 455[1]	530[3]/474[2]/ 509[2]	-	21,41, 45, 46, 66
CLA	-	390, 450/445	460[3]/467[5]	467	21, 24, 59, 60
MCLA	-	407/ 394	465[3]/471[5]	472	21, 24, 59, 60
8	470	411	523[2]	-	21, 41
9	472,575 [d]	570	597[2]	-	
10	472/460	421	545[2]	-	62
11	471	415	530[2]	-	
12	472	411	543[2]	-	
13	-	470	480[3]	-	21
14	-	412	473[3]	-	
15	-	470	470[3]	-	
16	-	381, 460	463[3]	-	
17	-	415	470[3]	-	
18	-	404	468[3]	-	
19	475	-	-	-	43, 44
20	480	405, 445[1]	-	-	45, 46
21	467	400	-	-	
22	470	-	-	-	48
23	465	-	-	-	
24	454	-	-	-	49
25	467	-	-	-	
26	462	-	-	-	
27	473	-	-	-	
28	479	-	-	-	
29	-	-	535, 590[6,5]	-	50
30	-	-	583	-	
31	480	-	-	-	53
32	-	-	580[4]	-	55
33	-	-	461[4]	-	
34	-	-	519[4]	-	
35	-	-	520[4]	-	
36	-	-	525[4]	-	
37	-	-	534[4]	-	
38	-	491	474[5]	472	59, 60, 23
39	-	390, 453	467[5]	467	
40	-	453	474[5]	468	
41	-	432	474[5]	476	
42	-	437	473[5]	-	

Table 2. (Continued)

Compounds	λmax (nm)				Ref.
	DMSO[a]	DGM-acetate[b]	DMSO-base[c]	CH3CN-TMG	
43	-	389, 450	470[5]	-	
44	-	401, 450	467[5]	-	
45	454				22
46	457 (480)	-	486[7]	-	62 (67) 68
47	497	-	-	-	
48	456	-	-	-	
49	499	-	-	-	
50	507	-	-	-	
53	-	521	-	-	63
54	-	524	-	-	
55	-	523	-	-	
56	-	518	-	-	
57	553	-	-	-	64
58	525	-	-	-	
59	513	-	-	-	
60	-	-	451[4]	-	65
61	-	-	551[4]	-	
62	-	-	582[4]	-	
63	-	-	456[4]	-	
64	-	-	458[4]	-	
65	479	-	538[2]	-	66
66	478	-	484[2]	-	
67	526	-	520[2]	-	
68	470	-	-	-	67
69	522	-	-	-	
70	-	-	489[7]	-	68
71	-	-	495[7]	-	
72	-	-	481[7]	-	
73	-	-	502[7]	-	
74	-	-	499[7]	-	
75	-	-	500[7]	-	
76	-	-	564[7]	-	
77	-	-	487[7]	-	
78	-	-	496[7]	-	
79	-	-	490[7]	-	
80	-	-	500[7]	-	
81	-	-	494[7]	-	

[a] aerated DMSO at 25°C; [b] DGM containing acetate buffer (0.10 M, pH 5.6, 0.66% v/v), except other indication (1) 0.20M acetate buffer (pH 7.6, 0.66% v/v); [c] different base or conditions: (2) 0.5% of 0.1M NaOH, (3) 0.1 N NaOH, (4) 0.1 M acetate buffer pH 6.5, (5) 0.1M TMG, (6) 1M KOH, (7) 0.05% (v/v) 1 M NaOH; [d] weak and barely detected.

Other studies were focused in the development of novel CLA derivatives. New *Cypridina* luciferin analogues were prepared by Goto and collaborators to test their application as superoxide anion detection in physiological media. 6-(3-indolyl)-2-methylimidazo[1,2-α]pyrazine-3(7H)-one (ICLA) and 2-methyl-6-(2naphthyl)imidazo [1,2-α]pyrazine-3(7H)-one (NCLA) were compared with CLA and MCLA (Figure 3). In DGM, RLY resulted much higher than within CLA (ICLA 10000, NCLA 1900, MCLA 3400 and CLA 430) (Table 4). However, in aqueous solution containing the xanthine-xanthine oxidase system (X-XOD system), NCLA emitted weaker light (RYL: 30) and ICLA gave almost no light (RYL: 2). These observations corroborated the superior ability of MCLA as superoxide probe due to its higher RLY (140 for MCLA, and 110 for CLA) and its emission maximum at 460 nm in visible region [42].

Table 3. Chemiluminescent properties (emission maxima) of imidazopyrazinone derivatives in protic solvents

Compounds	λmax (nm)			Ref.
	% H₂O or MeOD	H₂O/DMF	MeOH/DMF	
CLA	0	475	-	25
	50	387	-	
	60	387	-	
	70	387	-	
	90	387	-	
MCLA	0	460	-	
	50	420, 460	410	
	60	420, 460	410	
	70	420, 460	420	
	90	420, 465	420	
FCLA	0	463	-	
	50	387	378	
	60	387	378	
	70	387	378	
	90	387	380	

Between 1994 and 1997, Ohashi and coworkers described two novel derivatives. A photolabile analogue of CLZ was synthetized, with a trifluoromethyl diazirine group (CF_3N_2) at the C-2 position instead de hydroxyl group (19, Table 1). The studies of CL showed that this novel molecule emitted light with a flash pattern identical to that of native CLZ (475 nm in DMSO) [43, 44]. The other reported molecules were compounds possessing an adamantyl group in the C-2 and C-8 positions, to study the effects of this hydrophobic substituent (20 and 21, Table 1). CL maxima values were like that of CLZ (480 nm and 467 nm in DMSO, for C-2 and C-8 positions, respectively) (Table 2), but it was found a slight decrease in the relative chemiluminescent efficiencies (0.3, 0.6 in DMSO). When CL was measured in DGM-acetate buffer, the spectra were different from that of CLZ: with the adamantyl group at C-2 position, two maximum peaks were observed (405 and 455 nm), related to the excited states of both the neutral and amide anion forms [45, 46].

Table 4. Relative light yields of imidazopyrazinone derivatives

Compounds	R.L.Y.				Ref.
	DMSO[a]	DMG-acetate[b]	DMS-base[c]	Aqueous solution X-XOD	
CLZ	1	1	-	-	41, 46
CLA	-	430	-	110	42
MCLA	-	3400	-	140	
ICLA	-	1900	-	30	
NCLA	-	10000	-	2	
8	-	1.72	-	-	41
9	-	0.004	-	-	
10	-	1.44	-	-	
11	-	1.46	-	-	
12	-	1.72	-	-	
20	0.3	0.8	-	-	46
21	0.6	0.9	-	-	
29	-	-	100, 19[1,2]	-	50
30	-	-	9,19[1,2]	-	
68	0.77	-	-	-	67
69	0.81	-	-	-	

[a] aerated DMSO at 25°C; [b] DGM containing acetate buffer (0.10 M, pH 5.6, 0.66% v/v); c different base conditions: (1) 0.1M TMG (2) 1M KOH.

In 1997, Teranishi and Shimomura described some imidazopyrazinone derivatives with higher sensitivity, decreased background luminescence, reduction in the influence of environmental factors, and increased water solubility. A higher luminescent intensity was reached with a bridge between the phenyl group and pyrazine ring, in C-2 methyl substituted CLZ analogue. Two MCLA analogues with a cyclodextrin (CD) in C-2 position were soluble in water, and the intensity of the β-CD analogue was about five times greater than MCLA [47].

In 1998, Hirano et al. prepared analogues of CLZ in the C-6 position, consisting in the substitution of H for a fluoro group at *ortho-* or *meta-* position of the 4-hydroxyphenyl moiety, and they investigated the electronic substituent effect (22 and 23, Table 1). CL reactions in DMSO showed emission maxima at 470 and 465 nm, respectively, and relative CL efficiency of 1.6 and 1.8. The light emitters were assigned as the singlet excited states of the amide anions [48].

At the same time, Ohashi and coworkers investigated the effect of substituents in the CL of imidazopyrazinone derivatives with a methyl group in C-2 position. They substituted the C-6 position, from electron-donating groups (OCH_3 or $N(CH_3)_2$) to electron-withdrawing groups (CF_3 or F) at the *para-* position of the phenyl (24-28, Table 1). Their results indicated that the variation of the electronic properties of these analogues caused the small change of the efficiency of chemical generation of a singlet excited light-emitter (Φ_S) [49].

In addition, Nakamura et al. reported imidazopyrazinone derivatives, with C-2 methyl substituted, modifications in C-6 (-C≡C-Ph) and different styryl groups in C-8 (CH=CHPh, CH=CClPh) positions (29 and 30, Table 1). It was revealed that the C-8 substituted CH=CClPh product had bimodal CL, emitting orange-colored luminescence under acidic to neutral conditions and yellow-colored luminescence under basic conditions. Besides that, they could confirm the large bathochromic shift caused by the chlorostyryl group at the C-8 position [50].

Moreover, Teranishi et al. designed novel chemiluminescent systems, in which MCLA as a light-producing chromophore was covalently bonded to a CD molecule. This strategy provided the enhancement of CL, having a better efficiency in aqueous solution with MCLA-bounded CD at C-2 position (Φ_{CL}= 1.9) than MCLA (Φ_{CL}= 0.17) or MCLA with CD in the media (Φ_{CL}= 0.29) (Table 5) [51]. One year later, in 1999, other systems were designed, but this time special attention was paid to the length of the spacer that bounded a CD to MCLA. The results demonstrated that a short spacer gave the highest CL efficiency (44 times greater than MCLA). The CL spectra indicated a dependence of the light-emitting efficiency with the type of bound CD, the spacer length between the MCLA and CD, and the binding site in CD [52].

During this year, Teranishi and coworkers also reported a chemiluminescent study in protic solvents of CLA analogues, with a trifluoromethoxy derivate and its comparison with both CLA and MCLA. To study the influence of the electronic characteristics on the phenyl group at C-6 position, a trifluoromethyl substituent was inserted in the *para-* position of the phenyl moiety. CL was tested in mixtures water/DMF and MeOH/DMSO. These results coincided with previous reports showing that the substituent effect on the 6-phenyl group of 6-arlyimidazo[1,2-a]pyrazin3(7H)-one is extremely weak [25].

In 2000, Ohashi and coworkers reported a new CLZ containing a photoreactive azido group (N_3) as phenyl substituent at C-2 position (31, Table 1). Comparing with the native CLZ, the CL results showed similar behavior (λ_{max}= 480 in DMSO) [53].

In 2001, Teranishi reported analogues of MCLA with structural changes at the C-6 position, consisting in 4-methoxyphenyl rings with restrictions in the dihedral angle, introduced by the presence of an alkyl bridge, to investigate the effects of the conformation between the pyrazine and 4-methoxyphenyl rings on their CL efficiencies in aqueous solutions. The results showed that a decrease of the dihedral angle caused the enhancement of the fluorescence efficiency, and vice versa, the increase of the angle dramatically decreased the chemiluminescence efficiency. Two alkyl bridge derivatives (CH_2, CH_2CH_2) exhibited the highest CL efficiencies [54].

At the same time, Wu et al. prepared novel CLZ analogues with substitutions at the C-6 and C-8 positions, consisting in different aromatics groups at C-6, and conjugated olefins and/or aromatic groups in the C-8 position (32-37, Table 1). The analogue having a styryl group at C-8 showed the largest bathochromic shift under neutral conditions giving a peak at 580 nm. A comparison of the analogue with a hydroxyphenyl at C-8 position (peak at 519 nm) and the analogue with phenyl at C-6 and C-8 (peak at 520 nm) suggested that the electron-rich function group did not enhance the displacement to a longer wavelength (Table 2) [55].

Two years later, in 2003, a group of researchers prepared 3,7-dihydroimidazo[1,2-α]pyrazine-3-ones derivatives to study their ability as flash chemiluminescent labels. The difference between the common glow luminescence assays and a flash luminescence reaction is that the latter occurs quickly, in a matter of seconds or minutes, giving off a very bright signal, whereas glow luminescence assays can last for hours but they are not as bright as flash assays. However, all the performed tests resulted in a shorter efficiency [56].

Teranishi et al. continued with the design of chemiluminescent systems with CD to enhance the CL efficiency. MCLA was attached to the secondary hydroxyl face of δ-CD, which consists of nine D-glucose units. The results demonstrated that δ-CD was too large to include the singlet-excited amide moiety in the course of the CL reaction, suggesting that light was produced in a roughly 90% aqueous environment [57]. A year later, these authors reported a green luminescent probe, in which MCLA and fluorescein molecules were bound at the secondary and primary faces of γ-CD, respectively. This novel probe exhibited high CL intensity in comparison to FCLA, using as a green probe before [58].

It was 2005 when Hirano and coworkers decided to verify the reaction mechanism involved in CL by investigating the kinetics of a series of CLA analogues, namely its *para*-substituted phenyl derivatives: $(CH_3)_2NC_6H_4$, $CH_3OC_6H_4$, ClC_6H_4, NCC_6H_4 and 3-indolyl. The mechanism suggested successive single electron transfer (SET) (38-41, Table 1) [59]. One year later, they reported the influence of the nature of the group in C-6 position in the chemiluminescent properties, extending the studies with three novel derivatives, 3-(N-methyl)indolyl, 3-benzofuranyl and 3-benzothienyl substituted (42-44, Table 1). Particularly, an electron-donating group permitted the generation of a high Φ_S value via the charge transfer-induced luminescence (CTIL) mechanism (Table 5) [60]. In 2008, it was demonstrated the electronic effects of the C-6 substituted groups on the elementary reaction's steps. Electron-donating groups increased the basicity of the corresponding anionic dioxetanone giving neutral dioxetanone by rapid protonation. They also gave the transition state (TS) and the singlet-excited acetamidopyrazine (S1) a strong intramolecular charge transfer (ICT) character, and their similar ICT character leaded to the ICT TS → S_l route in the charge CTIL mechanism for efficient chemiexcitation. The electron-donating group also played an essential role in increasing the Φ_S value for the thermal decomposition (Table 5) [24].

Back to 2007, Teranishi synthesized two chemiluminescent probes that emitted red CL (λ_{max} around 610 nm) and their intensities were more intense than those of MCLA and FCLA when tested with superoxide anions produced by the hypoxanthine-xanthine oxidase system. These compounds consisted in moieties of MCLA and sulphorhodamine [61].

After a few years, in 2011, the research teams of Ohashi and Saito used five methyl analogues of CLZ described before and a novel CF_3 derivative of CLA (24-28, 45 Table 1) [22, 49] to quantitatively investigate their chemiluminescent properties in DMSO. They found that an electron-donating substituent (CF_3, F, H, OMe, and NMe_2) at *para-* position of the C-6 phenyl group, is not required for the efficient generation of a singlet-excited light emitter, but is required for the high fluorescence quantum yield (Φ_F) of the emitter [22].

In 2012, Guiliani et al. designed and synthesized new CLZ analogues with modification at the C-6 position, replacing the methylene group bounded to the phenyl at C-8 position for a S atom (46-50, Table 1) to obtain better bathochromic emissions shifts. The results showed that the introduction of the sulfur heteroatom made a new electron-rich structure, which had a significant bathochromic shift of the CL spectrum measured in neutral DMSO (about 40 nm) (Table 2) [62].

Two years later, in 2014, Hirano and coworkers reported CLA derivatives with modifications at the C-2 and C-6 positions. First, it was studied the effect of a phenyl group at C-2 position, varying the substituents at the *para-* position of the phenyl in C-6 (NMe_2, OMe and H) (51-53, Table 1). In DMSO, these compounds showed low CL because the amide anions were not fluorescent. In DGM-acetate, only the electron-donating 4-(dimethylamino)phenyl group showed CL. With these results, it was fixed $PhNMe_2$ at C-6, and changed the C-2 substituents (4-MeOPh, 3,4,5-$(MeO)_3$Ph and 2,3,4-$(MeO)_3$Ph) (54-56, Table 1), and measured their CL. The results demonstrated better Φ_S than the corresponding to the CLA (Table 5) [63].

Simultaneously, Saito el al. studied the derivatives of CLA having a substituent (CF_3, H, and OMe) at *para-* position of the phenyl group at C-8 position (57-59, Table 1). The CL efficiencies of these imidazopyrazinones were improved by the introduction of a phenyl substituted group but this fact did not change the value of Φ_S, concluding the small contribution of these groups (Table 5) [64].

Focused in the design of CLZ analogues that could produce a red shift of light emission with CLZ-dependent bioluminescent proteins, Dodd and coworkers reported the chemiluminescent properties of novel derivatives. Two different type of modifications were studied. First, a 4-fluorobenzyl group replaced the 4-hydroxybenzyl group at C-2 position of CLZ because it had been shown that the fluorine atom at this position produced a notable red shift in BL (60-62, Table 1). A second modification was introduced substituents in *meta-* to the phenyl ring at C-6 position of CLZ. The last strategy was replacing the C-8 benzyl group of native CLZ by a α-styryl moiety or by a 1-naphthyl group (63 and 64, Table 1). The results suggested that only compounds having an extended resonance system at the C-8 position (1-naphthyland and α-styryl analogues) showed a significant red shift of light emission (Table 2) [65].

In 2015, Suzuki and coworkers reported three novel CLZ derivatives modified at C-6 position. They investigated the effect of introducing a styryl group, resulting styryl-Ph-OH, styryl-Ph or styryl-Ph-Ph (**65-67**, Table 1). The spectral similarity between the first two analogues and native CLZ shows that the C-6 substituent has little influence on the CL emission wavelength but surprisingly, the styryl-Ph-Ph compound showed a 50 nm red-shifted emission compared to that of native CLZ (Table 2) [66].

One year later, a team of researchers prepared imidazopyrazinone derivatives with an oxygen or sulfur atom in the place of the methylene group at C-8 position (68 and 69, Table 1). The replacement with an extended electronic conjugation was chosen to develop red-shifted CLZ analogues. The results supported their hypothesis [67]. Following the same idea, in 2017, it was reported more derivatives varying substituents at the C-6 position of the imidazopyrazinone (70-81, Table 1) [68].

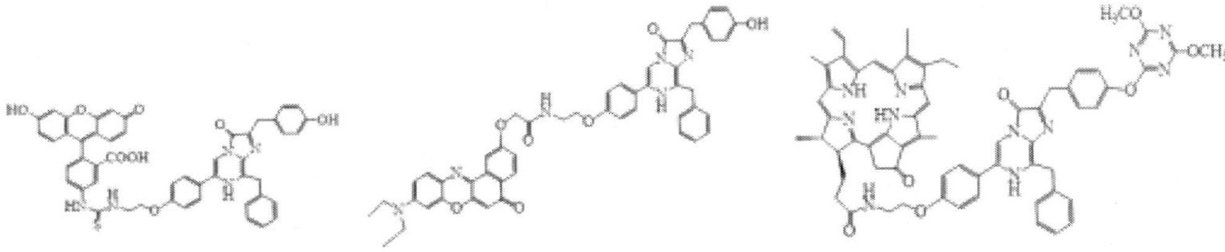

6-FITC-CLZ   6-Nile-Red-CLZ   6-Chlorin-2-DMT-CLZ

Figure 4. Selected fluorescent dye-conjugated CLZ analogues.

The most recent study was focused in the development of fluorescent dye-conjugated analogues for the creation of a molecular imaging platform. The CLZ analogues were dye-bridged at C-2 and/or C-6 positions, being the dye-conjugated at the C-6 position the best candidates. In fact, a fluorescein isothiocyanate derivative (6-FITC) showed green CL with a $\lambda_{max}= 538$ in DMSO, and 6-Nile-Red and 6-Chlorin-2-(4,6-dimethoxy1,3,5-triazin-2-yl) (6-Chlorin-2-DMT) derivatives had maximal peaks at ca. 650, and 680 nm, respectively. Then, this study provided new red shifted molecules (Figure 4) [69].

4. CURRENT AND POTENTIAL FUTURE APPLICATIONS

CL measurements have become extremely popular in recent years due to their high sensitivity, simplicity of operation, effectiveness, inexpensive instrumentation, and the emergence of novel chemiluminescent assays. Measurement of light from a chemical reaction is highly useful because the concentration of an "unknown" can be inferred from the rate at which light is emitted. The rate of light output is directly related to the amount of light emitted and accordingly, proportional to the concentration of the luminescent molecule present.

Table 5. Quantum yields and relative quantum yields of imidazopyrazinone derivatives

Compounds	Φ_{CL}					$R\Phi_{CL}$	Ref.
	DMSO[a] (x10⁻³)	DMSO-NaOH (x10⁻⁴)	DMSO-TMG (x10⁻³)	DMG-acetate[b] (x10⁻³)	CH₃CN-TMG (x10⁻⁴)	DMSO	
CLZ	-	7.1	-	2.3	-	100	41, 44 45
CLA	-	1.10	-	0.75	1	-	59, 60
MCLA	-	2.10	-	3.1	2	-	59, 60
8	-	10.10	-	0.041	-	-	41, 62
9	-	0.004	-	0.00009	-	-	
10	-	7.2	-	0.034	-	0.96	
11	-	8.8	-	0.034	-	-	
12	-	8.9	-	0.04	-	-	
19	-	-	-	-	-	10	44
22	-	-	-	-	-	1.6	48
23	-	-	-	-	-	1.8	
24	0.6	-	-	-	-	-	49
25	1.5	-	-	-	-	-	
26	1.5	-	-	-	-	-	
27	1.8	-	-	-	-	-	
28	1.5	-	-	-	-	-	
31	-	-	-	-	-	49	53
38	-	-	1.2	1.5	3.8	-	59, 60
39	-	-	0.34	0.55	1.1	-	
40	-	-	0.31	0.86	0.8	-	
41	-	-	0.48	9	1.2	-	
42	-	-	0.43	7.2	-	-	
43	-	-	0.5	1.1	-	-	
44	-	-	0.24	0.55	-	-	
46	-	-	-	-	-	1	62
47	-	-	-	-	-	1.83	
48	-	-	-	-	-	1.80	
49	-	-	-	-	-	0.96	
50	-	-	-	-	-	0.42	
53	-	-	-	2.5	-	-	63
54	-	-	-	12	-	-	
55	-	-	-	6.2	-	-	
56	-	-	-	14	-	-	
57	5.2	-	-	-	-	-	64
58	5.3	-	-	-	-	-	
59	4.5	-	-	-	-	-	

[a] aerated DMSO at 25°C; [b] DGM containing acetate buffer (0.10 M, pH 5.6, 0.66% v/v); TMG: 1,1,3,3-tetramethylguanidine.

The following section aims to provide an overview of the most popular applications of imidazopyrazinone-based compounds. Besides that, two different concepts will be introduced, to enable a better understanding of the processes involved in the described techniques: reactive oxygen species and photodynamic therapy.

Reactive oxygen species (ROS) are molecules with unpaired electrons in their structure, which are produced by living organisms because of normal cellular metabolism. These highly reactive intermediates are best known for their ability to react with several biological macromolecules in cell, inducing cellular damage because these species alter the functions of proteins, lipids, carbohydrates and/or nucleic acids. The most important ROS found in cells is, undoubtedly, superoxide anion O_2^-. It is produced during normal cellular respiration and plays key roles in cellular physiology with its dysregulation being associated with a variety of chronic diseases including cancer, inflammation and aging [70].

On the other side, photodynamic therapy (PDT) is a minimally invasive therapeutic modality for cancer therapy. For its application, three requirements must be present: molecular oxygen, a non-toxic photosensitizer (PS) and light of a specific wavelength. PS is a molecule only activated under an appropriated wavelength within the irradiation of light. This allows a great selectivity and reduce the number of side effects of this therapy when comparing with others. However, the depth of light penetration (less than 1 cm in tissues) limits this treatment, being nowadays in wide clinical use of skin, esophageal and non-small cell lung cancers [71].

4.1. Chemiluminescent Probes for Detecting Superoxide Anion

Besides molecular oxygen, the imidazopyrazinone scaffold can also react with other oxidizing species to generate visible light, being the O_2^- one the most interesting species for detecting applications [72].

Since CLA was first applied as CL probe, the use of CLA and its derivative MCLA for the detection of O_2^- have been widely tested in biological systems because of their low toxicity and high sensitivity. Different studies were described investigating the ability of O_2^- generation by macrophages during phagocytosis, the estimation of superoxide dismutase (SOD) levels in human erythrocytes or determining Cu-Zn and Mn-Zn in SOD [38, 73-77].

Moreover, in the development of chemiluminescent probes for O_2^-, MCLA has been described as the best molecule due to its higher sensitivity.

The first application of MCLA for detecting O_2^- production *in vivo* was published in 1993. The experiment was performed in a rat by stimulating Kupffer cells of the liver, which is a macrophage. Under experimental conditions in the perfused liver, a stimulator of O_2^- was added, and this MCLA method allowed identifying these species better than others used before [78].

MCLA was also described as a good candidate in the direct measurement of O_2^- produced by leucocytes adherent to tissues *in vivo*. In this study, it was induced a thrombosis in a femoral arteria of a guinea pig. An intravenous administration of MCLA detected the accumulation of leucocytes, suggesting that O_2^- was produced at the site of thrombolysis [79].

Five compounds derived from MCLA were presented as novel superoxide probes in 1998, more sensitive even than the MCLA. The measurement of the O_2^- generation was examined with the bacteria L. *monocytogenes*, a species of pathogenic bacteria that causes listeriosis. Two compounds, both with a triple bond in the C-6 position, appeared particularly useful not only due to their high sensitivity, but also because one of them gave the highest luminescence response and a relatively low background luminescence, and the other showed an exceptionally high signal-background ratio [80].

A few years later, it was described other example of MCLA used as specific probe to investigate the time course of generating O_2^- in ischemia-reperfusion in the *in vivo* rat lung. CL induced by MCLA was continuously monitored and results showed that the increase of oxygen radical formation leading to ischemia occurred after a short period of occlusion of the pulmonary artery alone in vivo [81].

Despite the interesting results obtained before, the successful detection of O_2^- in biological systems requires not only neutral conditions, but also luminescence at long wavelengths. The analogue of MCLA with a fluorescein molecule, FCLA, has its emission of light with the longest wavelength under neutral conditions but it was observed a low superoxide-induced CL intensity. Thereby, it was necessary the development of different approaches to produce a better molecular probe.

To overcome this problem, a cyclodextrin (CD) was introduced into the system, because it was well known that this cyclic oligosaccharide confers solubility [82]. Typical CDs contain several glucose monomers units in a ring, creating a cone shape. In all investigated experiences with CD, the source of O_2^- was the hypoxanthine–xanthine oxidase system. In 1997, two CD-bound analogues, α-CD (with six glucose subunits) and β-CD (with seven glucose subunits), at C-2 position of the imidazopyrazinone core were prepared. Both compounds were easily solubilized in water but showed different behavior, since the intensity CL of the β-CD analogue was about five times greater than MCLA, meanwhile α-CD was practically the same that MCLA [47]. One year later, Teranishi et al. reported the first example of the synthesis of MCLA covalently bound to a CD molecule. These chemiluminescent probes were achieved by the formation of an amide bond between MCLA and a CD. This strategy allowed an increase in the chemiluminescent quantum efficiency in the β-CD-bound MCLA, about eleven times greater that MCLA, whereas the quantum efficiency of α-CD-bound MCLA was practically the same as that of MCLA. The study also showed that the cavity of β-CD might be most suited to MCLA, but that very large amounts of CD were needed for the aqueous CL system [51].

In a further investigation, the C-2 position of MCLA was bound to the primary face of γ-CD (8 glucose subunits) or at the secondary face of an α-,β-,γ-CD through a short spacer, one more time by the formation of an amide bond. The probe of MCLA bound at the secondary face of γ-CD gave the highest CL efficiency, 44 times greater than that of MCLA [52].

Continuing with the studies of CD-bound MCLA compounds with enhanced luminescence, MCLA was attached to the secondary hydroxyl face of δ-cyclodextrin, which consists of nine D-glucose units. Although the oxygen-induced CL efficiency was 12 times greater than that of

MCLA, this efficiency was hardly lower than γ-CD-bound MCLA studied before, and singlet-excited state formation efficiency for δ-CD-bound MCLA was also lower than that for γ-CD bound MCLA [57].

One more step was made to obtain probes with longer wavelengths; a series of MCLA bounded with fluorescein through γ-cyclodextrin were designed. In the spectra of the superoxide-induced CL, and in contrast with the results obtained before (λ_{max} around 460nm), it was observed maxima at around 515-527 nm along with the absence of luminescence due to the MCLA moiety. A probe in which MCLA and fluorescein molecules were bound at the secondary and primary faces of γ-cyclodextrin, respectively, generated green light, possessed high sensitivity to O_2^- and exhibited high CL intensity, in comparison to FCLA, which has been used as the green luminescent probe in the past [58].

Despite the chemiluminescent probes reported until now, CL at longer wavelengths is still desirable. In a new attempt to obtain red CL probes, MCLA and sulphorhodamine 101 (texas red) moiety were covalently attached on the secondary and primary hydroxyl faces, respectively, of the γ-cyclodextrin molecule. These probes emitted red light (λ_{max} 610 nm) when induced by O_2^-, improving the detection. In particular, the probe with the modification at the C-2 position of MCLA exhibited strong intensity and it is a commercially available as "red-CLA" (Figure 5a) [61].

A new approach to enhance CL was described in 2009, when a novel type of functional chemiluminescent probe KEIO-BODIPY-imidazopyrazine (KBI) was developed (Figure 5b). The molecule consists of two parts connected through a direct bond: the imidazopyrazinone moiety as a CL emitter, and the BODIPY as a fluorophore. KBI showed yellow-green emission (545 nm) under neutral pH conditions, at a longer wavelength than the CL of MCLA, and its CL intensity was much higher than the intensities of both MCLA and FCLA, assumed to be due to the high fluorescence quantum yield (Φ_F) of BODIPY [83].

Tang and coworkers described other strategies. In 2016, this team described a supersensitive imaging nanoprobe called PCLA, which contained two moieties linked covalently, namely CLA that is capable of CL triggered by O_2^- as the energy donor, and conjugated polymers with light-amplifying property as the energy acceptor. One interesting aspect was that for the first time it was possible to *in situ* visualizes O_2^- level differences between normal and tumor tissues in mice, without an external excitation source. PCLA revealed concentration differences of O_2^- between normal and tumor tissues without exogenous stimulation [84].

A year later, CLA was included in a biological sensor with simultaneous turn-on signals of fluorescence (FL) and CL triggered by one single species. CLA was used as the reactive motif, and it was covalently linked to tetraphenylethene (TPE) as the prototype motif of aggregation-induced emission (AIE). TPE-CLA was successfully applied in mice, in imaging native O_2^- in Raw264.7 cells, and in endogenous O_2^- in HL-7702 cells, which was achieved non-invasively with increased accuracy, based on complementary information obtained simultaneously from two independent signals. This conjugated TPE-

CLA probe was presented as an example of FL/CL dual sensing platform for real-time and continuous monitoring of endogenous O_2^- in live cells [85].

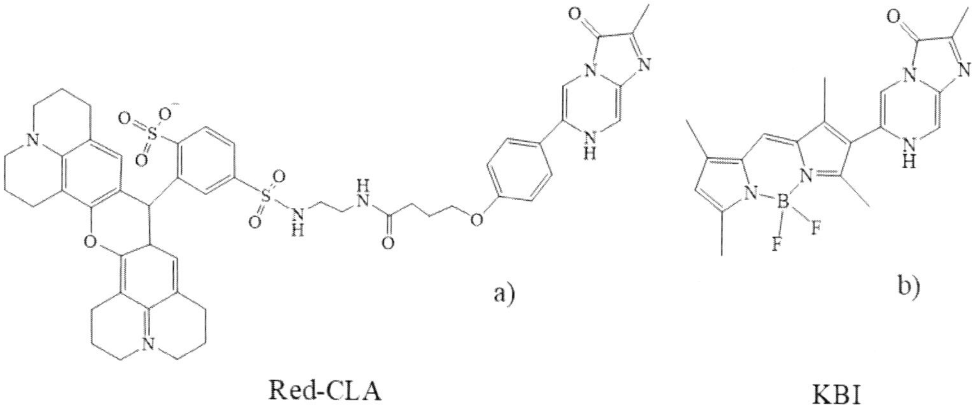

a)

b)

Red-CLA

KBI

Figure 5. Examples of imidazopyrazinone functional systems with a) a fluorescent dye (sulphorhodamine 101) and b) a fluorophore (BODIPY), respectively.

Regarding the use of CLZ in CL detection probes, the first attempt to test its potential was in 1992. It was reported the first example of CLZ use as chemiluminescent probe in an assay of respiratory burst (sometimes called oxidative burst), that consist in the rapid release of ROS from different types of cells. In immune cells, monocytes and neutrophils release these chemical molecules as they are exposed to bacteria or fungi. When this happens, NADPH oxidase produce O_2^-. CLZ assay was sensitive in detecting reactive oxygen metabolites produced by human neutrophils. In addition, it was shown some advantages in the experimental design to follow the activity of NADPH in normal cells, i.e., CLZ did not require extra reagents and it seemed more selective than luminol [86].

In 1999, a study compared the effect of CLZ as chemiluminescent probe for detecting O_2^- with the classical lucigenin in vascular tissues. Tests were performed with cultured bovine aortic endothelial cells and isolated rings from rat aorta. The results showed that the use of lucigenin to estimate rates of O_2^- formation were not exact because its ability to undergo cycles of reduction and oxidation, which contributes to O_2^- production. However, the properties of oxidant-dependent CL in CLZ suggested its utility as a sensitive and more reliable alternative probe in biological systems [87].

In 2004, a new study compared the ability of CLZ against lucigenin as O_2^- indicators *in vivo*. The effects of bacteria *Paracoccus denitrificans* were studied in submitochondrial particles and cytoplasmic membranes from a rat liver. CLZ resulted more reliable for quantitative analysis because a redox chemistry is not required for measuring O_2^- production [88].

In 2010, a group of researchers presented a CL assay to quantify total O_2^- scavenging activity (SOSA) in which CLZ was chosen as indicator to create this inhibition assay. It allowed the differentiation of enzymatic and non-enzymatic superoxide scavengers that can be used as biomarkers to characterize metabolic shifts and cellular oxidative stress [89].

The most recent reports involving a more biological context were made by Contag and coworkers in 2016. They have validated the use of CLZ as a chemiluminescent probe for O_2^- both *in vitro* and *in vivo*, in situations of pathology such as cancer, diabetes mellitus and chronic inflammation.

CLZ was evaluated as a potential reporter of cancer associated O_2^- in cell culture and in mice. In vitro, various cancer cell lines were quantified by CLZ CL, showing different concentrations with a signal range of 3 to 12 times that of background. For *in vivo* studies, it was selected a mouse model with a breast adenocarcinoma, and it was obtained significantly higher CL than that of surrounding normal tissues. These results indicated that CLZ can be used to assay O_2^- concentrations in cultured cancer cells and in tumors growing in mice [90].

For *in vivo* studies between the relation of ROS and diabetes, β cells were selected because they are a type of cell found in pancreatic islets that synthesize and secrete insulin. In patients with type I or type II diabetes, β-cell mass and function are diminished, leading to insufficient insulin secretion and hyperglycemia. In vivo imaging was used for non-obese diabetic (NOD) mouse model, and it was revealed the production of CL in a concentration dependent manner [91].

In an effort of identified appropriate chemiluminescent reagents associated with ROS and inflammation, CLZ was tested in a mouse model of inflammatory bowel disease (IBD), term used to describe disorders that involve chronic inflammation of digestive tract. Results showed a dynamic quantification with fine temporal resolution of ROS in populations of cells, making CLZ a useful tool to detect early pre-symptomatic inflammation for studies involving early disease pathogenesis or intervention [92].

4.2. Light Delivery in PDT of Cancer

As we mention before, CL detection methods provides clinicians efficient tools for early detection and accurate diagnosis of cancer due to the possibility of *in-vivo* monitoring the short-lived ROS in cells. The next step would be killing cancer cells, and the PDT technique appears as one solution in which CL systems can act as the source of light. This treatment is today a reality but not widely used due to its limits, for example, it only treats areas where light can reach and do not work for metastatic cancers [93, 94].

Singlet oxygen ($^1O_2$), other of the most interesting and versatile molecules within the ROS family, is the most important cytotoxic agent in PDT. $^1O_2$ is a metastable state of normal triplet oxygen, and its potential for the control of oncogenesis and in a variety of therapeutic antitumor approaches has begun to be recognize [94]. Although it is very short lived, $^1O_2$ is highly reactive and can oxidize a variety of biological molecules, able to cause direct tumor-cell death via apoptosis or necrosis damage. In PDT, $^1O_2$ is generated after the excited PS molecule transfer their energy to the nearby oxygen molecules dissolved in the cell. Therefore, a direct monitoring of $^1O_2$ production may provide a more direct indicator of PDT efficacy.

Researchers have been proposed chemiluminescent compounds as an excitation source in PDT in cancer to improve its limitation into deeply localized tissues. This is a very interesting and advantageous characteristic since no light excitation source being required and given the ability of these molecules to excite the PS by intracellular generation of light, that could overcome the limitation.

The first attempt to test the potential use of a CL probe in PTD was made by Wang et al. in 2002. They selected a Hematoporphyrin derivative (HpD) as a photosensitizer and FCLA as a CL probe. The study demonstrated that the main product from the photosensitization reaction, the $^1O_2$, was detected with FCLA. In fact, the results of photosensitized CL imaging *in vivo* performed with a tumor-bearing nude mouse, showed a clear tumor image. This achievement indicated the potential application of this localization method in clinics for tumor diagnosis [95].

In 2005, Qin et al. successfully demonstrated the applicability of FCLA *in vitro* for real-time monitoring of $^1O_2$, produced during PDT. They showed that FCLA-CL emission during PDT was linearly proportional to the corresponding cytotoxicity, regardless of the treatment protocol. The CL reagent was absorbed by cells with minimal cytotoxicity and minimal interference with the outcome of PDT treatment [96]. In a further investigation, these authors evaluated the lifetime of the FCLA and the results showed that it is much longer than that of direct luminescence of $^1O_2$, suggesting the feasibility of using delayed CL to detect singlet oxygen *in vivo* [97].

4.3. Commercially Available Compounds

Since CLZ analogues have been synthesized during decades, showing different properties in terms of emission wavelength, quantum efficiency or cell membrane permeability, as well as they can function as substrates for Renilla luciferase, an enzyme used as a reporter gene for luminescence-based assays, nowadays it is easy to access to purchasable compounds. Several companies offer different molecules, and among the main suppliers, we can highlight Tokyo Chemical Industry, Co., ltd. (TCI), Thermo Fischer Scientific Corporation (Invitrogen brand), Prolume ldt (NanoLight Technology division) and Biotium [98, 99, 100, 101].

TCI Co., ltd. is a leading global manufacturer of fine chemicals. Since its foundation in Japan, it has also established affiliate companies, manufacturing sites and overseas offices in the USA and Europe. The chemiluminescent compounds provided e are Cypridina Luciferin analogues: CLA, MCLA, FCLA and Red-CLA (Figures 3 and 5) Other company, which offers molecular probes for detection of ROS is Invitrogen, with CLZ and MCLA among their own products.

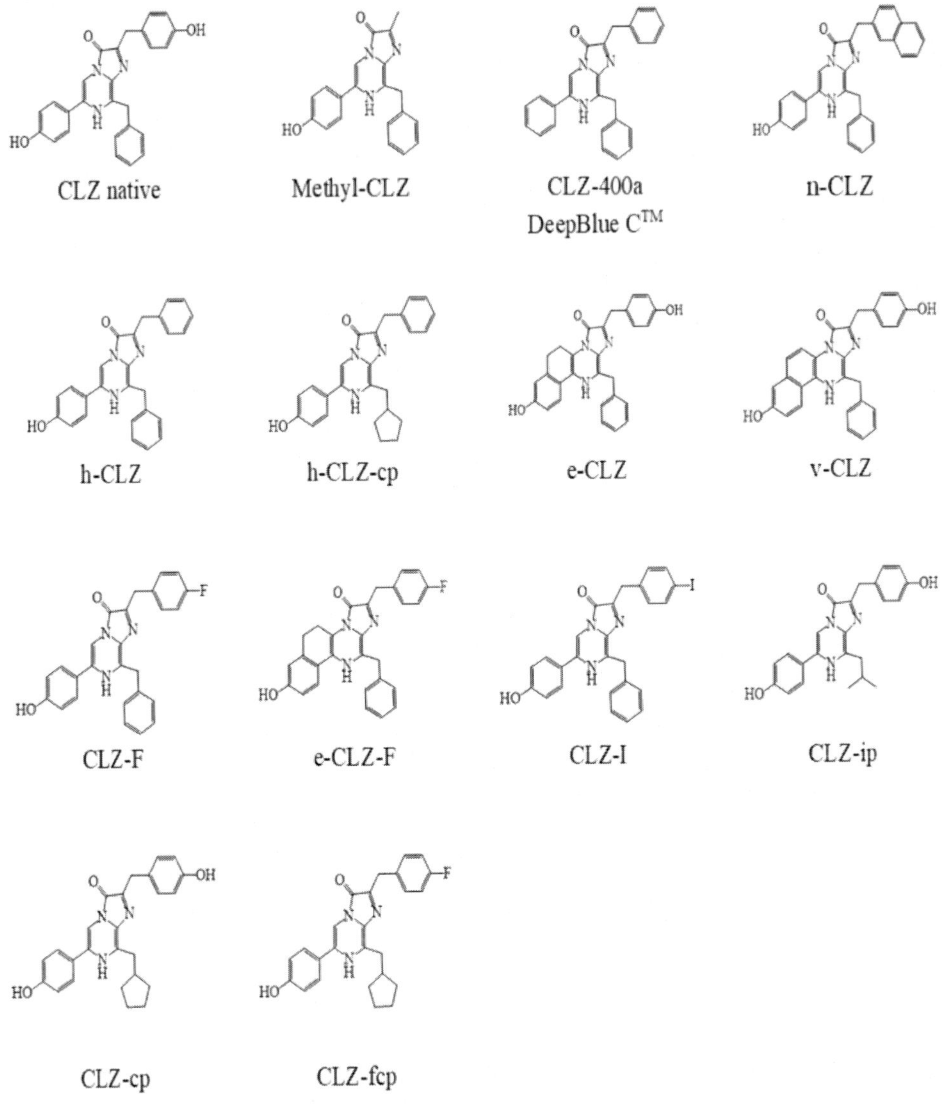

Figure 6. CLZ analogues commercially available.

NanoLight Technology is a division of Prolume Ltd., a biotechnology company focusing upon broad based applications of marine bioluminescence organisms. They supply *Cypridina* Luciferin, native CLZ and a number of analogues, such as h-CLZ, CLZ-400a (also known as DeepBlue C™), e-CLZ, CLZ-F, e-CLZ-F, v-CLZ, CLZ-I. Biotium is a leading provider of fluorescent products, putting in the market CLZ, CLZ-cp, CLZ-fcp, h-CLZ, h-CLZ-cp, CLZ-400a, methyl-CLZ, e-CLZ, CLZ-F, CLZ-I, CLZ-ip, CLA (Figure 6).

CONCLUSION

Chemiluminescence offers great advantages in biomedical research and diagnostic thanks to its imaging potential and the needless of photoexcitation, which diminishes the possibility of autofluorescence arising from the background signal. As a result, it has been widely used due to the high specificity of the chemiluminescent probes in different detection methods. The ready availability of synthetic imidazopyrazinone-based compounds as molecular probes has tremendous potential.

We can draw some conclusions after reviewing the numerous imidazopyrazinone derivatives reported over the last years.

First, the modifications are focused on the C-2, C-6 and C-8 positions of the imidazopyrazinone core with the aim to improve the CL properties. Secondly, we can learn about how CL activity is influenced by the relation between the substituents and efficiency.

Native CLZ emits light with a weak intensity due to the presence of a hydroxyl group on the 6-phenyl group in the structure. In fact, electron donation from the phenyl group is not necessary for the high CL efficiency. The conformational effects have more influence in CL, for example, rigidizing bridges between position 5 and the 6-phenyl group increase the efficiency, and the incorporation of a hydroxyl group capable of bonding to nitrogen atom, decrease the luminescence intensity. A fluoro group at *ortho-* or *meta-* position of the 4-hydroxyphenyl moiety also improve the Φ_{CL}.

Modifications in C-2 (methyl mostly) together with substitutions on the 4-position of 6-phenyl moiety, from electron-donating group, OCH_3 or $N(CH_3)_2$, to electron-withdrawing groups, CF_3 or F, have a small effect. Besides that, the CL efficiencies of the analogues having electron-donating groups are only slightly lower than those having electron-withdrawing groups.

The addition of an extended resonance system at the C-8 position (1-naphthyl or α-styryl) adds a significant red shift of light emission. The same effect is observed with derivatives with an oxygen or sulfur atom in place of the methylene group at C-6 or C-8 position.

Cypridina luciferin analogues show better Φ in aqueous media than native CLZ. The most sensitive chemiluminescent probes are the blue MCLA and the green FCLA. MCLA has the better efficiency, and its Φ is about three times that of CLA. FCLA has a slow reaction rate compared with the other compounds.

The substituent effect at C-6 position in MCLA has also a small influence of electron donating and electron withdrawing. About the conformational rigidity, analogues with a decreased dihedral angle between the pyrazine ring and the 4-methoxyphenyl ring in their structures show an enhancement in CL efficiency.

Different approaches have been done to achieve substrates with high light-emitting efficiency in aqueous solution. MCLA in a covalent attachment with a cyclodextrin, contributes to the enhanced of the CL. Particularly, it was demonstrated that γ-CD was the best cyclodextrin, suggesting that CL efficiencies of the CD-bound MCLA compounds depend on the efficiencies of singlet-excited state formation: the entrance and cavity of δ-CD and β-CD were too large and too small, respectively, to include the singlet-excited amide moiety. Other approaches consist in the introduction of fluorescent moieties in the molecular probe, such as BODIPY or dyes.

The development of useful probes for detecting ROS species with bathochromic shift of the CL is important for the practical application. Longer wavelengths than blue or green are desirable. Recent efforts are concentrated in the development of novel red-chemiluminescent probes with higher intensities. Although imidazopyrazinone-based compounds have been widely used for superoxide monitoring, even demonstrated their ability to selectively visualize O_2^- in normal vs. inflammation tissues, prospects in imaging and real time monitoring are expected.

Moreover, CL methods are powerful with the capability of detecting cancerous cells and cancer biomarkers, what means an increase in the chance of patient's survival.

The detection of cancer in the incipient stages of the disease is possible since these chemiluminescent molecules offer the opportunity to dynamically assess the global production of O_2^- related to cancer development and progression. As we described in this chapter, a method for multimodality imaging with CLZ as an indicator of cancer-associated superoxide was demonstrated both *in vitro* and *in vivo*. As a reminder, the absence of photoexcitation in chemiluminescent systems eliminates problems associated with light penetration into biologic tissues.

And the ROS-mediated signaling in cancer cells can be a target for developmental therapeutics. In PDT a promising alternative approach is monitoring the production of singlet oxygen in real-time imaging. In addition, it will be important for further development the exploration of both novel excitation sources and new photosensitizers with high fluorescence quantum yields and tunable absorption-emission.

ACKNOWLEDGMENTS

This work was made in the framework of the project Sustainable Advanced Materials (NORTE-01-00145-FEDER-000028), funded by "Fundo Europeu de Desenvolvimento Regional (FEDER)," through "Programa Operacional do Norte" (NORTE2020). Acknowledgment to project POCI-01-0145-FEDER-006980, funded by FEDER through COMPETE2020, is also made. The Laboratory for Computational Modeling of Environmental Pollutants-Human Interactions (LACOMEPHI) is acknowledged. Luís Pinto da Silva also acknowledges funding from FCT in the framework of the Scientific Employment Stimulus (CEECIND/01425/2017). Projects PTDC/QEQ-QFI/0289/2014 is also acknowledge, which is co-funded by FCT/MEC (PIDDAC) and by FEDER through "COMPETE – Programa Operacional Fatores de Competitividade" (COMPETE-POFC).

REFERENCES

[1] Widder, E. A. (2010). Bioluminescence in the Ocean: Origins of Biological, Chemical, and Ecological Diversity. *Science* 328: 704-708.

[2] Markova, S. W., Vysotski, E. S. (2015). Coelenterazine-Dependent Luciferases. *Biochem. (Mosc.)* 80: 714-732.

[3] Shimomura, O., Musickit, O., Kishi, Y. (1989). Semi-synthetic aequorins with improved sensitivity to Ca^{2+} ions. *Biochem. J.* 261: 913-920.

[4] Hosoya, T., Iimori, R., Yoshida, S., Sumida, Y., Sahara-Miura, Y., Sato, J., Inouye, S. (2015). Concise Synthesis of v-Coelenterazines. *Org. Lett.* 17: 3888-3891.

[5] Shimomura, O. (2006). *Bioluminescence: Chemical Principles and Methods*. 91-185. World Scientific Publishing: Singapore.

[6] Coutant, E. P., Janin, Y.L. (2015). Synthetic Routes to Coelenterazine and Other Imidazo[1,2-a]pyrazin-3-one Luciferins: Essential Tools for Bioluminescence-Based Investigations. *Chem. Eur. J.* 21: 17158-17171.

[7] Jiang, T., Du, L., Li, M. (2016). Lighting up bioluminescence with coelenterazine: strategies and applications. *Photochem. Photobiol. Sci.* 15: 466-480.

[8] Kaskova, Z. M., Tsarkovaab A. S., Yampolsky, I. (2016). 1001 lights: luciferins, luciferases, their mechanisms of action and applications in chemical analysis, biology and medicine. *Chem. Soc. Rev.* 45: 6048-6077.

[9] Teranishi, K., Goto, T. (1989). Effects of conformational rigidity and hydrogen bonding in the emitter on the Chemiluminescence efficiency of Coelenterazine (Oplophorus Luciferin). *Chem. Lett.* 1423-1426.

[10] Teranishi, K. (2007). Luminescence of imidazo[1,2-a]pyrazin-3(7H)-one compounds. *Bioorg. Chem.* 35: 82-111.

[11] Shimomura, O., Teranishi, K. (2000). Light-emitters involved in the luminescence of coelenterazine. *Luminescence* 15: 51-58.

[12] Isobe, H., Takano, Y., Okumura, M., Kuramitsu, S, Yamaguchi, K. (2005). Mechanistic insights in charge-transfer-induced luminescence of 1,2-dioxetanones with a substituent of low oxidation potential. *J. Am. Chem. Soc.* 127: 8667-86.79.

[13] Isobe, H., Yamanaka, S., Kuramitsu, S., Yamaguchi, K. (2008). Regulation mechanism of spin-orbit coupling in charge-transfer-induced luminescence of imidazopyrazinone derivatives. *J. Am. Chem. Soc.* 130: 132-149.

[14] Min, C., Ferreira, P. J. O., Pinto da Silva, L. (2017). Theoretically obtained insight into the mechanism and dioxetanone species responsible for the singlet chemiexcitation of Coelenterazine. *J. Photochem. Photobiol. B* 174: 18-26.

[15] Chen, S., Navizet, I., Roca-Sanjuán, D., Lindh, R., Liu, Y., Ferré, N. (2012). Chemiluminescence of coelenterazine and fluorescence of coelenteramide: a systematic theoretical study. *J. Chem. Theory Comput.* 8: 2796-2807.

[16] Pinto da Silva, L., Magalhães, C. M., Esteves da Silva, J. C. G. (2016). Interstate crossing-induced chemiexcitation mechanism as the basis for imidazopyrazinone bioluminescence. *Chemistry Select* 1: 3343-3356.

[17] Pinto da Silva, L., Magalhães, C. M., Crista, D. M. A., Esteves da Silva, J. C. G. (2017). Theoretical modulation of singlet/triplet chemiexcitation of chemilumine scent imidazopyrazinone dioxetanone via C_8-substitution. *Photochem. Photobiol. Sci.* 16: 897-907.

[18] Pinto da Silva, L, Pereira, R. F. J., Magalhães, C. M., Esteves da Silva, J.C.G. (2017). Mechanistic Insight into Cypridina bioluminescence with a combined experimental and theoretical chemiluminescent approach. *J. Phys. Chem. B* 121: 7862-7871.

[19] Magalhães, C. M., Esteves da Silva, J. C. G., Pinto da Silva, L. (2018). Study of coelenterazine luminescence: Electrostatic interactions as the controlling factor for efficient chemiexcitation. *J. Lumin.* 199: 339-347.

[20] Pinto da Silva, L., Magalhães, C.M. (2019). Mechanistic insights into the efficient intramolecular chemiexcitation of dioxetanones from TD-DFT and multireference calculations. *Int. J. Quantum Chem.* 119: e25881.

[21] Hirano, T., Gomi, Y., Takahashi, T., Kitahara, K., Qi, C. F., Mizoguchi, I., Kyushin, K. Ohashi, M. (1992). Chemiluminescence of Coelenterazine Analogues-Structures of Emitting Species. *Tetrahedron Lett.* 33: 5771-5774.

[22] Saito, R., Hirano, T., Maki, S., Niwa, H., Ohashi, M. (2011). Influence of Electron-Donating and Electron-Withdrawing Substituents on the Chemiluminescence Behavior of Coelenterazine Analogs. *Bull. Chem. Soc. Jpn.* 84: 90-99.

[23] Takahashi, Y., Kondo, H., Maki, S., Niwa, H, Ikeda, H., Hirano, T. (2006). Chemiluminescence of 6-aryl-2-methylimidazo[1,2-α]pyrazine-3(7H)-ones in DMSO/TMG and in diglyme/acetate buffer: support for the chemiexcitation process to generate the singlet-excited state of neutral oxyluciferin in a high quantum yield in the Cypridina (Vargula) bioluminescence mechanism, *Tetrahedron Lett.* 47: 6057-6061.

[24] Hirano, T., Takahashi, Y., Kondo, H., Maki, S., Kojima, S., Ikedab, H., Niwa, H. (2008). The reaction mechanism for the high quantum yield of Cypridina (Vargula) bioluminescence supported by the chemiluminescence of 6-aryl-2-methylimidazo [1,2-α]pyrazin-3(7H)-ones (Cypridina luciferin analogues). *Photochem. Photobiol. Sci.* 7: 197-207.

[25] Teranishi, K., Hisamatsu, M., Yamada, T. (1999). Chemiluminescence of 2-methyl-6-arylimidazo [1,2a]pyrazin-3(7 H)-one in protic solvents: electrondonating substituent effect on the formation of the neutral singlet excited-state molecule. *Luminescence* 14: 297-302.

[26] Lourenço, J. M., Esteves da Silva, J. C. G., Pinto da Silva, L. (2018). Combined experimental and theoretical study of Coelenterazine chemiluminescence in aqueous solution. *J. Lumin.* 194: 139-145.

[27] Kishi, T., Goto, T., Hirata, Y., Shimomura, O., Johnson, F. (1966). Cypridina bioluminescence I. Structure of Cypridina luciferin. *Tetrahedron Lett.* 7: 3427-3436.

[28] Kishi, Y., Goto, T., Inoue, S., Sugiura, S., Kishimoto, H. (1966). Cypridina bioluminescence III total synthesis of Cypridina luciferin. *Tetrahedron Lett.* 7: 3445-3450.

[29] Hori, K., Cormier, M. (1973). Structure and Chemical Synthesis of a Biologically Active Form of Renilla (Sea Pansy) Luciferin. *Proc. Natl. Acad. Sci. U.S.A.* 70: 120-123.

[30] Kishi, Y., Tanino, H., Goto, T. (1972). The structure confirmation of the light-emitting moiety of bioluminescent jellyfish Aequorea. *Tetrahedron Lett.* 13: 2747-2748.

[31] Shimomura, O., Johnson, F. H. (1975). Chemical nature of bioluminescence systems in coelenterates. *Proc. Natl. Acad. Sci. U.S.A.* 72: 1546-1549.

[32] Inoue, S., Sugiura, S., Kakoi, H., Hasizume, K. (1975). Squid bioluminescence ii. Isolation from watasenia scintillans and synthesis of 2-(p-hydroxybenzyl)-6-(p-hydroxyphenyl)-3,7-dihydroimidazo[1,2-α]-pyrazin-3-one. *Chem. Lett.* 141-144.

[33] Inoue, S., Taguchi, H., Murata, M., Kakoi, H., Goto, T. (1977). Squid bioluminescence IV. Isolation and structural elucidation of Watasenia dehydro-preluciferin. *Chem. Lett.* 259-262.

[34] Keenan, M., Jones, K., Hibbert, F. (1997). Highly efficient and flexible total synthesis of coelenterazine. *Chem. Commun.* 0: 323-324.

[35] Adamczyk, M., Johnson, D. D., Mattingly, P. G., Pan, Y., Reddy, R.E. (2001). Synthesis of Colenterazine. *Org. Prep. Proced. Int.* 33: 477-485.

[36] Goto, T., Takagi, T. (1980). Chemiluminescence of a Cypridina Luciferin analogue, 2-methyl-6-phenyl-3,7-dihydroimidazo[1,2-α]pyrazin-3-one, in the presence of the xanthine-xanthine oxidase system. *Bull. Chem. Soc. Jpn.* 53: 833-834.

[37] Nishida, A., Kimura, H., Nakano, M., Goto, T. (1989). A sensitive and specific chemiluminescence method for estimating the ability of human granulocytes and monocytes to generate O_2^-. *Clin. Chim. Acta* 179: 177–181.

[38] Nakano, M., Sugioka, K., Ushijima, Y., Goto, T. (1986). Chemiluminescence probe with Cypridina luciferin analog, 2-methyl-6-phenyl-3,7-dihydroimidazo[1,2-α]pyrazin-3-one, for estimating the ability of human granulocytes to generate O_2^-. *Anal. Biochem.* 159: 363-369.

[39] Suzuki, N., Suetsuna, K., Mashiko, S., Yoda, B., Nomoto, T., Toya, T., Inaba, H., Goto, T. (1991). Reaction rates for the chemiluminescence of Cypridina Luciferin analogues with superoxide: a quenching experiment with superoxide dismutase. *Abric. Biol. Chem.* 55: 157-160.

[40] Teranishi, K., Goto, T. (1990). Synthesis and chemiluminescence of Coelenterazine (Oplophorus Luciferin) analogues. *Bull. Chem. Soc. Jpn.* 63: 3132-3140.

[41] Qi, C. F., Gomi, Y., Hirano, T., Ohashi, M., Ohmiya, Y., Tsuji, F.I. (1992). Chemi- and bio-luminescence of coelenterazine analogues with phenyl homologues at the C-2 position. *J. Chem. Soc., Perkin Trans 1*: 1607-1611.

[42] Toya, Y., Kayano, T., Sato, K., Goto, T. (1992). Synthesis and chemiluminescence properties of 6-(4-methoxyphenyl)-2-methylimidazo[1,2-a]pyrazin-3(7h)-one and 2-methyl-6-(2-naphthyl)imidazo[1,2-α]pyrazin-3(7H)-one. *Bull. Chem. Soc. Jpn.* 65: 2475-2479.

[43] Chen, F. Q., Hirano, T., Hashizume, Y., Ohmiya, Y., Ohashi, M. (1994). Synthesis and preliminary chemi- and bio-luminescence studies of a novel photolabile coelenterazine analogue with a trifluoromethyl diazirine group. *J. Chem. Soc., Chem. Commun.* 2405-2406.

[44] Chen, F. Q, Zheng, J. L., Hirano, T., Niwa, H., Ohmiya, Y., Ohashi, M. (1995). A potential photoaffinity probe for labelling the active site of aequorin: a photolabile coelenterazine analogue with a trifluoromethyldiazirine group. *J. Chem. Soc., Perkin Trans 1*: 2129-2134.

[45] Hirano, T., Negishi, R., Yamaguchi, M., Chen, F. Q., Ohmiya, Y., Tsuji, F., Ohashi, M. (1995). Chemi- and bio-luminescent properties of coelenterazine analogues possessing an adamantyl group. *J. Chem. Soc., Chem. Commun.* 1335-1336.

[46] Hirano, T., Negishi, R., Yamaguchi, M., Chen, F. Q., Ohmiya, Y., Tsuji, F., Ohashi, M. (1997). Chemi- and Bioluminescence of Coelenterazine Analogues Possessing an Adamantylmethyl Group. *Tetrahedron* 53: 12903-12916.

[47] Teranishi, K., Shimomura, O. (1997). Coelenterazine analogs as chemiluminescent probe for superoxide anion. *Anal. Biochem.* 249: 37-43.

[48] Hirano, T., Ohmiya, Y., Maki, S., Niwa, H., Ohashi, M. (1998). Bioluminescent properties of fluorinated semi-synthetic Aequorins. *Tetrahedron Lett.* 39: 5541-5544.

[49] Saito, R., Hirano, T., Niwa, H., Ohashi, M. (1998). Substituent effects on the chemiluminescent properties of coelenterazine analogues. *Chem. Lett.* 27: 95-96.

[50] Nakamura, H., Wu, C., Takeuchi, D., Murai, A. (1998). Bimodal Chemiluminescence of 8-Chlorostyryl- 6-phenylethynylimidazopyrazinone: Large Bathochromic Shift Caused by a Styryl Group at 8-Position. *Tetrahedron Lett.* 39: 301-304.

[51] Teranishi, K., Komoda, A., Hisamatsu, M., Yamada, T. (1998). Synthesis and enhanced chemiluminescence of new cyclomaltooligosaccharide (cyclodextrin)-bound 6-phenylimidazo[1,2-α]pyrazin-3(7H)-one. *Carbohydr. Res.* 306: 177-187.

[52] Teranishi, K., Tanabe, S., Hisamatsu, M., Yamada, T. (1999). Investigation of cyclomaltooligosaccharide-bound 6-(4-methoxyphenyl)imidazo[1,2-]pyrazin-3(7H)-one for enhanced chemiluminescence. *Luminescence* 14: 303-314.

[53] Zheng, L., Chen, F. Q., Hirano, T., Ohmiya, Y., Maki, S., Niwa, H., Ohashi, M. (2000). Synthesis, chemi- and bioluminescence properties, and photolysis of a coelenterazine analogue having a photoreactive azido group. *Bull. Chem. Soc. Jpn.* 73: 465-469.

[54] Teranishi, K. (2001). Effect of conformation on the chemiluminescence efficiency of light-producing 2-methyl-6-(4-methoxyphenyl)imidazo[1,2-α]pyrazine-3(7H)-ones. *Luminescence* 16: 367-374.

[55] Wu, C., Nakamura, H., Muraib, A., Shimomura, O. (2001). Chemi- and bioluminescence of coelenterazine analogues with a conjugated group at the C-8 position. *Tetrahedron Lett.* 42: 2997-3000.

[56] Adamczyk, M., Akireddy, S. R., Johnson, D. D., Mattingly, P. G., Pan, Y., Reddy, R.E. (2003). Synthesis of 3,7-dihydroimidazo[1,2-α]pyrazin-3-ones and their chemiluminescent properties. *Tetrahedron* 59: 8129-8142.

[57] Teranishi, K., Nishiguchi, T., Ueda, H. (2003). Enhanced chemiluminescence of 6-(4-methoxyphenyl)imidazo[1,2-α]pyrazin-3(7H)-one by attachment of cyclomal - tooligosaccharide (cyclodextrin). Attach-ment of cyclomaltononaose (δ-cyclodextrin). *Carbohydr. Res.* 338: 987-993.

[58] Teranishi, K., Nishiguchi, T. (2004). Cyclodextrin-bound 6-(4-methoxyphenyl)imidazo[1,2-α]pyrazin-3(7H)-ones with fluorescein as green chemiluminescent probes for superoxide anions. *Anal. Biochem.* 325: 185–195.

[59] Kondo, H., Igarashi, T., Maki, S., Niwa, H., Ikedab, H., Hirano, T. (2005). Substituent effects on the kinetics for the chemiluminescence reaction of 6-arylimidazo[1,2-α]pyrazin-3(7H)-ones (Cypridina luciferin analogues): support for the single electron transfer(SET)–oxygenation mechanism with triplet molecular oxygen. *Tetrahedron Lett.* 46: 7701-7704.

[60] Takahashi, Y., Kondo, H., Maki, S., Niwa, H., Ikeda, H., Hirano, T. (2006). Chemiluminescence of 6-aryl-2-methylimidazo[1,2-α]pyrazine-3(7H)-ones in DMSO/TMG and in diglyme/acetate buffer: support for the chemiexcitation process to generate the singlet-excited state of neutral oxyluciferin in a high quantum yield in the Cypridina (Vargula) bioluminescence mechanism. *Tetrahedron Lett.* 47: 6057-6061.

[61] Teranishi, K. (2007). Development of imidazopyrazinone red chemiluminescent probes for detecting superoxide anions via a chemiluminescence resonance energy transfer method. *Luminescence* 22: 147-156.

[62] Giuliani, G., Molinari, P., Ferretti, G., Cappelli, A., Anzini, M., Vomero, S., Costa, T. (2012). New red-shifted coelenterazine analogues with an extended electronic conjugation. *Tetrahedron Lett.* 53: 5114-5118.

[63] Ishii, Y., Hayashi, C., Suzuki, Y., Hirano, T. (2014). Chemiluminescent 2,6-diphenylimidazo[1,2- α]pyrazin-3(7H)-ones: a new entry to Cypridina luciferin analogues." *Photochem. Photobiol. Sci.* 13: 182-189.

[64] Saito, R., Hirano, T., Maki, S., Niwa, H. (2014). Synthesis and chemiluminescent properties of 6,8-diaryl-2-methylimidazo[1,2-α]pyrazine-3(7H)-ones: systematic investigation of substituent effect at para-position of phenyl group at 8-position. *J. Photochem. Photobiol. A* 293: 12-25.

[65] Gealageas, R., Malikova, N. P., Picaud, S., Borgdorff, A. J., Burakova, L. P., Brûlet, P., Vysotski, E. S., Dodd, R. H. (2014). Bioluminescent properties of obelin and aequorin with novel coelenterazine analogues. *Anal. Bioanal. Chem.* 406: 2695-2707.

[66] Nishihara, R., Suzuki, H., Hoshino, E., Suganuma, S., Sato, M., Saitoh, T., Nishiyama, S., Iwasawa, N., Citterio, D., Suzuki, K. (2015). Bioluminescent coelenterazine derivatives with imidazopyrazinone C-6 extended substitution. *Chem. Commun.* 51: 391-394.

[67] Yuan, M., Jiang, T., Du, L., Li, M. (2016). Luminescence of coelenterazine derivatives with C-8 extended electronic conjugation. *Chin. Chem. Lett.* 27: 550-554.

[68] Jiang, T., Yang, X., Zhou, Y., Yampolsky, I., Dua, L., Li, M. (2017). New bioluminescent coelenterazine derivatives with various C-6 substitutions. *Org. Biomol. Chem.* 15: 7008-7018.

[69] Nishihara, R., Hoshino, E., Kakudate, Y., Kishigami, S., Iwasawa, N., Sasaki, S., Nakajima, T., Sato, M., Nishiyama, S., Citterio, D., Suzuki, K., Kim, S. (2018). Azide- and dye-conjugated coelenterazine analogues for a multiplex molecular imaging platform. *Bioconjugate Chem.* 29: 1922-1931.

[70] Schieber, M., Chandel, N. (2014). ROS function in redox signaling and oxidative stress. *Curr. Biol.* 24: 453-462.

[71] Mroz, P., Yaroslavsky, A., Kharkwal, G., Hamblin, M. (2011). Cell death pathways in photodynamic therapy of cancer. *Cancers* 3: 2516-2539.

[72] Hayyan, M., Hashim, M., AlNashef, I. (2016). Superoxide ion: generation and chemical implications. *Chem. Rev.* 116: 3029-3085.

[73] Sugioka, K., Nakano, M., Kurashige, S., Akuzawa, Y., Goto, T. (1986). A chemiluminescent probe with a Cypridina luciferin analog, 2-methyl-6-phenyl-3,7-dihydroimidazo[1,2-α]pyrazin-3-one, specific and sensitive for O_2^- production in phagocytizing macrophages. *FEBS J.* 197: 27-30.

[74] Kimura, H., Nakano, M. (1988). Highly sensitive and reliable chemiluminescence method for the assay of superoxide dismutase in human erythrocytes. *FEBS J.* 239: 347-350.

[75] Nakano, M., Kimura, H., Hara, M., Kuroiwa, M., Kato, M., Totsune, K., Yoshikawa, T. (1990). A highly sensitive method for determining both Mn. And Cu-Zn superoxide dismutase activities in tissues and blood cells. *Anal. Biochem.* 187: 277-280.

[76] Nakano, M. (1990). Assay for superoxide dismutase based on chemiluminescence of luciferin analog. *Methods Enzymol.* 186: 227-232.

[77] Uehara, K., Hori, K., Nakano, M., Koga, S. (1991). Highly sensitive chemiluminescence method for determining myeloperoxidase in human polymorphonuclear leukocytes. *Anal. Biochem.* 199: 191-196.

[78] Uehara, K., Maruyama, N., Huang, C-K., Nakano, M. (1993). The first application of a chemiluminescence probe, 2-methyl-6-[p-methoxyphenyl]-3,7-dihydroimidazo

[1,2-α]pyrazin-3-one (MCLA), for detecting O_2^- production, in vitro, from Kupffer cells stimulated by phorbol myristate acetate. *FEBS J.* 335: 167-170.

[79] Wada, K., Umemura, K., Nishiyama, H., Saniabadi, A. R., Takiguchi, Y., Nakano, M., Nakashima, M. (1996). A chemiluminescent detection of superoxide radical produced by adherent leucocytes to the subendothelium following thrombolysis: studies with a photochemically induced thrombosis model in the guinea pig femoral artery. *Atherosclerosis* 122: 217-224.

[80] Shimomura, S., Wu, C., Murai, A., Nakamura, H. (1998). Evaluation of five imidazopyrazinone-type chemiluminescent superoxide probes and their application to the measurement of superoxide anion generated by Listeria monocytogenes. *Anal. Biochem.* 258: 230-235.

[81] Midorikawa, J., Maehara, K., Yaoita, H., Watanabe, T., Ohtani, H., Ushiroda, S., Maruyama, Y. (2001). Continuous observation of superoxide generation in an in-situ ischemia –reperfusion rat lung model. *Jpn. Circ. J.* 65: 207-212.

[82] Mitani, M., Sakaki, S., Koinuma, Y., Toya, Y., Kosugi, M. (1995). Enhancement Effect of 2,6-O-Dimethyl-β-cyclodextrin on the chemiluminescent detection β-D-Galactosidase using a Cypridina Luciferin analog. *Anal. Sci.* 11: 1013-1015.

[83] Sekiya, M., Umezawa, M., Sato, A., Citterio, D., Suzuki, K. (2009). A novel luciferin-based bright chemiluminescent probe for the detection of reactive oxygen species. *Chem. Commun.* 3047-3049.

[84] Li, P., Liu, L., Xiao, H., Zhang, W., Wang, L., Tang, B. (2016). A new polymer nanoprobe based on chemiluminescence resonance energy transfer for ultrasensitive imaging of intrinsic superoxide anion in mice. *J. Am. Chem. Soc.* 138: 2893-2896.

[85] Niu, J., Fan, J., Wang, X., Xiao, J., Xie, X, Jiao, X., Sun, C., Tang, B. (2017). Simultaneous fluorescence and chemiluminescence turned on by aggregation-induced emission for real-time monitoring of endogenous superoxide anion in live cells. *Anal. Chem.* 89: 7210-7215.

[86] Lucas, M., Solano, F. (1992). Coelenterazine is a superoxide anion-sensitive chemiluminescent probe: its usefulness in the assay of respiratory burst in neutrophils. *Anal. Biochem.* 206: 273-277.

[87] Tarpey, M. M., White, C. R., Suarez, R., Richardson, G., Radi, R., Freeman, B. A. (1999). Chemiluminescent detection of oxidants in vascular tissue Lucigenin but not Coelenterazine enhances Superoxide formation. *Circ. Res.* 84: 1203-1211.

[88] Kervinen, M., Pätsi, J., Finel, M., Hassinen, I. (2004). Lucigenin and coelenterazine as superoxide probes in mitochondrial and bacterial membranes. *Anal. Biochem.* 324: 45-51.

[89] Saleh, L., Plieth, C. (2010). A coelenterazine-based luminescence assay to quantify high-molecular-weight superoxide anion scavenger activities. *Nat. Protoc.* 5: 1635-1641.

[90] Bronsart, L., Stokes, C., Contag, C. (2016). Multimodality imaging of cancer superoxide anion using the small molecule Coelenterazine. *Mol. Imaging Biol.* 18: 166-171.

[91] Bronsart, L., Stokes, C., Contag, C. (2016). Chemiluminescence imaging of superoxide anion detects beta-cell function and mass. *PLoS One* 11: e0146601.

[92] Bronsart, L., Nguyen, L., Habtezion, A., Contag, C. (2016). Reactive oxygen species imaging in a mouse model of inflammatory bowel disease. *Mol. Imaging Biol.* 18: 473-478.

[93] Magalhães, C., Esteves da Silva, J., Pinto da Silva, L. (2016). Chemiluminescence and bioluminescence as an excitation source in the photodynamic therapy of cancer: a critical review. *Chem Phys Chem* 17: 2286-2294.

[94] Agostinis, P., Berg, K., Cengel, K., Foster, T., Girotti, A., Gollnick, S., Hahn, S., Hamblin, M., Juzeniene, A., Kessel, D., Korbelik, M., Moan, J., Mroz, P., Nowis, D., Piette, J., Wilson, B., Golab, J. (2011). Photodynamic therapy of cancer: an update. *Cancer J. Clin.* 61: 250-281.

[95] Wang, J., Xing, D., He, Y., Hu, C. (2002). Localization of tumor by chemiluminescence probe during photosensitization action. *Cancer Lett.* 188: 59-65.

[96] Qin, Y., Xing, D., Luo, S., Zhou, J., Zhong, X., Chen, Q. (2005). Feasibility of using fluoresceinyl cypridina luciferin analog in a novel chemiluminescence method for real-time photodynamic therapy dosimetry. *Photochem. Photobiol.* 81: 1534-1538.

[97] Wei, Y., Zhou, J., Xing, D., Chen, Q. (2007). In vivo monitoring of singlet oxygen using delayed chemiluminescence during photodynamic therapy. *J. Biomed. Opt.* 12: 014002.

[98] TCI Co., ltd. 2019. Accessed January 10. https://www.tcichemicals.com.

[99] Invitrogen, a brand from Thermo Fischer Scientific Corporation 2019. Accessed January 10. https://www.fishersci.com/us/en/brands/IIAM0WMR/invitrogen.html.

[100] Prolume Ltd. 2019. Accessed January 10. http://www.prolume.com.

[101] Biotium. 2019. Accessed January 10. https://biotium.com.

In: A Comprehensive Guide to Chemiluminescence
Editor: Luís Pinto da Silva
ISBN: 978-1-53616-170-0
© 2019 Nova Science Publishers, Inc.

Chapter 2

EMITTER STRUCTURE IN THE FIREFLY LUCIFERIN-LUCIFERASE SYSTEM: A HISTORICAL PERSPECTIVE

Natalia N. Ugarova[1,], PhD and Galina Y. Lomakina[1,2], PhD*
[1]Department of Chemistry, Lomonosov Moscow State University, Moscow, Russia
[2]Bauman Moscow State Technical University, Moscow, Russia

ABSTRACT

Bioluminescence of firefly luciferases is characterized by complex changes of the emission spectra and bioluminescence λ_{max} occurring with changing pH, temperature, and the enzyme structure. Analysis of the literature data and the result of our own research on the nature of the emitter in the firefly luciferin-luciferase system allows concluding that keto-enol tautomerization of the excited oxyluciferin molecule provides the most adequate explanation of the observed complex spectral changes. Only one emitter molecule can be present in the active center of the luciferase molecule, hence, the emitter can be considered as an intramolecular probe reporting on the properties of its microenvironment in the enzyme active center. Superposition of two or three forms of the emitter recorded in the bioluminescence spectra implies that equilibrium of different conformations of the enzyme exist in the reaction medium with each generating its own form of the emitter.

1. INTRODUCTION

Scientists have been interested in the phenomenon of bioluminescence since ancient times. The terms "luciferin" and "luciferase" were first introduced in the publication authored by the French scientist R. Dubois in 1875, who investigated bioluminescence of

* Corresponding Author's Email: nugarova@gmail.com.

the extracts of several luminous organisms and established that the enzyme (luciferase) contained in the extract produced at low temperature and substrate (luciferin) produced by extraction at boiling point were required for generation of light. The modern period of investigation of the firefly luciferin-luciferase system started in the 1940s. In 1947 McElroy demonstrated that the intensity of luminescence from the extract of *Photinus pyralis* fireflies was proportional to concentration of ATP and that oxygen was required for generation of luminescence [1]. In 1953, it was shown that four components were needed in the system for bioluminescence generation: enzyme – luciferase (Luc), substrate – luciferin (LH$_2$), ATP$\cdot$ Mg^{2+}, and oxygen [2]. In 1956 Green and McElroy detected pyrophosphate (PP$_i$) and AMP among the reaction products [3]. The scheme of the firefly luciferase reaction was proposed to include two steps.

$$Luc + ATP\bullet Mg^{2+} + LH_2 \rightarrow Luc\bullet LH_2 - AMP + PPi\bullet Mg^{2+} \tag{1}$$

$$Luc\bullet LH_2 - AMP + O_2 \rightarrow Luc + CO_2 + AMP + oxyluciferin (LO) \tag{2}$$

In the first step, the enzyme catalyzes the interaction of ATP$\bullet$Mg^{2+} and LH$_2$ resulting in the formation of luciferyl-adenylate (LH$_2$–AMP), independently of the presence of oxygen. In the second step LH$_2$–AMP is oxidized by air oxygen and emission of light is observed. Luciferin was isolated from the extract of fireflies in 1957, and it was established that the molecule contained carboxyl and phenol groups. The carboxyl group is essential for formation of LH$_2$–AMP, which does not occur if the luciferin methyl ether is used [4]. The adenylation step was confirmed by the fact that the chemically-synthesized LH$_2$–AMP also generated bright luminescence in the presence of luciferase [5]. The chemical structure of LH$_2$ was suggested in 1961 [6], and was later validated by the complete chemical synthesis of the substrate [7]. McElroy and Seliger discovered in 1962 that one oxygen molecule was consumed during oxidation of one LH$_2$ molecule and that was required for formation of the linear hydroperoxide of LH$_2$–AMP [8]. In 1960s dipolar aprotic solvents were successfully employed for investigation of the mechanism of luminol chemiluminescence, and later the same approach was used for elucidation of the mechanism of LH$_2$ oxidation. Bright chemiluminescence was observed for LH$_2$–AMP as well as for the methyl ether of luciferin in DMSO in the presence of strong bases. It was concluded that the main function of ATP was the increase of acidity of the C$_4$ proton in the thiazoline cycle that facilitated its subsequent removal with formation of carbanion [9]. The mechanism of the luciferase reaction was suggested [10] based on the use of luciferin chemiluminescence as models [8, 9]. Schematic representation of the luciferin oxidation is presented in Figure 1.

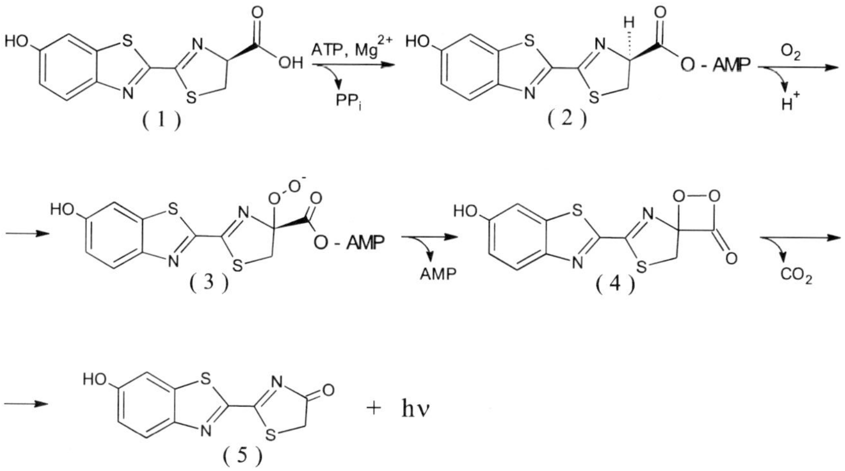

Figure 1. Schematic representation of enzymatic oxidation of firefly luciferin.

According to this scheme, in the first step the enzyme binds luciferin (1) and ATP that results in producing LH_2–AMP (2) and release of PP_i. Interaction of the intermediate carbanion with O_2 leads to production of a hydroperoxide (3) at position C_4 of the thiazoline cycle [8]. The subsequent removal of AMP, as a good leaving group, results in formation of the cyclic peroxide (dioxetanone) (4), decomposition of which is accompanied by the release of CO_2 and formation of the electronically excited state of oxyluciferin (5). The reaction product in this scheme—oxyluciferin (LO)—is presented as a ketone. However, this is only an arbitrary representation of the emitter of the bioluminescence reaction. The scheme and mechanism of enzymatic oxidation of luciferin was the research goal of numerous studies. However, this topic is outside of the scope of this review. We focused our attention in this review on the consideration of the emitter (or emitters) structure based on the available literature data and our own research as, so far, there is no consensus among researchers regarding this matter. Historical perspective of this issue will be provided with main emphasis on the experimental studies that produced most adequate, to our opinion, validation of the hypothesis suggested by authors.

2. INVESTIGATION OF THE EMITTER STRUCTURE IN THE FIREFLY LUCIFERASE REACTION USING MODEL SYSTEMS

Great interest has been expressed by researchers to investigation of the bioluminescence spectra and emitter structure in the luciferin-luciferase system already in the first studies on the mechanism of the luciferase reaction. In 1960 Seliger and McElroy studied bioluminescence spectra in the reaction catalyzed by the luciferase from *Photinus pyralis* fireflies. The dependence of emission spectra on pH was demonstrated for the first time: at pH 7.6 the *in vitro* bioluminescence spectrum was identical to *in vivo* bioluminescence. A single band in the yellow-green region of the spectrum was observed at neutral and slightly basic pH *in vitro*. The emission maximum shifted to 616 nm at pH 6.3, but the shoulder in the yellow-green region (approximately 70% of the maximum) was still maintained. Only one band with maximum at 620 nm was observed at pH 5.4 and below. The emission quantum yield also depended on pH: it was ~88% at 562 nm (pH 7.6), and it decreased to (20 – 40)% at acidic pH. These measurements were carried out with luciferin isolated from fireflies, which could be a mixture of enantiomers [11]. The measurements conducted later demonstrated that the quantum yield in the firefly luciferin-luciferase system was ~41%. Nevertheless, this system is considered as the most effective among the all known bioluminescent systems [12]. Considering that the firefly bioluminescence maximum in nature varied in a wide range, the conclusion made by Seliger and McElroy [13] that the spectra were defined to a great extent by microenvironment of the emitter in the enzyme active center seemed entirely plausible.

It was shown in a number of studies that the chemiluminescence spectra of luciferin and its analogues in DMSO in the presence of bases were very similar to the fluorescence spectra of the products of these reactions. This provided a means for extensive use of these spectral methods for identification of the structure of feasible emitters in the firefly bioluminescence. In 1967 Hopkins et al. [14] suggested based on the investigation of luciferin chemiluminescence in DMSO that the red chemiluminescence of luciferin in DMSO solution in the presence of bases could serve as a model of red bioluminescence observed *in vitro* at acidic pH. The authors assigned the keto-anion (1) formula to this product, which had two resonance structures (Figure 2).

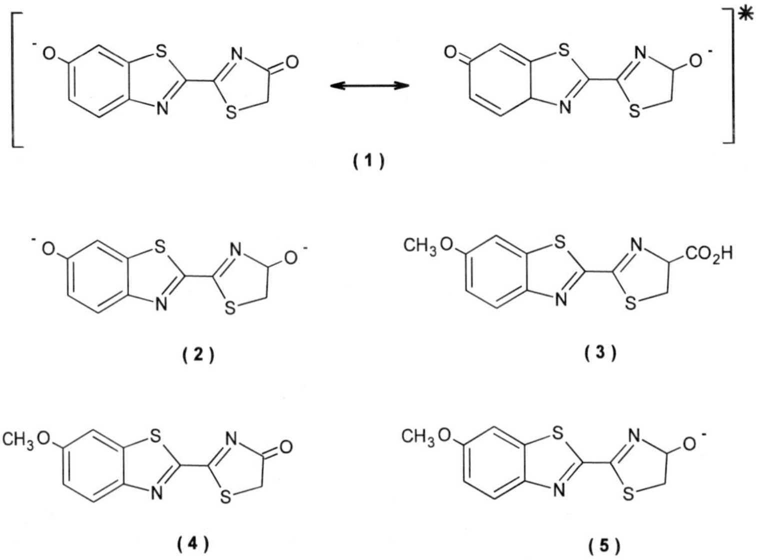

Figure 2. Suggested structures of emitter in chemiluminescence of luciferin and its analogues in DMSO in the presence of strong bases as models of bioluminescence emitters. Structure designations:

(1) – keto-anion as red emitter; (2) – dianion as yellow-green emitter; (3) - 6'-methoxy-luciferin; (4) - 6'-methoxy-keton as blue emitter; (5) - 6'-methoxy-enolate as weak orange emitter.

In 1969 White et al. investigated the effect of the base concentration in DMSO on the chemiluminescence spectra of luciferin, its methyl- and dimethyl-derivatives at the C_5 atom in thiazoline cycle and suggested that the dianion (2) was the yellow-green emitter in this system. Only red chemiluminescence independent on the base concentration was observed in the case of 5, 5-dimethylluciferin that was not capable of forming enol structure. 6'-Methoxy-luciferin (3) with phenol group not capable of ionization demonstrated only low-intensity blue chemiluminescence in the presence of low base concentration, which was assigned to the neutral form of the ketone (4). Higher base concentrations induced weak orange luminescence, which was likely related to the product with ionized enol group (5); nevertheless, it was the authors' opinion that this product was an unlikely emitter for the bioluminescence system due to the low efficiency of its chemiluminescence and its spectral characteristics [15].

Morton et al. investigated fluorescent properties of luciferin, luciferyl adenylate, and several of its analogues in aqueous solutions [16] and showed that 6'-methoxyluciferin was practically non-fluorescent in aqueous solutions (quantum yield below 0.03). Moreover, it was impossible to produce quantitative data for these compounds due to their low stability in aqueous solutions. The pK value of the phenolic group of luciferin in the excited state was found to be around −1. This implied that luciferin existed in the excited state as a phenolate ion. The blue fluorescence shoulder was observed in the green fluorescence spectrum of luciferin at acidic pH. Similarly to luciferin, luciferyl adenylate produced fluorescence with quantum yield of approximately 0.45 at pH 4.5. This indicated that the emitter assumed the form of phenolate ion during emission. Bioluminescence spectra for the luciferases from 45 species of fireflies were examined by the time this paper was published (1969). The authors mentioned that the bioluminescence spectra for different firefly luciferases were rather broad and unstructured and did not correspond to a one single band with $\lambda_{max.em}$ varying from 546 to 594 nm. Morton et al. very reasonably suggested that the observed differences in the luciferase bioluminescence spectra could be explained by the three different mechanisms (or their combination):

1) Emitter is bound to the enzyme active center, and its microenvironment produces a non-specific solvent effect on the spectral properties of the emitter. The non-specific interactions result in the change of $\lambda_{max.em}$ in the bioluminescence spectrum not affecting significantly the spectrum shape. Thus, reflecting the combined effect of all solvent molecules surrounding the emitter. The total effects are determined by electron polarizability of the solvent, which is characterized by the refractive index, and by molecular polarizability, which is the result of re-orientation of the dipoles of solvent molecules and depends on permittivity of a medium.

2) Emitter environment in the enzyme active center specifically affects properties of the chromophore and its bioluminescence spectrum. The specific interactions comprise formation of hydrogen and acid-base bonds, or charge transfer complexes between the functional groups of the emitter and compounds in its vicinity. Such interactions could result in a significant shift of the $\lambda_{max.em}$ and change of the emission spectrum shape. The specific interaction of the emitter with solvent can occur either in the ground or excited state. If this interaction occurs only in the excited state it does not affect the absorption spectra.

3) Interaction between the benzothiazole and adenine cycles in the enzyme active center could affect properties of the excited chromophore. Moreover, the energy of such interaction could be different for different luciferases. Under *in vitro* conditions the shift of the bioluminescence λ_{max} is observed from yellow-green to red region of the spectrum depending on pH and presence of various additives. This indicates the change in the emitter structure. In this case light emission could occur from two excited species, that is, from two ionic forms of the emitter molecule. This is not the gradual shift of the maximum position that is observed in the case when the solvent – emitter interactions change the energy of the excited state of the same emitter species. The problems of interrelation between the emitter structure and its microenvironment have been discussed later in the review [17].

Hence, the results presented in [14-16] suggest that it is very likely that the red emission is related to the keto-form of the bioluminescence emitter, while the yellow-green emission – to the enol/enolate form of the emitter. The authors emphasize that the enzyme acts not only as a catalyst of the bioluminescence reaction, but also a) ensures formation of the phenolate ion, b) creates microenvironment of the product molecule, which determines $\lambda_{max.em}$ of the electronically excited state of the product via general and/or specific interactions with the emitter, and c) capable of establishing the pH-dependent tautomeric equilibrium between the keto- (red emitter) and enol (yellow-green emitter) forms of the product molecule.

In 1970s the Suzuki group significantly improved the method for oxyluciferin synthesis [18-19], which facilitated the synthesis of the product with high purity and validation of its structure using IR and NMR spectroscopy. The fluorescence $\lambda_{max.em}$ of oxyluciferin in DMSO (556 ± 3) nm was shown to be practically identical to the spectrum of yellow-green chemiluminescence of luciferin ($\lambda_{max.em}$ =555 nm) reported in [15] that was assigned to the dianion (2) (Figure 2) as an emitter of the yellow-green bioluminescence. These data once again corroborated the hypothesis on keto–enol tautomerism and on the structure of red and yellow-green fluorophores as plausible bioluminescence emitters. The authors demonstrated that the oxyluciferin molecule existed in DMSO and acetone in a form of enol. Special consideration was given by authors to oxyluciferin stability. Its synthesis required anaerobic conditions. Even the attempts of re-crystallization of the synthesized oxyluciferin resulted in its partial decomposition. It revealed the reason why previously many researches failed in isolation of the luciferase reaction products from the reaction medium. Oxyluciferin degradation occurred in aqueous solutions in the presence of oxygen. The authors isolated the product of degradation and demonstrated that this product comprised amide that was formed via the C–S cleavage in the thiazoline cycle (Figure 3). This product demonstrated fluorescence in the yellow-green region of the spectrum with $\lambda_{max} = 498$ nm [19].

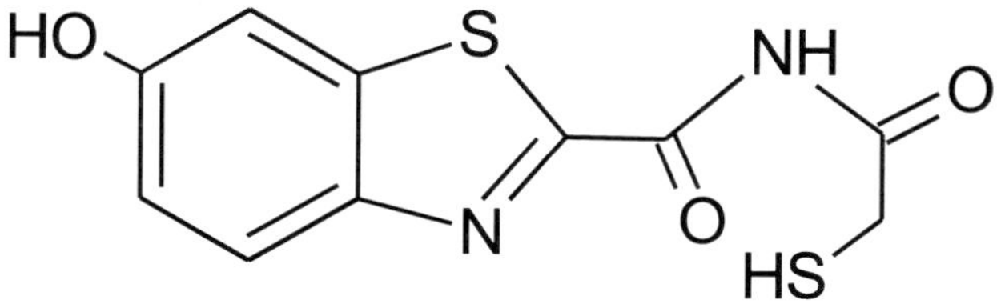

Figure 3. Structure of the amide - product of degradation of oxyluciferin in aqueous solution in the presence of oxygen.

The absorption, stationary and sub-nanosecond time-resolved fluorescence spectra of oxyluciferin and its analogues (luciferin, 6'-methoxyluciferin and 2-cyano-6-hydroxybenzothiazole (BT)) in aqueous solutions in the pH range 1–10 were obtained for the first time in our laboratory in 1993 [20-21]. The spectra were recorded under strictly anaerobic conditions to prevent oxyluciferin decomposition in the presence of air oxygen. The spectra of these compounds in ethanol were recorded for comparison. Oxyluciferin in ethanol fluoresces only in the blue region of the spectrum and demonstrates very short fluorescence lifetime (0.72 ns), while in aqueous solutions the fluorescence lifetime increases to 3.34 ns. Three oxyluciferin fluorescence emitters were identified: blue – phenol, yellow-green – enol-phenolate, and red – keto-phenolate. It was shown that the rate of dissociation of the 6'-OH phenol group in the electronically excited luciferin was lower than the rate of emission of the phenolic form itself. Hence, despite the fact that the pK values of the 6'-OH phenol group dissociation in the ground and excited states differ significantly (8.5 and –0.5, respectively), fluorescence in the blue region of the spectrum is observed in nonpolar and strongly acidic aqueous solution although of low intensity. It was concluded that the aqueous solutions of oxyluciferin represent more realistic and adequate model of firefly bioluminescence than the ethanol ones. The pH-dependence of the fluorescence spectra in aqueous solutions was similar to the pH-dependence of the bioluminescence spectra [20].

The absorption and fluorescence spectra of oxyluciferin derivatives with substitutions at the position C_5 (monomethyl-oxyluciferin, M-LO, and dimethyl-oxyluciferin, DM-LO) in aqueous solutions in the pH range 6.5 -9.6 were obtained in our laboratory in 2006 [22]. The scheme of possible equilibria for oxyluciferin and its derivatives in aqueous solutions was confirmed (Figure 4).

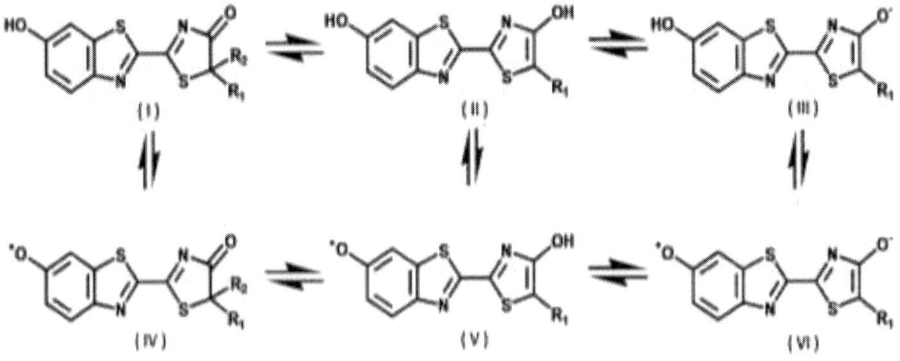

Figure 4. Different form of oxyluciferin and its 5, 5-substituted analogues: oxyluciferin $R_1 = R_2 = H$; monomethyl-oxyluciferin (M-LO) $R_1 = CH_3$, $R_2 = H$; dimethyl-oxyluciferin (DM-LO) $R_1 = R_2 = CH_3$.

Phenolic form (I) ($\lambda_{max.abs} = 383$ nm) and phenolate form (IV) ($\lambda_{max.abs} = 485$ nm, $\lambda_{max.em} = 639$ nm with pK = 7.8) were detected for DM-LO. Excitation of the phenolate form ($\lambda_{max.ex} = 485$ HM) generated only one fluorescence peak with $\lambda_{max.em} = 639$ nm in the entire investigated pH range (from 5.7 to 10.5). Three forms were identified based on the absorption spectra of M-LO: neutral, form (II) ($\lambda_{max.abs} = 375$ nm), mono-anion, form (III) ($\lambda_{max.abs} = 390$ nm), and dianion, form (VI) ($\lambda_{max.abs} = 440$ nm). The fluorescence spectra of M-LO displayed one maximum at 550 nm independent on the excitation wavelength, which corresponded to the emission of dianion (form (VI)). Hence, the DM-LO fluorescence corresponds to the red luminescence, and the M-LO fluorescence – to the yellow-green.

Although DM-LO and M-LO were more stable in aqueous solution than oxyluciferin, nevertheless we detected degradation of these substituted oxyluciferins during incubation in aqueous solutions at room temperature in the presence of oxygen based on the changes of their absorption spectra from which the rate constants of decomposition were calculated. The half-lives of M-LO and DM-LO at pH 7.8 in buffer solutions were 65 min and ~ 8.6 h, respectively. Excitation of the DM-LO phenol form ($\lambda_{max.ex} = 383$ nm) at pH ≤ 8.0 produced one fluorescence peak at $\lambda_{max.em} = 639$ nm. Yellow-green fluorescence ($\lambda_{max.em} = 500$ nm) emerged in the alkaline solutions with its intensity increasing with increasing pH. Moreover, the peak remained constant with the decreasing pH. It was suggested that this new peak is related to the product of DM-LO decomposition. In the case of M-LO, excitation at 375 nm resulted in gradual shift of the fluorescence spectrum towards shorter wavelength with the increasing pH. It seemed likely that this effect could be also explained by the M-LO decomposition, and the observed fluorescence spectrum comprised superposition of two spectra: dianion ($\lambda_{max.em} = 550$ nm) and product of degradation ($\lambda_{max.em} = 500$ nm). The products of degradation of M-LO and DM-LO were isolated from the reaction mixtures that demonstrated $\lambda_{max.abs}$ at 310 and 350 nm and $\lambda_{max.em} = 500$ nm, respectively. These spectral characteristics were found to be similar to the spectra of the product of oxyluciferin degradation in aqueous solutions in the presence of oxygen - amide

(Figure 3), which was described in [19]. Moreover, emission maximum of these products corresponded to the emission maximum of the BT compound described in [20], which also did not contain ionogenic groups.

Hence, already at the start of 21st century substantial experimental evidence has been accumulated corroborating contribution of the keto-enol tautomerization mechanism to the firefly bioluminescence reaction. It must be mentioned that the studies described above were conducted with model systems. However, the observed similarity of the chemiluminescence and fluorescence spectra in these systems with the firefly luciferase bioluminescence spectra under *in vitro* conditions allowed using the same approaches for deconvolution of the bioluminescence spectra of different luciferase and explanation of their pH-dependence.

3. ALTERNATIVE MECHANISMS DEFINING EMITTER STRUCTURE IN THE FIREFLY LUCIFERIN-LUCIFERASE SYSTEM

Despite the large number of publications considering the structure of bioluminescence emitter based on keto-enol tautomerization, several well-known authors suggested alternative hypothesis on the emitter structure. McCapra et al. suggested in 1994 [23] that oxyluciferin in its keto-form represents the only emitter in this system, which could exist as two stereo-conformations with the planar and perpendicular mutual arrangement of the benzothiazole and thiazole rings. Quantum-mechanical calculations conducted by the authors showed that the perpendicular conformation in the excited state corresponded to the minimum of potential energy of the excited oxyluciferin associated with emission of the red light, while the planar conformation corresponded to the saddle point of potential energy surface with higher emission energy. The authors suggested that conformation of the excited oxyluciferin in the rigid microenvironment of the enzyme active center would remain planar as in the ground state, and, hence, would generate emission of green light, while if the rigidity of the enzyme active center is compromised the perpendicular conformation would be formed accompanied by the emission of red light. The quantum-chemical calculations conducted later by another group of authors revealed that the planar conformation was the optimal one [24 - 27], which is why the mechanism of McCapra has been not considered as valid since.

In 2002 Branchini et al. reported that the synthetically produced luciferin analogue - 5, 5-dimethylluciferin - that was incapable of enolization, generated in the reaction with luciferase not the expected red, but green bioluminescence as natural luciferase substrate, luciferin [28]. Based on these experimental data in 2004 Branchini et al. suggested the model, in which different resonance structures of the oxyluciferin keto-anion generated green and red bioluminescence [29] (Figure 5).

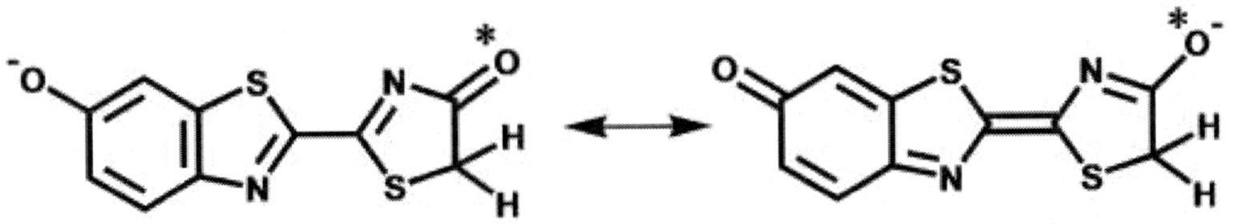

Figure 5. Phenolate (*a*) and quinoid (*b*) resonance forms of oxyluciferin keto-anion.

It is worth reminding that the equilibrium between the quinoid and phenolate forms of oxyluciferin keto-anion was reported first in the study of White et al. [15] in 1969 as indicated above in Figure 2 (form 1). White et al. considered the forms (a) and (b) as possible resonance forms of the same bioluminescence red emitter. According to the mechanism suggested by Branchini et al. on the other hand, the quinoid form of keto-anion contains the system of conjugated bonds, and its precisely this form is responsible for the typical long wavelength (red) fluorescence of the ketone in polar solvents [30, 31] and bioluminescence of the firefly luciferase. Formation of the quinoid anion also explains the strong long wavelength shift of the absorption spectrum during deprotonation of the neutral phenolic keto-form (1) (Figure 4): from 385 nm to 485 nm in water [22] and to 580 nm in DMSO [31]. Branchini et al. suggested that phenolate structure of the product was associated predominately with the green bioluminescence, and the color of the luciferase bioluminescence was controlled by the resonance stabilization of the phenolate form of the keto-anion via Coulomb interaction with the negatively charged phosphate groups of AMP and R220 cation [29]. Distortion of the enzyme structure with changes of pH and/or other external conditions, as well as on introduction of mutations could disrupt rigid orientation of AMP and R220, which could result in formation of the regular quinoid form of the product and red bioluminescence. The authors argued that theoretically the resonance mechanism could explain the entire set of the bioluminescence spectra, however, this hypothesis was difficult to validate experimentally [29]. It must be also mentioned that the resonance mechanism is in good agreement with the gradual change of the spectra but cannot explain the discrete shift observed with the pH decrease, which more likely corresponds to the availability of two separate molecular forms of the emitter. The fact that in the case of amino-luciferin only red bioluminescence is observed [32] also requires explanation as the stabilization mechanism suggested by Branchini is not applicable in this case. A number of studies on theoretical modeling of emission spectra of oxyluciferin in the complex with protein reveal that oxyluciferin interaction with the surrounding protein groups theoretically can cause switch from red to green emission spectrum of the oxyluciferin keto-anion [24, 25]. However, these results seem questionable and must be treated with caution. Firstly, not all the presented calculation results were in agreement with the experimental results for the fluorescence spectra of various forms of oxyluciferin [25, 33]. Secondly, oxyluciferin environment during calculations was selected rather arbitrary: for example, the R220 residue was positioned in [34] as counter-ion of the 6'-O⁻ -group, which did not correspond to the data on the structure of the luciferase complexes with DLSA and reaction products [35].

On the other hand, careful consideration of experimental data presented in [29] showed that the relative activity of 5,5-dimethylluciferyl adenylate was only 2.9% of the natural luciferyl adenylate. This suggested that a certain non-specific mechanism of green bioluminescence initiation was involved in the process, likely related to the low stability of 5,5-dimethylluciferyl adenylate. One cannot rule out the possibility that in the Branchini's experiments degradation of 5,5-dimethylluciferyl adenylate took place resulting in formation of the products similar to the ones mentioned above for oxyluciferin and its analogues [19, 22], which were transformed during incubation with luciferase into the products with $\lambda_{max.em}$ in the yellow-green range of the spectrum.

Another explanation suggested by Orlova et al. based on the quantum-chemical calculations is the so-called effect of microenvironment [26]. According to these authors, the entire variety of the firefly bioluminescence spectra theoretically can be explained by polarization of the OH-bond in the protein microenvironment in any form of the emitter [27]. In 2015 Pinto da Silva et al. also employed theoretical approach for investigation of the process of oxyluciferin deprotonation in the enzyme-like environment via proton transfer from the chemiexcited oxyluciferin to AMP. The scheme suggested by authors includes AMP as an acceptor of the proton from the 5-CH_2 - group of oxyluciferin. The conducted calculations demonstrated that the spectrum of enolate ion covered the entire range of yellow-green and red bioluminescence and explained variety of colors in firefly bioluminescence. The suggestion seems too categorical as it excludes keto-form from the list of plausible emitters in the firefly luciferase reaction. Although at the end of the paper, the authors point to the possibility of co-existence of various forms of oxyluciferin [36].

Despite the large number of alternative suggestions on the mechanisms describing the nature of emitter in the firefly luciferin-luciferase system, it is our opinion that the most substantiated mechanism is the mechanism of keto-enol tautomerization based on co-existence of various forms of oxyluciferin (Figure 4), spectral characteristics of which are defined by their molecular structure and microenvironemt in the luciferase active center. Furthermore, it is commonly accepted that independent on the nature of particular molecular forms of the emitter, the bioluminescence spectrum is defined to a considerable degree by the interaction of the emitter with protein microenvironment. A rigid microenvironment of the emitter and strict orientation of the key amino acid groups of the enzyme are required for realization of the green bioluminescence [17, 35, 37]. It is precisely the different conformation of the luciferase active center that stabilizes one or another form of the emitter.

4. Novel Experimental Evidence for Existence of Keto-Enol Tautomerization of Oxyluciferin in Solution and in Complex with Luciferase

Proton transfer is a well-studied and fundamental chemical process that plays an important role in multiple chemical and biological reactions. The most important examples include keto-enol tautomerization, proton transport, and proton-conducting systems in membrane proteins and enzymes. Heterocyclic aromatic molecules represent a special class of compounds for which proton transfer is typical. These molecules change their acidity on electron excitation. The acidity of photoacids and basicity of photo bases increases on excitation. The intramolecular excited state proton transfer (ESPT) is observed when the acidic and basic forms co-exist in one electronically excited molecule. The ESPT process can be also realized between molecules as in the case of proton transfer to a solvent. Oxyluciferin belongs to the family of fluorophores and represents an example of photoacids participating in the intermolecular ESPT to solvent molecules. Investigation of this class of compounds resulted in discovery of numerous interesting and unexpected photo-induced phenomena based on ESPT: efficient fluorescence quenching, strong acidity of enol group, photo-induced keto-enol tautomerization catalyzed by bases, and others [36].

New interesting results on the processes of proton transfer were reported recently based on the investigation of spectral properties and structure of oxyluciferin and its analogues in various systems: in crystals, in solution, as well as in complex with the *Luciola cruciata* firefly luciferase. Crystal structures of the unsubstituted oxyluciferin (LO) and and its 5-methyl analogue (M-LO) were elucidated for the first time by Naumov et al. in 2009 [38], according to which both molecules exist in crystals as true *trans*-enol forms — enol-LO and enol-M-LO —with head-to-tail orientation via hydrogen bonds. Their stationary absorption and emission spectra (in solution and in solid state) and nanosecond time-resolved fluorescence spectra (in solution) were obtained and assigned to the six plausible *trans* forms of the emitter and its anions (Figure 4). It was shown that the wavelengths in $\lambda_{\text{max.em.}}$ of the six chemical forms of LO were in good agreement with the theoretically predicted energy of S_0-S_1 transfers, which were in the range from blue to red region of the spectrum. The corresponding $\lambda_{\text{max.abs}}$ varied from ultraviolet to green range of the spectrum. It was confirmed that both neutral forms — phenol-enol and phenol-ketone — were blue emitters, while the phenolate-enolate form was a yellow-green emitter. The phenol-enolate form was suggested to exist only in the mixture with other forms, and the phenolate-enolate dianion was either yellow or orange emitter with very close $\lambda_{\text{max.em.}}$. The ratio of concentrations of various chemical forms in the oxyluciferin solution depends of several factors, which affect triple equilibrium with most important among them being pH, solvent polarity, hydrogen bonds, addition of various ions, and π-π stacking interactions. Owing to stabilization of the enol group in the 4-hydroxythiazole ring by the hydrogen bond with proton acceptors in its vicinity, which according to theoretical calculations is close in energy to dimerization of two enol molecules observed in crystals, the phenolate-enol is the dominating species in the ground state in buffer solutions at pH 7.44–8.14 and the most probable emitter of yellow-green bioluminescence in the majority of wild type firefly luciferases. [38].

In 2012 Naumov et al. published the results of first systematic study examining oxyluciferin and two of its chemically modified analogues by picosecond time-resolved spectroscopy, which allowed establishing that the enol group in the excited state was more acidic than the phenol one. It is worth mentioning that this fact was described in earlier publications devoted to investigation of the fluorescence spectra of oxyluciferin analogues [15, 20, and 22]. The pKa values of the 4-enol hydroxyl proton in the excited state of 0.9 in DMSO and –0.3 in water were measured experimentally. Moreover, it was demonstrated that even in the nonpolar environment in the presence of bases the keto-form in the excited state could be transformed into an enol as a result of tautomerization, which next formed the enolate-ion via the excited state proton transfer. These data provided strong evidence of keto-enol tautomerization of the emitter in the firefly luciferase reaction [39].

Individual emission spectra of all chemical forms of the oxyluciferin emitter and its analogues in aqueous buffer solutions and equilibrium constants in the excited state were determined by Naumov et al. in 2015 using methods of stationary and time-resolved fluorescence. The previously proposed hypothesis was confirmed suggesting that the excited-state proton transfer from the enol group was the predominate process in comparison with the proton transfer from the phenol group. In water the phenol-keto form is the strongest photoacid among all other isomers, and its conjugated base (phenolate-keto) displayed the lowest emission energy (634 nm). Authors suggested a complex scheme of equilibria existing in aqueous solutions for different forms of oxyluciferin both in the ground and excited states, which could be used for interpretation of the bioluminescence spectra in the firefly luciferin-luciferase system [40].

Although investigation of oxyluciferin and its derivatives in aqueous solution in the absence of enzyme provided new insights to understanding photochemistry of these compounds, certain information on the polarity and proteolytic properties of the enzyme active center is required in order to extrapolate these conclusions to real biological systems. With this in mind, subnanosecond spectroscopy was used in [41] for investigation of the dynamics of the processes occurring in the excited-state oxyluciferin in complex with the *L. cruciata* luciferase. It allowed characterizing all photochemical processes that were plausible in the luciferase active center at least in principle. Due to the complex organization of the oxyluciferin chemical structure in the ground state, the complexes of luciferase with different chemical species of oxyluciferin exist, which are excited simultaneously, and, hence, the produced results more likely refer to the possible processes that may occur in the enzyme active center in the moment of emission rather than most probable ones. It was shown that the emission of the excited fluorophore comprised a complex cascade of photo-induced processes of proton transfer and could be interpreted as a fluorescence pH-dependence. Furthermore, the process of proton transfer is the central point in spectrochemistry of this system, and assignment of the pH-dependent emission to any single chromophore would represent significant simplification. As has been mentioned above [38], in the crystal and in most of the polar solvent, where oxyluciferin molecules are stabilized, it exists as a neutral enol forming dimers via hydrogen bond. Moreover, the analysis of distances of the emitter bonds in the crystal structure of the oxyluciferin-AMP complex with *Luciola cruciata* luciferase [35] revealed that in the ground state oxyluciferin also existed in the neutral enol form likely due to its stabilization by the closely located AMP residue.

Excitation and emission spectra of the complexes of oxyluciferin with the *Luciola cruciata* luciferase were recorded at pH 6.8, 7.8 and 8.3 [41]. The normalized absorption spectra were derived from these data. It was shown that the bathochromic shift of the absorption maximum with the pH increase was observed for the complex similarly to the oxyluciferin aqueous solution due to oxyluciferin deprotonation. The fluorescence spectrum of the complex at pH 6.8 is significantly broader than in aqueous solution, which indicates significant contribution of other forms of ground state oxyluciferin bound to the luciferase. A minor shoulder at 450 nm was observed in the fluorescence spectrum of the complex, which could be assigned to the neutral from of the emitter. The emission spectrum with maximum at 550 nm was the dominating one, which could be assigned to the conjugated base emitter. Contribution of the shoulder at 450 nm decreases with the increase of pH due to the decrease of the fraction of the respective form in the ground state. The abovementioned data on tautomeric equilibrium of oxyluciferin in aqueous solution [39] implied that several different forms absorbed light in the wavelength range 350−400 nm, and precisely these conditions were used for excitation of emitter during recording of the stationary and time-resolved fluorescence of the luciferase complex with oxyluciferin. Deconvolution of the stationary emission spectra into individual components conducted by the authors is presented at Figure 6, according to which the main emitters in this system are phenolate-enolate, phenol-enolate, and phenol-keton forms. Phenolate-enolate becomes dominating emitter with the increase of pH.

The accuracy of assignment of the oxyluciferin fluorescence spectra to its phenol forms seems problematic (Figure 6). It is known that the phenol group both in luciferin and oxyluciferin becomes strong acid in the excited state, and emission is generated from the electronically excited phenolate. The fluorescence quantum yield for the luciferin derivatives not containing phenol group (such as 6'-methoxyluciferin) is extremely low, and fluorescence is observed only in the blue region of the spectrum [16]. Blue fluorescence was observed for the oxyluciferin under anaerobic conditions only at pH below 2, when the phenol group was not ionized [20]. It is not clear why the phenol-keto form (Figure 6) at pH 6.8 can contribute ∼ 40% to the fluorescence intensity of oxyluciferin in the complex with the enzyme. This requires further examination. Consideration of the crystal structure of the *L. cruciata* luciferase with oxyluciferin and AMP demonstrates that the residues R220 and R337 are located in the vicinity of the oxyluciferin phenol group as well as several water molecules, which could serve as proton acceptors [35]. This creates conditions for stabilization of the charged phenolate form of the emitter. In regard to the phenol-enolate form (Figure 6), this form can be also represented as a phenolate-enol depending on where the stronger proton acceptor is located – closer to the phenol or to the enol proton.

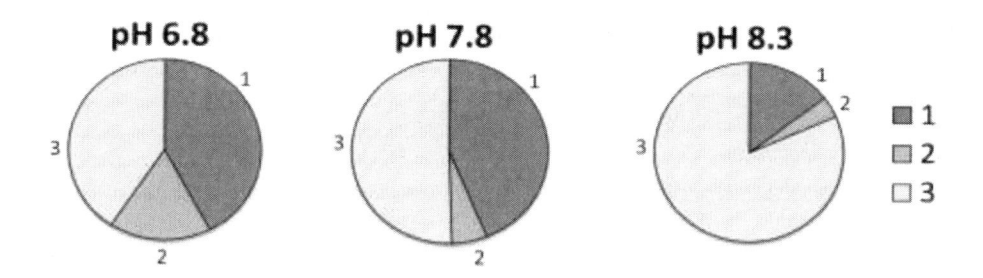

Figure 6. Contribution of main emitters to the stationary fluorescence spectra of oxyluciferin in complex with *L.cruciata* luciferase at different pH (%) . Structure designations: 1 – phenol-keto form; 2 – phenol-enolate form; 3 - phenolate-enolate form.

Despite the fact that the data obtained for the complex are in good agreement with the data produced for oxyluciferin in aqueous solutions [40], there is no certainty that they reflect the processes occurring in the course of bioluminescence adequately. Broad bioluminescence spectra were obtained in the same study for the live fireflies [41]. The spectrum was characterized by a single Gauss distribution and one kinetic decay curve with lifetime of 18 ms. Although the luciferase structure was unknown, the bioluminescence maximum was at 550 nm and the spectrum width at half height was 60 nm, which was significantly narrower than in the case of complex in aqueous solution at any pH.

5. EMITTER AS A PROBE OF CONFORMATIONAL STATE OF THE LUCIFERASE ACTIVE CENTER

As was mentioned above, the results on the investigation of the fluorescence spectra of oxyluciferin in aqueous solutions and in complex with luciferase provide essential information on the complex photophysical processes occurring in these systems in a wide range of pH, but this cannot be the full description of the processes occurring in the luciferase active center in the moment of bioluminescence event. First of all, this is related to the fact that only one single electronically excited molecule of the emitter is formed in the active center of each luciferase molecule in the course of oxidative decarboxylation of dioxetanone and depending on its microenvironment it can assume only one definite form. And the cases that present not monomodal but more complex (bi- and ever three-modal) bioluminescence spectra imply that under those conditions different enzyme conformers are present in the reaction mixture with each mediating formation of different forms of the emitter. Hence, the emitter molecule is in essence an intramolecular probe that characterizes microenvironment of the emitter at the moment of light emission. By now a large number of studies were published presenting bioluminescence spectra of native and mutant luciferases often at different pH or temperatures. But in the majority of publications the authors focused only on the position of the bioluminescence λ_{max}, but not on the spectrum shape and its change with varying pH or temperature. Only few examples of analysis of the bioluminescence spectra in a wide range of pH or temperature are available. Some of them are presented below.

In 2005 pH-dependencies of the bioluminescence spectra of the WT recombinant *Luciola mingrelica* firefly luciferase and its mutant form with His433Tyr substitution were investigated in our laboratory [42]. The amino acid sequence of *Luciola mingrelica* firefly luciferase has more than 80% homology with the luciferases from the Japanese fireflies of the *Luciola* genus [43]. The indicated mutation resulted in the shift of the bioluminescence λ_{max} from 566 nm to 606 nm at pH 7.8 (pH optimum of luciferase activity). The bioluminescence spectra of the WT luciferase and its mutant form were obtained in the pH range 5.6–10.2. Only yellow-green bioluminescence was observed for the WT luciferase at pH $\geq$ 7.0, at pH 5.6 only red bioluminescence was detected, and two bioluminescence peaks were detected at intermediate pHs. The luciferase mutant demonstrated red bioluminescence at pH $\leq$ 6.1. A shoulder in the yellow-green range of the spectrum emerged with the pH increase, its intensity increased with the increase of pH, and only at pH ~10.2 the intensity of yellow-green bioluminescence overcame the red bioluminescence intensity. Hence, the shift of the bioluminescence λ_{max} for the mutant form at the pH-optimum of activity can be explained by the change of the ratio between the different forms of oxyluciferin. Deconvolution of the bioluminescence spectra into individual Gaussian components allowed establishing that the observed spectra represented superposition of three spectra of three different forms of electronically excited oxyluciferin. Assuming that all emitters in the excited state are in the phenolate form, we identified the three species revealed from the spectra as enolate (LO-O⁻, $\lambda_{max.em}$ = 556 nm), enol (LO-OH, $\lambda_{max.em.}$= 587 nm), and ketone (LO=O, $\lambda_{max.em.}$= 618 nm). The relative content of each form was determined for different pH, which varied with the pH change due to the shift of equilibria between the three forms of emitter (Table 1):

ketone $\leftrightarrow$ enol $\leftrightarrow$ enolate   (3)

The change of relative content of different emitter forms with pH resulted both in the change of the position of bioluminescence λ_{max} and of the shape of the bioluminescence spectrum [42]. Identification of the three emitter forms indicates co-existence of three luciferase conformers in the reaction medium with the content ratio between these conformers changing with varying pH [44]. Different conformers of luciferase exist in solution:

$$E_1 \leftrightarrow E_2 \leftrightarrow E_3 \qquad (4)$$

Each luciferase conformer contains one form of electronically exited oxyluciferin:

$$E_1\,(LO=O)^* \leftrightarrow E_2\,(LO\text{-}OH)^* \leftrightarrow E_3\,(LO\text{-}O^-)^* \qquad (5)$$

The relative content of each form of the enzyme-emitter complex was determined at different pH, which varied with the pH change in accordance with the shift of equilibrium between the three forms of emitter as shown in Table 1.

Table 1. Distribution of luciferase conformers for the WT and mutant luciferases at different pH (in %)

pH	$E_1\,(LO=O)^*$		$E_2\,(LO\text{-}OH)^*$		$E_3\,(LO\text{-}O^-)^*$		Sum	
	WT	Mutant	WT	Mutant	WT	Mutant	WT	Mutant
6.8	38	63	28	28	27	6	93	97
7.8	22	50	32	40	50	12	104	102
8.3	21	45	26	42	54	16	101	103

It is interesting to note that the similar results (Figure 6) were obtained from the analysis of the fluorescence spectra of the complexes of oxyluciferin with the native *Luciola cruciata* firefly luciferase [47], although the spectra assignment was different in these two studies. Hence, the bioluminescence emitters are intramolecular probes for the emitter microenvironment in the luciferase active center, which depends on micro pH, polarizability, orientation, and mobility of the key amino acid residues [17].

This brings up the question: how the single-point mutation, replacement of the residue located at a distance of 12 A° from the active center, affected the bioluminescence spectra? Computer modeling of the change of the luciferase 3D structure on substitution of His433 with Tyr allowed revealing significant changes in conformation of the residue 433. The Tyr phenyl ring in the structure of the mutant luciferase was found to be rotated by almost 60° relative to the imidazole ring of His in the WT structure [42]. It is know based on the X-ray diffraction data for luciferases and other adenylating enzymes to which firefly luciferases belong that the His433 residue is located in the mobile loop formed by the Tyr427–Phe435 residues, which connects N- and C-domains of the luciferase [35, 45]. This loop can be considered as a "hinge" connecting two luciferase domains. The imidazole cycle of the His433 residue forms hydrogen bond with the carboxyl group of the Asp431 residue, which increases rigidity of the hinge and decreases the amplitude of thermal fluctuation of the N- and C-domains relative each other. This provides sufficiently strong anchoring of the amino acid residues Thr529 and Lys531 of the C-domain, which are in contact with the oxygen atom of the oxyluciferin enol group via water molecule [35] and likely participate directly in formation of the yellow-green emitter – enolate. The His433Tyr substitution makes the hydrogen bond between His433 and Asp431 impossible, which decreases rigidity of the Tyr427–Phe435 hinge and increases the amplitude of thermal oscillations of the domains relative to each other. The emitter microenvironment becomes loose, interaction of the enol group with water is disrupted, and this causes equilibrium shift from enolate toward enol and ketone. The significant effect of C-domain on the shape and bioluminescence λ_{Max} has been corroborated by the literature data: the mutant luciferase with removed small C-domain generated only very weak red bioluminescence [46]. Hence, the detailed investigation of the bioluminescence spectra in a wide pH range, quantitative evaluation of the contribution of different forms of emitter to the observed bioluminescence spectra, and consideration of the structure data of the luciferases and their complexes with substrates and reaction products, as well as employment of computer modeling allow elucidating interconnections between the spectral characteristics of bioluminescence and luciferase structure.

Site-directed or random mutagenesis of the luciferase protein globule allows modulating physicochemical properties of the emitter microenvironment in the enzyme active center, and thus changing the ratio between the different enzyme conformers and shift the equilibrium towards green or vice versa red bioluminescence. It was shown in our laboratory that the bioluminescence λ_{Max} for the WT *L. mingrelica* firefly luciferase shifted from 564 to 610 nm with the pH decrease from 7.8 to 6.0. The single Tyr35His mutation resulted in formation of the mutant *L. mingrelica* luciferase for which the $\lambda_{Max} = 564$ nm was maintained in the bioluminescence spectra in the entire pH range 6.0–7.8. The monomodal spectrum with λ_{Max} at 564 nm indicates that this mutant exists as a single conformer. The Tyr35His replacement produced the effect similar to the one described above for the *L. mingrelica* luciferase with the His433Tyr substitution [42]. The enzyme containing His residue emitted green light at pH 7.8, while the one containing Tyr residue – red. This fact can be explained as follows. The comparative analysis of the crystal structures of the complexes of WT and mutant (Ser286Asn) *L. cruciata* luciferases with the luciferyl adenylate analogue (DLSA) (complexes LucDLSA and mLucDLSA) allowed concluding that the active center in the LucDLSA complex was in the "closed" conformation, which ensured green bioluminescence. An "open" conformation with less rigid substrate microenvironment was realized in the mLucDLSA due to the change of orientation of the Ile288 residue, which generated red bioluminescence [35]. Although the 3D-structures of these complexes are almost identical, orientation of Asn and Ser at position 286 and of the Ile288 residue as well as orientation of the 233–237 loop are different. The Tyr35 residue in the LucDLSA is in the vicinity of the 233–237 loop. It can be suggested that such localization is required for generation of the enzyme conformer mediating green bioluminescence. It is quite likely that the pH decrease leads to the "open" less rigid conformation of the enzyme active center, which results in the shift of the bioluminescence spectra towards the red region. The replacement of the bulky aromatic Tyr35 residue for the smaller His residue stabilizes the dense packing in the vicinity of the 35th residue. The 233–237 loop maintains its position with the pH decrease, hence, the "closed" conformation is not disturbed. The Tyr35 residue is conserved in all firefly luciferases. For example, the click beetle luciferases contain His at the position 35 [48]. It is possible that this is one of the sites in the luciferase structure that contain amino acid residues responsible for formation of pH-insensitivity of the bioluminescence spectra of these luciferases. The easiness of generating of mutations decreasing sensitivity of bioluminescence to pH indicates that it is likely that this property in fireflies is under control stabilizing selection. It must be mentioned that formation of the oxyluciferin microenvironment in the active center and realization of green (or red) bioluminescence is defined by the balance of interactions between numerous amino acid residues in the entire protein globule of the enzyme. In particular, the network of hydrogen bonds between the residues Arg220, Asn231, Ser286, Glu313, Lys339 [37] in the region of the luciferin binding site, hydrophobic packing of internal residues [49], interactions in the ATP binding site [50] and between the two mobile luciferase domain [42, 51] play an important role. Conformation of the 233–237 loop could be only one of several required factors. Hence, there are no specific residues explaining the pH-dependence of the bioluminescence spectra of luciferases. Numerous remote mutation has been identified, which disrupt directly or indirectly the required interactions and result in the increase of the fraction of red bioluminescence [17, 42, 52], as well as mutations stabilizing the structure of active center thus reducing dependence of the bioluminescence spectra on external conditions [49, 51].

Another interesting result was obtained during investigation of the bioluminescence spectra of the thermostable *L. mingrelica* luciferase and its single-point mutants at the Glu457 residue (Glu457Asp/Gln/Val/Lys) at temperatures 10, 25, and 42°C [53]. All these luciferases displayed the net bioluminescence spectra represented by superposition of the spectra of green ($\lambda_{Max.em.}$ = 554±3 nm) and red emitter ($\lambda_{Max.em.}$ = 595±5 nm) with the ratio shifting with the temperature increase towards the red emitter. The fraction of red emitter reaches 90% at 42 °C for the initial thermostable luciferase, and for mutants – to 100%. The initial luciferase and its Glu457Asp mutant demonstrate similar temperature dependencies of the bioluminescence spectra. In the Glu457Gln/Val mutants the green emitter provides only ~ 20% of bioluminescence at 10°C and ~ 10% at 25°C. The Glu457Lys luciferase mutant displays a monomodal bioluminescence spectrum with maximum at 600 nm in the entire investigated temperature range. The analysis of the spectra shows that $\lambda_{Max.em}$ for the green and red components of the emission are the same as in the case of initial luciferase, and only the ratio between these components is changing with the increase of temperature. These results indicate that the emitter structure does not depend on the temperature, but the ratio between the different protein conformers is temperature-dependent. Hence, loosening of the protein structure on heating is the cause for the shift of the bioluminescence spectra towards the red emitter.

ACKNOWLEDGMENTS

This work was financially supported by the Lomonosov Moscow State University, State-Registered theme No. AAAA-A16-116052010081-5.

REFERENCES

[1] McElroy, W. D. (1947). The energy source for bioluminescence in an isolated system. *Proc. Nat. Acad. Science*, 33: 342-345.

[2] Hastings, J. W., McElroy, W. D. and Coulombre, J. (1953). The effect of oxygen upon the immobilization reaction in firefly luminescence. *J. Cell. Comp. Physiol.*, 42: 137–150.

[3] McElroy, W. D. and A. Green, A. (1956). Function of adenosine triphosphate in the activation of luciferin. *Arch. Biochem. Biophys.*, 46: 399–416.

[4] Bitler, B. and McElroy, W. D. (1957). The preparation and properties of crystalline firefly luciferin. *Arch. Biochem. Biophys.*, 72: 358–368.

[5] Rhodes, W. C., W. D. McElroy, W. D. (1958). The Synthesis and Function of Luciferyl-adenylate and Oxyluciferyl-adenylate. *J. Biol. Chem.* 233: 1528-1537.

[6] White, E. H., Field, G. F., McElroy, W. D. and McCapra, F. (1961). Structure and synthesis of firefly luciferin. *J. Am. Chem. Soc.*, 83: 2402–2403.

[7] White, E. H., Field, G. F. and McCapra, F. (1963). Structure and synthesis of firefly luciferin. *J. Am. Chem. Soc.*, 85: 337–343.

[8] McElroy, W. D. and Seliger, H. H. (1962). Mechanism of action of firefly luciferase. *Fed. Proc.*, 21: 1006–1012.

[9] Seliger, H. H. and McElroy, W. D. (1962). Chemiluminescence of firefly luciferin without enzyme. *Science*, 132: 683–685.

[10] McCapra, F., Chang, Y. C. and Francëois, V. P. (1968). Chemiluminescence of a firefly luciferin analogue. *Chem. Commun.*, 22–23.

[11] Seliger, H. H. and McElroy, W. D. (1960). Spectral Emission and Quantum Yield of Firefly Bioluminescence. *Archiv. Biochem. Biophys.* 88: 136-141.

[12] Ando, Y., Niwa, K., Yamada, N., Enomoto, T., Irie, T., Kubota, H., Ohmiya, Y. and Akiyama, H. (2008). Firefly bioluminescence quantum yield and color change by pH-sensitive green emission. *Nat. Photonics*, 2: 44–47.

[13] Seliger, H. H. and McElroy, W. D. (1964). The colors of firefly bioluminescence: enzyme configuration and species specificity. *Proc. Natl. Acad. Sci. USA*, 52: 75–81.

[14] Hopkins, T. A., Seliger, H. H., White, E. H. and Cass, M. W. (1967). Chemiluminescence of firefly luciferin. A model for bioluminescent reaction and identification of product excited state. *J. Am. Chem. Soc.*, 89: 7148–7150.

[15] White, E. H., Rapaport, E., Hopkins, T. A. and Seliger H. H. (1969) Chemi and bioluminescence of firefly luciferin. *J. Am. Chem. Soc.*, 91: 2178–2180.

[16] Morton, R. A., Hopkins, T. A. and Seliger, H. H. (1969), The Spectroscopic Properties of Firefly Luciferin and Related Compounds. An Approach to Product Emission. *Biochemistry*, 8(4), 1598 – 1607.

[17] Ugarova, N. N. and Brovko, L. Yu. (2002) Protein structure and bioluminescent spectra for firefly bioluminescence. *Luminescence*. 17: 321-330.

[18] Suzuki, N. and Goto, T. (1971) Firefly bioluminescence II. Identification of 2-(60-hydroxybenzothiazol-20-yl),-4-hydroxythiazole as a product in the bioluminescence of firefly lanterns and as a product in the chemiluminescence of firefly luciferin in DMSO. *Tetrahedron Lett.* 12: 2021–2024.

[19] Suzuki, N., Sato, M., Okada, R. and Goto, T. (1972). Studies of firefly bioluminescence. I. Synthesis, and spectral properties of firefly oxyluciferin, a possible emitting species in firefly bioluminescence. *Tetrahedron*, 28: 4065–4074.

[20] Gandelman, O. A., Brovko, L. Y., Ugarova, N. N., Chikishev, A. Y. and Shkurimov, A. P. (1993). Oxyluciferin fluorescence is a model of native bioluminescence in the firefly luciferin--luciferase system. *J. Photochem. Photobiol., B*. 19: 187-191.

[21] Gandelman, O. A., Brovko, L. Yu., Chikishev, A. Y., Shkurinov, A. P. and Ugarova, N. N. (1994). Investigation of the interaction between firefly luciferase and oxyluciferin or its analogues by steady state and subnanosecond time-resolved fluorescence. *J. Photochem. Photobiol., B*, 22: 203–209.

[22] Leontieva, O. V., Vlasova T. N. and Ugarova, N. N. (2006). Dimethyl- and monomethyloxyluciferins as analogs of the product of the bioluminescence reaction catalyzed by firefly luciferase. *Biochemistry (Moscow)*, 71: 51–55.

[23] McCapra, F., Gilfoyle, D. J., Young, D. W., Church, N. J. and Spencer, P. (1994). The chemical origin of color differences in beetle bioluminescence. In: *Proceedings of the 8th International Symposium on Bioluminescence and Chemiluminescence*. (Eds. Campbell, A. K., Kricka, L. J. and Stanley, P. E.) John Wiley & Sons. Chichester.: 387-391.

[24] Liu, Y. J., De Vico, L. and Lindh, R. (2008). Ab initio investigation on the chemical origin of the firefly bioluminescence. *J. Photochem. Photobiol., A*. 194: 261-267.

[25] Nakatani, N., Hasegawa, J., Nakatsuji, H. (2007). Red light in chemiluminescence and yellow-green light in bioluminescence: color-tuning mechanism of firefly, Photinus pyralis, studied by the symmetry-adapted cluster-configuration interaction method. *J. Am. Chem. Soc.* 129: 8756-8765.

[26] Orlova, G., Goddard, J. D., Brovko, L. Y. (2003). Theoretical study of the amazing firefly bioluminescence: the formation and structures of the light emitters. *J. Am. Chem. Soc.* 125: 6962-6971.

[27] Yang, T. and Goddard, J. D. (2007) Predictions of the geometries and fluorescence emission energies of oxyluciferins. *J. Phys. Chem. A.* 111: 4489-4497.

[28] Branchini, B. R., Murtiashaw, M. H., Magyar, R. A., Portier, N. C., Ruggiero, M. C. and Stroh, J. G. (2002) Yellow-green and red firefly bioluminescence from 5,5 dimethyloxyluciferin. *J. Am. Chem. Soc.* 124: 2112-2113.

[29] Branchini, B. R., Southworth, T. L., Murtiashaw, M. H., Magyar, R. A., Gonzalez, S. A., Ruggiero, M., C. and Stroh, J. G. (2004) An alternative mechanism of bioluminescence color determination in firefly luciferase. *Biochemistry*, 43: 7255-7262.

[30] White, E. H., Steinmetz, M. G., Miano, J. D., Wildes, P. D. and Morland, R. (1980) Chemi- and bioluminescence of firefly luciferin. *J. Am. Chem. Soc.,* 102: 3199-3208.

[31] White, E. H., and Roswell, D. F. (1991) Analogs and derivatives of firefly oxyluciferin, the light emitter in firefly bioluminescence. *Photochem. Photobiol.,* 53: 131-136.

[32] White, E. H., Worther, H., Seliger, H. H. and McElroy, W. D. (1966) Amino analogs of firefly luciferin and biological activity thereof. *J. Am. Chem. Soc.* 88: 2015-2019.

[33] Ren, A. M. and Goddard, J. D. (2005). Predictions of the electronic absorption and emission spectra of luciferin and oxyluciferins including solvation effects. *J. Photochem. Photobiol., B.* 81: 163-170.

[34] Nakatani, N., Hasegawa, J. and Nakatsuji, H. (2007). Red light in chemiluminescence and yellow-green light in bioluminescence: color-tuning mechanism of firefly, *Photinus pyralis,* studied by the symmetry-adapted cluster-configuration interaction method *J. Am. Chem. Soc.,* 129: 8756-8765.

[35] Nakatsu, T., Ichiyama, S., Hiratake, J., Saldanha, A., Kobashi, N., Sakata, K. and Kato, H. (2006). Structural basis for the spectral difference in luciferase bioluminescence. *Nature*, 440: 372-376.

[36] Pinto da Silva, L. and Esteves da Silva, J. C. G. (2015). Chemiexcitation Induced Proton Transfer: Enolate Oxyluciferin as the Firefly Bioluminophore. *J. Phys. Chem. B.,* 119: 2140−2148.

[37] Viviani, V. R., Arnoldi, F. G. C., Neto, A. J. S., Oehlmeyer, T. L., Bechara, E. J. H. and Ohmiya, Y. (2008) The structural origin and biological function of pH-sensitivity in firefly luciferases. *Photochem. Photobiol. Sci.*, 7: 159-169.

[38] Naumov, P., Ozawa, Y., Ohkubo, K. and Fukuzumi, S. (2009). Structure and Spectroscopy of Oxyluciferin, the Light Emitter of the Firefly Bioluminescence, *J. Am. Chem. Soc.,* 131: 11590–11605.

[39] Solntsev, K. M., Laptenok, S. P. and Naumov, P. (2012). Photoinduced Dynamics of Oxyluciferin Analogues: Unusual Enol "Super"photoacidity and Evidence for Keto−Enol Isomerization. *J. Am. Chem. Soc.* 134: 16452−16455.

[40] Ghose, A., Rebarz, M., Maltsev, O. V., Hintermann, L., Ruckebusch, C., Fron, E., Hofkens, J., Mély, Y., Naumov, P., Sliwa, M. and Didier, M. (2015). Emission Properties of Oxyluciferin and Its Derivatives in Water: Revealing the Nature of the Emissive Species in Firefly Bioluminescence. *J. Phys. Chem. B,* 119: 2638−2649.

[41] Snellenburg, J. J., Laptenok, S. P., DeSa, R. J., Naumov, P. and Solntsev, K. M. (2016). Excited-State Dynamics of Oxyluciferin in Firefly Luciferase. *J. Am. Chem. Soc.* 138: 16252−16258.

[42] Ugarova, N. N., Maloshenok, L. G., Uporov, I. V. and Koksharov, M. I. (2005). Bioluminescence Spectra of Native and Mutant Firefly Luciferases as a Function of pH. *Biochemistry (Moscow),* 70 (11): 1262-1267.

[43] Devine, J. H., Kutuzova, G. D., Green, V. A., Ugarova, N. N. and Baldwin, T. O. (1993). Luciferase from the East European firefly Luciola mingrelica: Cloning and nucleotide sequence of the cDNA, overexpression in Escherichia coli and purification of the enzyme. *Biochim. Biophys. Acta, Gene Struct. Expression.* 1173: 121-132.

[44] Ugarova, N. N. (2009). Mechanism responsible for the spectral differences in firefly bioluminescence. *Proceedings of the 15th International Symposium on Bioluminescence and Chemiluminescence. Light Enission: Biology and Scientific Applications.* (Ed. Xun Shen et al.). World Scientific, Publishing Co., Singapore: 75-78.

[45] Conti, E., Franks, N. P. and Brick, P. (1996) Crystal structure of firefly luciferase throws light on a superfamily of adenylate-forming enzymes. *Structure,* 4: 287-298.

[46] Zako, T., Ayabe, K., Aburatani, T., Kamiya, N. A., Ueda, H., and Nagamune T. (2003). Luminescent and substrate binding activities of firefly luciferase N-terminal domain. *Biochim. Biophys. Acta.* 1649: 183-189.

[47] Koksharov, M. I. and Ugarova, N. N. (2008). Random Mutagenesis of *Luciola mingrelica* Firefly Luciferase. Mutant Enzymes with Bioluminescence Spectra Showing Low pH Sensitivity. *Biochemistry (Moscow),* 73(8): 862-869.

[48] Wood, K. V., Lam,Y. A., McElroy, W. D. and Seliger, H. H. (1989) Bioluminescent click beetles revisited. *J. Biolum. Chemilum.* 4: 31-39.

[49] Kajiyama, N. and Nakano, E. (1991) Isolation and characterization of mutants of firefly luciferase which produce different colors of light. *Protein Eng.* 4: 691-693.

[50] Tisi, L. C., Law, G. H., Gandelman, O., Lowe, C. R. and Murray, J. A. (2002) The basis of the bathochromic shift in the luciferase from *Photinus pyralis*. In:

Proceedings of the 12th International Symposium on Bioluminescence and Chemiluminescence: Progress and Current Applications. (Eds. Stanley, P. E. and Kricka, L. J.) World Scientific, Singapore: 57-60.

[51] Law, G. H., Gandelman, O. A., Tisi, L. C., Lowe, C. R. and Murray, J. A. (2006) Mutagenesis of solvent-exposed amino acids in *Photinus pyralis* luciferase improves thermostability and pH tolerance. *Biochem. J.,* 397: 305–312.

[52] Branchini, B. R., Southworth, T. L., Khattak, N. F., Murtiashaw, M. H. and Fleet, S. E. (2004) Rational and random mutagenesis of firefly luciferase to identify an efficient emitter of red bioluminescence. *Proc. SPIE.,* 5329: 170-177.

[53] Modestova, Y., Koksharov, M. I. and Ugarova, N. N. (2014). Point mutations in firefly luciferase C-domain demonstrate its significance in green color of bioluminescence. *Biochim. Biophys. Acta,* 1844: 1463–1471.

In: A Comprehensive Guide to Chemiluminescence
Editor: Luís Pinto da Silva
ISBN: 978-1-53616-170-0

Chapter 3

CHEMILUMINESCENCE OF FULLERENES AND THEIR DERIVATIVES

Ramil G. Bulgakov[1,*] and Dim I. Galimov[2]
[1]Institute of Molecule and Crystal Physics,
Ufa Federal Research Center RAS, Ufa, Russia
[2]Institute of Petrochemistry and Catalysis,
Ufa Federal Research Center RAS, Ufa, Russia

ABSTRACT

In this chapter we present the first review regarding the liquid-phase chemiluminescence (CL) arising in the reactions of fullerenes C_{60} and C_{70}, as well as some of their derivatives. The main attention in this review will be given to the identification of the nature and characteristics of the CL emitters, the mechanism of their generation, and also to the effect of quenchers. Peculiarities and regularities of CL generated in the following reactions will be considered. The ozone oxidation of the fullerenes C_{60} and C_{70}, fluorides $C_{60}F_{36}$ and $C_{60}F_{48}$ in organic solvents, as well as the CL which occurs during the ozonation of C_{60} aqueous dispersions. The CL emitters in ozone addition to the C_{60} or C_{70} cage are the excited molecules of the fullerene polyketones $^*O = C_{60}(= O)_{1\text{-}5}$ or $^*O = C_{70}$ $(= O)_{1\text{-}5}$. The reactions of the hydrolysis of secondary ozonides C_{60} and C_{70} will also be discussed. Today, only one example is known of CL by oxidation of fullerene derivatives by oxygen in solution; it consists on the oxidation of the $C_{60}H_{36}$ hydride which gives two CL emitters: the $C_{60}H_{18}^*$ hydride and the singlet-excited fullerene $^1C_{60}^*$. The features of the transformations of carbon-centered radicals $EtC_{60}^{\bullet}$ in the chemiluminescent system $(C_{60}\text{-}Et_3Al)\text{-}O_2$, which determine the mechanism for the generation of CL emitters, are also considered. The other new CL occurs in the dimerization of the $EtC_{60}^{\bullet}$ fullerenyl radicals with generation the excited fullerene dimer $(EtC_{60}C_{60}Et)^*$. Attention was also given to the CL generated by UV-illumination, gamma-irradiation, or sonication of solutions of

* Corresponding Author's Email: profbulgakov@yandex.ru.

fullerene C_{60} in dimethylformamide. It is also discussed the electrochemiluminescence (ECL) arising in electron transfer reactions between the $C_{60}^{-\bullet}$ radical anion (or di- (C_{60}^{2-}) and tri- (C_{60}^{3-}) anions) with some organic radical cations, caused by the emission of singlet-excited fullerene $^1C_{60}^*$. Surprisingly, but in a purely chemical reaction of the $C_{60}^{-\bullet}$ radical anion (or dianion C_{60}^{2-}) with the Ce^{4+} ion, the CL emitter is a triplet of $^3C_{60}^*$.

1. INTRODUCTION

At present, the phenomenon of light emission by excited chemical reaction products or luminescent additives i.e., chemiluminescence (CL), finds wide application in the process of converting the energy of a chemical reaction into light emission [1-9], in analytical chemistry [10-15], in the manufacture of powerful chemical lasers [16 and references therein], as well as in medicine [9, 10, 12, 13, 15, 17]. For almost a hundred years, the emitters of CL were known excited states of organic [1-10] and inorganic substances, including metal ions [14, 15 and references therein] and even free radicals [18]. With the discovery of fullerenes in 1985 [19], new types of potential participants in chemiluminescent reactions have been discovered, namely the C_{60} and C_{70} molecules and their derivatives, both as light-emitting emitters and initial reagents, whose transformations are accompanied by CL. Fullerenes are the third allotropic form of carbon in addition to the already well-known diamond and graphite. The most widely available fullerenes to date are C_{60} and C_{70}, hollow molecules of a spherical and ellipsoidal shape, respectively. Despite the fact that more than thirty years have passed since 1985, the study of chemical, physical, photophysical and other properties, as well as the possibilities of practical application of fullerenes, evoke extraordinary interest [see for example 20-38].

The unique structure and unusual properties of fullerenes also attracted the attention of scientists engaged in the study of chemiluminescence. The first reports of the discovery and study of CL of fullerene C_{60} [38, 39] and its derivatives NaC_{60}, Na_2C_{60}, Na_3C_{60} [40-42] appeared a few years after the publication by Kratschmer et al. [43] regarding a new method for large scale fullerene production.

2. CHEMILUMINESCENCE IN ELECTRON TRANSFER REACTIONS OF FULLERENE C_{60} ANIONS WITH ORGANIC AND INORGANIC CATIONS

2.1. Electrochemiluminescence in Electron Transfer Reactions between the $C_{60}^{-\bullet}$ Radical Anions and the C_{60}^{2-}, C_{60}^{3-} Anions with Some Organic Radical Cations

F. Gupta and K. S. V. Santhanam were the first to discover the ability of ionic fullerene derivatives to generate in the electrochemical reactions the excited states of the fullerene molecules, namely, the singlet excited molecule $^1C_{60}*$ [41-43]. The authors called the detected light emission as chemiluminescence. However, in essence this emission is a classic example of electrochemiluminescence (ECL), by nature similar to the previously discovered ECL [45], which occurs due to electron transfer reactions between radical anions and radical cations derived from the same parent aromatic molecules during electrolysis. The ECL, detected by Gupta and Santhanam, originates at electrolysis with a controlled potential of C_{60} solution (10^{-3} M) in CH_2Cl_2 containing aromatic hydrocarbon (Ar) and supporting electrolyte such as tetra-n-butyl ammonium fluroborate (TBAF) or tetra-n-butylammonium perchlorate (TBAP) [40, 41]. As aromatic-precursors were taken thianthrene (Th), tri-p-tolylamine (TPTA), 9,10-diphenylanthracene (9,10-DPA) [40, 41], and also carbazole (Car) [42]. The ECL was associated with the following successive processes. When the working Pt-electrode was cycled in the anode regime from 0 to +1.4 V (during from 0.1 to 5.0 s), the Ar molecules were oxidized to the radical cations $Ar^{+\bullet}$. These radical cations live for a few minutes due to their relative stability. Then, the Pt-electrode was switched to the cathode regime (-0.45 V), as result C_{60} was reduced to the radical anion $C_{60}^{-\bullet}$ (or until the C_{60}^{2-}, C_{60}^{3-} anions). The formation of these radical anions (or molecular anions) in the Pt-electrode diffusion layer is followed by an annihilation reaction and light emission in the form of frequent flashes of glow [40]. Each flash of light occurs at the time of switching the polarity of the electrode. The maximum intensity of ECL is observed at the first switching from the anode regime to the cathode one, after which the intensity decreases to some more constant level. The brightness of the ECL grows in the series $C_{60}^{-\bullet} < C_{60}^{2-} < C_{60}^{3-}$ [41].

The maximum of ECL spectrum (Figure 1 and 2) is located at 720 nm and coincides with the maximum of the photoluminescence (PL) spectrum (Figure 2) of the singlet excited fullerene $^1C_{60}*$ [41, 42]. The obtained spectral characteristics of the ECL and the enthalpy of separate redox stages, calculated from the electrochemical data, suggested the following scheme of dark and light processes (Scheme 1).

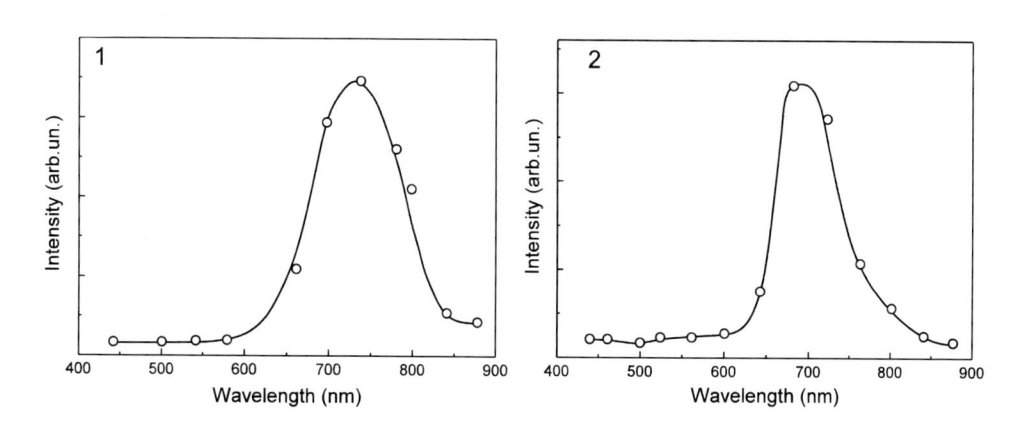

Figure 1. ECL spectra in reactions between $C_{60}^{-\bullet}$ with $Th^{+\bullet}$ (1) and $DPA^{+\bullet}$ (2) recorded using a running interference filter and R2066 Hamamatsu photomultiplier.

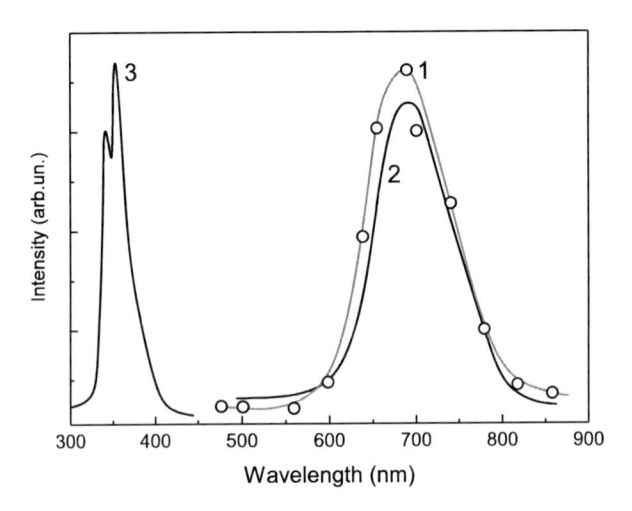

Figure 2. Emission spectrum obtained during the electron-transfer reaction between $C_{60}^{-\bullet}$ and Car (1). Fluorescence spectra of C_{60} (2) and Car (3). $\lambda_{exc} = 532$ (1, 2) and 290 (3) nm. Taken from [42].

At the Pt-electrode:

$$Ar \longrightarrow Ar^{+\bullet} + e^- \text{ (anode reaction)} \qquad (1)$$

$$C_{60} + e^- \longrightarrow C_{60}^{-\bullet} \text{ (cathode reaction)} \qquad (2)$$

In solution:

$$Ar^{+\bullet} + C_{60}^{-\bullet} \longrightarrow Ar + {}^3C_{60}^* \qquad (3)$$

$${}^3C_{60}^* + {}^3C_{60}^* \longrightarrow {}^1C_{60}^* + C_{60} \qquad (4)$$

$${}^1C_{60}^* \longrightarrow C_{60} + h\nu \text{ (720 nm)} \qquad (5)$$

Scheme 1.

As a result of electrolysis, the radical cations of an aromatic hydrocarbon (1) and radical anions of fullerene (2) are generated. The annihilation of these species gives the triplet excited state of fullerene ${}^3C_{60}^*$ (3). The energy of the lowest excited singlet state of C_{60}, $E_{S1} = 2.01$ eV and that of the first triplet level, $E_{T1} = 1.56$ eV.

The energetics involved in the electron transfer reactions (3), ΔG^0 (eV) were: 1.75 (Ar = Th), 1.66 (Ar = DPA), 1.50 (Ar = TPTA) [41], and 1.78 (Ar = Car) [42].

Hence, for all these aromatic hydrocarbons, ΔG^0 is less than $E_{S1} = 2.01$ eV. Because of the energy deficit, the annihilation reaction (3) cannot lead to the excitation of C_{60} into the singlet state ${}^1C_{60}^*$. When 2,5-diphenyl-1,3,4-oxadiazole (PPD) or N,N,N`,N`-tetrametyl-1,4-phenilendiamin (TMPD), are used as Ar, a ECL does not arise, because the enthalpies of reactions of their radical cations PPD$^{+\bullet}$ and TMPD$^{+\bullet}$ (0.76 and 0.78 eV) are not sufficient for the generation of triplets of ${}^3C_{60}^*$ (1.56 eV). The ECL efficiencies calculated as the ratio of the moles of photons which were radiated to the moles of the radical ion generated are $9.2 \cdot 10^{-7}$, $3.4 \cdot 10^{-7}$ and $1.85 \cdot 10^{-8}$ in the case of reactions of $C_{60}^{-\bullet}$ with DPA$^{+\bullet}$, Th$^{+\bullet}$ and TPTA$^{+\bullet}$, respectively.

The formation of the $^3C_{60}^*$ triplets is also confirmed by the increasing in the ECL intensity in the presence of an applied external magnetic field due to reducing the quenching rate of the triplet by radical ion in the magnetic field and to increasing in triplet-triplet annihilation rate [41]. For example, in case of DPA, in the magnetic field the ECL intensity is increased by 40% (at $H = 500$ G).

Another argument in favor of the formation of triplets $^3C_{60}^*$ is the quenching of ECL by molecular oxygen due to energy transfer from triplets $^3C_{60}^*$ to O_2; so, after removal of O_2 from the solution by vacuum, the intensity of the ECL is 0.133 and in the solution saturated with O_2 it is 0.045.

2.2. Chemiluminescence in Electron Transfer Reaction between Sodium Fulleride Na_2C_{60} with Complex $(NH_4)_2Ce(NO_3)_6$

Another example of the chemical glow that appears in the electron transfer reaction is CL, occurring at interaction of the Na_2C_{60} and NaC_{60} fullerides with a complex of tetravalent cerium $(NH_4)_2Ce(NO_3)_6$ (hereinafter Ce^{4+}) in THF [45]. In contrast to works of Gupta and Santhanam [40-42], where active radical ions were generated electrochemically, in this case the active species are already present in the composition of the initial compounds. Thus, the mild reducing agents, the fullerides Na_2C_{60} and NaC_{60}, contain the dianion C_{60}^{2-} and radical anion $C_{60}^{-\bullet}$, respectively, and the cerium complex contains the Ce^{4+} cation, which is a strong oxidant with $E(Ce^{4+}/Ce^{3+}) = 1.6$ V (*vs.* SCE). At mixed THF solutions of red Na_2C_{60} and colorless Ce^{4+}, rapidly decaying CL (40 s, $I_{max} = 1.1 \cdot 10^8$ photon·s^{-1}·mL^{-1}) appears accompanied by a decrease in the intensity of the absorption bands of the initial reagents: the C_{60}^{2-} dianion at 829 and 945 nm and Ce^{4+} at 322 nm (Figure 3). At the same time, the absorption maxima of the Ce^{3+} ions appear at 248, 252, 257, 265, 269 and 272 nm (Figure 3).

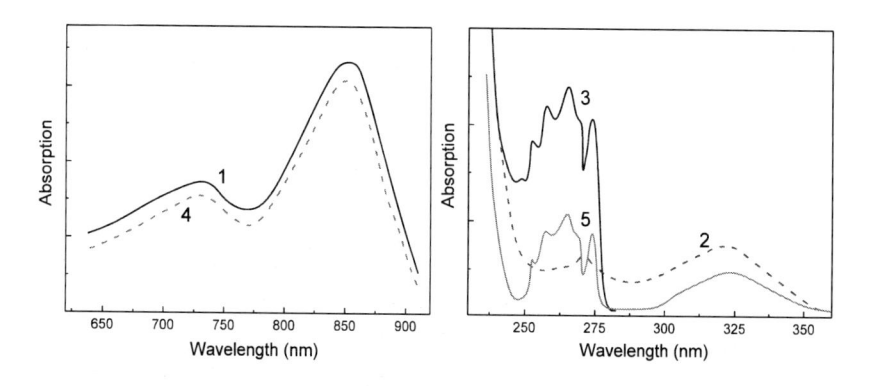

Figure 3. Absorption spectra: 1 and 2 – the initial THF solutions of Na_2C_{60} ($4\cdot10^{-3}$ M) and Ce^{4+} ($2\cdot10^{-2}$ M); 3 – the THF solution of $Ce(NO_3)_3\cdot6H_2O$ ($6\cdot10^{-3}$ M); 4 – the DMF solution of Na_2C_{60}; 5 – the THF solution after the oxidation of Na_2C_{60} with Ce^{4+}.

In addition to a compound of Ce^{3+} ion, as a second reaction product fullerene C_{60} which was precipitated due to poor solubility in THF was revealed. It manifested itself in solid as the absorption bands at 527, 577, 1183 and 1429 cm⁻¹ characteristic of C_{60}. The maximum of the CL spectrum is at 790 nm and matches well with the maximum of phosphorescence (PS) of the C_{60} solution in the methylcyclohexane/2-methyltetrahydrofuran/ethyl iodide mixture in 2:1:1 volume ratio at 77 K [46].

A comparison of the CL spectrum with the PS and fluorescence spectra of C_{60} in solutions (Figure 4) made it possible to exclude the singlet-excited fullerene $^1C_{60}^*$ as an emitter of CL and attributed this CL to the emission of a triplet-excited fullerene $^3C_{60}^*$. The triplet nature of the CL emitter is confirmed by the quenching of the CL by molecular oxygen. Indeed, when the O_2 flow is passed through the reaction solution take place a sharp drop in the intensity of CL, and the subsequent shutdown of the O_2 current causes the enhance of CL.

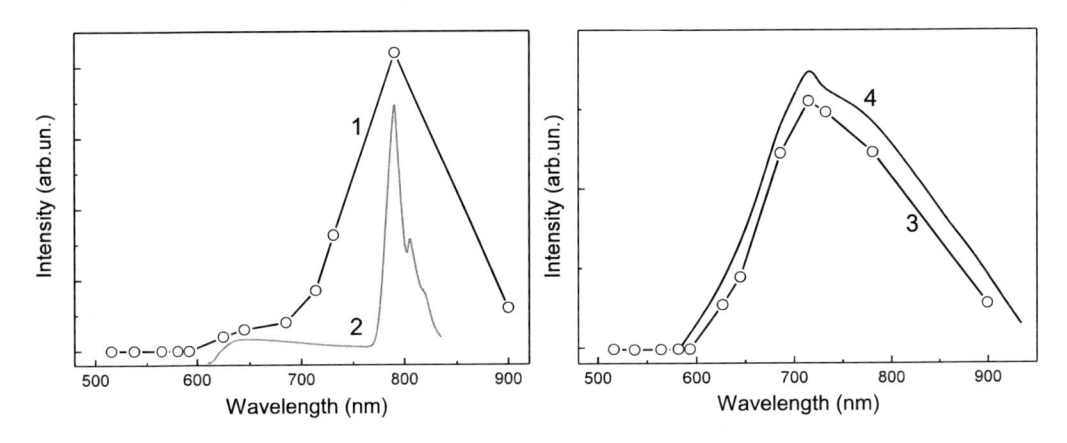

Figure 4. 1 - CL spectrum in the reaction of Na_2C_{60} ($2 \cdot 10^{-3}$ M) with Ce^{4+} (10^{-2} M) in THF measured by the stop-flow method at 300 K; 2 – PS spectrum of the C_{60} solution in the mixture of methyl-cyclohexane/2-methyltetrahydrofuran/ethyl iodide at 77 K [46]; 3 and 4 – PL spectra of C_{60} in toluene ($1.6 \cdot 10^{-4}$ M) measured on a microfluorimeter and an Aminko-Bowman spectrofluorimeter, respectively (77 K, λ_{exc} = 337 nm). Taken from [45]. Spectra 1 and 3 were measured by cut-off filters.

When the solution of Ce^{4+} in THF is added to a greenish-blue solution containing only the radical anion $C_{60}^{-\bullet}$, in the NaC_{60} form (whose absorption maxima are at 910 and 1075 nm) more weaker CL is observed. This CL is located in the same region as the CL with Na_2C_{60}. This experience indicates that the radical anion $C_{60}^{-\bullet}$ is the intermediate of the reaction of Na_2C_{60} with Ce^{4+}. In accordance with the results obtained, in the first stage the electron transfer from dianion C_{60}^{2-} to the Ce^{4+} cation occurs to give radical anion $C_{60}^{-\bullet}$ and Ce^{3+} (1, Scheme 2). The emitter of the CL, the triplet excited state of fullerene ($^3C_{60}^*$), is generated on the second stage (2) *via* electron transfer from intermediate $C_{60}^{-\bullet}$ to another ion Ce^{4+}.

$$C_{60}^{\cdot -} + Ce^{4+} \longrightarrow [C_{60}^{\cdot -} \cdots Ce^{4+}] \longrightarrow C_{60}^{\cdot -} + Ce^{3+} \tag{6}$$

$$C_{60}^{\cdot -} + Ce^{4+} \longrightarrow [C_{60}^{\cdot -} \cdots Ce^{4+}] \longrightarrow {}^3C_{60}^{\cdot} + Ce^{3+} \tag{7}$$

$${}^3C_{60}^{\cdot} \longrightarrow C_{60} + h\nu \ (790 \ nm) \tag{8}$$

Scheme 2.

The enthalpy of reaction (2) was estimated using known electrochemical redox potentials of cerium and fullerene, which are the following:

$$\Delta G_1^0 = E(Ce^{3+}/Ce^{4+}) - E(C_{60}^{2-}/C_{60}^{\cdot -}) = [1.7 \ eV - (-0.72 \ eV)] = 2.42 \ eV,$$

$$\Delta G_2^0 = E(Ce^{3+}/Ce^{4+}) - E(C_{60}^{\cdot -}/C_{60}) = [1.7 \ eV - (-0.44 \ eV)] = 2.14 \ eV.$$

The energy of reaction 2 (Scheme 2) is enough for the generation as the ${}^3C_{60}^*$ triplet (1.56 eV), as well as the ${}^1C_{60}^*$ singlet (2.01 eV). However, the fluorescence of the ${}^1C_{60}^*$ singlet in the CL spectrum (Figure 4) is absent and CL is only due to the phosphorescence of the ${}^3C_{60}^*$ triplet state. Interestengly, this CL is the only example of the emission of ${}^3C_{60}^*$ at room temperature. At first glance it seems surprising, but a similar situation take place, accompanying the initiated oxidation of hydrocarbons (300 K) [47]. Here, the CL due to the emission of triplet-excited ketones or aldehydes is reliably detected at 300 K, while the phosphorescence of these carbonyl emitters is often absent at 300 K and recorded only at 77 K.

3. CHEMILUMINESCENCE IN REACTIONS OF FULLERENES WITH OZONE

3.1. Chemiluminescence in Ozonolysis of the C_{60} and C_{70} Fullerenes in Organic Solvents

The oxidation by ozone is one of the first studied chemical reactions of fullerenes that attracted researchers' attention because it is interesting from the viewpoint of establishing fundamental chemical properties of fullerenes (structure and reactivity) and of using oxidative derivatization for the preparation of water soluble and other derivatives of fullerenes, which hold promise for practical application [see for example 48-54 and references therein].

The first example of a CL in oxidation of fullerenes by ozone was reported in the late nineties [38, 39]. This CL is also interesting as a efficient method for studying the kinetics of formation and nature of the stable and intermediate products of the C_{60} [55, 56] and C_{70} [57] fullerenes ozonolysis.

During puffing O_3 gas onto the surface of a solid sample of fullerite C_{60} at room temperature, CL was not observed because of the fullerite chemical inertness toward ozone [55]. In contrast, while bubbling an O_3/O_2 mixture (3–7% O_3) through solutions of C_{60} [38, 39, 55] or C_{70} [57], it was observed a quite bright CL (Figure 5).

This CL is much brighter than CL in the oxygen oxidation of the sodium fullerides [40-42] described above, that allows one to register its CL spectra using not light filters, but a monochromator (Figure 6) [57].

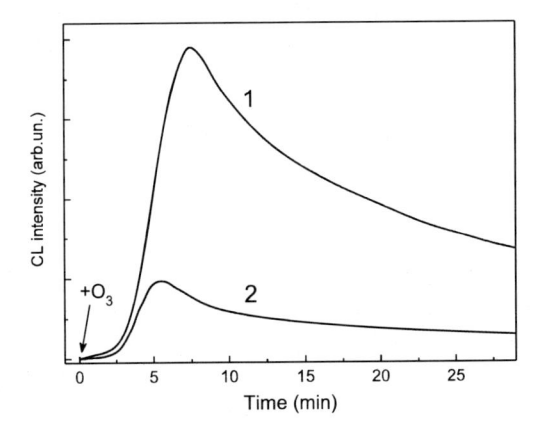

Figure 5. The CL kinetics in the ozonolysis of C_{60} (1) and C_{70} (2) in CCl_4. $[C_{70}]_0 = [C_{60}]_0 = 1.6 \cdot 10^{-4}$ M, $W(O_3) = 1.4$ mmol$\cdot$h^{-1}, 298 K. Taken from [57].

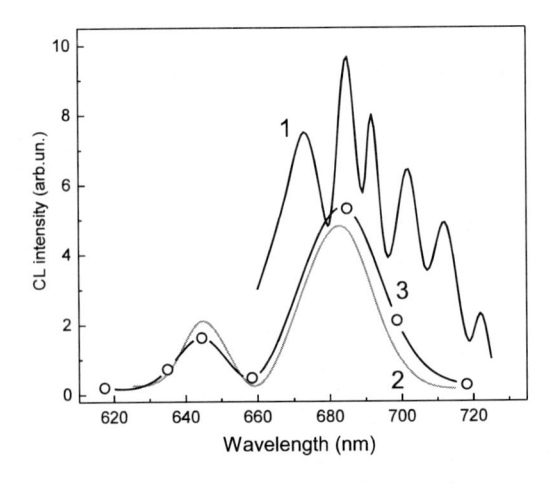

Figure 6. The PL spectra of a C_{70} solution in CCl_4 at 77 K (1) and solution of the solid products of C_{70} ozonolysis in an EtOH-Et$_2$O (2:1) mixture at 77 K (2). The CL spectrum in ozonolysis of the C_{70} solution in CCl_4 (3). $\lambda_{exc} = 530$ (1) and 400 (2) nm.

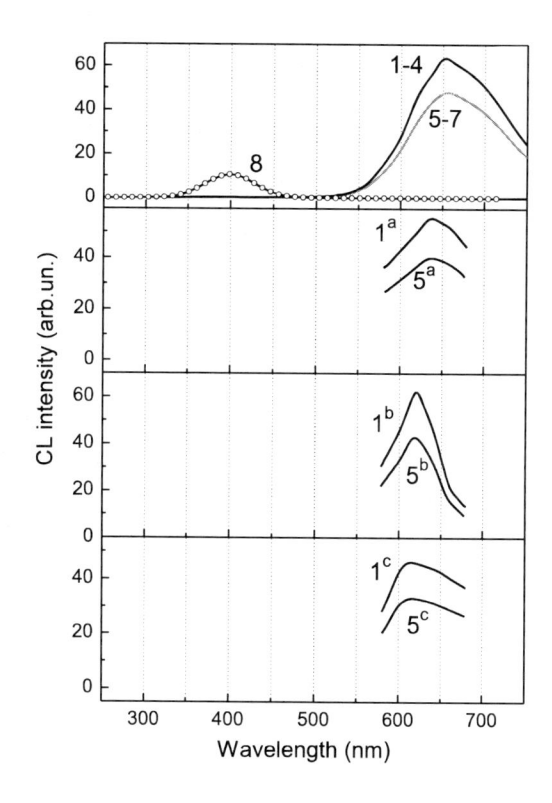

Figure 7. The CL spectra at 300 K in the ozonolysis of C_{60} solutions ($1.6 \cdot 10^{-4}$ M) in toluene (1), benzene (2), methylnaphthalene (3), tetrachloroethane (4) and CCl_4 (5), scanned by the Specol monochromator. Spectra 1 and 5; 1^a and 5^a; 1^b and 5^b; 1^b and 5^c were measured at the 2-, 5-, 7- and 9-minute of ozonolysis, respectively. The sweep time of spectra 1, 1^a-1^c is 65 s. 6, 7 – PL spectra of reaction samples in ozonolysis of C_{60} in toluene and CCl_4, selected at the moment of maximum CL brightness (λ_{exc} = 337 nm); 8 – CL spectra in ozonolysis of toluene.

The ozone CL of fullerenes is observed in solvents of different nature: benzene [39], toluene [38, 39, 57], methylnaphthalene [38, 39], 1,2-dichlorobenzene [57], tetrachloroethane [39], and CCl_4 [55-57]. The CL does not appear when pure oxygen is bubbled through the solution of fullerenes. It is well known that the reactions of ozone with many organic compounds are accompanied by CL [58]. However, the bubbling of ozone through the above solvents gives a much weaker CL, which is also located in the shorter wavelength region (<450 nm) than CL with fullerene (>600 nm) (Figure 7) [39, 55].

For example, the CL intensity in the reaction of C_{60} + O_3 in a solution of toluene is $I_{max} = 6.2 \cdot 10^9$ photon·s^{-1}·mL^{-1}, whereas in ozonation of pure toluene, it is only $I_{max} = 7.6 \cdot 10^5$ photon·s^{-1}·mL^{-1} [39]. Therefore the CL under study originates only from ozone oxidation of fullerenes.

The study of CL is complicated by the heterogeneous character of this reaction. In bubbling O_3 through the solutions of C_{60} or C_{70} (hereinafter referred to as $C_{60/70}$), after the CL appears in 2 to 6 minute, a brown precipitate is formed, which is a mixture of nonreacted parent fullerene and its oxyderivatives [39, 57]. In addition, in solvents containing hydrogen, as a result of oxidation by ozone, byproducts containing active oxygen are formed, which makes it difficult to identify the stable products of ozonolysis of fullerenes and emitters of the CL. Therefore, the most correct data on the composition of the products of ozonolysis and the nature of the CL emitters have been obtained for solutions of $C_{60/70}$ in the aprotic solvent CCl_4 [56, 57].

The $C_{60/70}$ cages are oppened during ozonolysis because their $C = C$ bonds are cleavaged to form two $C = O$ groups at the ends of the open hexagon. As stable ozonolysis products, a complex mixture of oxyderivatives of $C_{60/70}$ are formed. It consist of the epoxides $C_{60/70}O_{n/m}$ (n = 1-6, m = 1-4) [48-55], polyesters and polyketones $C_{60/70}(= O)_{n/m}$ (n = 2-6, m = 2-6) [56, 57], and secondary fullerene ozonides [60]. The latter are very unstable and destroyed by traces of moisture [59, 60].

Among all oxyderivatives, only polyketones are capable of being formed not only in the ground $O = C_{60/70}(= O)_{n/m-1}$ but also in the electronically excited state $*O = C_{60/70}(= O)_{n/m-1}$. These polyketones have been identified as the only emitters of a complex ozonolysis reaction of fullerenes. Polyketones were identified using the UV absorption band at 490 nm of fullerenic 2,4-dinitrophenylhydrazone formed by the treatment of an oxidate sample dissolved in MeOH with 2,4-dinitrophenylhydrazine [56, 57] and also using the IR absorption band (1736 cm^{-1}), which disappeared after the treatment of the precipitate isolated after ozonolysis of fullerene with a solution of $NaBH_4$ in MeOH [57]. These polyketones are also responsible for the photoluminescence, which is manifested by solid products of ozonolysis of fullerenes and their solutions in polar solvents [39, 57]. In addition, this is confirmed by the disappearance of PL of glassy solutions of precipitates formed upon ozonolysis along with the IR absorption band of $C = O$ groups at 1736 cm^{-1} after the treatment of the oxidate with an alcoholic solution of $NaBH_4$ [57].

The effect of hypsochromic shift of the maximum position in the CL spectra was observed with increasing duration of ozonolysis of C_{60}. Thus, after 2 min of ozonization, the maximum is recorded at 660 nm, and after 9 minutes at 610 nm ($\Delta = 50$ nm) (Figure 7). Regardless of the duration of ozonolysis, the position of the maximum of the CL spectrum lies in an essentially longer wavelength region of the spectrum, in comparison with the luminescence of the $>C = O^*$ group of hydrocarbons, usually recorded at 400-480 nm. Such long-wave arrangement of the maxima of CL and PL appears to be due to the unique electronic structure of fullerene molecule containing 60 active π-electrones. This electronic ensemble as a powerful inductor perturbs the excited electron of the radiating carbonyl group of polyketones, which leads to a shift of the luminescence maximum to the long-wavelength region in comparison with luminescence of conventional organic ketones. The reason above described for the hypsochromic shift of the CL spectrum peak is a disturbance of this electronic structure as a result of the splitting and opening the double $>C = C<$ bonds to form the $>C = O$ groups on the ends of the disclosed hexagons. The resemblance and difference in the structures of fullerenes C_{70} and C_{60} are reflected in the close composition and luminescence properties of their ozonolysis products, but the differences in the luminescence properties are more pronounced. For example, the products of C_{60} and C_{70} ozonolysis have practically the same IR spectra, but their CL and PL spectra are different by position and number of maxima. The CL spectrum (λ_{max} = 645 and 685 nm) in ozonolysis of C_{70} in CCl_4 (Figure 6) is practically overlapped with that of C_{60} (λ_{max} = 685 nm) [55, 57] but contains a greater number of maxima. This was explained by the fact that the C_{70} molecule (characterized by the lower symmetry D_{5h}) contains nonequivalent carbon atoms. Therefore, the carbonyl groups formed at ozonolysis are also nonequivalent, which is manifested as an additional maximum in the CL spectrum. The CL emitters are stable ozonolysis products of C_{70}, because the CL spectra coincide with the PL spectra (77 K, λ_{exc} = 400 nm) of glassy ether-alcohol solutions (Et_2O:EtOH = 1:2) of the solid ozonolysis products isolated from the reaction solutions (Figure 6).

The faster formation of the kinetic CL maximum for C_{70} compared to C_{60} is related to the higher rate constant of C_{70} oxidation with ozone [61, 62]. The CL intensity (I_{CL}) of C_{70} in CCl_4 ($I_{CL} = 1.8 \cdot 10^8$ photon$\cdot s^{-1} \cdot mL^{-1}$) is several times lower than that during ozonolysis of C_{60} solutions ($I_{CL} = 8 \cdot 10^8$ photon$\cdot s^{-1} \cdot mL^{-1}$) (Figure 5). It is known [63] that the I_{CL} is determined by Eq. (1)

$$I_{CL} = W \cdot \eta_{exc} \cdot \eta_{lum} \qquad (9)$$

were W is the reaction rate, η_{exc} and η_{lum} are the excitation and luminescence yields of the emitters, respectively.

The ozone oxidation rate constant for C_{70} is 6 times higher than that for C_{60}. The PL intensity of glassy solution of the C_{70} ozonolysis products is also higher by an order of magnitude than that for C_{60}. Therefore the lower brightness for C_{70} is due to the lower excitation yield of the luminescence emitter.

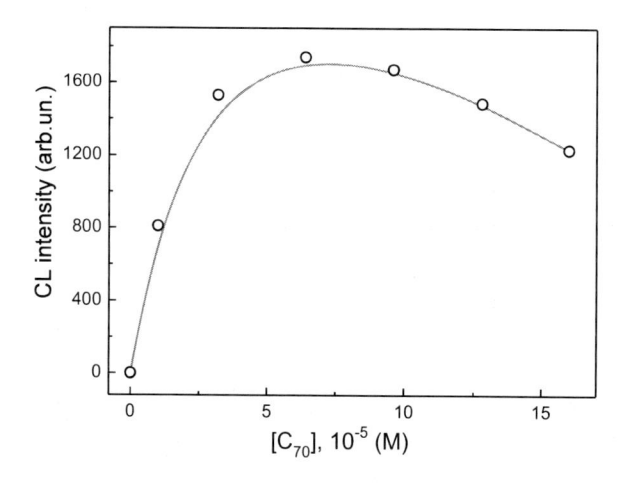

Figure 8. Dependence of the CL intensity at ozonolysis of C_{70} solution in CCl_4 on the initial concentration of C_{70}.

An increase in the initial concentration of fullerenes above a certain value leads to quenching of CL. For example, at the ozonolysis of C_{70}, the maximum CL brightness is observed at $[C_{70}]_0 = 6.4 \cdot 10^{-5}$ M (Figure 8), and at $[C_{70}]_0 = \text{const} = 1.6 \cdot 10^{-4}$ M after the 5 min ozonolysis (Figure 5).

This decrease in the CL intensityis caused by the self-quenching of luminescence of fullerene polyketones, whose content is continuously increasing during ozonolysis [57]. Similar self-quenching is also typical for PL of hydrocarbon ketones [64]. A confirmation of this explanation for the quenching effect of CL is the decrease in the intensity of CL by adding to the reaction mixture a solution of polyketones obtained in an independent experiment in another reactor (Figure 9).

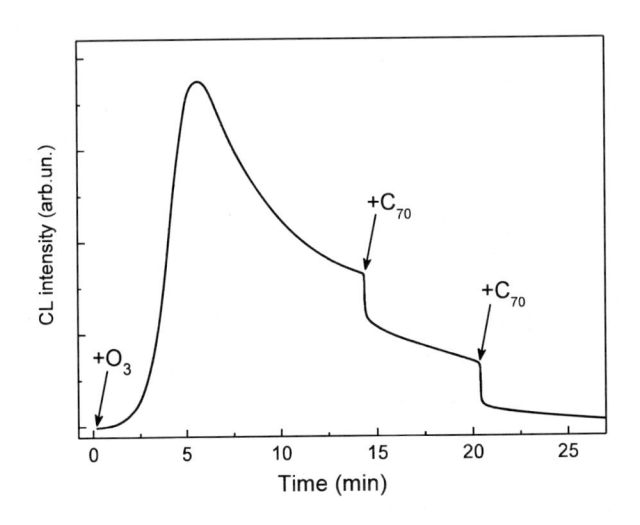

Figure 9. Quenching of CL in ozonolysis of C_{70} in CCl_4 by the addition of polyketones obtained in another reactor by ozonation of C_{70} in CCl_4 (20 ml, 10 min, $[C_{70}]_0 = 1.6 \cdot 10^{-4}$ M).

However, it should come to mind that the parent C_{60} and C_{70} fullerenes are supereffective quenchers of excited states of molecules and ions of different chemical nature [65-68]. Therefore, non-oxidized still fullerenes, along with polyketones, can contribute to quenching of CL (i.e., quenching of excited polyketones), and a quantitative evaluation of the relative contribution to quenching of polyketones and not oxidized fullerenes is necessary.

It is well known that CL in the free-radical reaction of the oxidation of hydrocarbons with ozone is quenched by inhibitors [69]. However, with the introduction of the ionol inhibitor, a decrease in the brightness of CL during ozonization of solutions of fullerenes is not observed, indicating a molecular mechanism of CL. The close composition of the stable products and the close regions of the location of the CL spectra of both fullerenes made it possible to propose the same scheme of the CL molecular mechanism for the ozonolysis of C_{60} and C_{70}, which for simplicity is represented by the example of C_{60} in the Scheme 3 [70].

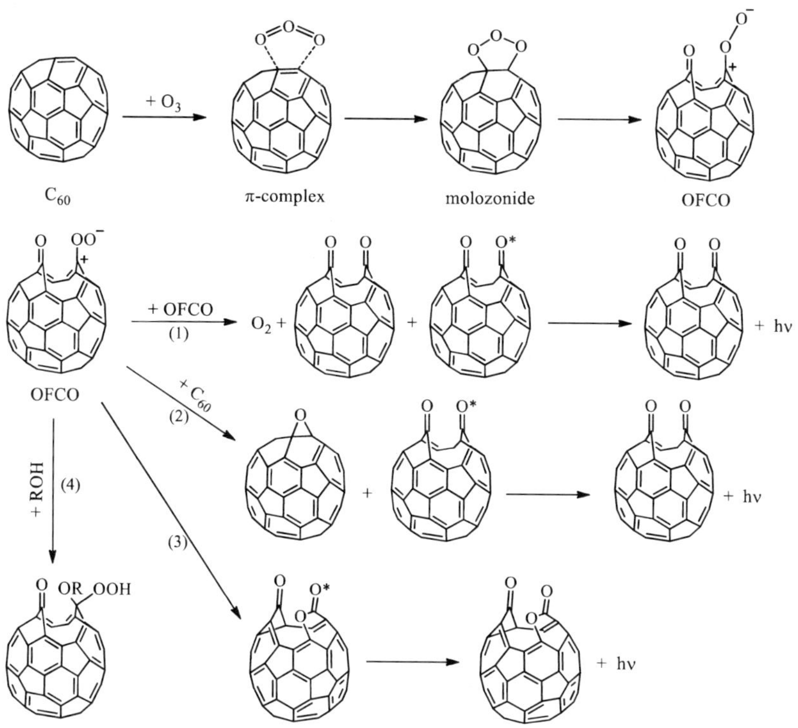

Scheme 3.

This mechanism is also based on well-known data on the study of dark ozonolysis of $C_{60/70}$, primarily on fundamental studies of Heymann [54, 71, 72] and Malhotra [51] groups.

According to Scheme 3, in the first stage a π-complex formed is transformed into 1,2,3-trioxolane (molozonide) which was identified experimentally by absorption spectra and HLPC [71, 72]. Destruction of molozonide gives an active intermediate, *i.e.*, oxofullerenecarbonyloxide (OFCO) [51]. The formation of OFCO confirms the quenching of CL by adding traps of oxocarbonyloxides of hydrocarbons, such as alcohols and others [73]. This test-reaction is completed by the formation of fullerene-containing hydroperoxides (4, Scheme 3). The bimolecular decay of OFCO led to formation of polyketones in the ground and excited states, which deactivated with emission of the CL (1). The another bimolecular reaction of OFCO also affords CL-radiating the excited polyketones along with epoxide (2). However, any contribution of epoxides to CL is excluded, since they do not have luminescence at room temperature, and besides, epoxides disappear from the reaction solution during of 3-6 min, while the emission of CL lasts much longer (about 1 hour) [55, 57]. The third potential light stage of ozonolysis of fullerenes is monomolecular isomerization of OFCO (3), the presence of which confirms the incomplete quenching of CL by additives of OFCO traps [73]. The calculated values of the enthalpies of light transformations of OFCO ($-\Delta H^0 = 203.9$ (1), 237.0 (2), 402.5 (3) kJ·mol^{-1}) are sufficient for the generation of CL emitters of excited fullerene polyketones (191.6 kJ·mol^{-1}).

3.2. Chemiluminescence in oxidation of the fullerenes fluorides $C_{60}F_x$ (x = 18, 36, 48) with ozone in solution

CL in ozone oxidation of fullerene fluorides $C_{60}F_x$ (x = 18, 36, 48) in CCl_4 solution for the first time was found in [74]. Bubbling pure O_2 in the same conditions did not give CL. Unlike CL at ozonolysis of parent C_{60} fullerene, the CL brightness of all fluorides $C_{60}F_x$ is much weaker. This is a consequence of the replacement of active C = C bonds of the fullerene cage with chemically inert C–F bonds. In accordance with this, the maximum CL intensity (in photon·s^{-1}·mL^{-1}) falls in the following row: C_{60} (5.2·10^8) > $C_{60}F_{18}$ (2.7·10^8) > $C_{60}F_{36}$ (1.1·10^7) > $C_{60}F_{48}$ (4.7·10^6). This is due to the lower reactivity of these fluorides with respect to ozone, due to the smaller number of active C = C bonds in fullerene fluorides as compared to C_{60}. In addition, the weaker CL intensity of fluorides $C_{60}F_{36}$ and $C_{60}F_{48}$ are due to the shielding of the C = C bonds adjacent to the F-atoms [75], which impede the attack of ozone to the C = C bond of these fluorides. With an increase in the fluorine content, the maxima of the CL spectra shift to a shorter wavelength region (Figure 11).

This is one more example that demonstrates the high ability of the electronic structure of the fullerene cage to shift the emission maxima of the >C = O group to the red side of the spectrum. Nevertheless, the CL spectra of fluorides $C_{60}F_{36}$ and $C_{60}F_{48}$ also contain long-wave maxima (at 560, 610, and 645 nm), which are close to the maxima of the CL spectrum in ozone oxidation of C_{60}.

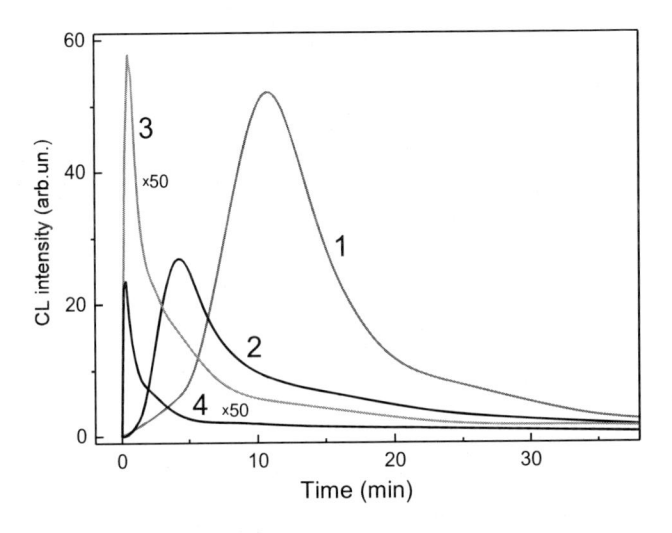

Figure 10. CL kinetics in the ozonolysis of C_{60} (1), $C_{60}F_{18}$ (2), $C_{60}F_{36}$ (3) and $C_{60}F_{48}$ (4) in CCl_4 solutions (293K, 1.6·10^{-4} M). Ozonation was carried out by a O_3/O_2 mixture bubbling (0.07 mmol O_3/min) through solutions.

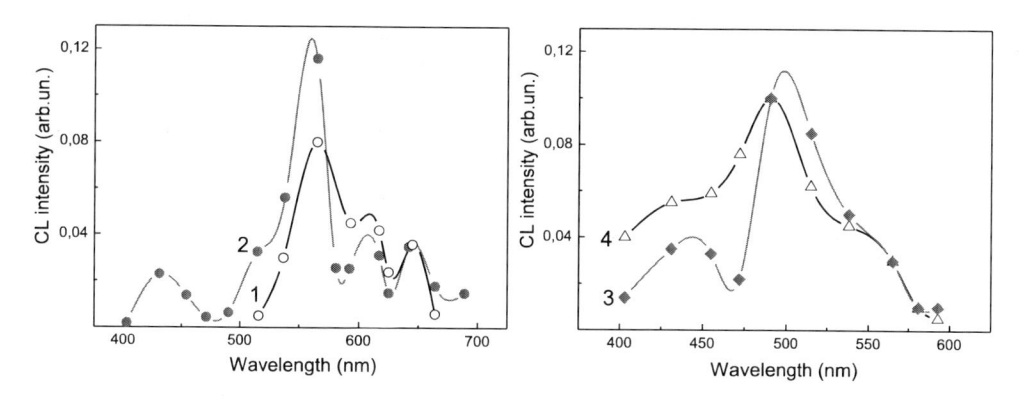

Figure 11. CL spectra in the ozonolysis of C_{60} (1) and $C_{60}F_{18}$ (2), $C_{60}F_{36}$ (3) and $C_{60}F_{48}$ (4) recorded using cut-off filters, FEU-79 (1, 2) and FEU-39 (3, 4) photomultipliers.

This made it possible to propose that CL emitters formed during ozonolysis of the fullerene fluorides are also excited polyketones resulting from the cleavage of the $C = C$ bonds of the fullerene cage (Scheme 4).

$$C_{60}F_x + O_3 \longrightarrow {}^*(O)C_{60}F_xO_y + (O)C_{60}F_xO_y \tag{10}$$

$${}^*(O)C_{60}F_xO_y \longrightarrow (O)C_{60}F_xO_y + h\nu \tag{11}$$

$x = 18, 36, 48.$

Scheme 4.

3.3. Chemiluminescence in Ozonation of C_{60} Aqueous Dispersions

Fullerene C_{60} can form aqueous dispersions (ADs) with different sizes of colloid particles and stabilities [76-78]. These ADs are colloid solutions that consist of associates of C_{60} donor-acceptor complexes with water, in which the C_{60} molecules are surrounded by a specific hydration shell. At low concentrations, these complexes correspond to the formula $C_{60}@nH_2O$, and at high dose to $[C_{60}@nH_2O]_y@zH_2O$. In the latter case, these clusters have a common hydration shell and no direct contacts between the C_{60} molecules. These ADs are of great interest, because they exhibit high biological activity, including antioxidant [79]. The reactivity of fullerene C_{60} in the composition of the ADs strongly depends on the method of their production. As it turned out, the chemiluminescent activity of ADs towards ozone is also influenced by the nature of ADs. This was established by the example of five ADs (AD-1–AD-5), different in particle sizes and obtained by various known methods [80]. The CL intensity at ozonation of all types of ADs is significantly less than that for organic C_{60} solutions (Figure 12) [80]. At the same time, the intensity of the ADs CL is noticeably higher than the intensity of the CL in ozonation of the pure water, which is the background CL (below called BCL). This type of CL was detected in ozonation of water or H_2SO_4 [81]. This CL originates due to the emission of singlet oxygen $^1O_2$ and its dimers $(^1O_2)_2$ formed during the catalytic decomposition of ozone on the surface of a glass reactor. Surprisingly, the concentration of ozone significantly affects the spectra of CL. So, when it is low ($N(O_3) = 2$ mmol·h^{-1}), the CL spectra recorded during the ozonation of all types ADs are the same as the BCL spectrum (Figure 13). This is a good argument for attributing this CL to the thermocatalytic decomposition of ozone. It is also confirmed by the high sensitivity of CL to the particle size of the ADs. For example, in the case of AD-1 and AD-5, which have a more developed surface of particles compared with the glass cell surface, the CL intensity is higher than in BCL.

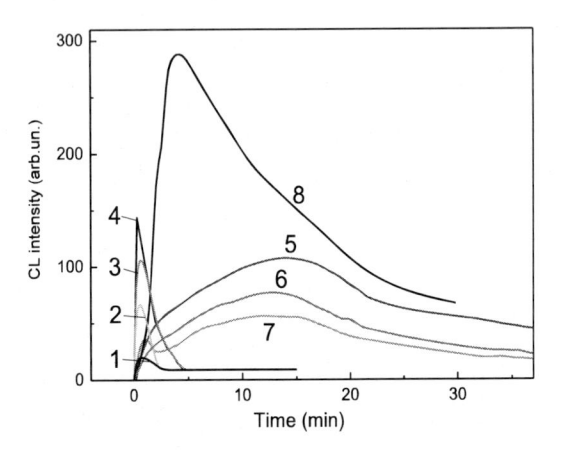

Figure 12. CL kinetics in the ozonation of ADs, H_2O and C_{60} solution. 1 – doubly distilled water; 2 – AD-1 ($[C_{60}]_0 = 5\cdot10^{-3}$ M); 3 – AD-2 ($[C_{60}]_0 = 1\cdot10^{-4}$ M); 4 – AD-3, AD-4, AD-5 ($[C_{60}]_0 = 2\cdot10^{-4}$ M); 5 – AD-5 ($[C_{60}]_0 = 2\cdot10^{-4}$ M); 6 – AD-3 ($[C_{60}]_0 = 2\cdot10^{-4}$ M); 7 – AD-4 ($[C_{60}]_0 = 1.77\cdot10^{-4}$ M); 8 – C_{60} in CCl_4 ($[C_{60}]_0 = 2\cdot10^{-4}$ M). The CL curves 1-4 increases to 2-fold. $T = 300$ K, $V = 5$ mL, $N(O_3) = 2.0$ (1-4), 33.4 (1-8) mmol·h^{-1}.

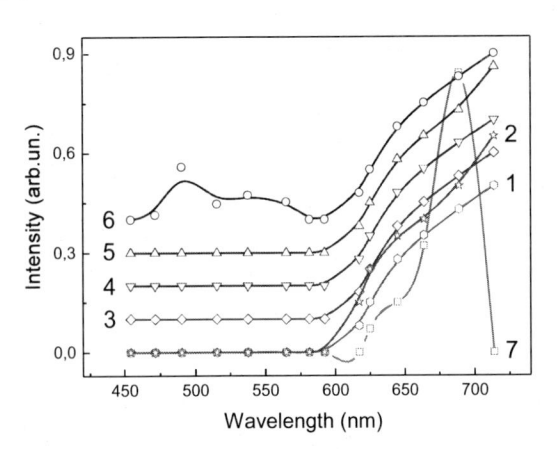

Figure 13. CL spectra in the ozonation of ADs, H_2O and C_{60} solution. 1 – doubly distilled water; 2 – AD-1 ($[C_{60}]_0 = 5\cdot10^{-3}$ M); 3 – AD-2 ($[C_{60}] = 1\cdot10^{-4}$ M); 4 – AD-5 ($[C_{60}]_0 = 2\cdot10^{-4}$ M), 5 – AD-3 ($[C_{60}]_0 = 2\cdot10^{-4}$ M); 6 – AD-4 ($[C_{60}]_0 = 1.77\cdot10^{-4}$ M); 7 – C_{60} in CCl_4 ($[C_{60}]_0 = 2\cdot10^{-4}$ M). $T = 300$ K, $V = 5$ mL, $N(O_3) = 2.0$ (1-7), 33.4 (1-3) mmol·h^{-1}. For the sake of clarity, the spectra 3-6 are shifted up along the Y-axis.

Due to the complete consumption of ozone in the catalytic decomposition reaction at $N(O_3) = 2.0$ mmol·h^{-1}, the C_{60} fullerene, on the contrary, is preserved. This is confirmed by not only identical the CL and BCL spectra, but also by identical the absorption spectra of the dispersions AD-1 to AD-5 before and after the ozonation. The increase in the CL brightness on going from AD-1 and AD-2 (particle size of up to 1.7 μm) to AD-3 and AD-4 (particle size of up to 700 nm) and to AD-5 (contains associates of hydrated C_{60} molecules) correlates well with an increase in the specific surface area of the catalyst, fullerene micro- and nanoparticles.

The most distinctly different nature of the ADs occurs at high ozone concentrations and the oxidation of C_{60} take place only at $N(O_3) = 33.4$ mmol·h^{-1}. In this case rather bright CL appears and the kinetic peaks ($I_{CL} = 2.1·10^8$ (AD-3), $1.4·10^8$ (AD-4), $2.8·10^8$ (AD-5) photon·s^{-1}·mL^{-1}) are formed much later (by ~15 min) compared with the CL observed in ozonation of other AD (1–2 min) or with a C_{60} solution in CCl_4 (5 min). In addition, the maxima of the CL spectra during the ozonation of AD-3, AD-4, and AD-5 are at shorter wavelengths (570 and 620 nm) than that during ozonation of C_{60} in CCl_4 (660 nm). In ozonation of these ADs take place the chemical reaction C_{60} with ozone, which affords C_{60} oxyderivatives in the ground and electronically excited states. For example, IR spectrum of the AD-3 product ozonolysis exhibits the following bands (cm^{-1}): 3400 (s, br) (OH), 1667 (s) (quinones), 1640 (s) (quinones), 1621 (s), 1429–1450 (m, br) and 1060–1110 (m, br) (C–O–C), 1196 (w), 1317 (w) (C–O), 805 (m), 527 (vw) (traces of C_{60}). The differences between the CL spectra during reaction of AD-3, AD-4, and AD-5 and the IR spectra of the products of ozonation of these ADs and analogous spectra at ozonolysis of C_{60} in CCl_4 show different nature of CL emitters and stable products of C_{60} oxidation. It is assumed that the most probable CL emitters during ozonolysis of these ADs are quinone derivatives of C_{60} as compounds with a carbonyl group (traditional CL emitters for organic compounds). In contrary, despite the high ozone concentration ($N(O_3) = 33.4$ mmol·h^{-1}), the CL spectra for the AD-1 and AD-2 ozonation are the same as for $N(O_3) = 2.0$ mmol·h^{-1} (Figure 13). The size of the particles in these ADs is such that they can only act as catalysts for the decomposition of ozone.

The above-described CL, which has signs of a two-faced Janus, has the prospect of practical application for testing the quality of biologically active preparations based on aqueous dispersions of fullerene.

3.4. Chemiluminescence in the Hydrolysis of Secondary Ozonides C_{60} and C_{70}

The CL found during in the hydrolysis of secondary ozonides of fullerenes C_{60} (CL-1) and C_{70} (CL-2) [59, 60] is a new type of CL because fullerene's molozonides, as well as organic compounds containing active oxygen (the –O–O– bond), such as peroxides, hydroperoxides and bisperoxides, do not produce CL in treatment by water. Another remarkable property of this CL is its selectivity. Indeed, of the entire mixture of oxyderivatives formed in the fullerene ozonolysis consisting of epoxides, polyketones, polyesters, carboxylic acids, and secondary ozonides of fullerenes (SOF), only SOF react with water to give the CL emission. This allows the CL in hydrolysis to be used for identification of SOF in a complex mixture of fullerene ozonolysis products. This CL arises as a narrow kinetic peak (I_{CL} = 2.65·10^8 and 1.63·10^8 photon·s^{-1}·mL^{-1} for C_{60} and C_{70}, respectively) when an aliquot of water (2 mL over 0.5 s) is added to the brown suspension of the ozonolysis products of both fullerenes [59, 60], containing fullerene epoxides, polyketones, polyesters and carboxylic acids [56, 57] (Figure 14).

Due to the polar nature of the fullerene oxyderivatives, the suspension was completely dissolved in the aqueous phase. According to iodometric-spectrophotometric analysis, the concentration of SOF increases during the whole ozonolysis time (Figure 15) [60] as well as that of another fullerene oxyderivatives [56, 57].

The most probable CL emitters during the hydrolysis of the SOF of both fullerenes are excited polyketones, as indicated by the coincidence of the spectra of CL-1 and CL-2 with the spectrum of CL during ozonolysis of C_{60} (Figure 16).

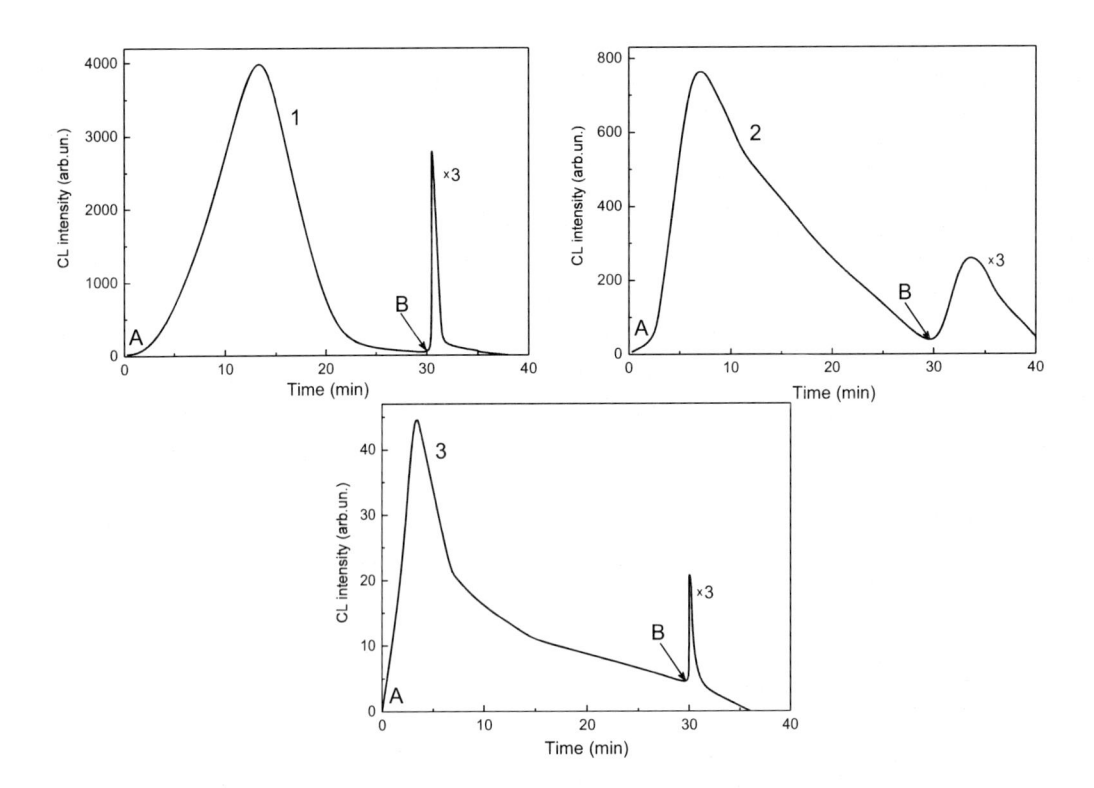

Figure 14. Kinetics of CLs: segment AB – CL curves during the ozonolysis of C_{60} (1), C_{70} (2) in CCl$_4$ and benzene (3); from B – CL curves during the hydrolysis of suspensions of SOF C_{60} (1), C_{70} (2) in CCl$_4$ and benzene ozonide (3). V = 10 mL, $V(H_2O)$ = 1 mL, $W(O_3)$ = 1.4 mmol·h^{-1}, $[C_{60}]_0$ = $[C_{70}]_0$ = 1.6·10^{-4} M, 294K.

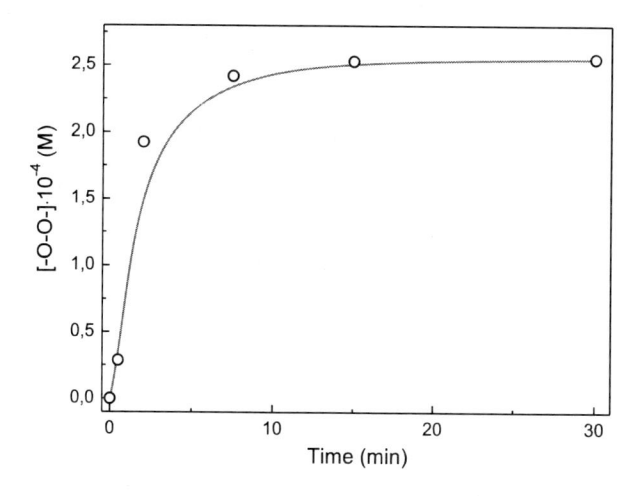

Figure 15. Change in the SOF content at different durations of ozonolysis of C_{60} solutions in CCl_4.

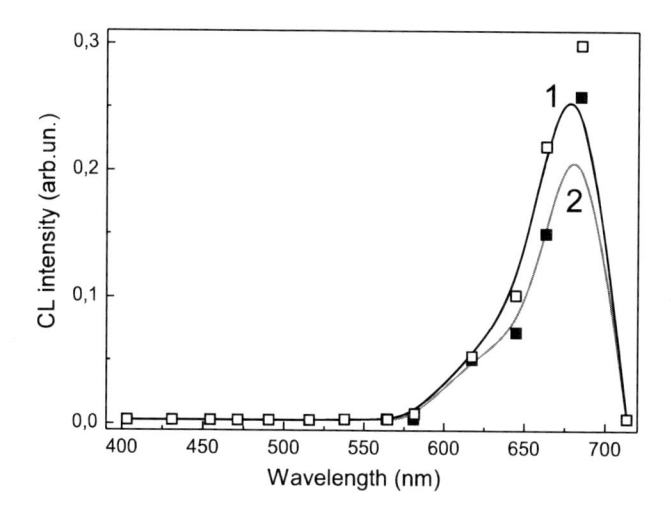

Figure 16. CL spectra in the hydrolysis of suspensions of solid products obtained by ozonation of solutions of C_{60} (1) and C_{70} (2) in CCl_4.

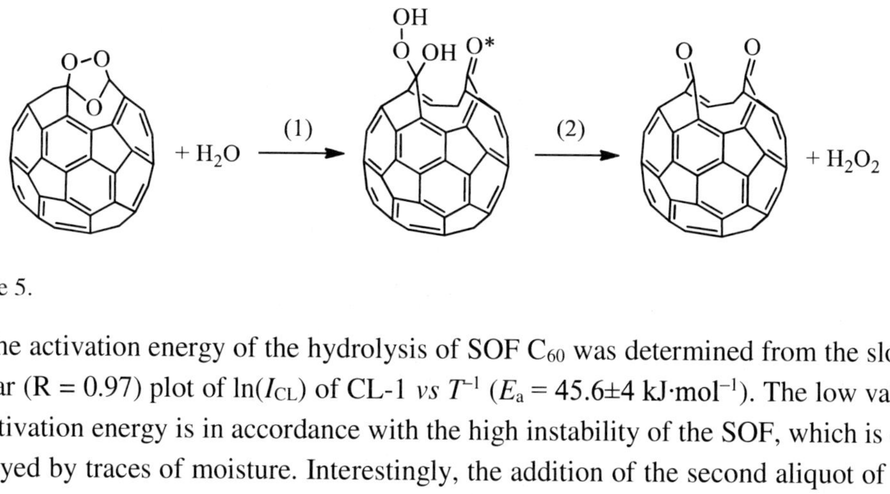

Scheme 5.

The activation energy of the hydrolysis of SOF C_{60} was determined from the slope of a linear (R = 0.97) plot of $\ln(I_{CL})$ of CL-1 *vs* T^{-1} (E_a = 45.6±4 kJ·mol^{-1}). The low value of the activation energy is in accordance with the high instability of the SOF, which is easily destroyed by traces of moisture. Interestingly, the addition of the second aliquot of water caused no CL. This means that SOF have completely disappeared in reaction with the first portion of water, which induces CL. However, the solution after hydrolysis contains reactive oxygen (KI test) and H_2O_2. The later was established due to the CL, which immediately arises at addition a solution of $FeSO_4 \cdot 9H_2O$ into the reaction slurry. Thus, polyketones and H_2O_2 are the main products formed at hydrolysis of SOF.

According to the quantum chemical calculations [60] of the heat effect by the semiempirical methods, the CL in hydrolysis of SOF is excited as shown on Scheme 5 [70].

For excitation into the radiating state of polyketones, energy of 214.4 $kJ \cdot mol^{-1}$ is necessary, corresponding to the maximum of the CL spectrum at 558 nm. This energy with a large excess can provide (even without considering the activation energy) the heat effect of reaction (1) (Scheme 5), equal 363.8 $kJ \cdot mol^{-1}$.

4. CHEMILUMINESCENCE IN OXIDATION AND REDUCTION OF C$_{60}$ SOLUTIONS AFTER THEIR PHOTOIRRADIATION, RADIATION EXPOSURE AND SONIFICATION

Under normal conditions, the initial, non-functionalized fullerene molecules are usually chemically inert with respect to oxidizers (O_2, H_2O_2, $KMnO_4$ and others) or reductants (Et_3Al, NaH, etc.) and therefore do not produce CL upon contact with them. This chemiluminescent passivity of fullerene can be overcome by pretreatment of a fullerene solution by different types of radiation. Such type of CL was registered after addition of the different chemical reagents into C_{60} solutions in *N,N*-dimethylformamide (DMF) saturated O_2 which were preliminary irradiated by light (2 min, Xe-lamp, 1000W) or γ-radiation (20 min, ^{60}Co source, 6500 Ci, 24 Gy min^{-1}) or ultrasound (180 min, Sultan 300, Pro-Sonic, 47 kHz) [82]. As chemical reagents were used solid NaH or solution of KOH in MeOH or Fenton's reagent (a solution of $FeSO_4 \cdot 8H_2O$ and H_2O_2). The most intense CL is given by the addition of NaH, probably due to more efficient decomposition of the intermediate peroxides by this reagent. However due to the low intensity and short exposure (a few seconds) of CL, recorded as flashes of light, the authors could not measure the spectra of CL. They also had problems identifying interaction products due to the formation of DMF-based polymers. Nevertheless, a simplified CL generation scheme (Scheme 6) was proposed.

$$C_{60}CH_2N(CH_3)CHO + Action \longrightarrow HC_{60}CH_2N(CH_3)CHO \qquad (12)$$

$$HC_{60}CH_2N(CH_3)CHO + Reagent \longrightarrow O=C_{60}CH_2N(CH_3)CHO + CL \qquad (13)$$

where Action is the photo-, radio- or sono-irradiation.

Scheme 6.

According to Scheme 6, as a result of different types of radiation, DMF is destroyed with the formation of the radical $^{\bullet}CH_2N(CH_3)CHO$, the attack of which on the fullerene gives a derivative of C_{60} containing a fragment of $CH_2N(CH_3)CHO$. The subsequent addition of a chemical reagent (Reagent) leads to the emission of CL and the formation of C_{60} derivatives containing a carbonyl group. The rationale for the proposed scheme was the next results of the analysis of the reaction products. The two groups of NMR signals between 140 and 125 ppm and also 65 and 50 ppm in the products of photolysis, radiolysis and sonolysis of the C_{60} solutions in DMF were attributed to the formation of C_{60} derivatives containing the DMF fragment $CH_2N(CH_3)CHO$. A photolysis product after registration of CL and after column chromatography showed peaks at m/z 809 and 810 in the mass spectrum and also maximum at 1700 cm^{-1} in IR spectrum which were attributed to the fullerene derivative which contains on the ends of an open hexagon the $CH_2N(CH_3)CHO$ and $>C = O$ fragments. However, the question regarding the identification of the CL emitter (emitters) remains open. The assignment of the absorption band of the products at 1680 cm^{-1} to the carbonyl group, the carbon atom of which belongs to the fullerene skeleton, also seems to be problematically. Indeed, carbonyl-containing fullerene polyketones usually manifest themselves in the IR spectra in the form of more shorter-wave maximum (1736 cm^{-1}) than at 1680 cm^{-1} [55]. Moreover, the band at 1680 cm^{-1} is characteristic of the keto-group DMF, and most likely belongs to the $>C = O$ group of the $CH_2N(CH_3)CHO$ fragment.

5. CHEMILUMINESCENCE IN OXIDATION OF THE FULLERENE HYDRIDE $C_{60}H_{36}$ BY OXYGEN

The study of the reactivity of fullerene hydrides (FH) towards oxygen is of theoretical and practical interest. The first aspect relates to the interest in the possible influence of the electronic structure of fullerene on the reactivity of the C–H bond attacked by oxygen. The second aspect is related to the potential possibility of using fullerene hydrides as hydrogen batteries [20, 83, 84].

Taking into account the significant success in applying the CL method for studying the reactivity of the C–H bond of hydrocarbons towards oxygen [3, 6, 9], a search was conducted and CL was found in the oxidation by O_2 of the $C_{60}H_{36}$ hydride, obtained by the action of the Zn–HCl reagent mixture on a solution of fullerene C_{60} under argon [85].

The CL was observed in two variants of the experiment: when O_2 was bubbling through a solution of $C_{60}H_{36}$ in toluene at 293K ($I_{CL} = 1.4 \cdot 10^6$ photon$\cdot$s$^{-1}\cdot$mL^{-1}) or when an aliquot of toluene saturated with O_2 was added to the $C_{60}H_{36}$ solution, followed by heating to 343K ($I_{CL} = 6.6 \cdot 10^7$ photon$\cdot$s$^{-1}\cdot$mL^{-1}) (Figure 17).

Since, under the experimental conditions, C_{60} and toluene are inert to O_2, the CL was assigned to the O_2 attack on the tertiary C–H bond of $C_{60}H_{36}$. Interestingly, the CL peculiarities in the oxidation of $C_{60}H_{36}$ and those for the known CL in oxidation of a tertiary C–H bond of hydrocarbons initiated by AIBN [9, 86] are drastically different. Indeed, it is well known that the addition of inhibitors quenches the CL of hydrocarbons [87]. However, even such strong inhibitors as ionol and galvinoxyl (10^{-4} M), have no influence on the CL of $C_{60}H_{36}$. Another feature of the CL hydride is that it is observed independently of the presence of AIBN, whereas the CL of hydrocarbons occurs only in the presence of AIBN. Also, the CL intensity of the $C_{60}H_{36}$ hydride is not affected by 9,10-dibromanthracene, the addition of which increases the intensity of the CL during the oxidation of hydrocarbons by several orders of magnitude.

The CL during the oxidation of $C_{60}H_{36}$ with oxygen is due to the emission of two emitters: the $*C_{60}H_{18}$ hydride ($\lambda_{max} = 495, 535$ nm) and singlet excited fullerene $^1C_{60}*$ (720-750 nm) (Figure 18).

These maxima correlate well with the known maxima of the PL spectra of $C_{60}H_{18}$ ($\lambda_{max} = 480, 520$ nm) [88] and fullerene C_{60} ($\lambda_{max} = 720$-750 nm) [89]. The formation of C_{60} during oxidation of $C_{60}H_{36}$ with oxygen confirms the appearance of a peak of C_{60} on the HPLC chromatogram of oxidation products [85].

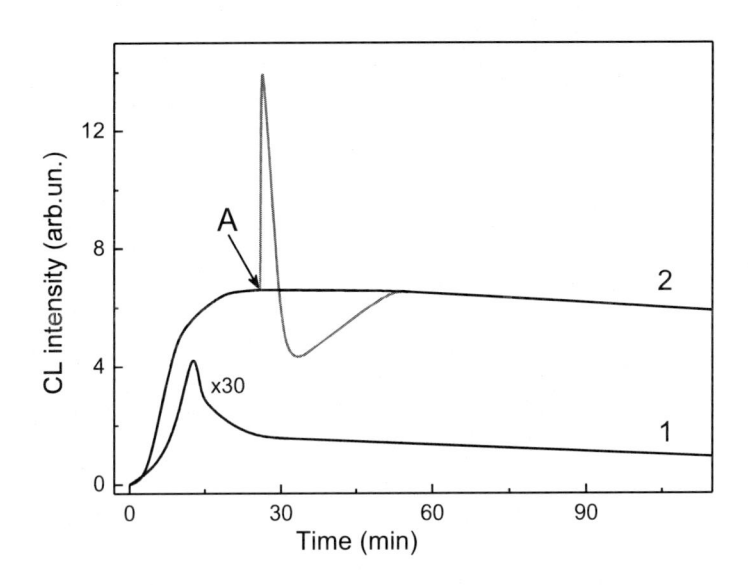

Figure 17. The CL kinetics in the oxidation of $C_{60}H_{36}$ ($4 \cdot 10^{-4}$ M) in toluene solution at 293K (1) and 343K (2). A is the moment of introduction of the Co(acac)$_2$ ($2 \cdot 10^{-4}$ M) solution.

Previously, among the products of the reaction $C_{60}H_{36} + O_2$, the formation of C_{60} and hydrides with a lower hydrogen content $C_{60}H_{18}$ was also found [90, 91]. This CL spectrum also contains an unidentified maximum at 610 nm (Figure 18), located in the emission region of fullerene polyketones. The formation of the excited carbonyl products in oxidation of $C_{60}H_{18}$ is possible by thermal decomposition of the intermediate C_{60}

hydroperoxide, formed by the insertion of oxygen into the C–H bond. This decomposition readily explains the prolonged emission of CL (Figure 17). A glow outburst above the kinetic curve of CL, which occurs when a complex Co(acac)$_2$ is added (Figure 17), serves as an argument in favor of the formation of C$_{60}$H$_{36}$ oxidation intermediates, containing active oxygen. This effect is also known for the CL in the oxidation of hydrocarbons [93].

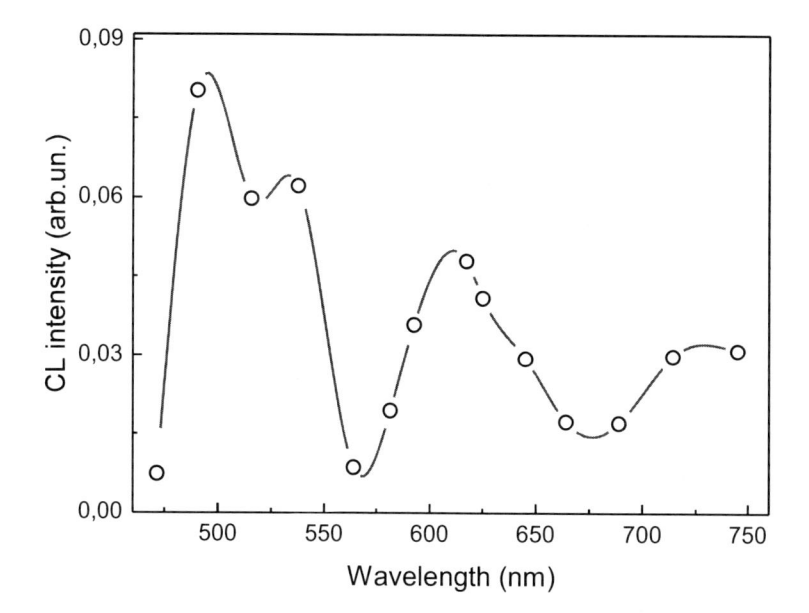

Figure 18. The CL spectrum in the oxidation of C$_{60}$H$_{36}$ ($4 \cdot 10^{-4}$ M) in toluene solution at 343K.

In accordance with the obtained results, only the most general scheme (Scheme 7) of the CL generation was proposed.

$$C_{60}H_{36} + O_2 \xrightarrow[\text{Sol}]{\Delta T} C_{60}H_{18}^{\cdot} + {}^1C_{60}^{\cdot} + P \qquad (14)$$

$$C_{60}H_{18}^{\cdot} \longrightarrow C_{60}H_{18} + h\nu_1 \ (495, 535 \text{ nm}) \qquad (15)$$

$${}^1C_{60}^{\cdot} \longrightarrow C_{60} + h\nu_2 \ (720\text{-}750 \text{ nm}) \qquad (16)$$

Sol = C$_6$H$_6$, C$_6$H$_5$CH$_3$, P - unidentified products.

Scheme 7.

6. CHEMILUMINESCENCE OF C$_{60}$ AND C$_{70}$ ALKYL DERIVATIVES IN THE (R$_3$AL-C$_{60/70}$)-O$_2$ SYSTEMS (R = ET, BU$^\text{I}$)

As is known, the highly exothermic oxidation of alkylaluminums (R$_3$Al) with dioxygen is accompanied by CL, caused by the emission of the excited carbonyl products such as aldehydes or ketones [5]. These emitters are generated in disproportionation of the intermediate peroxyl radicals ROO•. Along with ROO•, the R• and RO• radicals are also formed [92]. In connection with above data, an interest arose in studying the effect of fullerene C_{60} on the described CL, as well as the possibility of using the $R_3Al + O_2$ reaction in the presence of C_{60} as a new source of fullerenyl radicals of the form $XC_{60}•$, where X is an organic radical. In the research undertaken in this direction, success was achieved in both aspects. [94, 95].

The presence, or rather, active participation of fullerene in the reaction oxidation of Et_3Al with oxygen in solution substantially changes both the composition of the products and the characteristics of the CL. Indeed, when Et_3Al is oxidized in toluene in the absence of additives, the $Al(OEt)_3$ alkoxide is formed as the main product, and the content of other hydrocarbon products does not exceed ~5% [92]. In contrast, when interacting in the (C_{60}-Et_3Al)-O_2 system, the formation of fullerene derivatives was detected. Using the IR, UV-Vis, mass-spectroscopy and HLPS, it was established that fullerene ethylhydrides (FEH) and ethylfullerenes (EF), as well as the $EtC_{60}C_{60}Et$ dimer (hereinafter dimer) and unreacted C_{60} are the stable products of the interaction in this system. It means that of the series of the R•, RO•, and ROO• (R = Et) radicals which are formed in the reaction $Et_3Al + O_2$, only the R• radicals are added to the C_{60} molecule with the formation of FEH, EF and dimer. It confirms the mass spectrum of negative ions of the reaction products which contains the following lines, m/z (I_{rel}, %): 720 [C_{60}] (40), 750 [$C_{60}EtH$]⁻ (67), 778 [$C_{60}Et_2$]⁻ (40), 780 [$C_{60}Et_2H_2$]⁻ (40), 808 [$C_{60}Et_3H$]⁻ (50), 810 [$C_{60}Et_3H_3$]⁻ (53), 836 [$C_{60}Et_4$]⁻ (32), 838 [$C_{60}Et_4H_2$]⁻ (37), 840 [$C_{60}Et_4H_4$]⁻ (100), 866 [$C_{60}Et_5H$]⁻ (82), 868 [$C_{60}Et_5H_3$]⁻ (78), 870 [$C_{60}Et_5H_5$]⁻ (46), 894 [$C_{60}Et_6$]⁻ (32), 896 [$C_{60}Et_6H_2$]⁻ (18), 898 [$C_{60}Et_6H_4$]⁻ (12), and 900 [$C_{60}Et_6H_6$]⁻ (30). Among FEH, the $H_4C_{60}Et_4$ compound is formed in dominant quantities. Similar products were obtained in the oxidation of Et_3Al in the presence of fullerene C_{70}. In addition, the presence of a dimer among the products is clearly demonstrated by the appearance of the ESR signal of the $EtC_{60}•$ radical (g = 2.0037) or $EtC_{70}•$ radical (g = 2.0024) as a result of photo-irradiation of solutions obtained after interaction in the ($C_{60/70}$-Et_3Al)-O_2 system (Figure 19).

The characteristics of CL in the system (C_{60}-Et_3Al)-O_2 significantly differ from those for CL in oxidation of Et_3Al by O_2 (background CL or BCL) (Figure 20). First of all, the intensity of this CL ($I_{CL} = 3.2 \cdot 10^7$ photon·s⁻¹·mL⁻¹ at $[C_{60}]_0 = 7.4 \cdot 10^{-4}$ M) is much higher than that of BCL ($I_{CL} = 6.0 \cdot 10^6$ photon·s⁻¹·mL⁻¹). The spectrum of CL with C_{60} participation is located at a longer-wavelength region ($\lambda_{max} = 617$ and 664 nm) than the spectrum of BCL ($\lambda_{max} = 420$ nm) (Figure 21).

It was proposed two probable alternative mechanisms for the CL emitted in the (C_{60}-R_3Al)-O_2 system. According to the first of them, the $RC_{60}•$ radicals are oxidized by O_2 to peroxyl radicals i.e., $RC_{60}OO•$, which further disproportionate with the emission of CL (Scheme 8).

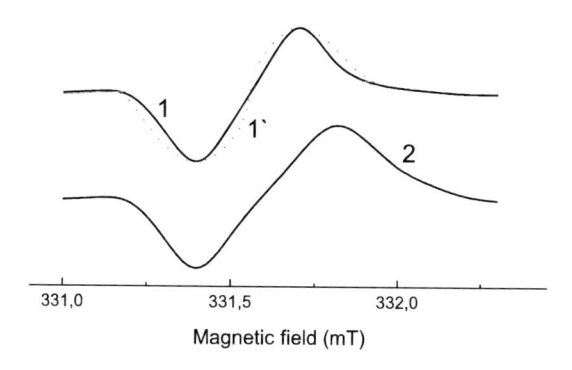

Figure 19. ESR spectra recorded during photo-irradiation of the hydrolysates of the reaction products in the (C_{60}-Et_3Al)-O_2 (1, 1`) and ($C_{70}$-$Et_3Al$)-$O_2$ (2) systems after three freezing–evacuation–thawing cycles (1, 2) and in air (1`).

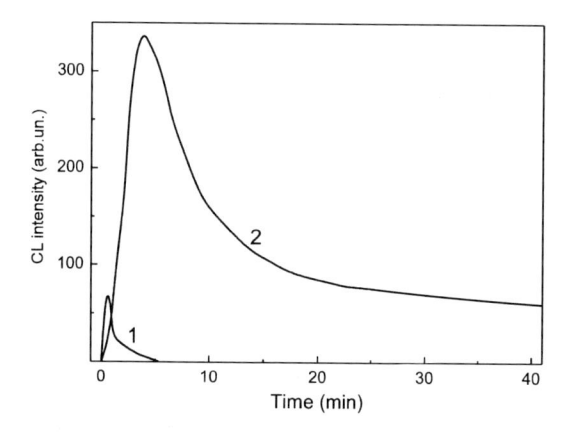

Figure 20. Kinetics of BCL (1) and CL (2) in toluene at 298 K. $[C_{60}]_0 = 7.4 \cdot 10^{-4}$ M, $[Et_3Al]_0 = 1.4 \cdot 10^{-1}$ M; O_2 bubbling through solutions of Et_3Al (1) and C_{60}-Et_3Al (2).

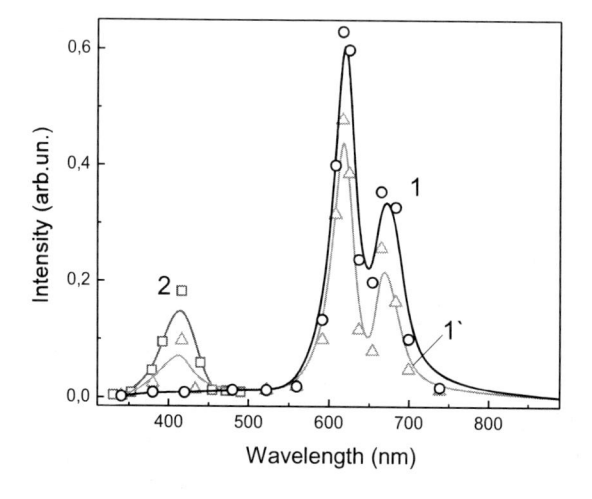

Figure 21. Spectra of CL (1, 1`) and BCL (2) at bubbling $O_2$ through solutions of $C_{60}$-$Et_3Al$ (1, 1`) and Et_3Al (2) in toluene. 298 K, $[Et_3Al]_0 = 0.14$ M, $[C_{60}]_0 = 7.4 \cdot 10^{-4}$(1) and $9.6 \cdot 10^{-6}$ M (1`).

The mechanism presented in Scheme 8 is similar to the mechanism of CL at the initiated oxidation of hydrocarbons, first proposed by Vasil'ev with colleagues [96], which in turn is based on the Russell's mechanism of the dark oxidation of hydrocarbons [97].

$$RC_{60}^{\cdot} + O_2 \longrightarrow RC_{60}OO^{\cdot} \qquad (17)$$

$$RC_{60}OO^{\cdot} + RC_{60}OO^{\cdot} \longrightarrow P + CL \qquad (18)$$

P - stable products

Scheme 8.

However, using the ESR method, it was found that the reaction of $RC_{60}^{\cdot}$ with O_2 (1, Scheme 8), unlike the $R^{\cdot} + O_2 \rightarrow ROO^{\cdot}$ reaction, which occurs for hydrocarbons with a constant close to k_{diff}, practically does not take place! Instead of oxidation, the $RC_{60}^{\cdot}$ radicals form with O_2 an unstable complex, which in the ESR spectrum manifested only in a broadening of the lines of the $RC_{60}^{\cdot}$ radicals (Figure 19), but not in their disappearance [94]. This means, that the CL is generated by a mechanism different from that accepted for oxidation of hydrocarbons. The second mechanism suggests that the triplet excited aldehyde ($^3CH_3CHO^*$) BCL emitter transfers energy to the interaction products in the system ($C_{60}-R_3Al$)-O_2, which emit more long-wavelength CL than BCL. The following arguments are in favor of the second mechanism. First, it was found the selective quenching of the short-wavelength maximum of the CL spectrum at 420 nm as a result of an increase in the concentration of C_{60} in the system ($C_{60}-R_3Al$)-O_2, up to its complete disappearance at $[C_{60}]_0 = 7.4 \cdot 10^{-4}$ M. It is important that in this case occurs simultaneously an increase in the intensity of the long-wavelength maxima at 617 and 664 nm in the CL spectrum. The combination of these two facts suggests that the CL is generated *via* the energy transfer from $^3CH_3CHO^*$ to the parent C_{60} or products, *viz.*, FEH, EF, and dimer. Not oxidized C_{60} as such an energy acceptor is excluded, since its fluorescence ($\lambda_{max} = 720$-750 nm [89]) and phosphorescence ($\lambda_{max} = 796$ and 812 nm [46]) spectra differ from the CL spectrum ($\lambda_{max} = 617$ and 664 nm). Moreover, fullerene as a luminescence enhancer (otherwise sensitizer) has not yet been known, while as a quencher of excited states fullerene has an extremely high efficiency [65-68]. This feature of fullerene also clearly manifests itself in a sharp drop in the intensity of CL when adding C_{60} solution ($[C_{60}]_0 > 2 \cdot 10^{-3}$ M) during registering BCL. This effect is due to quenching of the primary emitter $^3CH_3CHO^*$ by C_{60} (Figure 22).

Unlike C_{60}, a similar addition of the liquid hydrolysates of products (LH) formed in the ($C_{60}-Et_3Al$)-O_2 system enhances the CL (Figure 22). Hence, FEH, EF, and dimer are the CL enhancer in this CL system. The energy level of the $^3CH_3CHO^*$ donor ($\lambda_{max} = 420$ nm, $E = 2.95$ eV) is higher than the radiative levels of the energy acceptors emitted with the short wavelength maximum in the CL spectrum ($\lambda_{max} = 617$ nm, $E = 2.0$ eV). The luminescence spectrum of the $^3CH_3CHO^*$ donor well overlap with the absorption spectra of the acceptors, *viz.*, the noted above reaction products. The physical process of energy transfer in this case is complicated by a chemical reaction, that is manifested in a slower decrease in the intensity of CL compared to BCL due to the attack of the enhancer molecules by radicals Et• with formation of C_{60} derivatives with a high content of ethyl fragments.

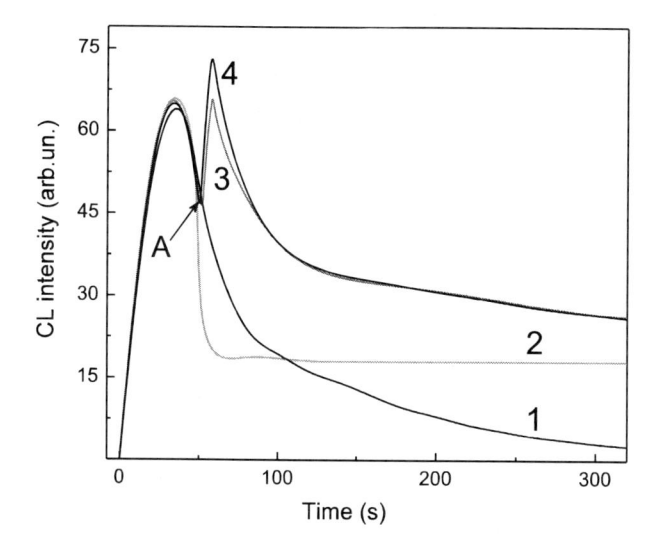

Figure 22. Effect of the quenching and activating additives on the BCL kinetics: without additives (1); with additives of $2 \cdot 10^{-3}$ M solution of C_{60} (2), LH (liquid hydrolysate) of the reaction products in the $(C_{60}\text{-}Et_3Al)\text{-}O_2$ system (3); FEH–EF mixture obtained in a non-dependent manner in the catalytic reaction of $C_{60} + Et_3Al$ (4). A is the moment of introduction of an additive. Solvent – toluene, 298 K, $[Et_3Al]_0 = 0.14$ M.

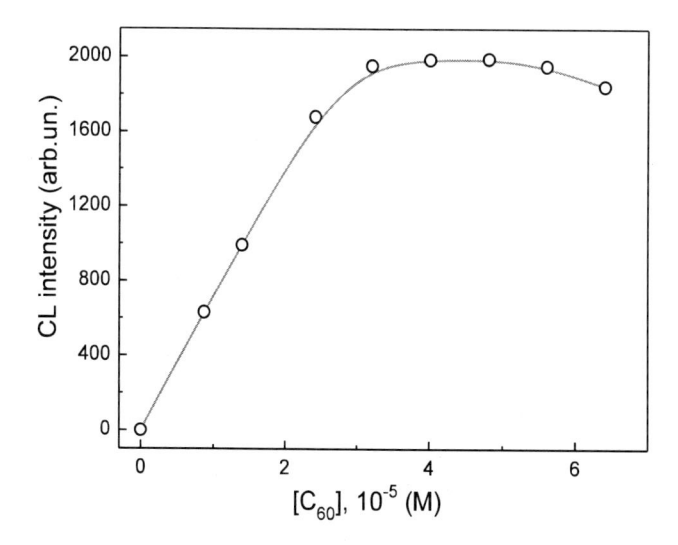

Figure 23. Plot of the maximum CL intensity *vs.* C_{60} concentration in the $(C_{60}\text{-}Et_3Al)\text{-}O_2$ system. Solvent – toluene, 298K, $[Et_3Al]_0 = 0.14$ M.

The dependence of the intensity of CL on the fullerene concentration is characterized by the presence of a plateau. This is because at some threshold concentration of the reaction products they accept all energy from primary emitters $^3CH_3CHO^*$. This leads to the disappearance of the maximum emission of $^3CH3CHO^*$ at 420 nm in the CL spectrum. The decrease in the intensity of the CL after this plateau is due to the trivial effect of the internal filter caused by the poor transparency of the reaction solution, which is 70% (for $\lambda_{max} = 617$ nm) and 30% (for $\lambda_{max} = 664$ nm) at $[C_{60}]_0 = 7.4 \cdot 10^{-5}$ M.

The CL characteristics obtained for Et_3Al are similar to those for the Bu^i_3Al in the system $(C_{60}\text{-}^iBu_3Al)\text{-}O_2$. Therefore, in the scheme of the mechanism of CL accompanying the oxidation of aluminum alkyls with oxygen with the participation of fullerene (Scheme 9), both radicals Et and ^{i}Bu are designated as R.

$$>AlR + O_2 \longrightarrow \quad >AlOO^\cdot + R^\cdot \tag{19}$$
$$\qquad\qquad\qquad\qquad >AlOOR \tag{20}$$

$$>AlOOR + >AlR \longrightarrow R^\cdot + RO^\cdot + P \tag{21}$$

$$R^\cdot + O_2 \longrightarrow RO_2^\cdot \tag{22}$$

$$RO_2^\cdot + >AlR \longrightarrow R^\cdot + >AlOOR \tag{23}$$

$$RO^\cdot + >AlR \longrightarrow R^\cdot + >AlOR \tag{24}$$

$$2C_2H_5OO^\cdot \longrightarrow CH_3C(H)=O^* + CH_3C(H)=O + C_2H_5OH + O_2 \tag{25}$$

$$CH_3C(H)=O^* \longrightarrow CH_3C(H)=O + h\nu_1 \text{ (BCL, 420 nm)} \tag{26}$$

$$R^\cdot + C_{60} \longrightarrow RC_{60}^\cdot \tag{27}$$

$$nRC_{60}^\cdot \longrightarrow R_mC_{60}H_o + R_pC_{60} + RC_{60}C_{60}R \tag{28}$$

$$CH_3C(H)=O^* + Em \longrightarrow CH_3C(H)=O + Em^* \tag{29}$$

$$Em^* \longrightarrow Em + h\nu_2 \text{ (CL, 617 and 664 nm)} \tag{30}$$

Scheme 9.

On Scheme 9, P (in reaction 3) is the sum of oxygen-containing products (aldehydes, carboxylic acids and esters); Em is the FEH, EF, dimer. Reactions (7), (8), and (11) are given for CL in the $(C_{60}\text{-}Et_3Al)\text{-}O_2$ system. The $R^\cdot$, $RO^\cdot$, and $ROO^\cdot$ radicals are formed in reactions (1) – (6). The reactions of these radicals afford the products of R_3Al oxidation: stable $Al(OR)_3$ and unstable $>AlOOR$ in which the $>Al$ atom is bonded to the R (closer to the beginning of the reaction) and RO (towards the end of the reaction) fragments. The emitter of BCL, the $^3CH_3CHO^*$ aldehyde, is formed in the disproportionation of the $ROO^\cdot$ radicals (7). The alkyl radicals, which avoid the reaction with O_2 (4), add to C_{60} to give the FR in reaction (9), and they are further transformed into the stable products P (FEH, EF, dimers) in reaction (10). The excited aldehyde $^3CH_3CHO^*$ transfers energy on the transformation products of FR denoted as Em (11). The excited states of Em^* are deactivated *via* the CL emission (12). The luminescence of $^3CH_3CHO^*$ in the CL spectrum take place only at low initial concentrations of C_{60}.

The results of the study of the CL in the system $(R_3Al-C_{60/70})-O_2$ can be useful to develop a new approach to fullerene functionalization with use the CL for control of fullerene conversion.

7. CHEMILUMINESCENCE IN THE DIMERIZATION OF ETHYL (ALKYL) FULLERENYL RADICALS

The ability of fullerenyl radicals of type $X_nC_{60}^•$ and $X_mC_{70}^•$ (X – organic radical) to dimerize with CL radiation was found on the example of alkyl derivatives of fullerenes [95].

The CL arised in the irradiation with a W-lamp (180 W, 500-600 nm) in argon of toluene solutions containing $EtC_{60}C_{60}Et$ (hereinafter, dimer), unreacted C_{60}, $Et_nC_{60}H_m$ ($n = m = 1$-6), and Et_nC_{60} ($n = 2$-6) (hereinafter, solution 1). This solution was obtained by oxidation of Et_3Al by O_2 in toluene in the presence of C_{60} followed by hydrolysis of the oxidant with 3% HCl for removing unreacted excess of Et_3Al [94]. The irradiated (2-3 min) samples of solution 1 emit a CL ($I_{CL} = 5.4 \cdot 10^7 photon \cdot s^{-1} \cdot mL^{-1}$) and exhibit an ESR signal with $g = 2.0023$, the intensity drop of which is shown in Figure 24 [95]. This ESR signal is characteristic of the $EtC_{60}^•$ radical. The slopes of linear anamorphous in the coordinates $[(I_0/I-1)^{-1}-1]-t$ afford the close rate constant values for attenuation of the CL and EPR signals: $k_{CL} = (0.62\pm0.04) \cdot 10^{-1} L \cdot mol^{-1} \cdot s^{-1}$ and $k_{ESR} = (0.59\pm0.02) \cdot 10^{-1} L \cdot mol^{-1} \cdot s^{-1}$ at 293K.

The short-wavelength boundary of the CL spectrum at 500 nm corresponds to energy of 57.04 $kcal \cdot mol^{-1}$, which correlates well with the binding energy ($\sim$ 53.8 $kcal \cdot mol^{-1}$) between the C_{60}–C_{60} fullerene cages in the $EtC_{60}C_{60}Et$ dimer.

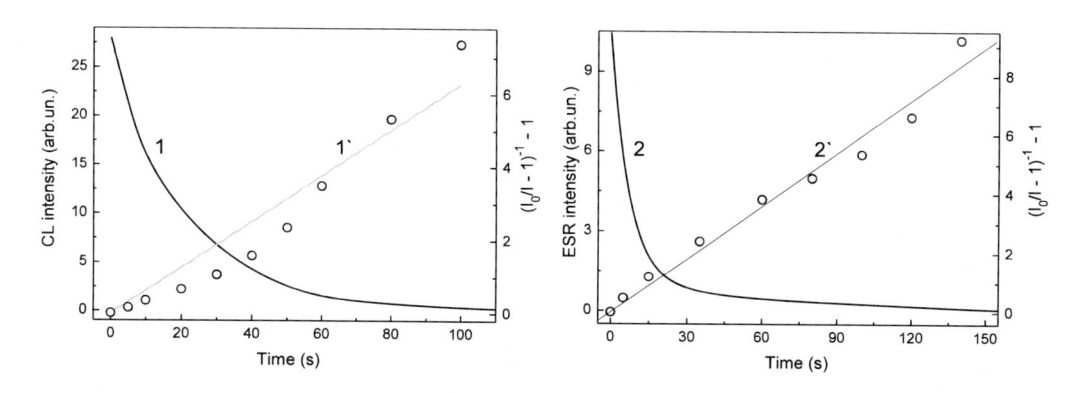

Figure 24. Changes in the intensity of CL (1) and the ESR signal (2) after the irradiation of solution 1 and their linear anamorphous (1` and 2`) in the coordinates [(I0/I-1)-1-1]–t. T = 273K.

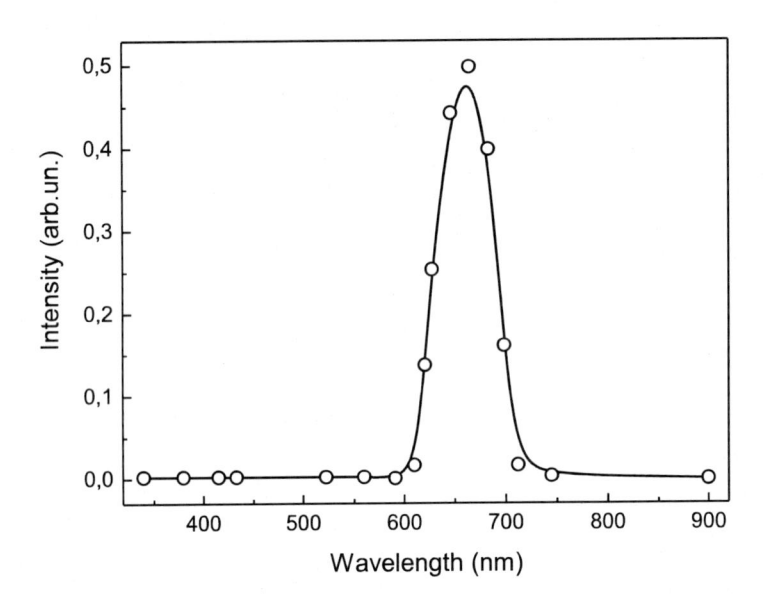

Figure 25. CL spectrum measured by the cut-off filters method during the decay of the kinetic curve 1 on Figure 24.

The bond energies in other compounds that are also present in the composition of solution 1 are much higher than the energy used for the photolysis. Therefore, it is clear that the generation of the CL and ESR-signal occurs by photodissociation of the dimer onto two $EtC_{60}^{\bullet}$ radicals (1, Scheme 10) followed by the recombination of $EtC_{60}^{\bullet}$ to form the dimer molecules, part of which are formed in the electronically excited state (2, Scheme 10).

$$EtC_{60}C_{60}Et + h\nu \ (500\text{-}600 \ nm) \longrightarrow EtC_{60}^{\bullet} + EtC_{60}^{\bullet} \qquad (31)$$

$$EtC_{60}^{\bullet} + EtC_{60}^{\bullet} \longrightarrow (EtC_{60}C_{60}Et)^{*} \longrightarrow EtC_{60}C_{60}Et + h\nu \ (664 \ nm) \qquad (32)$$

Scheme 10.

The maximum of the CL spectrum (Figure 25) at 664 nm is located at the long wavelength region characteristic of the luminescence of fullerene derivatives. The standard enthalpy (53.8 kcal·mol^{-1}) of the reaction 2 (Scheme 10) is enough for the population of the radiative level of the dimer. The latter was estimated from the short-wavelength boundary of the CL spectrum (λ_{max} = 612 nm, 47 kcal·mol^{-1}).

CONCLUSION

We have examined all the known works on the CL with participation of fullerenes C_{60}, C_{70} and some of their derivatives. These molecules and their compounds play the role of both the CL-emitters and initial compounds, the reaction of which are accompanied by CL.

In both cases, they exhibit luminescent properties that are also characteristic of organic and inorganic compounds. So, they participate in the same types of chemiluminescent reactions as hydrocarbons: in ion-radical reactions of electron transfer, in oxygen oxidation of the C–H bonds and oxidation of the C = C double bonds by ozone. However, due to the unusual structure of hollow spherical fullerene molecules rich in many electrons of double bonds, their CL differs significantly in several aspects from the CL of hydrocarbons. First of all, for the CL fullerenes and their derivatives is characteristic the emission in long-wavelength and even NIR region. For example, the excited molecules C_{60}^* and C_{70}^*, which are CL emitters of the ion-radical reactions have emission maxima at 610-660 nm and 645, 685 nm, respectively. The functionalization of fullerenes leads to a hypsochromic shift of emission maxima, which is clearly seen in CL, the emitters of which are fullerene derivatives, such as fullerene fluorides and polyketones. This hypsochromic shift of the CL maximum (from 660 to 610 nm) is most convincingly manifested at an increase in the exposure of C_{60} ozonolysis, leading to an increase in the number of C = O groups adding to the fullerene cage. At the same time, there is a significant red shift (up to 660 nm) for the CL of the fullerene polyketones compared with the emission of carbonyl products (400-460 nm), which are traditional emitters of chemiluminescent reactions of hydrocarbons. We believe that the cause of this red shift is the powerful influence of the electronic ensemble of 240 electrons of the fullerene cage perturbing the excited electron of this group, which leads to a decrease in the energy of the radiating level. Earlier, in a similar way, we explained the effect of abnormally strong quenching of the luminescence of organic molecules and Ln^{3+} ions by C_{60} and C_{70} molecules. In this case, the disturbing effect of the electronic structure of fullerenes was carried out at much larger distances than within the limits of one molecule of the fullerene polyketone. The influence of the original structure of fullerenes on the reactive ability of their alkyl fullerenyl radicals $RC_{60}^\bullet$ is also remarkable in the difference between the mechanisms of CL in the oxidation by oxygen of the C–H bond of the fullerene hydride $C_{60}H_{36}$ and hydrocarbons. The light stage of the CL of hydrocarbons is the disproportionation of the peroxyl radicals $ROO^\bullet$ formed in the reaction $R^\bullet + O_2 \rightarrow ROO^\bullet$. In the case of fullerene, a similar reaction ($RC_{60}^\bullet + O_2 \rightarrow RC_{60}OO^\bullet$) does not proceed, most likely because of the significant delocalization of the free spin along the spherical surface of the fullerene cage. As a result of this, the CL during the oxidation of $C_{60}H_{36}$ is excited by another mechanism, namely *via* the energy transfer from primary hydrocarbon emitters onto fullerene derivatives.

ACKNOWLEDGMENTS

The authors are grateful to Dr. Denis Sh. Sabirov (Institute of Petrochemistry and Catalysis, Ufa Federal Research Center RAS) for his help in the design of the review.

REFERENCES

[1] Vasil`ev, R. F. (1967). Chemiluminescence in liquid-phase reactions. *Prog. React. Kinet. 4:* 305-352.

[2] Adam, W., Cilento, G., Ed. (1982). Chemical and Biological Generation of Excited States. New York: Academic Press.

[3] Gunderman, K. D., McCapra, F. (1987). Chemiluminescence in organic chemistry. Berlin: Springer-Verlag.

[4] Campbell, A. K. (1988). Chemiluminescence: Principles and Applications in Biology and Medicine. Chichester: VCH Verlagsgesellschaft.

[5] Bulgakov, R. G., Kazakov, V. P., Tolstikov, G. A. (1990). Chemiluminescence of organometallics in solution. *J. Organomet. Chem., 387:* 11-64.

[6] Vasil`ev, R. F., Tsaplev, Yu. B. (2006). Light-created chemiluminescence. *Russ. Chem. Rev. 75:* 989-1002.

[7] Adam, W., Trofimov, A. V. (2006). Contemporary trends in dioxetane chemistry. In: The Chemistry of Peroxides, edited by Z. Rappoport, 1171-1209. Chichester: J. Wiley & Sons.

[8] Vacher, M., Fdez. Galvan, I., Ding, B.-W., Schramm, S., Berraud-Pache, R., Naumov, P., Ferre, N., Liu, Y.-J., Navizet, I., Roca-Sanjuan, D., Baader, W. J., Lindh, R. (2018). Chemi- and Bioluminescence of Cyclic Peroxides. *Chem. Rev. 118:* 6927-6974.

[9] Fedorova, G. F., Trofimov, A. V., Vasil`ev, R. F., Veprintsev, T. L. (2007). Peroxy-radical-mediated chemiluminescence: mechanistic diversity and fundamentals for antioxidant assay. *Arkivoc. viii:* 163-215.

[10] Roda, A. Ed. (2011). Chemiluminescence and Bioluminescence: Past, Present and Future. Cambridge: RSC Publishing.

[11] Roda, A., Guardigli, M. (2012). Analytical chemiluminescence and bioluminescence: latest achievements and new horizons. *Anal. Bioanal. Chem. 402:* 69-76.

[12] Büchel, G. E., Carney, B. T. M., Shaffer, M., Tang, J., Austin, Ch., Arora, M., Zeglis, B. M., Grimm, J., Eppinger, J., Reiner, T. (2016). Near-Infrared Intraoperative Chemiluminescence Imaging. *Chem. Med. Chem. 11:* 1978-1982.

[13] Roda, A., Mirasoli, M., Michelini, E., Di Fusco, M., Zangheri, M., Cevenini, L., Roda, B., Simoni, P. (2016). In situ controllable synthesis of cotton-like polyaniline nanostructures for a H_2O_2 sensor using an embedded three-electrode microfluidic chip. *Biosens. Bioelectron. 76:* 164-179.

[14] Elbanowski, M., Makowska, B., Staninski, K., Kaczmarek, M. (2000). Chemiluminescence of systems containing lanthanide ions. *J. Photochem. Photobiol. 130:* 75-81.

[15] Lis, S., Elbanowski, M., Makowska, B., Hnatejko, Z. (2002). Energy transfer in solution of lanthanide complexes. *J. Photochem. Photobiol. 150:* 233-247.

[16] Kompa, K. L. (1974). High Power Chemical Lasers: Problems and Perspectives. In: Laser Interaction and Related Plasma Phenomena, edited by H. J. Schwarz, H. Hora, 115-131. Boston: Springer.

[17] Rao, Y., Zhang, X. R., Luo, G. A., Baeyens, W. (1999). Chemiluminescence flow-injection determination of furosemide based on a rhodamine 6G sensitized cerium(IV) method. *Anal. Chim. Acta. 396:* 273-277.

[18] Bulgakov, R. G., Kuleshov, S. P., Valiullina, Z. S., Mustafin B. A. (1999). Red and green chemiluminescence of Na, Mg, and lanthanide triphenylmethyl derivatives during oxidation by dioxygen and cerium(IV). *Russ. Chem. Bull. 48:* 1091-1094.

[19] Kroto, H. W., Heath, J. R., O`Brien, S. C., Curl, R. F., Smalley, R. E. (1985). C_{60}: Buckminsterfullerene. *Nature. 318:* 162-163.

[20] Sokolov, V. I., Stankevich, I. V. (1993). The fullerenes – new allotropic forms of carbon: molecular and electronic structure, and chemical properties. *Russ. Chem. Rev. 62:* 419-435.

[21] Hirsch, A. (1994). The Chemistry of Fullerenes. Stuttgart: Georg Thieme.

[22] Hirsch, A. Ed. (1999). Fullerenes and Related Structures. Berlin: Springer.

[23] Reed, C. A., Bolskar. R. D. (2000). Discrete Fulleride Anions and Fullerenium Cations. *Chem. Rev. 100:* 1075-1120.

[24] Goldshleger, N. (2001). Fullerenes and fullerene-based materials in catalysis. *Fullerene Sci. Technol. 9:* 255-280.

[25] Anthopoulos, T. D., Tanase, C., Setayesh, S., Meijer, E. J., Hummelen, J. C., Blom, P. W. M., de Leeuw, D. M. (2004). Ambipolar Organic Field-Effect Transistors Based on a Solution-Processed Methanofullerene. *Adv. Mater. 16:* 2174-2179.

[26] Sidorov, L. N., Yurovskaya, M. A., Borshchevskii, A. Ya., Trushkov, I. V., Ioffe, I. N. (2005). Fullerenes. Moscow: Examen.

[27] Murayama, H., Tomonoh, S., Alford, J. M., Karpuk, M. E. (2005). Fullerene production in tons and more: from science to industry. *Fuller. Nanotub. Car. N. 12:* 1-9.

[28] Wang, Y., Seifert, G., Hermann, H. (2006). Molecular design of fullerene-based ultralow-k dielectrics. *Phys. Stat. Sol. A. 203:* 3868-3872.

[29] Piotrovskii, L. B., Kiselev, O. I. (2006). *Fullerenes in Biology.* St. Petersburg: Rostok.

[30] Troshin, P. A., Troshin, O. A., Lyubovskaya, R. N., Razumov, V. F. (2010). Functional fullerene derivatives: methods of synthesis and perspectives of using in organic electronics and biomedicine, 2nd edition. Ivanovo: Ivanovo State University.

[31] Thilgen, C., Diederich, F. (2006). Structural Aspects of Fullerene Chemistry – A Journey through Fullerene Chirality. *Chem. Rev. 106:* 5049-5135.

[32] Troshin, P. A., Lyubovskaya, R. N. (2008). Organic chemistry of fullerenes: the major reactions, types of fullerene derivatives and prospects for their practical use. Russ. *Chem. Rev. 77:* 305-349.

[33] Da Ros T. (2008). Medicinal Chemistry and Pharmacological Potential of Fullerenes and Carbon Nanotubes. In: *Carbon Materials: Chemistry and Physics*, Vol. 1, edited by F. Cataldo, T. Da Ros, 1-21. Dordrecht: Springer Netherlands.

[34] Cosseddu, P., Mattana, G., Orgiu, E., Bonfiglio, A. (2009). Ambipolar organic field-effect transistors on unconventional substrate. *Appl. Phys. A. 95:* 49-54.

[35] Tokunaga, K. (2011). Computational Design of New Organic Materials: Properties and Utility of Methylene-Bridged Fullerenes C_{60}. In: *Handbook on Fullerene: Synthesis, Properties and Applications,* edited by R. F. Verner, C. Benvegnu, 517–537. New York: Nova Science Publishers.

[36] Pinzón, J. R., Villalta-Cerdas, A., Echegoyen, L. (2011). Fullerenes, Carbon Nanotubes, and Graphene for Molecular Electronics. In: *Unimolecular and Supramolecular Electronics I. Topics in Current Chemistry,* Vol. 312, edited by R. Metzger. 127-174. Berlin: Springer.

[37] Tzirakis, M. D., Orfanopoulos, M. (2013). Radical reactions of fullerenes: From synthetic organic chemistry to materials science and biology. *Chem. Rev. 113:* 5262-5321.

[38] Bulgakov, R. G., Achmadieva, R. G., Musavirova, A. S. 1998. The first chemiluminescence fullerenes – oxidation C_{60} by ozone in solutions. Paper presented at the 12th International conference «Photochemical Conversion and Storage of Solar Energy», Berlin, August 9-14, 3W90.

[39] Bulgakov, R. G., Akhmadieva, R. G., Musavirova, A. S., Abdrakhmanov, A. M., Ushakova, Z. I., Sharifullina, F. M. (1999). The first example of chemiluminescence of fullerenes – oxidation of C_{60} by ozone in solution. *Russ. Chem. Bull. 48:* 1190-1190.

[40] Gupta, N., Santhanam, K. S. V. (1993). Electron-transfer chemiluminescence of buckminsterfullerene radical-anion and thianthrenecation. *Curr. Sci. 65:* 75-77.

[41] Gupta, N., Santhanam, K. S. V. (1994). Chemiluminescent electron transfer reactions of C_{60} anion radical – energetics and spectral features. *Chem. Phys. 185:* 113-122.

[42] Gupta, N., Santhanam, K. S. V. (1994). Exergonic Electron-transfer Reaction Between [60]Fullerene Anion and CarbazoleCation. *J. Chem. Soc., Chem. Commun.* 2409-2410.

[43] Kratschmer, W., Lamb, L. D., Fostiropoulos, K., Huffman, D. R. (1990). Solid C_{60}: a new form of carbon. *Nature. 347:* 354-358.

[44] Chandros, E. A., Sonntag, F. I. (1964). A Novel Chemiluminescent Electron-Transfer Reaction. *J. Am. Chem. Soc. 86:* 3179-3180.

[45] Bulgakov, R. G., Akhmadieva, R. G., Musavirova, A. S., Golikova, M. T. (2001). Chemiluminecsence in the oxidation of Na_2C_{60} by the $(NH_4)_2Ce(NO_3)_6$ complex in THF. *Russ. Chem. Bull. 50:* 731-733.

[46] Zeng, Y., Biszok, L., Linschitz, H. (1992). External heavy atom induced phosphorescence emission of fullerenes: the energy of triplet C_{60}. *J. Phys. Chem. 96:* 5237-5239.

[47] Belyakov, V. A., Vasil'ev, R. F. (1967). On some problems concerning investigation of liquid-phase chemiluminescence. *Photochem. Photobiol. 6:* 35-40.

[48] Chibante, P. F., Heymann, D. (1993). On the geochemistry of fullerenes: Stability of C_{60} in ambient air and the role of ozone. *Geochim. Cosmochim. Acta. 57:* 1879-1881.

[49] Heymann, D., Chibante, L. P. F. (1993). Reaction of C_{60}, C_{70}, C_{76}, C_{78}, and C_{84} with ozone at 23.5°C. *Rec. Trav. Chim. Pays. Bas. 112:* 639-642.

[50] Deng, J.-P., Ju, D.-D., Her, G.-R., Mou, C.-Y., Chen, C. J., Lin, Y. Y., Han, C. C. (1993). Odd-Numbered Fullerene Fragment Ions from C_{60} Oxides. *J. Phys. Chem. 97:* 11575-11577.

[51] Malhotra, R., Kumar, S., Satyam, A. (1994). Ozonolysis of [60]fullerene. J. *Chem. Soc. Chem. Commun.* 1339-1340.

[52] Deng, J.-P., Mou, C.-Y., Han, C.-C. (1995). Electrospray and Laser Desorption Ionization Studies of $C_{60}O$ and Isomers of $C_{60}O_2$. *J. Phys. Chem. 99:* 14907-14910.

[53] Cataldo, F. (2002). Polymeric fullerene oxide (fullerene ozopolymers) produced by prolonged ozonation of C_{60} and C_{70} fullerenes. *Carbon. 40:* 1457-1467.

[54] Heymann, D., Weismann, R. B. (2006). Fullerene oxides and ozonides. *C. R. Chimie. 9:* 1107-1116.

[55] Bulgakov, R. G., Musavirova, A. S., Abdrakhmanov, A. M., Nevyadovsky, E. Yu., Khursan, S. L., Razumovsky, S. D. (2002). Chemiluminescence in ozonolysis of solution of fullerene C_{60}. *J. Appl. Spectrosc. 69:* 220-225.

[56] Bulgakov, R. G., Nevyadovsky, E. Yu., Belyaeva, A. S., Golikova, M. T., Ushakova, Z. I., Ponomareva, Yu. G., Dzhemilev, U. M., Razumovsky, S. D., Valyamova, F. G. (2004). Water-soluble polyketones and esters as the main stable products of ozonolysis of fullerene C_{60} solutions. *Russ. Chem. Bull. 53:* 148-159.

[57] Bulgakov, R. G., Ponomareva, Yu. G., Muslimov, Z. S., Tuktarov, R. F., Razumovsky, S. D. (2008). Solution ozonolysis of C_{70}: stable products and chemiluminescence. *Russ. Chem. Bull. 57:* 2072-2080.

[58] Razumovskii, S. D., Zaikov, G. E. (1974). Ozone and its reactions with organic compounds. Moscow: Nauka.

[59] Bulgakov, R. G., Nevyadovsky, E. Yu., Ponomareva, Yu. G., Sabirov D. Sh., Budtov, V. P., Razumovsky, S. D. (2005). A new type of a chemiluminescent reaction: hydrolysis of ozonides of fullerenes C_{60} and C_{70}. *Russ. Chem. Bull. 54:* 2468-2470.

[60] Bulgakov, R. G., Nevyadovsky, E. Yu., Ponomareva, Yu. G., Sabirov, D. Sh., Razumovsky, S. D. (2006). Formation of secondary fullerene ozonides in the

ozonolysis of C_{60} solutions and chemiluminescence upon their hydrolysis. *Russ. Chem. Bull. 55:* 1372-1379.

[61] Razumovskii, S. D., Bulgakov, R. G., Nevyadovskii, E. Yu. (2003). Kinetics and Stoichiometry of the Reaction of Ozone with Fullerene C_{60} in a CCl_4 Solution. *Kinet. Catal. 44:* 229-232.

[62] Razumovskii, S. D., Bulgakov, R. G., Ponomareva, Yu. G., Budtov, V. P. (2006). Kinetics and stoichiometry of the reaction between ozone and C_{70} fullerene in CCl_4. *Kinet. Catal. 47:* 347–350.

[63] Dodeigne, C., Thunus, L., Lejeune, R. (2000). Chemiluminescence as diagnostic tool. A review. *Talanta. 51:* 415–439.

[64] Schuster, G., Turro, N. (1975). Energy migration. The energy hopping and self-quenching reaction involving carbonyl chromophores. *Tetrahedron Lett.* 2261-2263.

[65] Bulgakov, R. G., Galimov, D. I., Sabirov, D. Sh. (2007). New property of the fullerenes: the anomalously effective quenching of electronically excited states owing to energy transfer to the C_{70} and C_{60} molecules. *JETP Letters. 85:* 632-635.

[66] Bulgakov, R. G., Galimov, D. I. (2007). Fullerene C_{60} as a superefficient quencher of singlet exited states of polycyclic aromatic hydrocarbons. *Russ. Chem. Bull. 56:* 446-451.

[67] Bulgakov, R. G., Galimov, D. I., Ponomareva, Yu. G., Nevyadovskii, E. Yu., Gainetdinov, R. Kh. (2006). Quenching of electronically excited Ln^{3+*} ions by C_{60} fullerene. *Russ. Chem. Bull. 55:* 955-960.

[68] Bulgakov, R. G., Galimov, D. I. (2007). C_{60} fullerene, as ultraefficient quencher of singlet-excited adamantanone generated in photo- and chemiexcitation. *Russ. Chem. Bull. 56:* 1085-1087.

[69] Razumovskii, S. D., Zaikov, G. E. (1971). Kinetics and mechanism of the reaction of ozone with aromatic hydrocarbons. *Russ. Chem. Bull. 20:* 2524-2529.

[70] Bulgakov, R. G., Sabirov, D. Sh., Dzhemilev, U. M. (2013). Oxidation of fullerenes with ozone. *Russ. Chem. Bull. 62:* 304-314.

[71] Heymann, D., Bachilo, S. M., Weisman, R. B., Cataldo, F., Fokkens, R. H., Nibbering, N. M. M., Vis, R. D., Chibante, L. P. F. (2000). $C_{60}O_3$, a Fullerene Ozonide: Synthesis and Dissociation to $C_{60}O$ and O_2. *J. Am. Chem. Soc. 122:* 11473-11479.

[72] Heymann, D., Bachilo, S. M., Weisman, R. B. (2002). Ozonides, Epoxides, and Oxoannulenes of C_{70}. *J. Am. Chem. Soc. 124:* 6317-6323.

[73] Bulgakov, R. G., Sabirov, D. Sh., Khursan, S. L., Razumovskii, S. D. (2008). Chemiluminescent test for oxofullerenecarbonyl oxides generated in situ by C_{60} ozonolysis. *Mend. Comm. 18:* 307-308.

[74] Bulgakov, R. G., Galimov, D. I., Mukhacheva, O. A., Goryunkov, A. A. (2010). Chemiluminescence upon the oxidation of fullerene fluorides $C_{60}F_x$ (x = 18, 36, 48) with ozone in solution. *Russ. Chem. Bull. 59:* 1843-1845.

[75] Taylor, R. (2004). Why fluorinate fullerenes? *J. Fluorine Chem. 125:* 359-368.

[76] Andrievsky, G. V., Kosevich, M. V., Vovk, O. M., Shelkovsky, V. S., Vashchenko, L. A. (1995). On the Production of an Aqueous Colloidal Solution of Fullerenes. *J. Chem. Soc. Chem. Commun. 12:* 1281-1282.

[77] Andrievsky, G. V., Klochkov, V. K., Bordyuh, A. B., Dovbeshko, G. I. (2002). Comparative Analysis of Two Aqueous-Colloidal Solutions of C_{60} Fullerene with Help of FT-IR Reflectance and UV-Vis Spectroscopy. *Chem. Phys. Lett. 364:* 8-17.

[78] Andrievsky, G. V., Klochkov, V. K., Karyakina, E. L., Mchedlov-Petrossyan, N. O. (1999). Studies of aqueous colloidal solutions of fullerene C_{60} by electron microscopy. *Chem. Phys. Lett. 300:* 392-396.

[79] Andrievsky, G. V., Bruskov, V. I., Tykhomyrov, A. A., Gudkov, S. V. (2009). Peculiarities of the antioxidant and radioprotective effects of hydrated C_{60} fullerene nanostuctures in vitro and in vivo. *Free Radical Biol. Med. 47:* 786-793.

[80] Bulgakov, R. G., Sabirov, D. Sh., Andrievskii, G. V. (2012). Chemiluminescence in the ozonation of C_{60} aqueous dispersions. *Russ. Chem. Bull. 61:* 1093-1098.

[81] Bulgakov, R. G. (1975). The formation of excited states of the uranyl and lanthanide ions in chemical and electrochemical reactions in solution. PhD diss. Institute of Chemistry, Ufa.

[82] Papadopoulos, K., Triantis, T., Boyatzis, S., Dimotikali, D., Nikokavouras, J. (2001). Photo-, radio- and sonostorage chemiluminescence of buckminsterfullerene C_{60}. *J. Photochem. Photobiol. 143:* 93-97.

[83] Tarasov, B. P., Goldshleger, N. F., Moravsky, A. P. (2001). Hydrogen-containing carbon nanostructures: synthesis and properties. *Russ. Chem. Rev. 70:* 131-146.

[84] Goldshleger, N. F., Moravsky, A. P. (1997). Fullerene hydrides: synthesis, properties, and structure. *Russ. Chem. Rev. 66:* 323-342.

[85] Bulgakov, R. G., Galimov, D. I., Kinzyabaeva, Z. S. (2009). Chemiluminescence produced by oxidation of fullerene hydride $C_{60}H_{36}$ with oxygen in solution. *Russ. Chem. Bull. 58:* 857-858.

[86] Belyakov, V. A., Vasil`ev, R. F., Fedorova, G. F. (1978). *Dokl. Akad. Nauk SSSR. 239:* 344.

[87] Vasil`ev, R. F. (1962). Chemiluminescence kinetics and the study of reactions involved in the liquid phase oxidation of hydrocarbons. *Dokl. Akad. Nauk SSSR. 144:* 143-146.

[88] Dipak, K. P., Mohan, H., Mittal, J. P. (1998). Photophysical Properties of $C_{60}H_{18}$ and $C_{60}H_{36}$: A Laser Flash Photolysis and Pulse Radiolysis Study. *J. Phys. Chem. A. 102:* 4456-4461.

[89] Kim, D., Lee, M., Suh, Y., D. Kim, S. K. (1992). Observation of fluorescence emission from solutions of C_{60} and C_{70} fullerenes and measurement of their excited-state lifetimes. *J. Am. Chem. Soc. 114:* 4429-4430.

[90] Darwish, A. D., Abdul-Sada, A. K., Langley, G. J., Kroto, H. W., Taylor, R., Walton, D. R. M. (1995). Polyhydrogenation of [60]- and [70]-fullerenes. *J. Chem. Soc., Perkin Trans. 2. 12:* 2359-2365.

[91] Lobach, A. S., Perov, A. A., Rebrov, A. I., Roshchupkina, O. S., Tkacheva, V. A., Stepanov, A. N. (1997). Preparation and study of hydrides of fullerenes C_{60} and C_{70}. *Russ. Chem. Bull. 46:* 641-648.

[92] Davies, A. G., Roberts, B. P. (1968). Peroxides of elements other than carbon. Part XIV. The mechanism of the autoxidation of organic compounds of lithium, magnesium, zinc, cadmium, and aluminium. *J. Chem. Soc. B. 0:* 1074-1078.

[93] Belyakov, V. A., Vasil`ev, R. F., Fedorova, G. F. (1978). Chemiluminescence in liquid-phase oxidation of organic-compounds. 1. Methods of determining quantitative characteristics. *High Energ. Chem. 12:* 208-213.

[94] Bulgakov, R. G., Ponomareva, Yu. G., Muslimov, Z. S., Valyamova, F. G., Sadykov, R. A., Tuktarov, R. F. (2007). Generation of fullerenyl radicals and chemiluminescence in the $(C_{60}-R_3Al)-O_2$ system. *Russ. Chem. Bull. 56:* 211-217.

[95] Bulgakov, R. G., Ponomareva, Yu. G., Sadykov, R. A. (2008). Chemiluminescence upon the dimerization of $EtC_{60}^{\bullet}$ fullerenyl radicals. *Russ. Chem. Bull. 57:* 2028-2029.

[96] Belyakov, V. A., Vasil`ev, R. F. (1967). On some problems concerning investigation of liquid-phase chemiluminescence. *Photochem. Photobiol. 6:* 35-40.

[97] Russell, G. A. (1957). Deuterium-isotope effects in the autoxidation of aralkyl hydrocarbons. Mechanism of the interaction of peroxy. *J. Am. Chem. Soc. 79:* 3871-3877.

In: A Comprehensive Guide to Chemiluminescence   ISBN: 978-1-53616-170-0
Editor: Luís Pinto da Silva   © 2019 Nova Science Publishers, Inc.

Chapter 4

CHEMILUMINESCENCE OF LANTHANIDE IONS

Ramil G. Bulgakov[1],* and Dim I. Galimov[2]
[1]Institute of Molecule and Crystal Physics,
Ufa Federal Research Center RAS, Ufa, Russia
[2]Institute of Petrochemistry and Catalysis,
Ufa Federal Research Center RAS, Ufa, Russia

ABSTRACT

This chapter is devoted to studies of chemiluminescence (CL) with the participation of divalent (Ln^{2+}) and trivalent (Ln^{3+}) lanthanide ions. It was considered two main types of CL. In the first of these, the Ln^{2+} ions (which are part of the starting compounds) as a result of oxidation reactions, turn into electronically excited Ln^{3+*} ions, which emit light through the forbidden *f–f* transition. In the second CL type, Ln^{2+} ions, *via* accepting energy from the primary excited emitter, go to the electronically excited state and emit light due to the allowed *d–f* transitions. An analysis of the works under discussion also showed that a third, very interesting type of CL, the emitter of which – the excited divalent ion Ln^{2+*} – is generated in reducing of trivalent ion Ln^{3+}, still has not been found. The studies on CL, the emitter of which the Eu^{3+*} ion is formed during the oxidation of inorganic, organic, and organometallic compounds of the Eu^{2+} ion by UO_2^{2+} ions, O_2, H_2O_2, $H_2CO_4^-$ ions and XeF_2, are also discussed. Attention is paid to CL where the emitters of Dy^{3+*} and Nd^{3+*} ions are generated during the oxidation of DyI_2 and NdI_2 iodides with water. Unlike the traditional CL emitters ions Ln^{3+*}, the divalent Ln^{2+*} ions as CL emitters were discovered quite recently. This very bright CL, caused by emission of Eu^{2+*} or Sm^{2+*}, is generated in $LnCl_3 \cdot 6H_2O$-THF-Bu^i_2AlH-O_2 systems (Ln = Eu or Sm). The Eu^{2+*} and Sm^{2+*} emitters are formed due to the energy transfer to Eu^{2+} and Sm^{2+} ions from the primary CL emitter i.e., the triplet excited aldehyde ($^3Me_2CHC(H)=O^*$) which is a oxidation product Bu^i_2AlH by oxygen. The divalent Eu^{2+} ion is a much more efficient enhancer of CL in oxidation of Bu^i_2AlH than the trivalent lanthanide ion Tb^{3+}. Moreover, energy transfer *via* the Eu^{2+}

* Corresponding Author's Email: profbulgakov@yandex.ru.

mediator made possible the enhancement of the Tb^{3+} ion CL to a much greater extent, compared with direct transfer of energy from $^3Me_2CHC(H)=O^*$. Due to the high brightness of the CL caused by the Eu^{2+*} emission, this CL is promising for creating chemical light sources and analytically determining super-low concentrations of europium and oxygen.

1. INTRODUCTION

For the first time the CL with the participation of lanthanide divalent ions was discovered as early as 1975 by Bulgakov et al. [1, 2] during the oxidation of the europium ion Eu^{2+} by the uranyl ion UO_2^{2+} in aqueous solutions of $HClO_4$. Eight to nine years latter, Elbanowsky and his colleagues [3] have found CL in the oxidation of Eu^{2+} with H_2O_2 in water solutions, and Thomas and Ellis [4] have registered CL during the oxygen oxidation of the Yb^{2+} ion in the composition of the $(C_5Me_5)_2Yb$ compound in THF solution. These pioneering works stimulated other CL studies involving divalent lanthanides. However, compared with the CL in which divalent lanthanide ions take part, the CL of trivalent lanthanide ions is a much more developed area, which may be due to lower stability of divalent lanthanide ions. Indeed, the emission of Ln^{3+*} electronically excited states (*f-f* transitions) have been found in a wide variety of chemical reactions involving compounds of different nature [5-18]. Among them is the decomposition of dioxetanes [5-9], ozone decomposition in sulfuric acid [10], the oxidation of organic [11], biochemical [4, 6], and organometallic [4, 14-17] compounds, and also the condensation reaction of aniline with aldehydes [18]. To date, the patterns and mechanisms of CL of trivalent lanthanide ions, as well as the influence of the nature of the lanthanide, ligand (antenna effect), solvent, the Ln^{3+} concentration, temperature, and also of practical applications of this CL are well studied. More information on CL of Ln^{3+} ions and the prospects for its application in analysis, medicine and biochemistry, can be found in the studies [19-26]. In contrast, data on CL with the participation of divalent lanthanides are not so extensive and diversified. One of the reasons for this is the lack of literature reviews specifically devoted to this area of the luminescence of lanthanides. It should be noted that the closest in subject and content to this area is an excellent overview that Elbanowski with colleagues have devoted to CL with the participation of Ln^{3+} and Ln^{2+} ions [27]. However, most of this review deals with Ln^{3+} ions acting as CL enhancers, i.e., as sensitizers of CL. At the same time, CL during the oxidation of Ln^{2+} ions is described in detail only for systems that include Eu^{2+} and H_2O_2 as the main components, as well as a widely varied range of organic additives, including drugs. As concerning for the examples of CL in the oxidation of various compounds of the Eu^{2+} ion by oxidants of different nature, then they are only briefly listed without discussing the features of the interaction of chemical reagents and the mechanism of excitation of CL. In addition, after the publication of the review of Elbanowski and co-workers has been passed almost 20 years and new important data on CL with the participation of Ln^{2+} ions have appeared. So, very recently the ability of divalent lanthanide ions to act in the role of not only the initial reagents, but also as CL emitters was discovered [28-31]. Indeed, if Ln^{3+*} ions as CL emitters have been known for more than 30 years [6], then the first light reactions whose emitters are ions of divalent lanthanides (Eu^{2+*} [28] and Sm^{2+*} [29]) were discovered a little more than five years ago. Moreover, the first chemiluminescent reactions of other lanthanide ions (i.e., not only Eu^{2+}), namely of the ions Dy^{2+} and Nd^{2+} were also detected [32]. In addition, quite recently, using the reaction of $EuSO_4$ with XeF_2 as an example, the ability of divalent lanthanide ions to generate solid-phase CL was first discovered [33]. This is especially important given the well-known [34] instability of divalent ions, not only with respect to oxygen and moisture, but also to many organic solvents. In this connection, carrying out highly exothermic solid-phase reactions involving not only Eu^{2+}, Sm^{2+}, Yb^{2+} ions, but also much less stable Pr^{2+}, Nd^{2+}, Tm^{2+}, will expand the list of chemiluminescent reactions of Ln^{2+} ions. This represents not only of theoretical interest, but also practical for creating new chemical light sources.

In this review, all currently known chemiluminescent reactions involving divalent lanthanide ions were considered. These CLs are divided into two conditional types, depending on whether the Ln^{2+} ions are the original species or they act as CL emitters. We hope that this review will be an incentive for more active studies in the field of CL divalent lanthanides.

2. CHEMILUMINESCENCE IN WHICH DIVALENT LANTHANIDE IONS ARE INITIAL REAGENTS

2.1. Chemiluminescence in Oxidation of Eu^{2+} Ions by Uranyl Ions UO_2^{2+}

This CL is a remarkable example of the redox reaction of ions of 4f and 5f elements, accompanied by the emission of green light, which is clearly visible to unarmed eye in a dark room. Such CL was found in the reaction of Eu^{2+} and UO_2^{2+} ions in aqueous solutions [1, 2]. To study the chemiluminescent reaction of these ions the solutions of Eu^{2+} and UO_2^{2+} were prepared by dissolving the $EuCO_3$ and UO_3 samples, respectively, in $HClO_4$ of the required concentration. All experiments were carried out in an argon atmosphere to prevent the oxidation of Eu^{2+} and intermediate ions UO_2^+ with oxygen.

The CL intensity (I_{CL}) and duration (d) of CL depend on the concentration $HClO_4$. For example, in 0.13 M $HClO_4$, $I_{CL} = 2 \cdot 10^8$ photon$\cdot$s$^{-1}\cdot$mL^{-1}, $d = 1.5$ s, and in 8.5 M $HClO_4$ the CL is visible to the naked eye with $d = \sim 200$ s (Figure 1) [1, 2, 35]. The addition of $HClO_4$ during damping CL causes a jump in the intensity of CL (Figure 2).

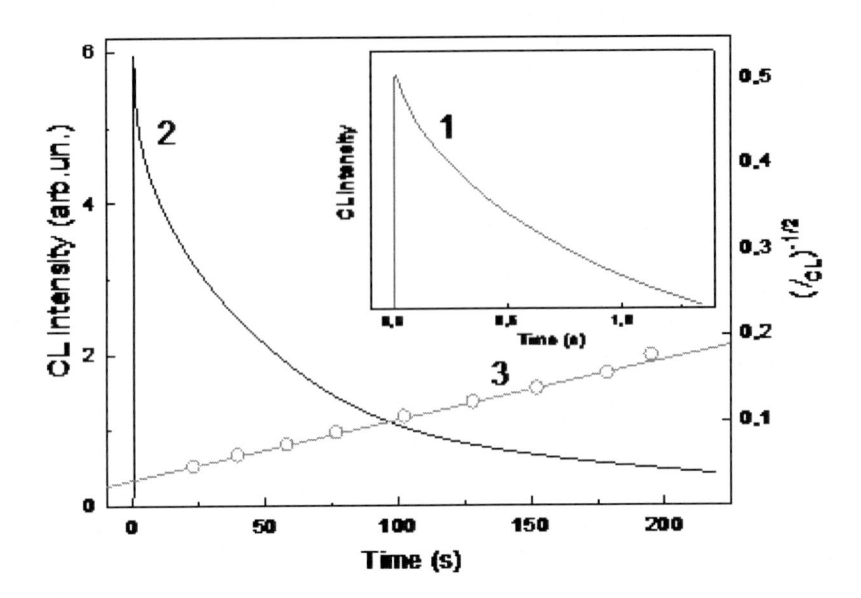

Figure 1. CL kinetics of the reaction of Eu^{2+} (10^{-3} M) and UO_2^{2+} (10^{-1} M) ions in aqueous solutions of $HClO_4$. $[HClO_4]_0 = 0.13$ M (1) and 8.5 M (2, 3), $T = 287$ K.

The increase in CL intensity in concentrated $HClO_4$ is due to two reasons: by an increase in the disproportionation reaction rate according to [36] as well as by an increase in the luminescence intensity of uranyl with an increase in the $HClO_4$ concentration (Figure 2, 2) [1, 35]. Well-structured CL spectrum, obtained by mixing of the flows of Eu^{2+} and UO_2^{2+} solutions, coincided with the PL spectrum of the ion UO_2^{2+} (Figure 3). Hence, the electronically excited state of the UO_2^{2+} ion is an emitter of CL. Still, before the discovery of the CL it was shown that the reaction of Eu^{2+} with UO_2^{2+} happens in two stages [36].

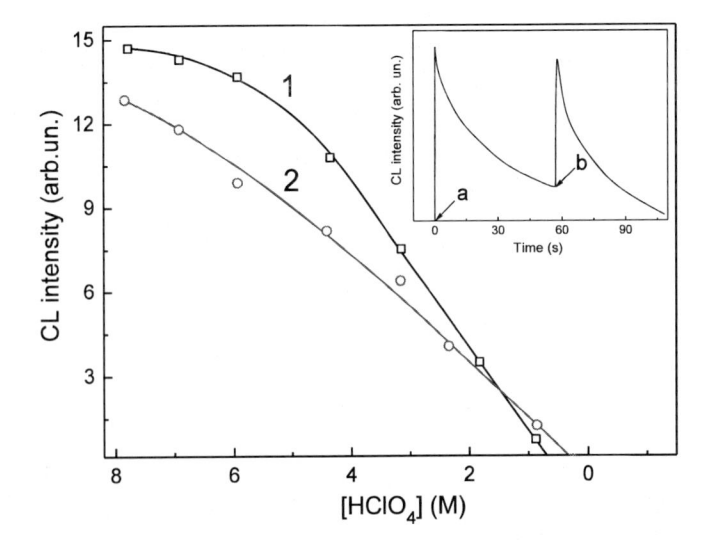

Figure 2. The dependences of the CL intensity of $Eu^{2+} + UO_2^{2+}$ reaction (1) and the PL of UO_2^{2+} (2) on $[HClO_4]$. On the inset – a sharp acidification from 0.8 to 3.0 M $HClO_4$: a – mixing $Eu^{2+} + UO_2^{2+}$; b – addition of $HClO_4$.

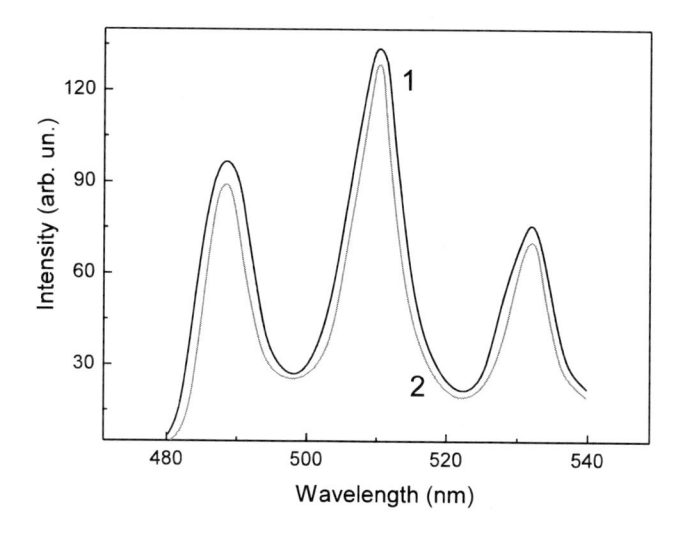

Figure 3. CL (1) and PL (2) spectra, measured during the reaction of Eu^{2+} and UO_2^{2+} ions. $[Eu^{2+}]_0 = [UO_2^{2+}]_0 = 10^{-3}$ M, $[HClO_4]_0 = 8.5$ M. CL spectrum was measured using the high-aperture monochromator Specol.

At the first fast stage ($k = 2 \cdot 10^3$ M^{-1} s^{-1}, 300 K), the pentavalent uranyl cation UO_2^+ is formed (1, scheme 1), which further disproportionate to the uranyl ion and ion of tetravalent uranium (2, scheme 1). The first stage (1, scheme 1) was excluded as the light reaction due to spectral and kinetic mismatches. First, the spectrum of the red emission of the Eu^{3+*} ion is completely different from the CL spectrum, and the singly charged ion UO_2^+ does not exhibit luminescence. Secondly, the duration of CL in 0.4 M $HClO_4$ is much longer than it should have been in accordance with the above-mentioned rate constant of the first stage. The generation of the CL emitter of the UO_2^{2+*} ion was assigned to the elementary act of the disproportionation reaction (3). The alternative excitation mechanism *via* energy transfer to the UO_2^{2+} ion from other primary excited products of the reaction, Eu^{3+*} and U^{4+*}, is excluded for the following reasons.

$$Eu^{2+} + UO_2^{2+} \longrightarrow Eu^{3+} + UO_2^+ \qquad (1)$$

$$UO_2^+ + UO_2^+ + 4H^+ \longrightarrow UO_2^{2+} + U^{4+} + 2H_2O \qquad (2)$$

$$UO_2^+ + UO_2^+ \longrightarrow U^{4+} + UO_2^{2+*} \longrightarrow UO_2^{2+} + CL \qquad (3)$$

Scheme 1.

The energy of the excited $^5D_0$ level of the Eu^{3+*} ion is less than that the radiating level of the UO_2^{2+*} ion, and the very weak PL of the U^{4+} ion is not quenched by the addition of UO_2^{2+*}. Finally, the mechanism of direct CL confirms the fact that the maximum intensity of CL is registered at the moment when the UO_2^{2+} ions are practically absent in the solution due to the consumption in a very fast reaction with Eu^{2+}.

Interestingly, in 0.4 M $HClO_4$, the products formed in the first and second reaction stages: the Eu^{3+} and U^{4+} ions, respectively, do not quench CL. This is established by the absence of quenching of the PL of the uranyl solution even at a concentration of U^{4+} equal to $2.5 \cdot 10^{-2}$ M. However, in 8.5 M $HClO_4$, the same concentration of U^{4+} quenches the PL of uranyl by more than two orders of magnitude.

When studying the peculiarities of the CL $Eu^{2+} + UO_2^{2+}$, it was first discovered that the ions of pentavalent uranium UO_2^+ effectively quench the luminescence of the ion UO_2^{2+} [1, 37]. This effect was discovered in an original way, by continuously measuring the PL of the uranyl solution, to which was added the Eu^{2+} solution. Immediately after the introduction of Eu^{2+}, the intensity of the PL decreases by more than two orders of magnitude (Figure 4), although the decrease in the concentration of uranyl (specially taken in excess) does not exceed 3.3%. At the moment of Eu^{2+} addition, the concentration of UO_2^+, and hence the quenching effect of uranyl, are maximal. Then, as the UO_2^+ ions are consumed in the disproportionation reaction, the PL intensity of uranyl increases. The Stern-Volmer quenching constant (K_{SV}), determined from the slope of the straight line 2 in Figure 4, is $(6.0\pm0.5) \cdot 10^3$ M^{-1}, and the bimolecular quenching rate constant (k_{bim}), calculated from the ratio $k_{bim} = K_{SV}/\tau_0$ is equal to $(5.0\pm0.5) \cdot 10^9$ $M^{-1} \cdot s^{-1}$, which is very close to the diffusion constant ($k_{diff} = 6.5 \cdot 10^9$ $M^{-1} \cdot s^{-1}$). Thus, the yield of CL during the reaction is a variable value due to a change in the concentration of the quencher UO_2^+. Since in the medium diluted with $HClO_4$, the UO_2^+ ion is the only quencher of uranyl, it is possible to study the CL kinetics in the disproportionation of UO_2^+ by measuring the fluorescence of uranyl. As the initial concentration of pentavalent uranium is known $[UO_2^+] = [Eu^{2+}]$, then $[UO_2^+]$ at any other time is determined from the kinetic curve PL (Figure 4, curve 1) according to the equation $[UO_2^+] = (I_0/I - 1) \cdot K_{SV}$, were I_0 and I are the PL intensities without and with the $[UO_2^+]$ quencher.

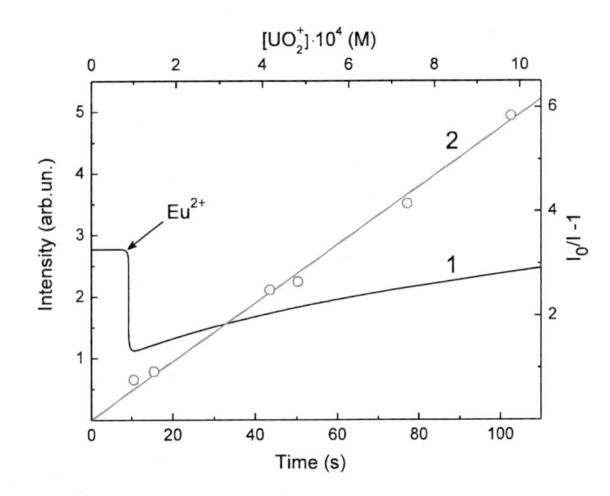

Figure 4. Kinetics of UO_2^{2+} ions PL intensity (1) before and after the Eu^{2+} solution addition. $[Eu^{2+}]_0 = [UO_2^{2+}]_0 = 8 \cdot 10^{-3}$ M, $[HClO_4]_0 = 0.4$ M. $[UO_2^+]_0 = 2.7 \cdot 10^{-4}$ M, $[LiClO_4] = 2.0$ M. 2 – dependence of the PL intensity of uranyl on the UO_2^+ ions concentration, $\lambda_{exc} = 365$ nm.

The straight-line 1 shown in Figure 5 indicates that the values of UO_2^+ concentrations obtained in this way, which change with time, are well described by a second-order law.

In the equation (I) of the dependence of the intensity of CL on the concentration of UO_2^+, the value of ϕ_q is also concentration dependent.

$$I_{CL} = k \cdot [UO_2^+]^2 \cdot \phi_{exc} \cdot \phi_q \cdot \phi_{temp} \tag{4}$$

The following notation is given in this equation: $k \cdot [UO_2^+]^2$ is the disproportionation reaction rate; ϕ_{exc} – excitation yield, that is, part of the elementary acts of the disproportionation reaction UO_2^+, generating excited ions UO_2^{2+*}; ϕ_q – quenching yield, i.e., part UO_2^{2+*} ions that have not been quenched by any reaction participants; ϕ_{temp} – is the part of the UO_2^{2+*} ions that avoided thermal quenching. The ϕ_{temp} is equal to the ratio τ/τ_{teor} [38], where τ is the lifetime of UO_2^{2+*} experimentally observed in a diluted $HClO_4$ medium, at 287 K in the absence of quenchers and is equal to $1.25 \cdot 10^{-6}$ s [39]; τ_{teor} – the theoretical maximum lifetime of UO_2^{2+*} in this environment was measured [1, 37] for uranyl solutions, glazed at 77 K. Starting from 170 K and up to 77 K inclusive, the value of τ_{teor} does not change, reaching a limit value of $3 \cdot 10^{-4}$ s. Hence, $\phi_{temp} = \tau/\tau_{teor} = 4.2 \cdot 10^{-3}$.

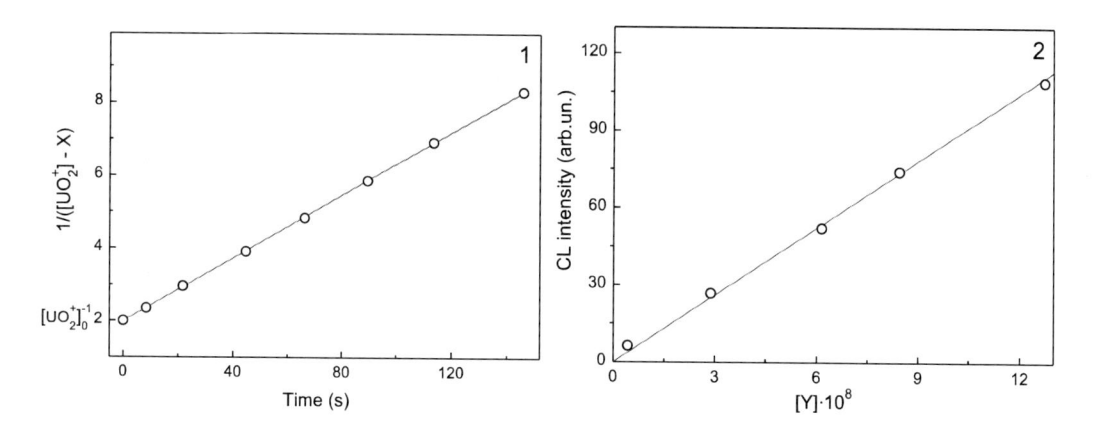

Figure 5. The time dependence of $1/([UO_2^+] - X)$ in the disproportionation reaction; $[UO_2^+]_0$ – the initial concentration; X – the $[UO_2^+]$ ions concentration, reacted at time t, and determined from the data on the uranyl quenching. $[HClO_4]_0 = 0.4$ M. 2 – dependence of the CL intensity in reaction Eu^{2+} + UO_2^{2+} on UO_2^+ ions concentration corrected according data on the quenching of UO_2^{2+*} by UO_2^+. Y = $[UO_2^+]^2/(K_{SV} \cdot [UO_2^+] + 1)$. $T = 287$ K. $[UO_2^{2+}] = const = 7.9 \cdot 10^{-3}$ M.

It is also fair to write that $\phi_q = I/I_0$, where I and I_0 are the CL intensities experimentally observed and in the absence of quenching, respectively. From the Stern-Volmer equation it follows that $\phi_q = 1/(K_{SV} \cdot [UO_2^+] + 1)$. Substituting this expression into the upper equation (I) gives (II):

$$I_{CL} = k \cdot [UO_2^+]^2 \cdot \phi_{exc} \cdot \phi_{temp}/(K_{SV} \cdot [UO_2^+] + 1) \tag{5}$$

As can be seen from Figure 5 (2), in the coordinates $I_{CL} - [UO_2^+]^2/(K_{SV}\cdot[UO_2^+] + 1)$ the linear dependence is well satisfied, which also confirms that the value ϕ_{exc} = const and does not change during the reaction. From the slope of this straight-line defined the value $\phi_{exc} = 10^{-7}$ for 0.4 M $HClO_4$ and molar ionic strength μ = 2.0 M.

According to the law $I_t = k\cdot\phi\cdot\phi_q\cdot[UO_2^+]^2$ (where $\phi = \phi_{exc}\cdot\phi_{temp}$) of the 2nd order, the concentration of UO_2^+ is related to its initial concentration by equation (III).

$$[UO_2^+] = [UO_2^+]_0/(k\cdot[UO_2^+]_0\cdot t + 1) \tag{6}$$

The substitution in (II) the concentration of UO_2^+ from (III) gives (IV).

$$I_t = k\cdot\phi\cdot\phi_q\cdot[UO_2^+]_0^2/(k\cdot[UO_2^+]_0\cdot t + 1)^2 \tag{7}$$

Immediately after mixing the reagents (t = 0), CL intensity is determined by equation (V). At this moment, the intensity of the CL is minimal, since the UO_2^+ quencher concentration is maximal.

$$I_{t0} = k\cdot\phi\cdot\phi_{q0}\cdot[UO_2^+]_0^2 \tag{8}$$

Dividing (IV) by (V) leads to expression (VI).

$$I_{t0}/I_t = (k\cdot[UO_2^+]_0\cdot t + 1)^2\cdot\phi_{q0}/\phi_q \tag{9}$$

Bearing in mind expressions (VII) and (VIII) and substituting the values of ϕ_q from (VII) into equation (VI), we obtain (IX)

$$\phi_q = [\{K_{SV}[UO_2^+]_0/(k[UO_2^+]_0 t + 1)\} + 1]^{-1} \tag{10}$$

$$\phi_q^0 = [K_{SV}[UO_2^+]_0 + 1]^{-1} \tag{11}$$

$$I_{t0}/I_t = (k[UO_2^+]_0 t + 1)^2[\{K_{SV}[UO_2^+]_0/(k[UO_2^+]_0 t + 1)\} + 1]\phi_q^0 \tag{12}$$

Opening the parentheses in (IX) gives a quadratic equation, the decision this equation with respect to (kt) gives the final expression (X), which reflects the dependence of the intensity of CL on time.

$$kt = -(\{[UO_2^+]_0^2 K_{SV} + 2[UO_2^+]_0\}\cdot 0.5[UO_2^+]_0^{-2}) + (\{[UO_2^+]_0^2 K_{SV} + 2[UO_2^+]_0\}^2 - 4[UO_2^+]_0^2)\{K_{SV}[UO_2^+]_0 + 1 - I_{t0}\cdot(\phi_q^0\cdot I_t)^{-1}\})^{0.5}\cdot 0.5[UO_2^+]_0^2 \tag{13}$$

The right-hand side of equation (X), containing the known constant terms is denoted as A, and the dependence is plotted in the coordinates $kt - A$, the linear correlation coefficient of which is R= 99.8% (Figure 6). Equation (X) allows one to determine the rate constant for the disproportionation of pentavalent uranium directly from the experimental curve of the decay of CL intensity.

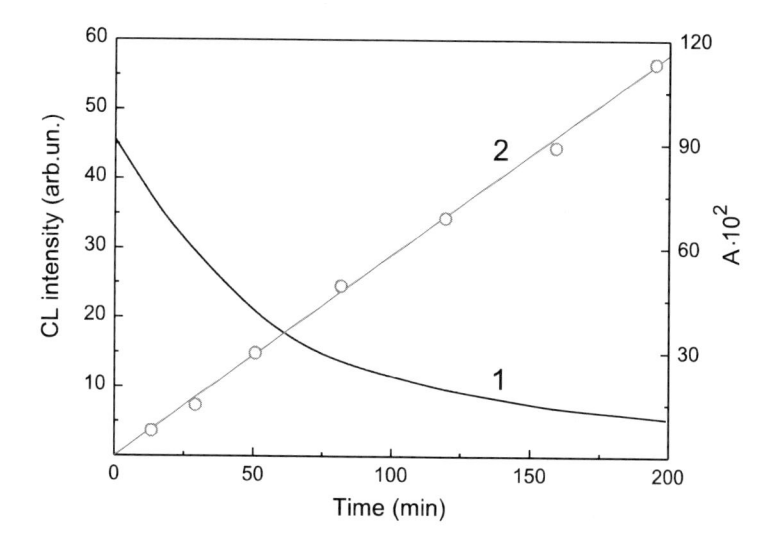

Figure 6. The CL kinetic curve (1) in the reaction $Eu^{2+} + UO_2^{2+}$ in 0.4 M $HClO_4$ at 287 K and linear anamorphosis (2) of curve 1 in coordinates A – time (A values are taken from equation X).

From the slope of the straight lines, similar to that shown in this figure, the values of disproportionation rate constants (k_{CL}) are determined for different concentrations of uranyl (the latter are shown in parentheses). Below are the values of similar constants, measured by spectrophotometric method (k_{SP}) in [36].

k_{CL}, $M^{-1} \cdot s^{-1} = 58.5 \pm 3.0$ ($8 \cdot 10^{-3}$), 17.3 ± 1.5 (78.7);
k_{SP}, $M^{-1} \cdot s^{-1} = 61.5$ ($8 \cdot 10^{-3}$), 15.2 (78.7).

Comparison of these data shows a good correlation between the rate constants determinate by CL and dark methods. The dependence of k on the concentration of uranyl according to in [36] is associated with the formation of the complex $UO_2^{2+} \cdot UO_2^{+}$. The effect of the concentration of the initial Eu^{2+} and UO_2^{2+} reagents has been studied in a wide range by changing them by several orders of magnitude from 10^{-1} to $3 \cdot 10^{-7}$ M (or $2 \cdot 10^{-8}$ M when cooling the photomultiplier up to 283 K) [1, 36]. Figure 7 shows the dependence of the intensity of CL $Eu^{2+} + UO_2^{2+}$ on the concentration of UO_2^{2+}. The right branch of curve 1 is the case when Eu^{2+} is taken in excess with respect to the stoichiometry of the reaction $Eu^{2+} + UO_2^{2+} \rightarrow Eu^{3+} + UO_2^{+}$. An increase in the intensity of CL in this area with an increase in the uranyl concentration is associated with an increase in the disproportionation reaction rate due to an increase in the concentration of UO_2^{+}.

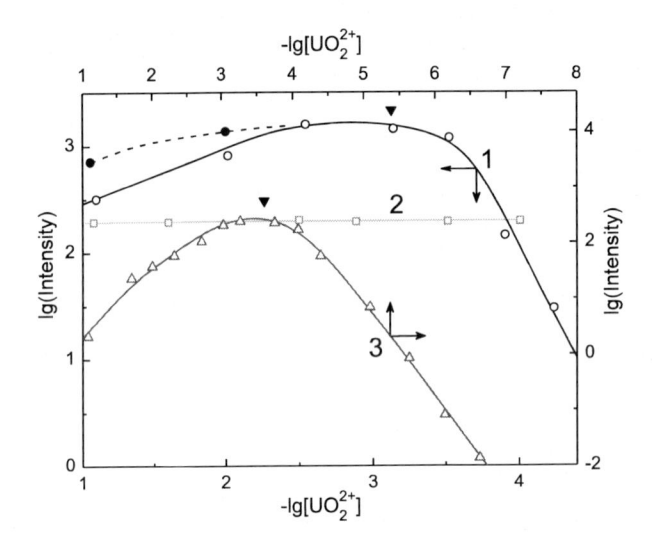

Figure 7. The dependence of the maximum CL intensity (at the moment of adding Eu^{2+} to UO_2^{2+}) on the concentration of UO_2^{2+}. 1 – $[HClO_4]_0 = 0.4$ M, $[LiClO_4]_0 = 2.0$ M, $T = 278$ K; dashed line – the effect of the formation of the UO_2^{2+} complex with UO_2^+ on the reaction rate is taken into account. 3 – $[HClO_4]_0 =$ 8.5 M; black triangle – $[UO_2^{2+}]/[Eu^{2+}] = 1.0$; 2 – the dependence of the ratio lg(intensity of the PL/$[UO_2^{2+}]$) on $[UO_2^{2+}]$.

Since in these experiments $[Eu^{2+}]_0 = \text{const}$, the maximum concentration of UO_2^+ formed in stage (1) is limited by the stoichiometry of this stage. Therefore, the concentration of UO^{2+} increases only to the equivalence point, where $[Eu^{2+}] = [UO_2^{2+}]$. As a result, the rate of disproportionation, and hence the intensity of CL increases only to the equivalence point. The decrease in the intensity of CL after the equivalence point is not related to the self-quenching concentration of uranyl, as evidenced by the independence of the fluorescence intensity of uranyl from its concentration (Figure 7, 2). The correction on slowing the rate of disproportionation due to the formation of the $UO^{2+} \cdot UO_2^+$ complex brings closer the dependence of the intensity of CL to the form of a curve with saturation (Figure 7, dashed curve). The kinetics of the disproportionation reaction of UO_2^+ is determined by a second-order law, that is fully consistent with the linear anamorphosis of the decay kinetics of CL in the coordinates: $(I_{CL})^{-1/2}$ *vs* time (Figure 1, 3).

It should be noted that the high brightness of the discussed green CL allows to determine the presence of uranyl ions in such low concentrations as $2 \cdot 10^{-8}$ M [1]. For comparison, it is known that the lowest detectable concentration of uranyl with the use of its brightest PL in the presence of sodium trimethyl phosphate is $1.7 \cdot 10^{-7}$ M [40].

2.2. Chemiluminescence in Oxidation of Eu^{2+} Ions by H_2O_2 in Aqueous Solution

CL in the oxidation of Eu^{2+} ions by H_2O_2 found by Elbanowski et al. [3] is one of the first light reactions involving divalent lanthanides. The CL was studied by use of the single-photon counting method (photomultiplier with a sensitive in the 200-800 nm region). This very weak CL occurs when mixing the aqueous solutions of Eu^{2+} (the parent form of $EuCl_2$) and H_2O_2 (pH = 1.5 or 5.3) in argon atmosphere. The kinetic curve of CL consists of two components (Figure 8). The fast decaying component with a lifetime of about 45 s is comparable with the time of Eu^{2+} oxidation. The second component is a broad maximum, which is formed at about 3-4 minutes. From this, the CL emitted during the prolonged component is not associated with the divalent europium ion, since Eu^{2+} at this time is already absent in the reaction solution (Figure 9) [41].

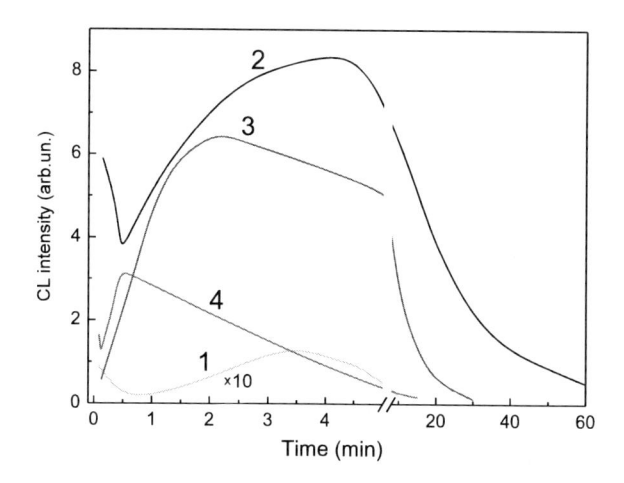

Figure 8. Kinetics of CL in interaction of Eu^{2+} with H_2O_2 in aqueous solutions at 295 K. 1 – the free Eu^{2+} ions ($1.6 \cdot 10^{-2}$ M) and H_2O_2 ($2.0 \cdot 10^{-2}$ M), pH=5.3; 2 – the complex of Eu^{2+} ($4 \cdot 10^{-3}$ M) with EDTA ($40 \cdot 10^{-3}$ M) and H_2O_2 ($20 \cdot 10^{-3}$ M), phosphate buffer ($30 \cdot 10^{-3}$ M), pH=7.5; 3 – the complex of Eu^{2+} ($4 \cdot 10^{-3}$ M) with EDTA ($40 \cdot 10^{-3}$ M), and H_2O_2 ($20 \cdot 10^{-3}$ M), phosphate buffer ($30 \cdot 10^{-3}$ M), and additionally 1% *n*-butanol, pH=7.5; 4 – the complex of Eu^{2+} ($4 \cdot 10^{-3}$ M) with EDTA ($40 \cdot 10^{-3}$ M) and H_2O_2 ($20 \cdot 10^{-3}$ M), phosphate buffer ($30 \cdot 10^{-3}$ M) in the presence of O_2, pH=7.5.

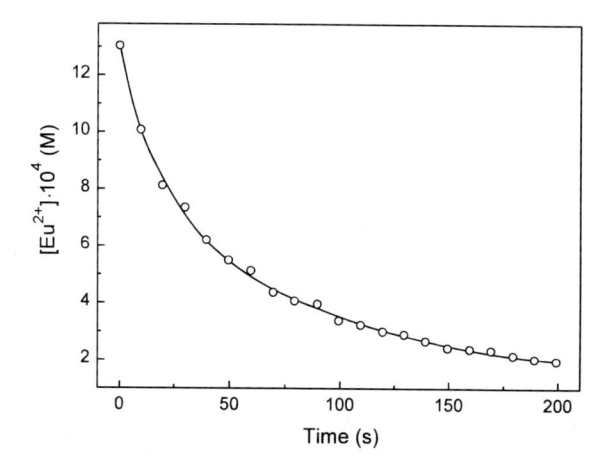

Figure 9. The decreasing of the Eu^{2+} ions concentration after the addition of H_2O_2. $[Eu^{2+}]_0 = [H_2O_2]_0 = 1.3 \cdot 10^{-3}$ M. The concentration of Eu^{2+} was determined from absorption spectra for $\lambda_{max} = 248$ and 328 nm.

Increasing the concentration of Eu^{2+} increases the intensity of CL, whereas the decay time decreases with growing H_2O_2 concentration. This CL has very low values of quantum yield (about 10^{-12}). Due to the very low intensity of this CL, its spectrum was not measured.

The brightness of CL increases significantly with the interaction of hydrogen peroxide with a complex of Eu^{2+} with ethylenediaminetetraacetic acid (EDTA): quantum yield of 10^{-10}. In this case the short-lived CL appears also immediately after addition of Eu^{2+}-complex to H_2O_2.

At higher pH values (5.0-8.5) also a long-lived component take place, similar to the uncomplexed solutions, but with much higher intensity and much longer decay time (Figure 8, 2). The brightness of the Eu^{2+}-complex CL was enough to register its spectrum by the cut-off filters, the strong maximum of which at about 610 nm testifies that the CL emitter is the electronically excited state $^5D_0$ of the Eu^{3+} ion.

Based on the above results, the short-lived CL caused by uncomplexed and complexed Eu^{2+} ions was associated with the occurrence of two consecutive reactions:

$$Eu^{2+} + H_2O_2 \longrightarrow Eu^{3+} + OH^- + {}^{\bullet}OH \tag{14}$$

$$Eu^{2+} + {}^{\bullet}OH \longrightarrow Eu^{3+} + OH^- \tag{15}$$

2.1 eV of energy is required for the excitation of Eu^{2+} in the $^5D_0$ state. The enthalpy (ΔH_0) of reaction (14) is 1.6 eV, while for the reaction (15), $\Delta H_0 = 2.6$ eV. The last value of ΔH_0 is quite enough to populate the excited level $^5D_0$, even without considering the activation energy.

As shown in this study, ion Eu^{3+*} can be generated by an oxidation reaction involving free radicals. Later in [27, 42, 43] the excitation reaction was presented in a more specific form of which it follows that the ions Eu^{3+*} are formed precisely in the elementary act of electron transfer from Eu^{2+} to the $^{\bullet}OH$ radical (3).

$$Eu^{2+} + {}^{\bullet}OH \longrightarrow Eu^{3+*} + OH^- \tag{16}$$

$$Eu^{3+*} \longrightarrow Eu^{3+} + CL \tag{17}$$

Also, thanks to digital data recording and processing was carry out a more precise spectral analysis of the CL in the reaction of Eu^{2+} with H_2O_2 (Figure 10) [42]. As can be seen from Figure 10, the CL spectrum of the $Eu^{2+} + H_2O_2$ reaction and the PL spectrum of Eu^{3+} solution coincide well. This means that CL is caused by radiation from an excited Eu^{3+} ion *via* transitions $^5D_0 \rightarrow {}^7F_1$ (594 nm) and $^5D_0 \rightarrow {}^7F_2$ (615 nm).

The study of the possibility of application of the chemiluminescent redox system $Eu^{2+}/Eu^{3+}-H_2O_2$ for investigation the characteristics of biologically active compounds is an important aspect of the research of Elbanowski et al. To this end, in a more complex system

Eu^{2+}/Eu^{3+}-H_2O_2-biomolecule, a wide variety of compounds were tested as biomolecules. Among them: the adenosine nucleotides – ATP, ADP, AMP, playing a very important role as biochemical, high-energy molecules [44], the reduced nicotinamide adenine nucleotide (NADH) molecule [45], deoxyribonucleic acid (DNA) [46] and others.

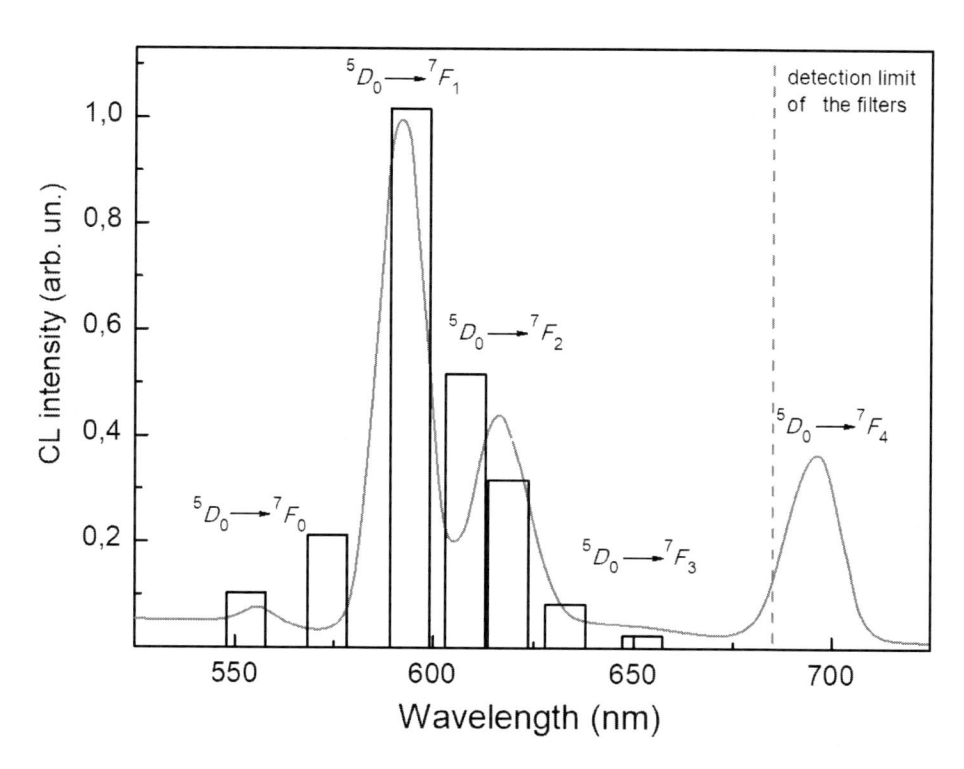

Figure 10. CL spectrum during oxidation of Eu^{2+} by H_2O_2. $[Eu^{2+}]_0 = [H_2O_2]_0 = 2\cdot10^{-3}$ M. The solid gray line is normalized spectrum of PL of Eu^{3+} ions, $\lambda_{exc} = 394$ nm.

We will not discuss these and others similar works in detail, because they are already reviewed in the excellent review by Elbanowski et al. and we will only note the most common features of these systems. First of all, we note that the Eu^{2+}/Eu^{3+}-H_2O_2-biomolecule CL systems are multi-emitter. The dominant emitters are the Eu^{3+*} ions which emit the "red" light (λ_{max} = 590-620 and 700 nm) from the Eu^{3+} complexes with biomolecules (for example with ATP). This "red" CL is caused by the magnetic dipole transition $^5D_0$–$^7F_1$ (λ_{max} = 594 nm), a hypersensitive transition $^5D_0$–$^7F_2$ (λ_{max} = 620 nm) and a forced electric dipole transition $^5D_0$–$^7F_4$ (λ_{max} = 705 nm). At the short-term CL stage (lasting about 40-60 s), the Eu^{3+*} ion is formed mainly due to the oxidation of Eu^{2+} ions by the $\cdot OH$ radicals, and then as a result of energy acceptance from other primary excited emitters. The position of the emission maxima of the Eu^{2+} ion varies insignificantly for different biomolecules coordinated with this ion, which is due to the effective shielding of the emitting $4f$ level of europium by the filled $^5S_2$ and $^5P_6$ shells. The donors of energy that excites the Eu^{3+} ions are assumed to be the excited dimers of singlet oxygen. These dimers are formed in radical reactions initiated by $\cdot OH$ radicals (Scheme 2). Singlet oxygen dimers are also responsible for emission of CL at 580 and 633 nm. In addition, CL in the region of 450-500 nm was also attributed to carbonyl fragments emission [27, 42, 45, 46].

$$HO^{\bullet} + H_2O_2 \longrightarrow HO_2^{\bullet} + H_2O;\ HO_2^{\bullet} \longleftrightarrow O_2^{\bullet-} + H^+$$

$$O_2^{\bullet-} + HO^{\bullet} \longrightarrow HO^- + {}^1O_2;\ O_2^{\bullet-} + O_2^{\bullet-} \longrightarrow HO_2^{\bullet} + {}^1O_2 \qquad (18)$$

$$HO_2^{\bullet} + HO_2^{\bullet} \longrightarrow H_2O_2 + {}^1O_2;\ O_2^{\bullet-} + H_2O_2 \longrightarrow HO^- + HO^{\bullet} + {}^1O_2$$

$$({}^1O_2)_2 + Eu^{2+} \longrightarrow 2\,{}^3O_2 + Eu^{2+*} \longrightarrow Eu^{2+} + CL\ (590\text{-}700\ nm) \qquad (19)$$

$$({}^1O_2)_2 \longrightarrow 2\,{}^3O_2 + CL\ (580\ and\ 633\ nm) \qquad (20)$$

Scheme 2.

2.3. Chemiluminescence in Oxidation of (Me5C5)2Yb by Oxygen in Solution

The CL of bis(η^5-pentamethylcyclopentadienyl)ytterbium(II), (Me$_5$C$_5$)$_2$Yb (wjere Me = CH$_3$) in THF solution is registered by oxygen bubbling or by addition of solution of O$_2$ in THF [4, 47].

Unfortunately, quantitative data on the intensity of CL are absent; however, it is recorded through a monochromator, which indicates its rather high intensity. The spectrum of the CL consists of a single broadband with a maximum at 990 nm, which coincides with the PL peak of the reaction product and is characteristic of the luminescence of the Yb^{3+} ion (configuration f^{13}) due to the $^2F_{5/2} \rightarrow {}^2F_{7/2}$ transition (Figure 11) [4]. This spectrum is significantly different from the PL spectrum of the initial solution (Me$_5$C$_5$)$_2$Yb showing a broad band with a maximum at 935 nm (Figure 11).

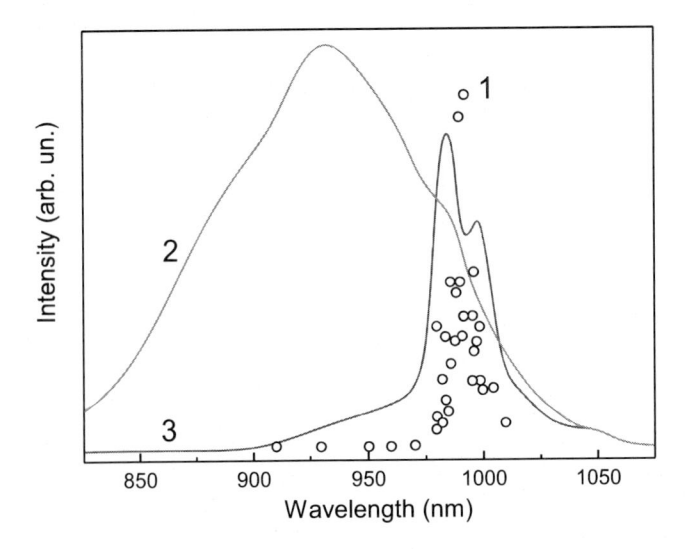

Figure 11. CL (1) and PL (2, 3) spectra of the reaction of (Me$_5$C$_5$)$_2$Yb with O$_2$ in THF solutions. PL spectra were obtained before (2) and after (3) addition of O$_2$, $\lambda_{exc} = 476.5$ nm.

It should be noted that the location of this maximum differed significantly from that for the maxima of Yb^{2+} ion luminescence in THF solutions at 431 nm for the $YbCl_2 \cdot (THF)_2\text{-Bui}_2AlH$ complex [29] and at 515 nm for the $YbI_2 \cdot (THF)_x$ complex [48]. The problem of a too short flash of CL for the correct registration of the CL spectrum was solved as follows. For obtaining CL spectra, 0.1 mL aliquots of 10-15 mM THF solutions of the $(Me_5C_5)_2Yb$ complex were added to THF saturated with O_2. Each addition was accompanied by a short flash of CL, which was recorded simultaneously through a monochromator installed at a specific wavelength and directly, i.e., in the form of a total stream of light. The ordinate of each point of the CL spectrum (Figure 11, 1) is the ratio of the light sums of the CL flashes recorded through the monochromator and without it. It should be noted that the establishment of the CL mechanism of organometallic compounds (OMC) of low-valent metal ions is complicated by the fact that both are prone to oxidation: the metal ion and ligand of OMC. Both routes are potentially chemiluminescent. As it turned out, the oxidation of OMC of trivalent ytterbium $(Me_5C_5)_2YbCl$ also gives CL due to emission of the analogous Yb^{3+} compound, although the maximum of the CL spectrum in the case of $(Me_5C_5)_2Yb$ is slightly blue-shifted. The higher CL intensity of $(Me_5C_5)_2Yb$ is related to the higher oxidation rate of this OMC. The use of the $O_2^{\bullet -}$ anion as an oxidizing agent, which is capable of oxidizing only a metal ion, results in CL only in the case of $(Me_5C_5)_2Yb$. The obtained results mean that in the case of $(Me_5C_5)_2YbCl$, only the ligand is oxidized by molecular oxygen to form an excited unidentified product that transfers its energy to the Yb^{3+} ion, converting it to the emitting excited state Yb^{3+*}. In contrast, CL during oxidation of $(Me_5C_5)_2Yb$ is caused by oxidation of both the Yb^{2+} ion and the ligand; the quantitative contribution of each has not been established. At the same time, it was found that for successful excitation of CL, the CL emitter containing Yb^{3+} must also contain oxygen. This important conclusion regarding "*oxophilicity*" for producing CL was made based on the absence of CL at the oxidation of $(Me_5C_5)_2Yb$ by oxidants contained no oxygen, such as $AgPF_6$ and the $Cp_2Fe\text{-}FeCl_4$ mixture. At the same time, Yb^{2+} was successfully oxidized to Yb^{3+} by both oxidizing agents.

In the paper [4, 47] under discussion, it was unambiguously proved that the radiating center of the CL is the excited ion Yb^{3+*}, although the final identification of a stable product-emitter based on Yb^{3+} is not given. At the same time, it was found that in a complex mixture of reaction products consisting of at least four species, hydroperoxide Me_5C_5OOH is dominant (60-70%). However, it should be remembered that Me_5C_5OOH is a secondary product, which is formed as a result of the hydrolysis of the product of complete oxidation $(Me_5C_5)_2Yb$ by O_2. With great confidence we can suggest as such a stable product only one species, namely, a complex organometallic peroxide $Me_5C_5OOYb^{3+}=O$. Therefore, it is most likely that $Me_5C_5OOYb^{3+*}=O$ is the CL emitter in the reaction $(Me_5C_5)_2Yb$ with O_2. Indeed, firstly, it contains the Yb^{3+} trivalent ion as an emitting center; secondly, the Yb^{3+} radiating ion is bound to oxygen atoms (corresponding to the requirement of "*oxophilicity*"); and, finally, the product of its hydrolysis will be hydroperoxide Me_5C_5OOH.

2.4. Chemiluminescence in Oxidation of Cp_2Eu and Cp_2Sm by Oxygen in Solution

Non-substituted cyclopentadinides of Eu^{2+} and Sm^{2+} also exhibit chemiluminescent activity with respect to oxygen. So, when O_2 is bubbling through Cp_2Eu or Cp_2Sm [15, 17] solutions in a mixture of THF-Et_2O at room temperature, reliably recorded CL appears (I_{CL} = ~10^5 photon·s^{-1}·mL^{-1}). In contrast to CL, the PL of the solution before and after the CL are observed only at 77 K (Figure 12) [17].

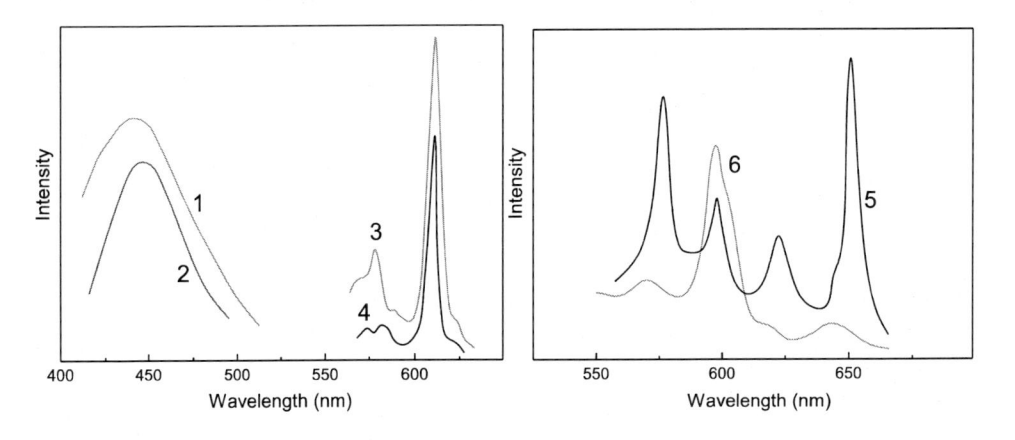

Figure 12. The PL spectra: 1 – Eu^{2+}_{aq} (10^{-2} M) in 2 M $HClO_4$; 2 – Cp_2Eu (10^{-3} M) in THF-Et_2O; 3 – solid $EuCl_3 \cdot 6H_2O$; 4 – Cp_3Eu (10^{-3} M) in THF-Et_2O; 5 – Cp_2Sm (10^{-3} M) in THF-Et_2O; 6 – Sm^{3+}_{aq} (10^{-3} M) in 1.0 N HCl. T = 77 K(1, 2, 4, 5), 300 K (3, 6); λ_{exc} = 365 nm.

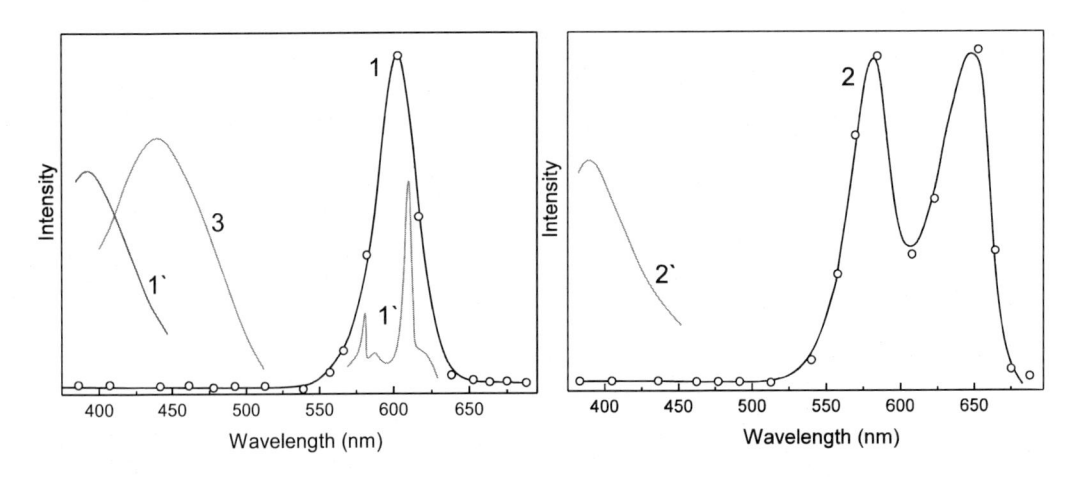

Figure 13. The CL (300 K) spectra of oxidation of Cp_2Eu (1) or Cp_2Sm (2) by O_2. The PL spectra (77 K) of Cp_2Eu (3) and products of the interaction Cp_2Eu (1`) and $Cp_2Sm$ (2`) with O_2 in THF. [Cp_2Eu]$_0$ = [Cp_2Sm]$_0$ = $2 \cdot 10^{-4}$ M.

The luminescent properties of the frozen solution Cp_2Eu is like those of Eu^{2+}_{aq} ion in 2 M $HClO_4$ studied in [49] (Figure 12). Indeed, they both show a broadband luminescence in the blue spectral region (430-450 nm) clearly visible to the naked eye and caused by the allowed $5d^1 \rightarrow 4f^7$ transition. After the emission of CL, the PL spectrum of the oxidized solution frozen at 77 K contains narrow maxima (580, 595, 613 and 640 nm) in the red region of the spectrum, characteristic of the Eu^{3+} ion (Figure 13) [17].

The PL spectra of the Cp_3Eu and Cp_3Sm solutions are recorded in the red region as narrow structural bands characteristic for shielded *f-f* transitions: 580, 595, 613, and 640 nm for Cp_3Eu ($\lambda_{exc} = 300$ nm) and 578, 600, 624, 656 nm for Cp_3Sm ($\lambda_{exc} = 365$ nm). The CL spectra in the oxidation of all lanthanidocenes were measured by cut-off filters due to the insufficient CL intensity. The CL spectrum of Cp_2Ln contains only bands characteristic of the luminescence of Eu^{3+} and Sm^{3+}. Thus, the radiating centers of CL $Cp_2Ln + O_2$ are the Eu^{3+*} or Sm^{3+*} ions, which are an ionic part of a stable lanthanidocenes oxidation product. The CL spectra at the long wavelength region are almost identical for the oxidation of lanthanidocenes of divalent and trivalent lanthanides ions (Figure 13). In the CL spectrum for trivalent lanthanide cyclopentadienides, Cp_3Eu, and Cp_3Sm, in addition to the emission of Ln^{3+*} ions, a short-wavelength emission at the 450-525 nm region is observed. The latter emission corresponds to the phosphorescence (PS) of triplet-excited 4-cyclopenten-1,3-dione which is a product of the Cp-ring oxidation. This ketone detects the characteristic band at 1750 cm^{-1} in the IR spectra of the oxidates of oxidation of Cp_3Eu and Cp_3Sm. The PL spectra of these oxidates coincide with the PS spectrum of 4-cyclopenten-1,3-dione (Figure 14), synthesized in a non-dependent manner [17].

If in the PL spectra of oxidized Cp_2Eu and Cp_3Eu solutions the luminescence of the Eu^{3+} ion is always clearly recorded (at 77 K), then the PL of the Sm^{3+} ion is recorded only with incomplete oxidation of Cp_2Sm. This is because the luminescence of Sm^{3+*} ions is significantly quenched by a not identified reaction product (noted as Pr). It is confirmed by the quenching the PL of Cp_3Sm at the addition of a solution of completely oxidized Cp_3Sm. The dependence of efficiency of the PL quenching on the quencher concentration [Pr] obeys the Stern-Volmer law: $I_0/I - 1 = K_{SV} \cdot [Pr]$, were I_0 and I are the PL intensity without and with a quencher, respectively [1]. In case Cp_3Eu similar quenching is much less effective.

Hydrolysates of the oxidation products Cp_2Ln and Cp_3Ln (Ln = Eu and Sm) contain as the main component the CpOOH hydroperoxide (IR-band at 850 cm^{-1}, KI-test reaction and H_2O-CL-test, the latter see in section 2.5). In combination with the CL and PL spectra containing the luminescence maxima of the Ln^{3+} ions, this means that the main product of oxidation of all lanthanideocenes contains a fragment of >LnOOCp, the hydrolysis of which gives CpOOH.

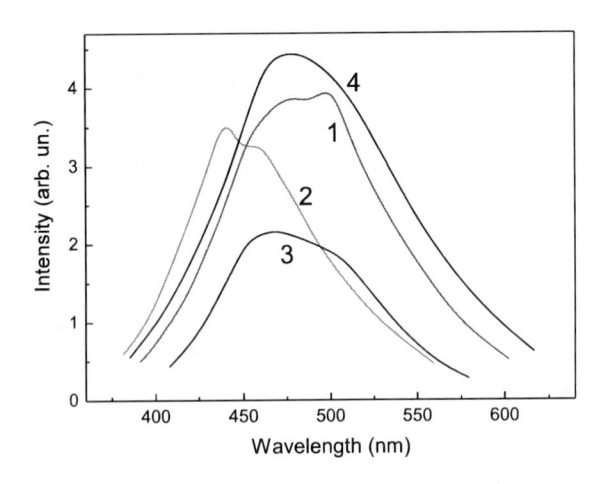

Figure 14. PS spectra of THF solutions of ketones (1-3) and the products after oxidation of Cp_nLn^{n+} (4); 1 – cyclopentenone, 2 – dimer of cyclopentadienone, 3 – cyclopenten-1,3-dione. [ketone] = 10^{-4} M, $T = 77$ K, $\lambda_{exc} = 300$ nm.

According to the proposed Scheme 3 [3], at the first fast stage of Cp_2Ln autoxidation, ketone and Ln(III) as its oxoderivatives are excited (1). Ketone is subjected to quenching by initial Cp_2Ln (1, Scheme 3). Further oxidation yields excited lanthanide organic peroxide, deactivation of which is responsible for the long wavelength "red" component of CL. In the oxidation of Cp_3Ln, except for the case with Ln^{3+*} (4, Scheme 3), in addition to the red luminescence, the emission of ketones is recorded, but at the short-wavelength region. Cyclopentadienone is formed, and very quickly disappears in the dimerization reaction as well as in its reactions with other reactive participants [3]. In addition, the excited cyclopentadienone can be quenching *via* energy transfer to the Ln^{3+} ions. Therefore, a luminescence of cyclopentadienone exhibits only in CL, but not in PL.

$$Cp_2Ln^{2+} + O_2 \longrightarrow Cp(O)Ln^{3+*} + C_5H_4O^* + C_5H_4(O)O^* \xrightarrow[\text{of ketones}]{Cp_2Ln^{2+} \text{ quenching}} \quad (21)$$

$$Cp(O)Ln^{3+*} \longrightarrow Cp(O)Ln^{3+} + LC\text{-}CL \quad (22)$$

$$Cp(O)Ln^{3+} + O_2 \longrightarrow (CpOO)(O)Ln^{3+*} \longrightarrow (CpOO)(O)Ln^{3+} + LC\text{-}CL \quad (23)$$

$$Cp_3Ln^{3+} + O_2 \longrightarrow Cp_2(CpOO)Ln^{3+*} \longrightarrow Cp_2(CpOO)Ln^{3+} + LC\text{-}CL \quad (24)$$

$$Cp_2(CpOO)Ln^{3+} + O_2 \longrightarrow CpOOLn^{3+*} < + C_5H_4O^* + C_5H_4(O)O^* \quad (25)$$

$$C_5H_4O^* + C_5H_4(O)O^* \longrightarrow C_5H_4O + C_5H_4O_2 + SC\text{-}CL \quad (26)$$

$$CpOOLn^{3+*} < \longrightarrow CpOOLn^{3+} < + P + LC\text{-}CL \quad (27)$$

C_5H_4O - cyclopenta-2,4-dienone ⬡=O; $C_5H_4O_2$ - 4-cyclopentene-1,3-dione ⬡ ;

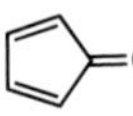

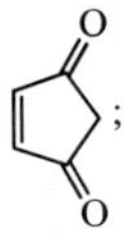

P stands for a complex mixture of polymers and metal oxides;

LC-CL and SC-CL - Long- and Short-wavelength Components of CL, respectively.

Scheme 3.

A lanthanide organic peroxide was identified as an emitter of "red" CL [3] by a good match of CL spectra of cyclopentadienyls of Eu and Sm with PL of the oxidation products each possessing a peroxide group. This cannot rule out a possible emission of an organometallic oxide lanthanide in the initial stages, when not all the Cp–Ln bonds are oxidized.

2.5. Chemiluminescence in Reaction of the Eu^{2+} and Sm^{2+} Cyclopentadienyls with Water

As it turned out, the hydrolysis of the oxidation products of cyclopentadienides of divalent and trivalent ions of europium and samarium, namely, peroxides $CpOOLn<$, is accompanied by CL [14, 17, 48, 50].

Adding an aliquot of water (0.5 mL) to an oxygen oxidized 10^{-4} M solution of Cp_nLn^{n+} (Ln = Eu, Sm, n = 2, 3) in THF (10 mL) at 300 K results in a rather bright CL ($I_{CL} = 5 \cdot 10^6$ photon$\cdot s^{-1} \cdot mL^{-1}$). Interestingly, a similar hydrolysis of even non-oxidized lanthanidocene Cp_3Sm in a THF solution ($2 \cdot 10^{-4}$ M) in an argon atmosphere also causes CL ($I_{CL} = 2 \cdot 10^6$ photon$\cdot s^{-1} \cdot mL^{-1}$). This CL fades by 90% within about 3 minutes. The interest of this CL is also connected with the numerous unsuccessful attempts that have been made to initiate the CL during the hydrolysis of a series of organometallic compounds, including Grignard reagents [51] and aluminum alkyls [52, 53]. The motive of these experiments is particularly understandable, bearing in mind the extremely high reactivity of these compounds with respect to water, characterized by the explosive nature of this reaction. From the above and considering the positive experience with CL in Cp_3Sm hydrolysis, it is clear that even a very high chemical exothermicity does not guarantee the generation of CL, if none of the reaction participants is able to emit luminescence. $Sm(OH)_3$ and C_5H_5 are identified as stable products of the hydrolysis reaction of Cp_3Sm; unidentified polymer products were also detected.

The spectra of CL in reactions with water of all Cp_nLn^{n+} oxygen oxidation products contain as a more intense component the red maxima: at 604 nm for europium and 590, 650 nm for samarium, which are due to intraconfigurational *f-f* transitions in the Ln^{3+} ions (Figure 15, 1 and 2).

The second less intense component (Figure 15, 1` and 2`) of the CL spectra is in the region of 400-550 nm. The positions of these shortwave CL components correlate well with those for the phosphorescence spectra of the Cp_2Eu and Cp_2Sm solutions oxidized by oxygen and then frozen at 77 K. This means that this part of the CL is caused by the emission of triplet excited carbonyl products. The coincidence of the CL and PL spectra during the hydrolysis of the oxidation products of divalent and trivalent lanthanide compounds indicates the similar nature of the CL emitters.

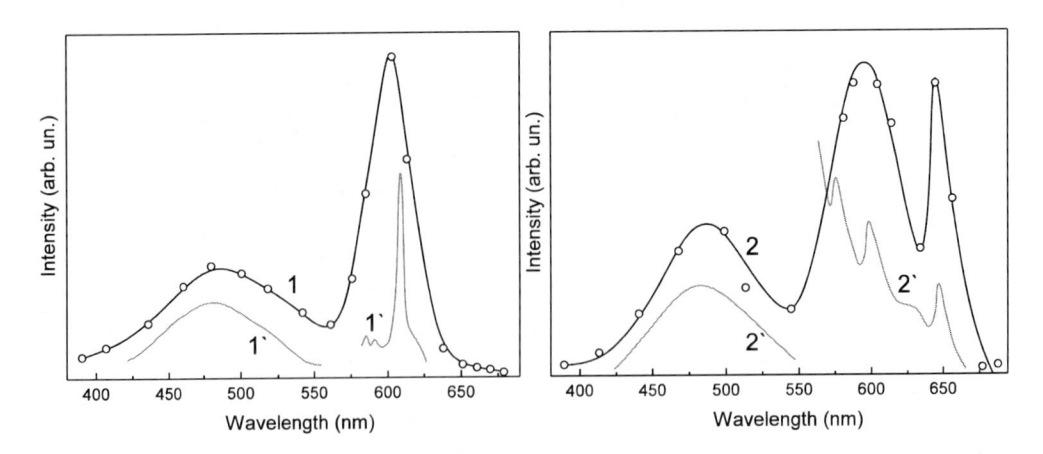

Figure 15. 1 and 2 – CL spectra in the hydrolysis (0.5 mL H_2O) of the products, obtained at the oxidation of Cp_3Eu (1) and Cp_3Sm (2) by oxygen in THF at 300 K. 1` and 2` – phosphorescence spectra of frozen (77 K) solutions, obtained in oxidation of Cp_3Eu and Cp_3Sm.

The detailed identification of the compounds which are CL emitters during hydrolysis is absent, although is undoubtedly that the excited Eu^{3+*} and Sm^{3+*} ions are the radiating centers of the long-wavelength part of CL. In the most general form, the reactions that cause CL during the hydrolysis of stable peroxides - the oxidation products of lanthanide cyclopentadienides by oxygen, are presented below (Scheme 4).

$$CpOOLn^{3+}< + H_2O \longrightarrow CpOOH + >Ln^{3+*}(OH) \qquad (28)$$

$$>Ln^{3+*}(OH) \longrightarrow >Ln^{3+}(OH) + CL \text{ (590, 604 and 650 nm)} \qquad (29)$$

$$CpOOH \xrightarrow{[Ln^{3+}]} Cp=O^* \longrightarrow Cp=O + CL \text{ (400-500 nm)} \qquad (30)$$

Scheme 4.

As noted above (in 2.4), the presence of CpOOH as the dominant compound in the hydrolysates of the oxidation products of Cp_nLn^{n+} is confirmed by the data of iodometric analysis and the presence of a band of 850 cm^{-1}. Two valence vacancies in the intended compound-emitter $>Ln^{3+}(OH)$ can be filled with moieties CpO-, O= and HO-, to give the next compounds: $(CpO)_2LnOH$, O=LnOH and $CpOLn(OH)_2$ or/and $Ln(OH)_3$. The formation of the compounds, contained the Cp-Ln bond as stable products can be completely excluded, given the extremely high reactivity of this bond in relation to O_2 and to water and other participants in the hydrolysis. As for short-wavelength CL emitters, it is most likely the excited ketones Cp=O*, the formation of which is quite possible as a result of decomposition of hydroperoxide under the influence of europium and samarium heavy metal ions. Similar generation of excited ketones occurs in many oxidative chemiluminescent systems involving hydrocarbons and metal complexes [54].

Thus, CL during the hydrolysis of the lanthanidocenes of luminescent lanthanides is observed both in the reaction with non-oxidized compounds of Cp_nLn^{n+} and with organometallic peroxides, which include the CpOO-Ln bond. In contrast, organometallic compounds of non-luminescent metals generate CL during hydrolysis only after their preliminary oxidation with oxygen.

2.6. Chemiluminescence in the Reaction of DyI_2 and NdI_2 with Water

Not only organometallic, but also inorganic compounds of divalent lanthanides, possess the ability to react with water emitting a CL. Such unusual CL was found at the hydrolysis of DyI_2 and NdI_2 in argon atmosphere [32]. Water was added to solid LnI_2 as a liquid or with an argon flow containing H_2O vapor. Because the hydrolysis reactions of these compounds are very exothermic, the water (5 mL) to the powders LnI_2 (~3 mmol) was added dropwise for 3 s at 273 K. Immediately after water addition, a rapidly decaying CL appears (Figure 16).

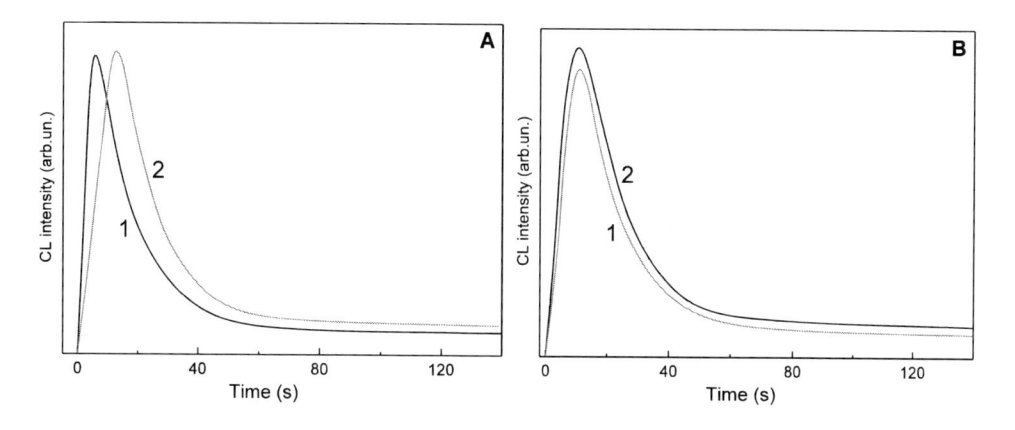

Figure 16. CL kinetics of the reactions of DyI_2 (A) and NdI_2 (B) with water (5 mL). A: 1 – the addition of liquid H_2O (2 mL), 2 – the treatment of solid DyI_2 by flow of argon with water vapor. B: with (1) and without (2) cut-off filter KS-19. $N(DyI_2) = N(NdI_2) = 0.04$ mmol.

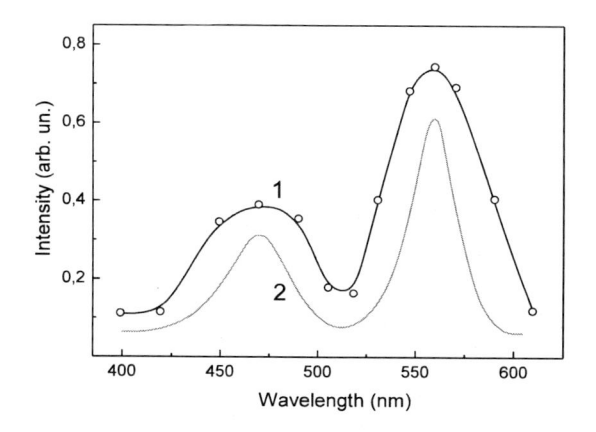

Figure 17. 1 – CL spectrum during the reaction DyI_2 (0.04 mmol) with H_2O [32]. 2 – PL spectrum of the solution of $DyCl_3 \cdot 3(BuO)_3PO$ in toluene, $\lambda_{exc} = 352$ nm [56].

The intensity of the CL was $I_{CL} = 5.4 \cdot 10^8$ and $6 \cdot 10^7$ photon$\cdot$s$^{-1}\cdot$mL^{-1} for DyI$_2$ and NdI$_2$, respectively. Since the emission intensity was low, the CL spectra were measured by the cut-off filters. The CL spectrum of hydrolysis of DyI$_2$ has maxima at 470 and 570 nm which close to the maxima of the PL spectra of solid salts DyCl$_3\cdot$6H$_2$O [55] and DyCl$_3\cdot$3(BuO)$_3$PO (Figure 17) [56]. This means that the emitter of this CL is the excited Dy^{3+*} ion.

In contrast to DyI$_2$, the CL during the hydrolysis of NdI$_2$ is at a longer wavelength region of the spectrum (700-1200 nm) characteristic [55] of the emission of the Nd^{3+*} ion. That was established by recording the CL through color filters: blue-green SZS-9, red KS-19, and violet FS-6 filters. The SZS-9 filter absorbs in the range of 400-600 nm, and CL no visible through this filter; the KS-19 and FS-6 filters quench the CL by 5 and 50%, respectively (Figure 18). Hence, the Nd^{3+*} is an emitter of CL in the hydrolysis of NdI$_2$. Unlike the products of hydrolysis reactions, the initial reactants DyI$_2$ and NdI$_2$ do not detect any PL at either 300 or 77 K. In this connection, it is appropriate to recall that quite recently stable PL of solid salts of divalent europium EuI$_2$ (432 nm), EuCl$_2$ (409 nm) and EuBr$_2$ (428 mm) was detected at room temperature [57]. As a result of the hydrolysis reaction of the LnI$_2$ salts, gaseous hydrogen is released, precipitates and solutions are formed.

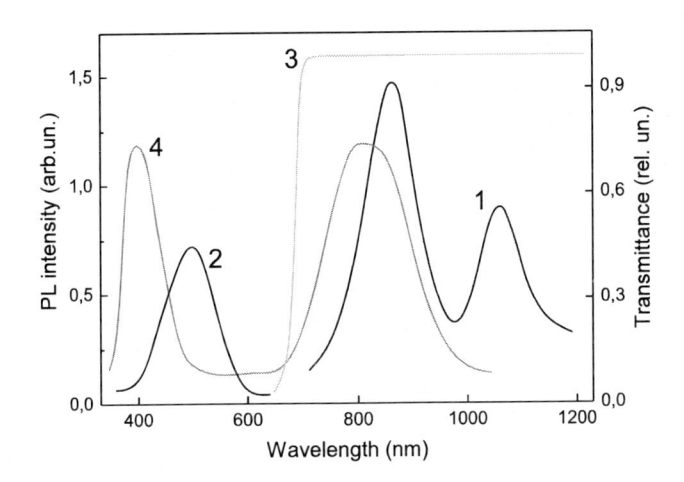

Figure 18. 1 – The PL spectrum of the NdCl$_3\cdot$3(BuO)$_3$PO complex in toluene solution ($\lambda_{exc} = 337$ nm) [56] and transmittances of the light filters SZS-9 (2), KS-19 (3) and FS-6 (4), used to determine the spectral region of the CL during the hydrolysis of NdI$_2$.

Triiodides DyI$_3$ and NdI$_3$ were found in solutions. Using complexometric titration, elemental analysis and IR-spectroscopy, it was established that the composition of the precipitates corresponds to the formula LnI(OH)$_2\cdot$H$_2$O (49-51% and 48-50% yield for DyI$_2$ and NdI$_2$, respectively). To these data on the composition of the products corresponds to the hydrolysis reaction (1).

$$2LnI_2 + 2H_2O \longrightarrow LnI_3 + LnI(OH)_2 + H_2 \tag{31}$$

When treating solid NdI_2 with a solution of water in THF (1.17 M), there is also a vigorous evolution of hydrogen (for 3 s) and staining the solution in a dark violet color characteristic of the Nd^{3+} ion. However, in this case, no precipitate is formed. The composition of the solid compound isolated from this solution corresponds to the formula $NdI_2(OH)\cdot(THF)_2\cdot 3H_2O$ (yield 88.62%), to which corresponds reaction (31).

$$NdI_2 + H_2O \xrightarrow{\text{THF}} NdI_2(OH)\cdot(THF)_2 + {}^1/_2H_2 \tag{32}$$

The mechanism of CL excitation during the hydrolysis of LnI_2 is reflected in Scheme 5.

$$Ln^{2+} + HOH \longrightarrow [Ln^{2+}\cdots HOH] \longrightarrow Ln^{3+*} + OH^- + H^\bullet \tag{33}$$
$$Ln^{3+*} \longrightarrow Ln^{3+} + h\nu \tag{34}$$

$$Ln = Dy \text{ and } Nd$$

Scheme 5.

When water comes into contact with solid LnI_2, water molecules enter the first coordination sphere of the lanthanide ion. This is followed by electron transfer from the Ln^{2+} ion to a water molecule with the oxidation of Ln^{2+} to Ln^{3+} and the formation of atomic hydrogen (30). The lanthanide ion turns into excited Ln^{3+*} ion, which is deactivated with the radiation of CL (31). As follows from the position of the short-wave maxima of the PL spectra (Figures 18), the excitation of Ln^{3+} ions require an energy of 1.44 eV (for Nd^{3+*}) and 2.68 eV (for Dy^{3+*}). The free Gibbs energy (ΔG^0) of reaction 30 (Scheme 5) was calculated by Weller`s equations (32) and (33). The calculated values of $\Delta G^0 = -4.87$ and -3.69 eV are more than enough for the excitation of Dy^{3+} (2.68 eV) and Nd^{3+} (1.44 eV) ions.

$$\Delta G^0 = E^0(Dy^{3+}/Dy^{2+}) - E^0(H_2O/H^\bullet,OH^-) - ke^2/(\varepsilon r) - E^*(Dy^{3+}) ==$$
$$-2.56 - (-0.41) - 0.04 - 2.68 = -4.87 \; eV \tag{35}$$

$$\Delta G^0 = E^0(Nd^{3+}/Nd^{2+}) - E^0(H_2O/H^\bullet,OH^-) - ke^2/(\varepsilon r) - E^*(Nd^{3+}) ==$$
$$-2.62 - (-0.41) - 0.04 - 1.44 = -3.69 \; eV \tag{36}$$

where $E^0(Ln^{3+}/Ln^{2+})$ and $E^0(H_2O/H^\bullet,OH^-)$ are the redox potentials; $E^*(Ln^{3+})$ is the energy of the excited level of the Ln^{3+*} ion; k is the Boltzmann constant; e is the electron charge; ε is the dielectric constant of the solvent (for water 78.5); r is the distance between the Ln^{2+} ion and H_2O (0.5 nm).

Thus, in the work [32], using the example of diiodides NdI_2 and DyI_2 for the first time, it was shown that the reaction of the divalent Dy^{2+} and Nd^{2+} ions with water can generate the emitting excited states of their trivalent ions.

2.7. Chemiluminescence in the Reaction of $NaLn(BH_4)_4 \cdot nDME$ with Oxygen ($Ln = Eu^{2+}$, Sm^{3+}, Pr^{3+} and Tb^{3+})

For the first time, CL and PL of lanthanide hydrides were detected on the example of the complexes $NaLn(BH_4)_4 \cdot nDME$, where $Ln = Eu^{2+}$, Sm^{3+}, Pr^{3+} (n = 4), and Tb^{3+} (n = 3); DME – dimethoxyethane [16]. To shorten, the $NaLn(BH_4)_4 \cdot nDME$ complexes for different lanthanides are designated as [Eu-comp], [Sm-comp], [Pr-comp] and [Tb-comp].

It is interesting to note that prior to studying the luminescent properties of these hydrides in [16], based on the data on the synthesis of $NaLn(BH_4)_4 \cdot nDME$ complexes, it was considered in [58] that europium, like other lanthanides, is in a trivalent state. However, as a result of the registration of bright blue PL of the [Eu-comp], it was found that europium is in this hydride as a divalent ion Eu^{2+}. This assignment found a clear additional confirmation in the form of the appearance of the red PL of the Eu^{3+} ion after the oxidation of [Eu-comp] with oxygen in THF solution (Figure 19).

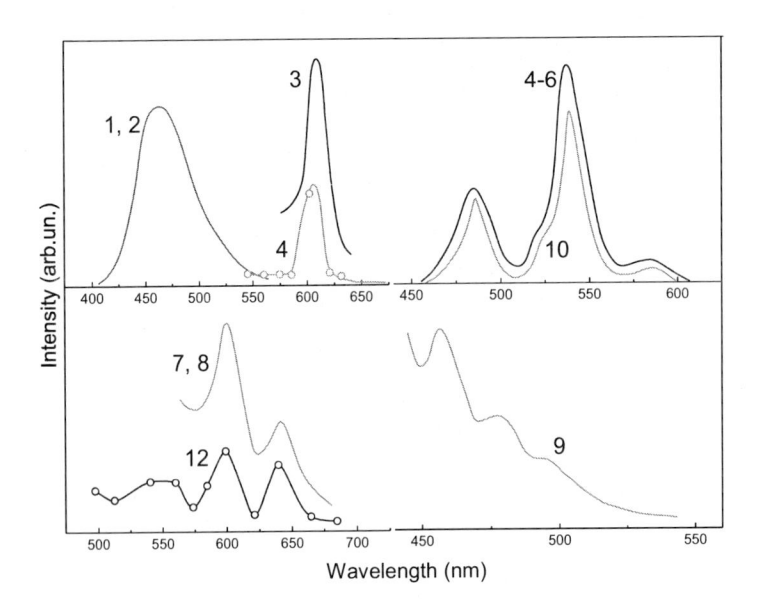

Figure 19. The PL (1-9) and CL (10-12) spectra. 1, 2 and 3 – the PL spectra of [Eu-comp] as crystal, $5 \cdot 10^{-3}$ M solution and after oxidation by O_2; 4, 5 and 6 – the PL spectra of [Tb-comp] as crystal, $5 \cdot 10^{-3}$ M solution and after oxidation by O_2; 7 and 8 – the PL spectra of [Sm-comp] as crystal and solution and after oxidation by O_2; 9 – PL spectrum of [Pr-comp] solution after oxidation by O_2. 10, 11 and 12 – the CL spectra at O_2 bubbling through THF solutions of [Tb-comp], [Eu-comp] and [Sm-comp]. Spectra 1-10 measured using MZD-2 monochromator, 11 and 12 – cut-off-filters. $T = 77$ K (4-9), 300 K (1-3, 10-12).

Apparently, the formation of Eu^{2+} occurs as a result of the reduction of Eu^{3+} during the synthesis of [Eu-comp] by the reaction of $EuCl_3$ with $NaBH_4$, according to the method developed in [58]. To such reduction contributes the greater value of reduction potential of Eu^{3+} compared to other Ln^{3+}, including Sm^{3+}, which are formed in a similar synthesis in the trivalent state, as can be seen from their PL spectra containing only bands characteristic of Ln^{3+} luminescence (Figure 19).

The brightest CL and PL before and after oxidation are observed for the [Tb-comp], which is characteristic of the brightly luminescent Tb^{3+} ion. Thanks to this, the PL of terbium hydride is recorded at room temperature. The initial solution of [Sm-comp] and [Pr-comp] do not have PL at 300 and 77 K. However, PL was detected after oxidation of these complexes by oxygen, and this PL belonged to the emission of the $Sm^{3+}*$ and $Pr^{3+}*$ ions (Figure 19). PL of the [Eu-comp] exhibits as a bright blue emission with maximum at 465 nm at 300 and 77 K. After oxidation of [Eu-comp] by oxygen the PL at 465 nm run disappears and the red PL characteristic for Eu^{3+} ion does appear. CL in oxygen oxidation take place for all complexes except of [Pr-comp].

2.8. Chemiluminescence in the Reactions of Cp₂SmCl and EuSO₄ with XeF₂

For the first time CL of lanthanide compounds with XeF_2 was detected by the example of the oxidation of Cp_2SmCl in solution [59, 60]. This CL occurs in interaction of Cp_2SmCl ($2.6 \cdot 10^{-3}$ M) and XeF_2 ($5 \cdot 10^{-3}$ M) in CH_2Cl_2 at 300 K in argon atmosphere in the form of a glow that fades in about 10 minutes. However, no further studies of this CL have been conducted.

The next CL of lanthanide compounds with XeF_2 was discovered only after 32 years already by the example of a divalent europium compound, namely, $EuSO_4$.

CL in oxidation of $EuSO_4$ by XeF_2 found in [33] is the first example of solid-phase reaction of divalent lanthanides accompanied by a light emission. This CL in the form of a wide kinetic maximum appears by mixing the solid samples of $EuSO_4$ and XeF_2; the CL is completely highlighted within one hour (Figure 20, 1). An increase in the number of loadable reagents causes an increase in the CL intensity, the formation time of kinetic maximum and the decay time of the glow (Figure 20, 2).

As a result of comparing the analysis of IR absorption spectra and PL spectra of the initial reactants and reaction product, as well as of the $Eu_2(SO_4)_3$ and EuF_3 reference salts, the $FEuSO_4$ compound was identified as the main solid reaction product. In addition, the second stable product – Xe gas – has been identified by gas chromatography-mass spectrometry. The PL spectra of the $FEuSO_4$ and the parent $EuSO_4$ differ drastically from each other.

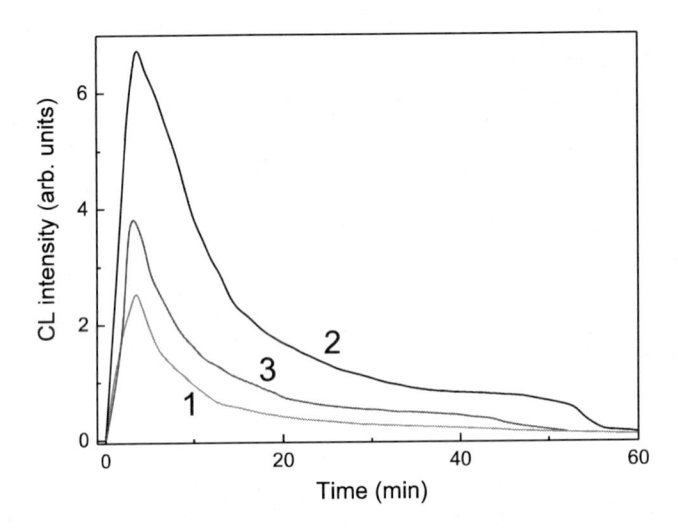

Figure 20. The CL kinetics in reaction powered samples of $EuSO_4$ (1, 2) and $Eu_2(SO_4)_3$ (3) with XeF_2 at 293 K. $[EuSO_4] = [XeF_2] = 5 \cdot 10^{-5}$ mol (1), $1 \cdot 10^{-4}$ mol (2); $[Eu_2(SO_4)_3] = 5 \cdot 10^{-5}$ mol (3), $[XeF_2] = 1 \cdot 10^{-4}$ mol (3).

The PL spectrum of $EuSO_4$ (Figure 21) consist of a broadband maximum at 378 nm caused by the allowed $4f^6 5d^1 \rightarrow 4f^7$ transition characteristic of the Eu^{2+*} ions.

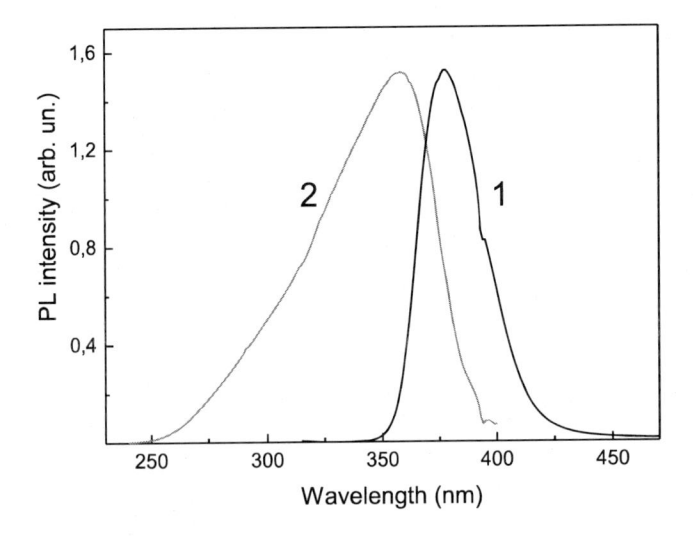

Figure 21. PL (1) and excitation (2) spectra of $EuSO_4$ were recorded at 300 K using the Fluorolog-3 spectrofluorimeter. $\lambda_{exc} = 300$ nm (1), $\lambda_{em} = 420$ nm (2).

The lifetime of the Eu^{2+*} ions (400 ns) in the parent $EuSO_4$ is in the nanoseconds time scale characteristic of the divalent Eu^{2+*} ions, while for $FEuSO_4$ that is equal to 200 μs which is typical for the trivalent europium ions Eu^{3+*}. Surprisingly, these luminescent characteristics of crystalline $EuSO_4$ were first obtained only recently in work [33]. The PL spectrum of $FEuSO_4$ (Figure 22, a) contains four narrow, well-structural maxima at 590, 612, 649 and 697 nm corresponding to the following transitions from the excited Eu^{3+*}

ion: $^5D_0$–$^7F_1$, $^5D_0$–$^7F_2$, $^5D_0$–$^7F_3$, and $^5D_0$–$^7F_4$, consequently. The PL excitation spectrum of FEuSO$_4$ also contains well-structured maxima which closed to the absorption maxima of the Eu^{3+} ion (Figure 22, a).

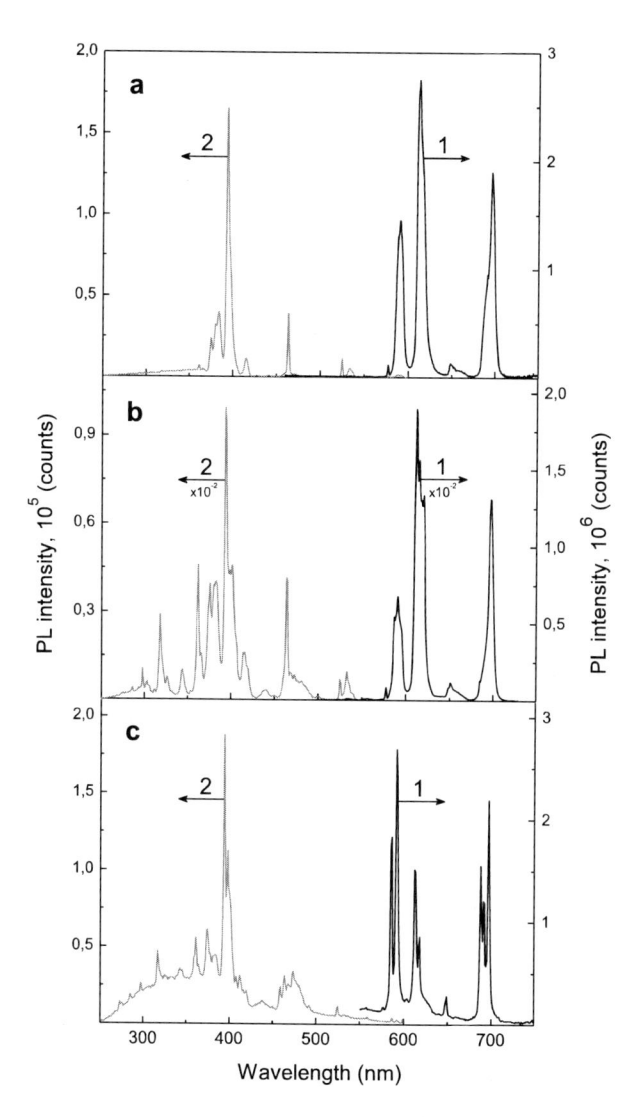

Figure 22. PL (1) and excitation (2) spectra of solid samples: a – FEuSO$_4$, obtained after 60 min at 293 K in reaction EuSO$_4$ + XeF$_2$ ([reactants] = 5·10^{-5} M); b – Eu$_2$(SO$_4$)$_3$ standard; c – EuF$_3$ standard. All spectra were recorded at 300 K using the Fluorolog-3. λ_{exc} = 393 nm (1), λ_{em} = 612 nm (2).

The PL spectra of FEuSO$_4$ and the Eu$_2$(SO$_4$)$_3$ reference are very close (Figure 22, b). Indeed, the location, shape and relative intensity of the most intense peaks of Eu^{3+*} at 590, 612 and 690 nm in the compared spectra are practically the same. This is because the ligand environment of the Eu^{3+} ion in the parent EuSO$_4$ and in the FEuSO$_4$ product is very similar i.e., a sulphate group linked to Eu ion in the reaction product is saved. This is confirmed also by the fact that the most intense emission peaks in the PL spectrum of FEuSO$_4$ are in the group of about 612 nm (Figure 22, a), and not at 590 nm, as is the case with the PL spectra (Figure 22, c) of the standard compound EuF_3. The peak intensity at about 612 nm is higher than that at about 590 nm in a PL spectrum of $FEuSO_4$, because the Eu^{3+} ion in $FEuSO_4$ has only one bond with F, whereas with the SO_4 group there are two bonds Eu–OS. As a result, the excited Eu^{3+*} ion in the $FEuSO_4$ is more influenced by the electron clouds of these two bonds, which manifest themselves in very close of the ratio of peak intensities at 590 and 612 nm in the spectrum of $FEuSO_4$ (Figure 22, a) and $Eu_2(SO_4)_3$ (Figure 22, b), and their large differences in the spectrum of EuF_3 (Figure 22, c).

The above PL data are in full agreement with the results of IR analysis of the participants of reaction under study. Indeed, the IR spectrum of the solid residue contains two strong bands at 609.53 and 639.43 cm^{-1}, in the region 598-670 cm^{-1}, which coincides with the 598-669 cm^{-1} absorption region of europium hydrosulfate $Eu_2(SO_4)_3\cdot4H_2O$. In addition, the S–O vibration regions (1050-1250 cm^{-1}) in the IR spectra of the solid product of the reaction $EuSO_4 + XeF_2$ and reference $Eu_2(SO_4)_3\cdot4H_2O$ (1050-1160 cm^{-1}) are very close. The presence of F^- anion in the reaction product confirms the presence of absorption bands in the region 400-500 cm^{-1}, which correlates with the bands at 410, 435 and 480 cm^{-1}, responsible for the Eu–F vibrations in the IR spectrum of reference EuF_3. In addition, the IR spectra of the reaction product and model sample EuF_3 (given in parentheses) comprise the following close bands: 389.69 (380.96), 398.28 (399.28), 419.54 (415.68), 432.07 (425.32), 457.15 (450.40), 471.62 (467.66), 479.33 (476.44), 497.66 (483.19) cm^{-1}. Thus, based on the analysis of the PL and IR spectra, it was concluded that the excited Eu^{3+*} ions emit from $FEuSO_4$ and responsible for the red PL of this product.

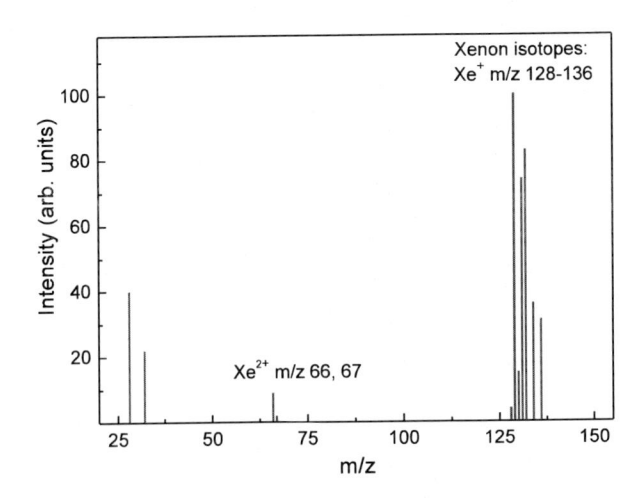

Figure 23. Mass spectrum of the gas formed in reaction $EuSO_4 + XeF_2$ during 5 min at 295 K, [reactants] = 0.1 mmol.

The formation of xenon – the second stable reaction product – was found on the signals of its isotopes in the mass spectrum (Figure 23) of the gas-probe taken from the gas phase of the reactor after the interaction of equimolar amounts of powder samples $EuSO_4$ and XeF_2.

The kinetics of Eu^{2+} reduction to Eu^{3+} was traced not only by CL, but also by an increase in the PL intensity of the solid reaction mixture at 612 nm (Figure 24).

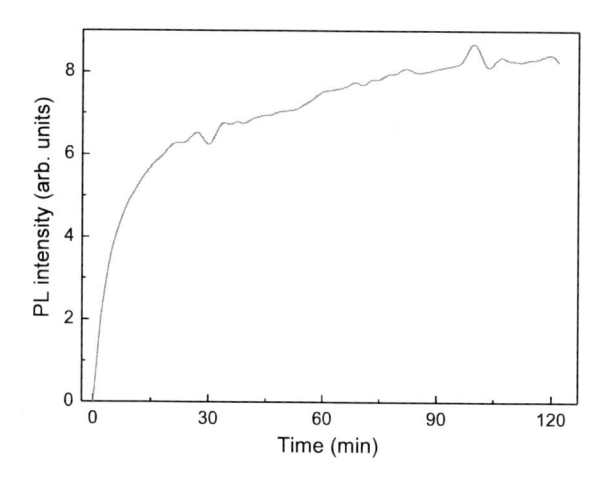

Figure 24. The change in PL intensity at 612 nm during the reaction $EuSO_4 + XeF_2$ at 293 K. $\lambda_{exc} = 395$ nm, $[EuSO_4] = [XeF_2] = 2.5 \cdot 10^{-4}$ mol. PL was recorded from a glass ampoule using a spectrofluorimeter Solar CM-2203.

Most part of the Eu^{2+} ions are oxidized relatively quickly (~20 min), while the oxidation of the remaining part takes about 2 hours due to poor contact of the reactants separated by a layer of the resulting product.

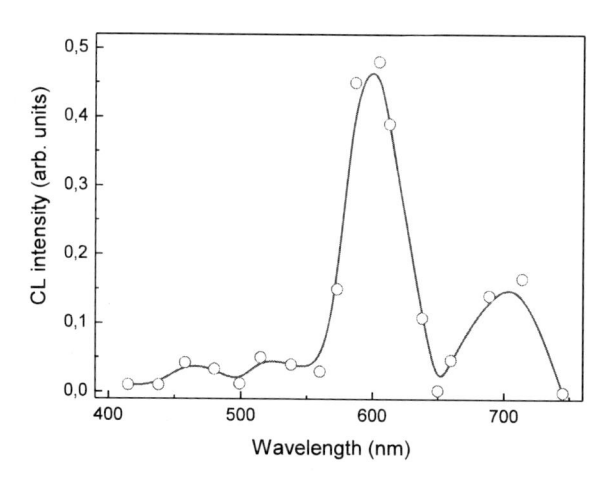

Figure 25. CL spectrum of the reaction $EuSO_4 + XeF_2$. $[reactants]_0 = 10^{-4}$ mmol.

The most intense peaks at ~608 and ~705 nm in the CL spectrum (Figure 25) are very close to the stronger peaks at 590, 612 and 697 nm in the PL spectrum of the reaction product (Figure 22). The broad and unresolved maximum in the CL spectrum at about 608 nm covers the range of PL peaks at 590, 612 nm. It means that the CL maxima at ~608 and ~705 nm are due to the emission of electronically excited state of the Eu^{3+*} ions. The red emission accounts for about 90% of the total CL (Figure 25), i.e., the Eu^{3+*} ion is a dominant emitter of the CL reaction of $EuSO_4$ and XeF_2. Two much weaker bands at around 475 and 530 nm are assigned to the excited reaction products not containing europium. Presumably they are attributed to the emission of an excited xenon atom Xe^* (475 nm) and $XeF^{\cdot *}$ radical (530 nm).

There are two possible types of excitation for the dominant CL emitter of the Eu^{3+*} ion in reaction of $EuSO_4$ with XeF_2. First, in the elementary act of electron transfer from Eu^{2+} to XeF_2 (direct CL) and second, *via* of energy transfer from primary emitters to Eu^{3+} ion. In favor of indirect CL indicates registration CL at the interaction of salt trivalent europium $Eu_2(SO_4)_3$ with XeF_2. The intensity and light sum of this CL is lower than that for $EuSO_4$ (Figure 20, 3). However, the emitter of this CL is also the Eu^{3+*} ion and the shape of the kinetic curves of CL for salt of two- and trivalent europium are the same (Figure 20, 3). These two circumstances are a clear sign that the indirect CL is quite possible. As an energy donor for energy transfer to Eu^{3+} ions can be serving the Xe^* atoms and $XeF^{\cdot *}$ radicals which have the higher energy excited states (475 and 530 nm or 2.61 and 2.34 eV, respectively) than that of Eu^{3+*} (590 nm, 2.10 eV). Moreover, it takes place the overlapping of absorption spectra of the energy acceptor (Eu^{3+}) and luminescence spectra of potential donors (Xe^* and $XeF^{\cdot *}$). In the region of the emission maxima of energy donors at 475 and 530 nm, there are weak maxima in the excitation spectrum of $FEuSO_4$ (Figure 22).

A possible mechanism of the CL arising in the reaction $EuSO_4 + XeF_2$ shown on the Scheme 6.

$$EuSO_4 + XeF_2 \xrightarrow{\ e^-\ } [Eu^{2+}SO_4\cdots F\text{-}XeF] \longrightarrow FEu^{3+*}SO_4 + XeF^{\cdot}\text{(possible and/or XeF}^{\cdot}) \qquad (37)$$

$$EuSO_4 + XeF^{\cdot} \xrightarrow{\ e^-\ } [Eu^{2+}SO_4\cdots F\text{-}Xe^{\cdot}] \longrightarrow FEu^{3+*}SO_4 + Xe \text{ (or } Xe^{\cdot}) \qquad (38)$$

$$Xe^{\cdot} \text{ (and possible } XeF^{\cdot\cdot}) + FEu^{3+}SO_4 \longrightarrow Xe \text{ (and possible } XeF^{\cdot}) + FEu^{3+*}SO_4 \qquad (39)$$

$$FEu^{3+*}SO_4 \longrightarrow FEu^{3+}SO_4 + CL \text{ (590, 612 nm)} \qquad (40)$$

$$Xe^{\cdot} \longrightarrow Xe + CL \text{ (475 nm)} \qquad (41)$$

$$XeF^{\cdot\cdot} \longrightarrow XeF^{\cdot} + CL \text{ (530 nm)} \qquad (42)$$

Scheme 6.

In the first step (37), is formed an intermediate reaction complex, in which electron transfer from the Eu^{2+} ion to XeF_2 molecule occurs with formation the stable $FEuSO_4$ molecule and the $XeF^{\cdot}$ radical. Then $XeF^{\cdot}$ radicals as an effective fluorine atom source and stronger oxidant than XeF_2 react with another molecules $EuSO_4$ like to reaction (37) with formation $FEuSO_4$ and the Xe atoms. The electron transfer with the formation of a new Eu^{3+}–F bond occurs much faster than the change in the position of the reacted atoms. Because of this, the energy released in this act can be localized in the form of electron excitation on both Eu^{3+} ion and $XeF^{\cdot}$ radical (for 37) or Xe atom (for 38) which did not leave to escape from the reaction area. This is especially possible for solid phase, which hinders the diffusion and exit of the reactants from the reaction zone. Since the $Eu^{3+}*$ ion is the dominant emitter it may be proposed that the probability of localization of energy on the Eu^{3+} ion is much higher. However, the dominance of europium as an emitter of CL, apparently, is also due to the large yield of its luminescence then that for Xe and XeF. A part of the elementary acts 34 and 35 can give products in an electronically excited state (which are indicated on the Scheme 6 in parentheses). It should be remembered that in one and the same acts (34) and (35) one or the other emitter is formed, and not two at the same time. The CL with formation of the radiating $Eu^{3+}*$ ions as a result of electron transfer from Eu^{2+} ion was previously observed (by M. Elbanowski et al. in 1983) only in the reaction $Eu^{2+} + HO^{\bullet} \rightarrow Eu^{3+}* + HO^{-}$ [3]. Based on the above results of experiments on CL in the interaction of XeF_2 with $Eu_2(SO_4)_3$, the $Eu^{3+}*$ emitter can be generated not only on the stages (34) and (35), but *via* the energy transfer from Xe* atoms (and possibly from $XeF^{\bullet}*$ radicals) followed by emission from the $Eu^{3+}*$ ion. So, in the reaction $EuSO_4 + XeF_2$, both direct and indirect CL are generated, with the formation in both cases of the electronically excited $Eu^{3+}*$ ions as the dominant emitter. The value of the contribution for the total glow of direct and indirect CL changes during the reaction. At the initial stages of the reaction (especially at the first contacts of the reagents), when the content of Eu^{2+} ions is substantially greater than that of Eu^{3+} ions, the contribution of the direct CL is also significantly higher. But with the further progress of the reaction (especially close to its completion), the content of Eu^{3+} ions dominate, and the contribution of indirect CL is the main one. The energy necessary for the generation of CL emitters in the $EuSO_4 + XeF_2$ reaction estimated from the positions of theirs emission maxima in the CL spectrum (Figure 25), are characterized by the following values: $Eu^{2+}*$ (612 nm, 2.02 eV, 46.50 kcal·mol^{-1}), Xe* (475 nm, 2.61 eV, 60.21 kcal·mol^{-1}), $XeF^{\bullet}*$ (530 nm, 2.34 eV, 53.98 kcal·mol^{-1}). In order to confirm the correctness of the identification of the stages of the CL emitters excitation (Scheme 6), a rough estimate of the enthalpies of reactions (34) and (35) has been carried out from the balance of the forming (Eu–F) and breaking (Xe–F) bonds:

$\Delta H_0(43) = E(\text{Xe–F in XeF}_2) - E(\text{Eu–F in FEuSO}_4) = 32.00\ [61] - 123.80\ [62] = -91.8$ kcal·mol^{-1};

$\Delta H_0(44) = E(\text{Xe–F in XeF}^{\bullet}) - E(\text{Eu–F in FEuSO}_4) = 11.00\ [63] - 123.80\ [62] = -112.8$ kcal·mol^{-1}.

As shown, the reaction (44) is even more exothermic than (43). Thus, the enthalpies of reactions (43) and (44) are more than enough for the generation of the $Eu^{3+}*$ ions, as well as for Xe* atoms and $XeF^{\bullet}*$ radicals.

3. CHEMILUMINESCENCE IN WHICH DIVALENT LANTHANIDE IONS ARE EMITTERS OF EMISSION

3.1. The Blue and Green Chemiluminescence of Eu2+* and Red Chemiluminescence of Sm^{2+}* Ions at the Autoxidation of Diisobutylaluminum Hydride

So far, we have considered chemiluminescent reactions in which divalent lanthanide ions were present in the starting compounds. The focus of this section, as its name implies, are reactions in which the Ln^{2+} ions are CL emitters. Surprisingly, this type of CL was discovered just over 5 years ago (*vs* 39 years for Ln^{3+}-emitters), first using the example of the CL of the Eu^{2+} [28] ion, and then Sm^{2+} [29] ion. Other Ln^{2+} ions as CL emitters have not yet been detected, although there are no fundamental limitations on the possibility of manifestation of such an attractive and important property of Ln^{2+} ions.

The CLs of Eu^{2+} and Sm^{2+} ions were generated in a rather complex system $(LnCl_3 \cdot 6H_2O\text{-}THF\text{-}O_2)\text{-}Bu^i_2AlH$, in which several processes take place almost simultaneously: gas formation, reduction and complexation of Ln-ions, as well as light emission. Therefore, it is rational to first consider the dark processes occurring in the system $(LnCl_3 \cdot 6H_2O\text{-}THF\text{-}O_2)\text{-}Bu^i_2AlH$, which are reflected in Scheme 7.

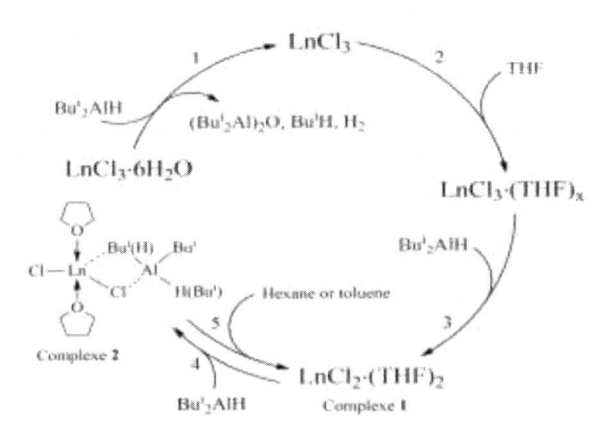

Scheme 7.

This scheme is implemented when loading into the CL-reactor (in an argon atmosphere) the crystalline hydrate $LnCl_3 \cdot 6H_2O$ and THF; O_2 is present in the reaction system in the form of difficult to remove impurities. The active process begins immediately after the subsequent addition of Bu^i_2AlH to the heterogeneous mixture $LnCl_3 \cdot 6H_2O\text{-}THF$. The Bu^i_2AlH attacks $LnCl_3 \cdot 6H_2O$ with removal of crystallization water in the form of alumoxane $(Bu^i_2Al)_2O$, as well as gaseous hydrogen and isobutane (1). The resulting anhydrous $LnCl_3$ immediately forms a soluble complex with THF, in the form of which

Eu^{3+} goes into a THF solution (2). This $LnCl_3 \cdot (THF)_x$ complex undergoes reduction by Bu^i_2AlH (3) to form the complex $LnCl_2 \cdot (THF)_2$ (I), in which europium (or samarium) is in the form of a divalent ion Eu^{2+} (or Sm^{2+}). Complex (I) forms with an excess of Bu^i_2AlH a bulkier complex (II) (4). Complex (II) in a simplified form is depicted as $LnCl_2 \cdot (THF)_2 - Bu^i_2AlH$. The complexes $LnCl_2 \cdot (THF)_2$ as individual compounds were isolated by addition of hexane to complex (II) solutions and were completely characterized (for Ln = Eu, Sm) by methods of complexometry, elemental analysis and IR spectroscopy [64]. For the complete removal of the crystallization water from the $LnCl_3 \cdot 6H_2O$ is required about 18 s. For example, in the case of europium, it is as evidenced from the gas volume $V(Bu^iH, H_2)_{exp} = 7.6$ mL, evolved in the reaction, which is 98% of the theoretical $V(Bu^iH, H_2)_{theor} = 7.75$ mL (Figure 26, 1). The reduction of Eu^{3+} to Eu^{2+} is almost finished only in ~400 s (6.6 min), that was established by controlling an increase in the intensity of the Eu^{3+} ion PL peak at 465 nm (Figure 26, 2). Of course, the reduction and dehydration processes of the Eu^{3+} ion are not perfectly consecutive as indicated by the simultaneous increase in the gaseous products volume and the Eu^{2+} PL intensity starting with the Bu^i_2AlH addition (Figure 26, 1 and 2).

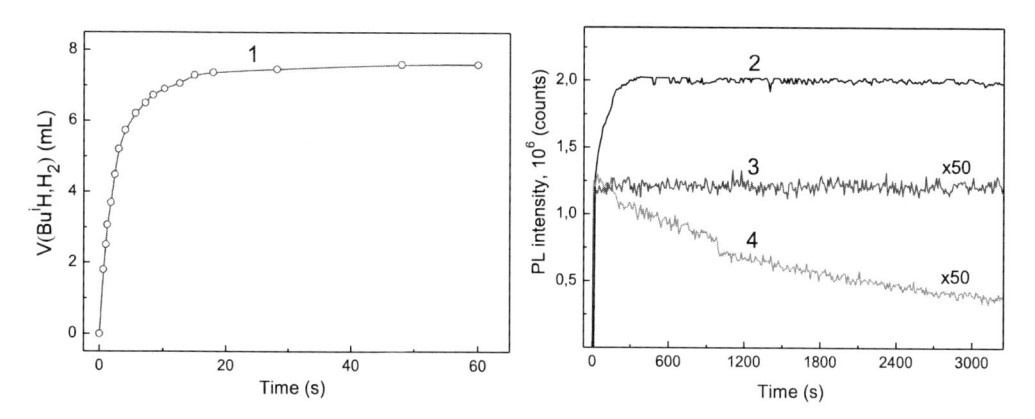

Figure 26. The kinetics of gas release (1) and reduction $Eu^{3+} \rightarrow Eu^{2+}$ (2-4). 1, 2 – at interaction of the slurry $EuCl_3 \cdot 6H_2O$ (0.02 mmol) with Bu^i_2AlH (0.8 mmol) in THF (2 mL); 3 – at addition of Bu^i_2AlH (0.8 mmol) in the solution $EuCl_3 \cdot 6H_2O$ (0.02 mmol) in MeOH (0.02 mL) in the presence of THF (2 mL); 4 – at interaction of the slurry $Eu(NO_3)_3 \cdot 6H_2O$ (0.02 mmol) with Bu^i_2AlH (0.8 mmol) in THF (2 mL). $\lambda_{max} = 447$ (1-3), 454 (4) nm; $\lambda_{exc} = 410$ (1-3), 400 (4) nm.

The addition of Bu^i_2AlH solution to the suspension of $EuCl_3 \cdot 6H_2O$ in THF causes a bright, easily visible to the unaided eye, blue CL (CL-Eu), which everyone can see with their own eyes in a video recording and in the real-color photograph in the section "Supplementary materials" in [28]. Although the reaction of solid $LnCl_3 \cdot 6H_2O$ (Ln = Eu, Sm) with Bu^i_2AlH in THF occurs in Ar atmosphere, the residual amount of oxygen in the reaction system is enough to generate the CL-Eu and CL-Sm. As can be seen from the kinetic curves shown in Figure 27, the CL intensities for both lanthanides are much higher than the intensity of CL during oxidation of Bu^i_2AlH with oxygen in the absence of any additives (hereinafter, background CL or BCL). For example, the intensity of BCL is weaker by more than six orders of magnitude ($5.3 \cdot 10^6$) than that of CL-Eu. The emitter of BCL is triplet excited isobutyric aldehyde $^3Me_2CHC(H)=O^*$, which is generated by disproportionation of peroxyl radicals formed in the oxidation of Bu^i_2AlH or aluminum alkyls R_3Al by oxygen [65]. The maxima of the CL-Eu and CL-Sm kinetic curves (Figure 27) are formed very quickly within 5-8 s and then the CL intensity decreases slowly because of the oxygen deficit in the reaction solution. The duration of this attenuation depends on the tightness of the CL cell, i.e., from the access of atmospheric oxygen to the reaction solution. In an argon atmosphere containing traces of oxygen, the CL-Eu is attenuated for more than a day. The CL-Eu is very sensitive not only to low concentrations of oxygen, but also europium. So, even repeatedly diluted europium solutions ($[Eu^{3+}]_0 = 10^{-11}$ M) exhibit CL-Eu (Figure 27, inset).

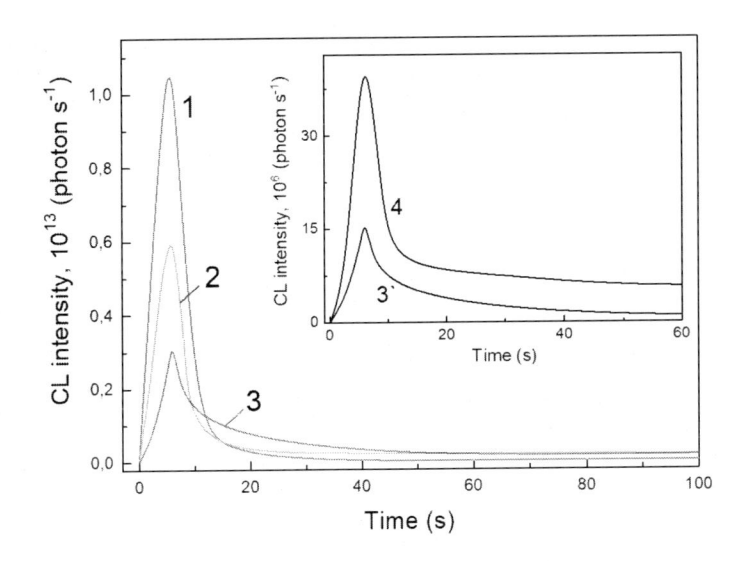

Figure 27. Kinetics of CL-Eu (1, 4), CL-Sm (2) and BCL (3, 3`) in THF, saturated by air. Intensity of curves 2 and 3 scaled up to 10^6 and $2 \cdot 10^6$ times, respectively. $[LnCl_3 \cdot 6H_2O]_0 = 10^{-2}$ M (1, 2) and 10^{-11} M (4); $[Bu^i_2AlH] = 10^{-1}$ M (1-4). V(THF) = 2 mL, $T = 298$ K.

The fast formation of the CL maxima is caused by the very fast free-radical oxidation of Bu^i_2AlH, which cannot be effectively inhibited even by such a strong inhibitor as galvinoxyl [65]. The high brightness of CL-Eu allowed to register its spectrum (Figure 28) using a spectrofluorimeter, while in the case of the weaker CL-Sm only the cut-off filters can be applied.

The broadband CL-Eu spectrum with $\lambda_{max} = 465$ nm (Figure 28) [28] is in a good agreement with the PL spectra of Eu^{2+} at 77 K aqueous solution of $EuCl_2$ [15]. Also, the CL-Eu spectrum is close to the PL spectra of Eu^{2+} complexes with crown ethers (433-448 nm) and cryptands (445-460 nm) [66, 67]. In addition, the spectrum of CL-Eu coincides with the PL spectrum of the reaction solution (Figure 28). Thus, the sole emitter of the CL-Eu is an electronically excited state of Eu^{2+*} ion, radiating through the $4f^6 5d^1 \rightarrow 4f^7$ transition. Surprisingly, with simultaneous measuring of the spectra of the CL-Eu and PL of the reaction solution in one reactor placed in the spectrofluorimeter, the contribution of the PL to the total luminescence did not exceed 10%.

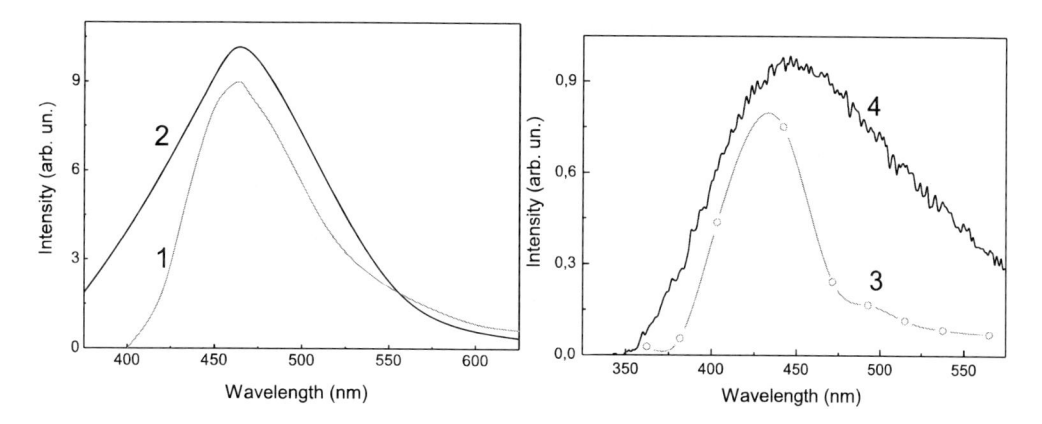

Figure 28. CL (1, 3) and PL (2) spectra, measured during the oxygen oxidation of Bu^i_2AlH (10^{-1} M) with (1, 2) and without (3) of the $EuCl_3 \cdot 6H_2O$ salt. $[Eu^{2+}]_0 = 10^{-2}$ M, $[Bu^i_2AlH]_0 = 10^{-1}$ M, V(THF) = 2.0 mL. 4 – the phosphorescence spectrum of $^3Me_2CHC(H)=O^*$ ($5 \cdot 10^{-4}$ M) in THF at 77 K. $\lambda_{exc} = 280$ (2), 310 (4) nm.

Thus, a very rare case is realized here, when the chemical generation of Eu^{2+*} ions is more effective than the photoexcitation. The spectrum of the CL-Sm contains two maxima, low-intensity at 643 nm and intense at 780 nm (Figure 29, 1 and 2). The weak maximum at 643 nm belongs to the emission of trivalent ion Sm^{3+}, which remains in very little amounts in solution due to the slow (unlike Eu^{3+} [64]) and incomplete reduction of $SmCl_3$. The maximum at 640 nm is very typical for the PL of the Sm^{3+} complexes [55, 68], and it differs from the most intense peak at 595 nm in the PL spectrum of the solid $SmCl_3 \cdot 6H_2O$ (Figure 30, 3). The maximum at 643 nm is present only in the CL-Sm spectrum and it is absent in the PL spectrum of the reaction solution, which is associated with the greater sensitivity of the CL method. The broad maximum at 780 nm in the CL-Sm spectrum was assigned to emission of Sm^{2+*} ions ($4f^55d^1 \rightarrow 4f^6$ transition) and it close to the broad maximum at 769 nm in PL spectrum of the SmI_2 solution in THF [48].

The forms of kinetic curves of CL-Eu, CL-Sm and BCL are the same and the intensities of CL-Eu and CL-Sm are increasing proportionally with an increase in the Eu^{2+} and Sm^{2+} concentrations. These experimental facts and the higher energy of the emitting level of aldehyde $^3Me_2CHC(H)=O^*$ ($\lambda_{max} = 430$ nm, 2.88 eV) than that of Eu^{2+*} ($\lambda_{max} = 465$ nm, 2.66 eV), and Sm^{2+*} ($\lambda_{max} = 780$ nm, 1.59 eV) ions indicate that the emitters of CL-Eu and CL-Sm are generated as a result of energy transfer from the primary emitter of $^3Me_2CHC(H)=O^*$ to Eu^{2+} and Sm^{2+} ions. This energy transfer is facilitated by the overlap integral between the luminescence spectrum of the donor $^3Me_2CHC(H)=O^*$ and absorption spectra of the acceptors Eu^{2+} and Sm^{2+} ions. Thus, the Eu^{2+} and Sm^{2+} ions are the enhancers of BCL.

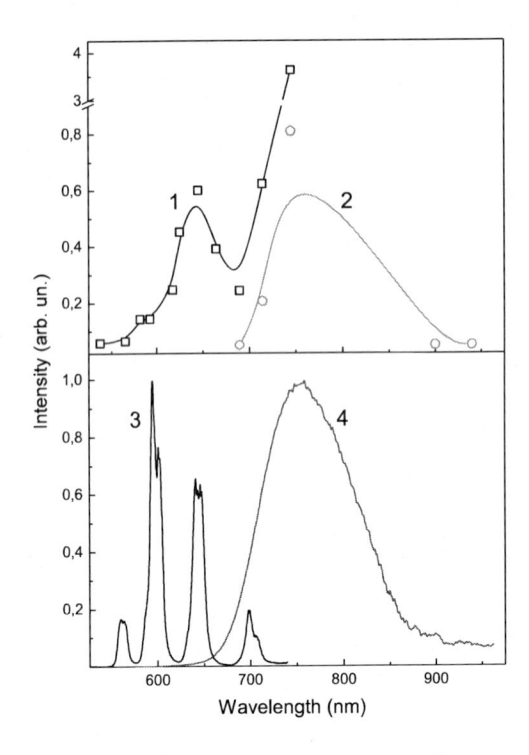

Figure 29. The CL-Sm spectra (1, 2) in reaction of $SmCl_3 \cdot 6H_2O$ (10^{-2} M) with $Bu^i{}_2AlH$ (10^{-1} M) in 5 mL THF in the presence of the oxygen impurity (295 K); The spectra were measured using photomultipliers «FEU-79» (1) and «FEU-83» (2). 3 and 4 – PL spectra of crystalline $SmCl_3 \cdot 6H_2O$ (λ_{exc} = 420 nm) and the solution after interaction in system ($SmCl_3 \cdot 6H_2O$-THF-O_2)-$Bu^i{}_2AlH$ (λ_{exc} = 488 nm), respectively.

The higher intensity of CL-Eu compared to CL-Sm is due to the fact that the Sm^{2+} ion has a much smaller quantum yield of luminescence. This is confirmed by the fact that the solution after interaction in the system ($LnCl_3 \cdot 6H_2O$-THF-O_2)-$Bu^i{}_2AlH$ in the case of samarium has a less PL intensity than that for europium. The higher efficiency of the Eu^{2+} ion compared to the Sm^{2+} ion as a CL enhancer is also due to the smaller energy gap between the radiative levels of the donor $^3Me_2CHC(H)=O^*$ and Eu^{2+} (0.22 eV) than Sm^{2+} (1.59 eV). This means that the Eu^{2+} acceptor is in greater resonance with the donor $^3Me_2CHC(H)=O^*$ than the Sm^{2+} acceptor.

On the Scheme 8 is shown the light and dark reactions happening in the interaction in the system ($LnCl_3 \cdot 6H_2O$-THF-O_2)-$Bu^i{}_2AlH$. The dark processes (1 and 2, scheme 8) leading to formation of the bulky complex II were discussed above and presented in detail in Scheme 7. The reactions 3-10 (Scheme 8) are characteristic of CL, which accompanies the free-radical autoxidation of aluminum alkyls (R_3Al) and their hydrides, which we studied earlier [59, 65, 69]. The light stage of oxidation of $Bu^i{}_2AlH$ is the disproportionation reaction of peroxyl radicals generating triplet excited aldehyde $^3Me_2CHC(H)=O^*$ (9), which is the emitter of a very weak BCL (10).

$$LnCl_3 \cdot 6H_2O^{solid} + Bu^i_2AlH \xrightarrow{THF} LnCl_2 \cdot (THF)_2^{solution} + P_1 \tag{45}$$
$$\text{Complex I}$$

$$LnCl_2 \cdot (THF)_2 + Bu^i_2AlH \longrightarrow LnCl_2 \cdot (THF)_2 \cdot Bu^i_2AlH \tag{46}$$
$$\text{Complex II}$$

initiation
$$Bu^i_2AlH + O_2 \begin{cases} \longrightarrow R^{\cdot} + RO^{\cdot} \tag{47} \\ \longrightarrow {>}AlOOR \tag{48} \end{cases}$$

$$R^{\cdot} + O_2 \longrightarrow RO_2^{\cdot} \tag{49}$$

$${>}AlOOR + {>}AlR \longrightarrow R^{\cdot} + RO^{\cdot} \tag{50}$$

propagation
$$RO_2^{\cdot} + {>}AlR \longrightarrow R^{\cdot} + {>}AlOOR \tag{51}$$

$$RO^{\cdot} + {>}AlR \longrightarrow R^{\cdot} + {>}AlOR \tag{52}$$

$$2RO_2^{\cdot} \longrightarrow O_2 + ROH + Me_2CHC(H){=}O \text{ (or } {}^3Me_2CHC(H){=}O^*) \tag{53}$$

$${}^3Me_2CHC(H){=}O^* \longrightarrow Me_2CHC(H){=}O + BCL \text{ (430 nm)} \tag{54}$$

$${}^3Me_2CHC(H){=}O^* + Ln^{2+} \longrightarrow Me_2CHC(H){=}O + Ln^{2+*} \tag{55}$$

$$Ln^{2+*} \longrightarrow Ln^{2+} + CL\text{-}Eu \text{ (or } CL\text{-}Sm) \tag{56}$$

$$P_1 = (Bu^i_2Al)_2O, Bu^iH, H_2; R = Bu^i, H.$$

Scheme 8.

The spectrum of BCL coincidences with PL spectrum of the frozen solution of the ${}^3Me_2CHC(H){=}O^*$ aldehyde in THF (Figure 28, 4). In the presence of europium or samarium in the reaction solution occurs much brighter CL-Eu with $\lambda_{max} = 465$ nm or CL-Sm with $\lambda_{max} = 780$ nm (12), the emitters of which are an electronically excited state of the Eu^{2+*} ($4f^65d^1{-}4f^7$ transition) and Sm^{2+*} ($4f^55d^1{-}4f^6$ transition). There is a simple and important argument in favor of the mechanism of the indirect CL: the absence of CL-Ln in the absence of oxygen in the reaction system under study.

The quantum yield of BCL is $4 \cdot 10^{-10}$ Einstein·mol^{-1}, which is comparable to 10^{-8}-10^{-10} Einstein·mol^{-1} for non-activated CL at initiated oxidation of hydrocarbons [70]. The quantum yield of CL-Eu is higher $1.4 \cdot 10^6$ times than that for BCL and equal to $5.85 \cdot 10^{-4}$ Einstein·mol^{-1} [31]. This is a relatively high quantum yield, which is primarily due to the high rate of oxygen oxidation by Bu^i_2AlH and a high PL quantum yield of the Eu^{2+} ions ($\phi_{PL}^{I} = 0.51$) in reaction solution [31]. The PL intensity and the maxima position of Eu^{2+} in PL spectra of the reaction solution and the solution of $EuCl_2 \cdot (THF)_2$ are 0.07 arb. un. (481 nm) and 12.17 arb. un. (447 nm), respectively (Figure 30). Quantum yields corresponding to these two solutions, ϕ_{PL}^{II} and ϕ_{PL}^{II} are equal to 0.51 and 0.07, respectively, i.e., are different by 7.3 times. Significant differences in the luminescent characteristics of complexes I and II are due to a high sensitivity to external influence of the Eu^{2+*} excited level $4f^65d^1$, not shielded by the fully filled $5s^2$ and $5p^6$-orbitals. Apparently, as part of complex II, the Eu^{2+*} ion is more protected from this effect due to the formation of additional new bonds with Bu^i_2AlH.

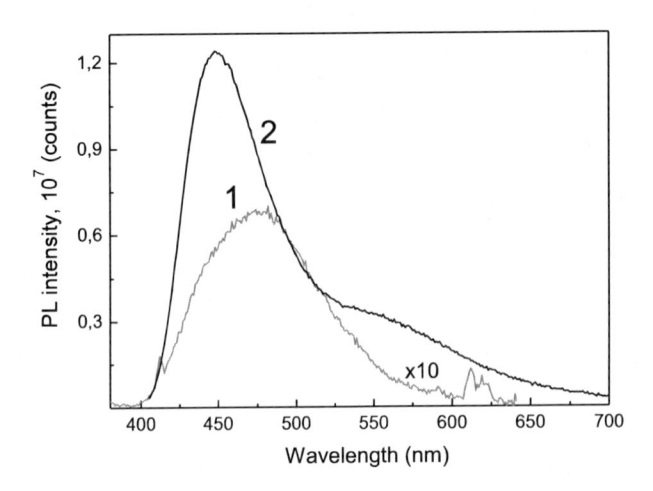

Figure 30. The PL spectra of complexes I (1) and II (2), $\lambda_{exc} = 410$ nm.

It is assumed that complex II has a four-center bridged structure containing the Eu–Bu^i–Al and Eu–Cl–Al bonds (see Complex II in Scheme 7), as the structure of lanthanide Ziegler–Natta catalysts. Similar structures have been previously identified as products of reactions of Ln^{3+} [71] and Ln^{2+} [72] complexes with trialkylaluminum compounds or their hydrides.

The formation of the bulky complex EuCl$_2$(THF)$_2$–Bu$^i_2$AlH is the main reason for the increase in the Eu^{2+} PL intensity of the reaction solution and, hence, of the intensity emission of the Eu^{2+*} ions in the blue CL-Eu. This increase in the intensity of Eu^{2+*} ion luminescence is due to the rigid bridge structure of complex II, which reduces the likelihood of non-radiative dissipation of electronic excitation energy ion Eu^{2+*} *via* vibrations of the atoms, coordinated with Eu^{2+}.

To study the effect of europium concentration ($[Eu^{2+}]_0$) on the CL-Eu intensity, the stock solutions of LnCl$_3\cdot$6H$_2$O in methanol were prepared, which were then diluted by THF and included in the reaction with Bu$^i_2$AlH. As result it was obtained a linear relationship of CL-Eu intensity *vs* $[Eu^{2+}]_0$ (Figure 31), which described by the equation y = 0.01 + 127.37x (R_1 = 0.996), varies in the range of five orders of magnitude. This linear relationship can serve as a calibration curve for determining the content of europium in alcohol-soluble salts. The rate constant for the energy transfer from the primary emitter ^{3}Me$_2$CHC(H)=O* to Eu^{2+} ion was obtained from a linear anamorphosis of dependence 1 (R_2 = 0.998) in coordinates of the reciprocals of the enhancement efficiency CL-Eu ($I/I_0-1)^{-1}$ of the reciprocals europium concentrations ($[Eu^{2+}]^{-1}$) according to the Eq. (I).

$$(I/I_0 - 1)^{-1} = [\alpha_1/\alpha \cdot (\phi_{En}/\phi_0 - 1)]^{-1} + [\alpha_1/\alpha(\phi_{En}/\phi_0 - 1)]^{-1}/K_{SV}[En] \quad (57)$$

where K_{SV} – Stern-Volmer constant, α and α_1 are the spectral sensitivity of a photomultiplier in a region of emission of the energy donor ^{3}Me$_2$CHC(H)=O* and the Eu^{2+}

enhancer, respectively; ϕ_0 and ϕ_{En} are the luminescence quantum yields of the primary emitter $^3Me_2CHC(H)=O^*$ and the CL enhancer Eu^{2+}.

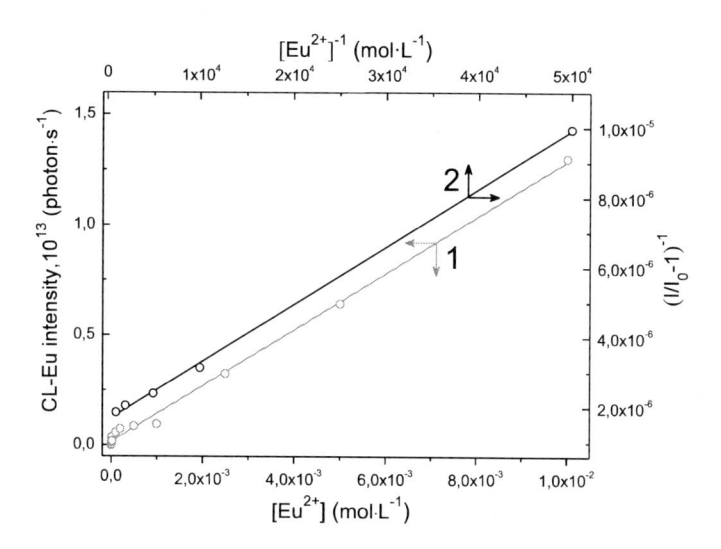

Figure 31. Dependence of the CL-Eu intensity from concentration of Eu^{2+} ions of equal $[Eu^{3+}]_0$ (1) in air oxidation of Bu^i_2AlH and its linearization (2) in coordinates $(I/I_0-1)^{-1}-[Eu^{2+}]^{-1}$. $[Bu^i_2AlH]_0 = 10^{-1}$ M, $[Eu^{3+}]_0 = 5 \cdot 10^{-4}-10^{-2}$ M, V(THF) = 2 mL.

The slope of the line segment (Figure 31, 2) give Stern-Volmer constant $K_{SV} = 9712 \pm 229$ M^{-1}. The real lifetime ($\tau_0 = 1.4 \cdot 10^{-7}$ s) of the $^3Me_2CHC(H)=O^*$ excited state under experimental conditions which was established from the study of the enhancement of BCL by of 9,10-dibromoanthracene. By the substitution the τ_0 value obtained in the formula $k_{bim} = K_{SV}/\tau_0$, was determined bimolecular rate constant for energy transfer $k_{bim} = 6.97\pm0.2 \cdot 10^{10}$ M$^{-1} \cdot$s^{-1} in the system ($EuCl_3 \cdot 6H_2O$-THF-O_2)-Bu^i_2AlH. The increase in the concentration of Bu^i_2AlH (up to $5.6 \cdot 10^{-1}$ M) in this system leads to an increase in the CL-Eu intensity and the CL-Eu light sum.

The CL-Eu intensity increases linearly with the concentration of dissolved oxygen (created before the introduction of Bu^i_2AlH) in the concentration range $1 \cdot 10^{-3}$-$8.5 \cdot 10^{-3}$ M (Figure 32). Maximum brightness CL was observed in THF saturated by pure oxygen ($[O_2]_0 = 8.5 \cdot 10^{-3}$ M). In general, CL decays after about 10 min from the start of the reaction. However, at increasing the sensitivity of chemiluminometer by 100-fold, CL is registered even on the next day!

If after the fall of the intensity of CL-Eu on about one order of magnitude, to add by a syringe an aliquot of gaseous oxygen into a reaction solution, it is registered a surge of CL-Eu that can be called multiple times (Figure 33). After a certain CL-Eu intensity level is reached ($\sim$270 s), the inclusion of oxygen bubbling causes the presence of a wider kinetic maximum, after which the glow completely disappears (Figure 33). The maximum intensity of the CL-Eu has been observed by the eye for 5-6 s, then its brightness falls slowly and 5-10 min from the start of the reaction, the CL-Eu is not visible to the naked eye. At the same time, at a moderate supply of oxygen (for example, by sucking oxygen from atmospheric air through the rubber stopper) the CL-Eu is visible in the form of narrow strip of the blue light about 2-3 days (!) in a darkened room. The minimum artificially created oxygen concentration to obtain the dependence in Figure 32 was not lower than 10^{-3} M. However, the estimates given in [31] allow us to hope that the minimum detectable concentration of oxygen by using CL-Eu is 10^{-13} M.

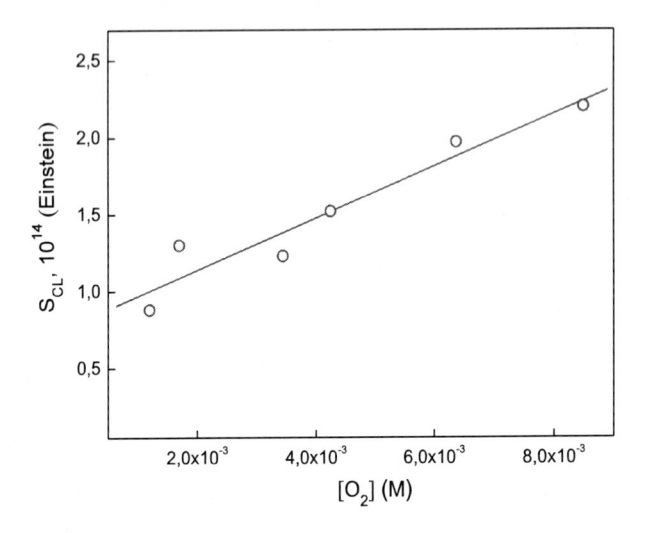

Figure 32. The effect of the concentration of $[O_2]_0$ dissolved in THF on the light sum of CL-Eu. $[Eu^{3+}]_0=1\cdot10^{-2}$ M, $[Bu^i_2AlH]_0=4\cdot10^{-1}$ M, V(THF) = 2 mL, T = 298 K.

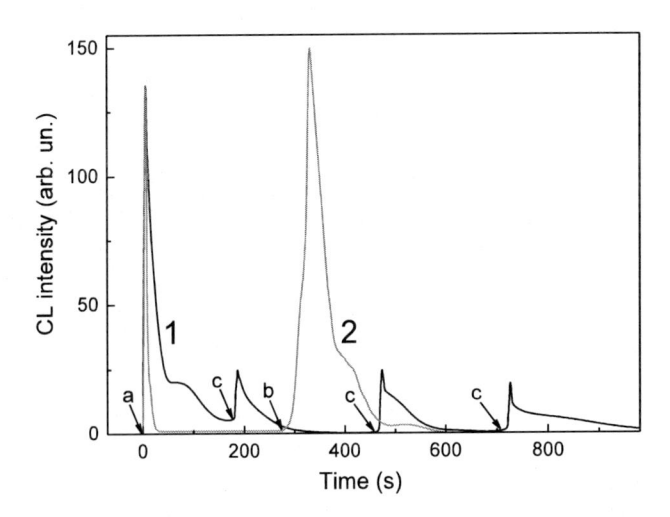

Figure 33. Kinetics of CL-Eu at addition of Bu^i_2AlH into the slurry of $EuCl_3\cdot6H_2O$ in THF saturated by argon (with O_2 impurities). The arrows indicate the moments: a (1 and 2) – addition of Bu^i_2AlH; b – switching O_2 bubbling; c – injection of 1 mL O_2 gas. $[Eu^{3+}]_0=1\cdot10^{-2}$ M, $[Bu^i_2AlH]_0=4\cdot10^{-1}$ M, V(THF) = 2 mL, T = 298 K.

The other aluminum alkyls Me_3Al, Et_3Al and Bu^i_3Al can be successfully used instead of Bu^i_2AlH for the efficient one-pot synthesis of the complexes of divalent europium soluble in THF from available crystalline hydrate $EuCl_3 \cdot 6H_2O$. The reactions of these aluminum alkyls with $EuCl_3 \cdot 6H_2O$ are also accompanied by bright blue CL, which originates from emission of the Eu^{2+} ions; the PL spectra of these reaction solutions are also exhibit the broadband emission characteristic of Eu^{2+} ion (Figure 34).

The use instead of Bu^i_2AlH the other hydrides such as B_2H_6 or 9-BBN affords at room temperature soluble complexes of only trivalent europium.

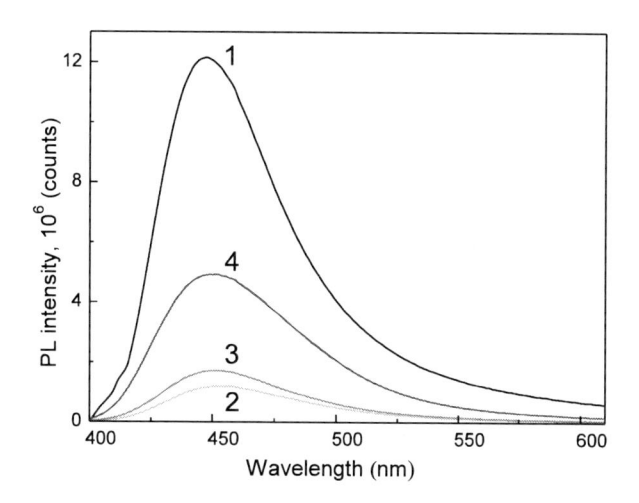

Figure 34. The PL spectra during the interaction in the systems ($EuCl_3 \cdot 6H_2O$-THF-O_2)-aluminum alkyl, were aluminum alkyl ($4 \cdot 10^{-1}$ M) – Bu^i_2AlH (1), Bu^i_3Al (2), Et_3Al (3) and Me_3Al (4). $[Eu^{3+}]_0 = 1 \cdot 10^{-2}$ M, V(THF) = 2 mL, T = 298 K. λ_{exc} = 410 (1), 389 (2), 385 (3), 365 (4) nm. All spectra were measured on Fluorolog-3.

When the $Eu(NO_3)_3 \cdot 6H_2O$ salt is used instead of $EuCl_3 \cdot 6H_2O$, a CL also appears, however its intensity is much lower (on six orders) than for CL-Eu, but it is still an order of magnitude higher than that of BCL. The spectrum of this new CL (λ_{max} = 475 nm) was close to the PL spectrum (λ_{max} = 453 nm) of reaction solution (Figure 35). Diffuse spectra of CL and PL of reaction solutions have the shape characteristic of the Eu^{2+*} ion luminescence and due to interconfiguration the allowed $4f^6 5d^1 \rightarrow 4f^7$ transition. It means that the emitter of CL and PL in reaction of $Eu(NO_3)_3 \cdot 6H_2O$ is the $5d^1$-excited state of the Eu^{2+*} ion have a one maximum in the blue region and these are completely different from the PL spectrum of solid $Eu(NO_3)_3 \cdot 6H_2O$ (Figure 35, 3).

Indeed, the PL spectrum of latter compound contains several of structural peaks characteristic [73] of the Eu^{3+*} ion and due to *f-f* intraconfiguration transitions.

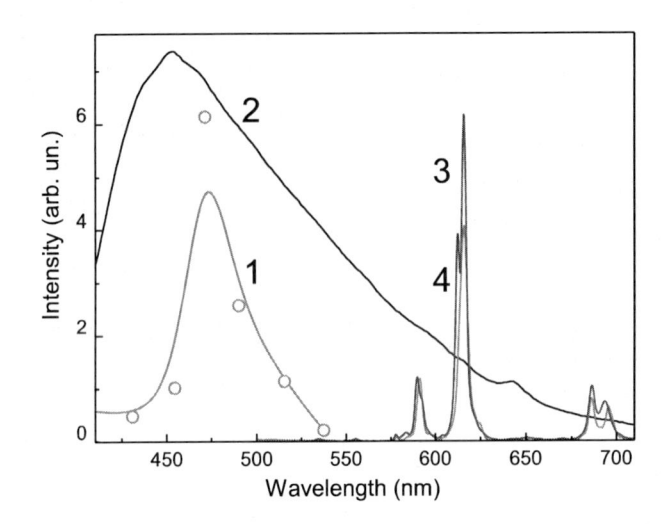

Figure 35. The CL (1) and PL (2) spectra registered during in reaction of Eu(NO$_3$)$_3$·6H$_2$O (0.02 mmol) with Bu$^i_2$AlH (0.8 mmol) in THF (2 mL) saturated by air at 298 K. 3 and 4 – the PL spectra of solid Eu(NO$_3$)$_3$·6H$_2$O (3) and a solid product obtained after reaction of Eu(NO$_3$)$_3$·6H$_2$O with Bu$^i_2$AlH (4) in THF. $\lambda_{exc} = 400$ nm.

The lifetime of the Eu^{2+*} ions in this reaction solution ($\tau = 377$ ns) is much shorter in comparison with $\tau = 100$ ms measured for the Eu^{3+*} ions in crystalline salt Eu(NO$_3$)$_3$·6H$_2$O. The ratio of quantum yields of the CL-Eu (5.85·10^{-4}) and "nitrate CL" (9.0·10^{-9}) is close to the ratio of the intensities of these CL and equal to 0.65·10^5. In Table 1 are shown some spectral characteristics of the reaction solutions formed after interaction in the systems (EuX$_3$·6H$_2$O-THF-O$_2$)-Bu$^i_2$AlH (X = Cl$^-$ or NO$_3^-$).

Table 1. The spectral characteristics of the reaction solutions after interaction of EuX$_3$·6H$_2$O with Bu$^i_2$AlH in THF at 298 K

X	λ_{PL}, nm	λ_{UV-VIS}, nm	λ_{CL}, nm	ϕ_{PL}	ϕ_{CL}
Cl$^-$	447	254, 333	465	0.51	5.85·10^{-4}
NO$_3^-$	454	257, 264, 347	475	0.05	9·10^{-9}

It is interesting to note that the maxima positions of the CL and PL spectra given in Table 1 depend little on the nature of the inorganic anions (ligands) of Cl$^-$ and NO$_3^-$. Indeed, these differences do not exceed 7-10 nm and, moreover, they are given for very wide maxima. At the same time, the PL yields for Eu^{2+} compounds with these anions differ by an order of magnitude and the CL yields even by 5 orders of magnitude. The noted features of the effect of anions on CL-Eu allow us to conclude that the significant difference between the CL-Eu yields is largely due to different yields of the excited state of the Eu^{2+} ion for different anions.

Thus, at the transition from the "chloride" to a "nitrate" system remains the same CL emitter, namely the exited $4f^6 5d^1$ configuration of Eu^{2+*} ion.

However, an attempt to isolate the estimated complex of $Eu(NO_3)_2 \cdot (THF)_x$ from the reaction solution, similar to chloride complex I, was unsuccessful due to the oxidation of Eu^{2+} to Eu^{3+} by the ligand NO_3^-.

Thus, to replace the chloride ligand with preservation of high intensity of CL-Eu, ligands that do not have strong oxidizing properties are suitable. As such inorganic ligands, the I^- and Br^- anions are promising, which also form well-luminescent compounds with europium ion EuI_2 and $EuBr_2$ [48, 57].

3.2. The Green Chemiluminescence of the Divalent Eu^{2+*} Ions at Interaction in the System $(EuL_3 \cdot xH_2O\text{-}THF\text{-}O_2)\text{-}Bu^i_2AlH$ (L = acac, dpm, fod and CH_3COO; x = 0, 1, 6)

Abbreviated names of ligands in the title of this section are deciphered as follows: acac – acetylacetonate, dpm – 2,2,6,6-tetramethyl-3,5-heptanedionato, fod – 1,1,2,2,3,3-heptafluoro-7,7-dimethyl-4,6-octandionato.

In the previous section, it was shown that the replacement of the Cl^- anion in the initial compound $EuCl_3 \cdot 6H_2O$ with another inorganic ligand does not lead to a change in the spectral region of the blue CL. As it turned out, organic ligand-anions, such as β-diketones, have a more significant effect on the color of the CL emitting by Eu^{2+*} ions. A striking example of this is the interaction in the system $(EuL_3 \cdot xH_2O\text{-}THF\text{-}O_2)\text{-}Bu^i_2AlH$ (L = acac, dpm, fod, and CH_3COO; x = 0, 1, 6) causes a rather bright CL which spectrum lies in the green region and contains one broad maximum at 555±10 nm (Figure 36) [30]. Further we will label this green CL as GR-CL. In the case of the $Eu(fod)_3$ complex the GR-CL is visible to the naked eye. The high brightness and duration of this GR-CL in the system $(Eu(fod)_3\text{-}THF\text{-}O_2)\text{-}Bu^i_2AlH$, make it promising as a chemical source of green light.

The duration of GR-CL for all complexes is about 1 hour, although for $Eu(fod)_3$ it is somewhat longer (Figure 37), and it depends on the oxidation of Bu^i_2AlH. The brightest GR-CL is registered for $Eu(fod)_3$ complex (Figure 37). However, the intensity of this GR-CL is much lower than that for the blue CL-Eu. In the same time the GR-CL for $Eu(fod)_3$ is observed with the naked eye in a darkened room as a beautiful persistent glow having the salad color. Less bright CL of $Eu(acac)_3 \cdot H_2O$ complex can also be seen with the naked eye but in a more strongly darkened room and at some adaptation of the eyes. The intensity of the BCL is many times smaller than the RG-CL, but the shapes of kinetic curves in both cases are the same (Figure 27 and 37). The sum of above date is characteristic of an enhanced CL.

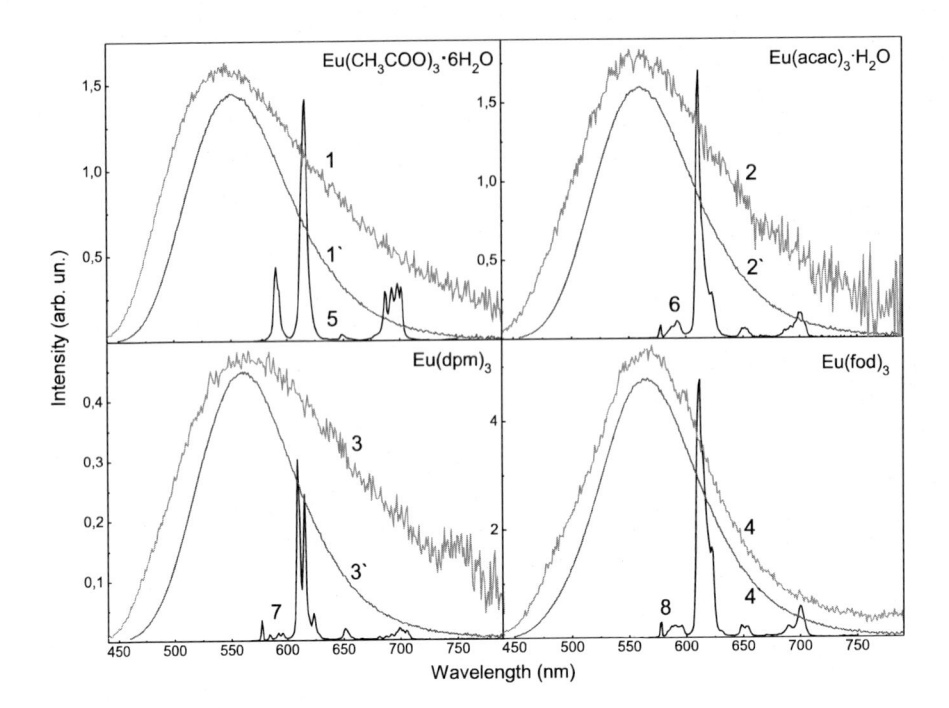

Figure 36. GR-CL (1-4) and PL (1'-4') spectra measured during the reaction of the β-diketonate complexes of Eu^{3+} with Bu^i_2AlH in THF in presence of oxygen. 5, 7 – PL spectra of crystalline samples of initial Eu^{3+} complexes; 6, 8 – PL spectra of the solutions of Eu^{3+} initial complexes in THF. $[Eu^{3+}]_0 = 10^{-2}$ M; V(THF) = 2 ml, $[Bu^i_2AlH]_0 = 4·10^{-1}$ M, 300 K.

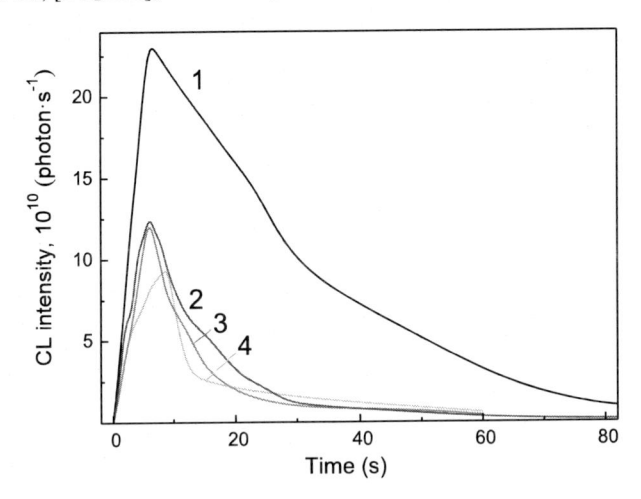

Figure 37. GR-CL kinetics (1-4) in THF solution at 300 K. 1 – $Eu(fod)_3$, 2 – $Eu(CH_3COO)_3·6H_2O$, 3 – $Eu(acac)_3·H_2O$, 4 – $Eu(dpm)_3$. The intensity of CL curves 2-4 increased to 3 times. $[Eu^{3+}]_0 = 1·10^{-2}$ M, $[Bu^i_2AlH]_0 = 4·10^{-1}$ M, V(THF) = 2.0 mL.

The reduction of Eu^{3+} to Eu^{2+} is fully finished by about ~13 min as this can be seen from the kinetics of increasing the intensity of the PL maximum of the reaction solution at 560 nm (Figure 38). The interaction in the system containing β-diketonate complexes with water of crystallization gives a series of the same products (aluminoxane $(Bu^i_2Al)_2O$, hydrogen and isobutane) as the interaction in the system $(EuCl_3·6H_2O-THF-O_2)-Bu^i_2AlH$.

But the reaction $Bu^i_2AlH + EuL_3$-moiety of the initial Eu^{3+} complexes give the previously no synthesized divalent europium complexes $EuL_2\cdot(THF)_2$, which were first isolated and characterized by elemental and spectral analysis in [30]. $EuL_2\cdot(THF)_2$ complexes formed in the reaction then form bulky complexes with an excess of Bu^i_2AlH, characterized by yellow-green color, and which are responsible for the PL of the reaction solutions, and whose spectra coincide with the spectra of GR-CL.

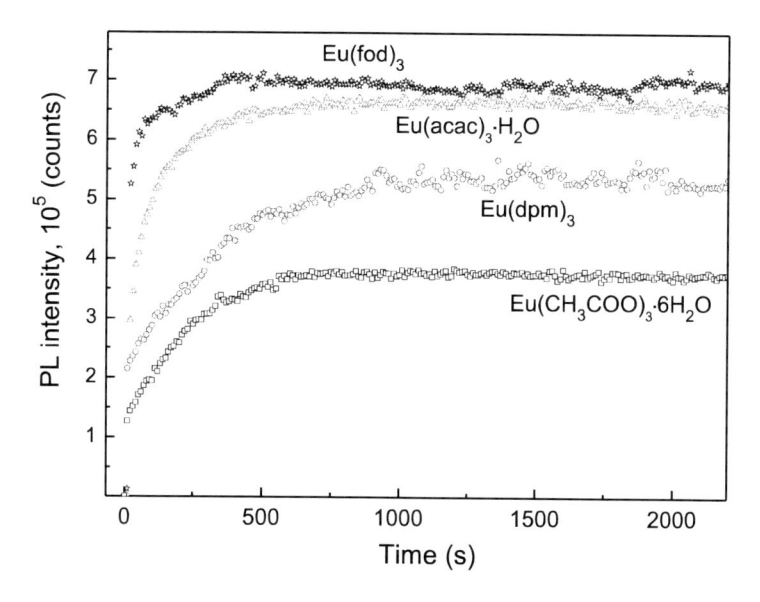

Figure 38. Kinetics of reduction of Eu^{3+} to Eu^{2+} in the reaction of the β-diketonate complexes of Eu^{3+} (10^{-2} M) with Bu^i_2AlH ($4\cdot10^{-1}$ M) in THF (2 mL) at 300 K, monitored by changes of intensity in PL maxima of Eu^{2+} at 560 nm. λ_{exc} = 450 nm.

Table 2. Luminescent characteristics of Eu^{3+} and Eu^{2+} ions before and after reaction of the β-diketonate (L) complexes of Eu^{3+} with Bu^i_2AlH in THF at 300 K

L	Crystalline samples of Eu^{3+} complexes		Solutions of Eu^{3+} complexes in THF		Yellow-green reaction solutions of Eu^{2+} complexes		
	λ_{PL}, nm	τ, μs	λ_{PL}, nm	τ, μs	λ_{PL}, nm	τ, ns	ϕ_{PL}
acac⁻	610[a]	273	611[a]	180.4	560	359.4	0.21
fod⁻	612[a]	144	612[a]	603.4	565	316.4	0.29
CH₃COO⁻	615[a]	404	–[b]	–[b]	560	308.2	0.19
DPM⁻	610[a]	40	–[b]	–[b]	550	336.8	0.28

[a] – dominant maxima luminescence of Eu^{3+} were given, [b] – not soluble in THF.

The lifetimes of the excited states of the PL emitters of the yellow-green reaction solutions are much shorter (ns time scale) than that for Eu^{3+} (μs time scale) in the parent crystals or solutions of complexes (Table 2). Registration lifetime in the nanosecond region is characteristic of Eu^{2+} *d-f* luminescence, for example, Eu^{2+} complexes with 15-crown-5 and its derivatives [67].

Thus, the emitters of GR-CLs and PL of the reaction solutions are Eu^{2+*} ions (the radiative $4f^65d^1 \rightarrow 4f^7$ transition), which are an ionic part of these bulky $EuL_2 \cdot (THF)_2$-Bu^i_2AlH complexes. Nature of the organic ligand in the parent Eu^{3+} complex has a little effect on the PL quantum yield of yellow-green solutions. So, the PL quantum yields of the Eu^{2+} complexes are decreased in the following series:

$Eu(fod)_2 \cdot (THF)_2$-Bu^i_2AlH ($\phi_{PL} = 0.29$) > $Eu(dpm)_2 \cdot (THF)_2$-Bu^i_2AlH ($\phi_{PL} = 0.28$) > $Eu(acac)_2 \cdot (THF)_2$-Bu^i_2AlH ($\phi_{PL} = 0.21$) > $Eu(CH_3COO)_2 \cdot (THF)_2$-$Bu^i_2AlH$ ($\phi_{PL} = 0.19$).

The presence of excess Bu^i_2AlH in the yellow-green solution, as well as in the Eu^{2+} coordination sphere in the $Eu(L)_2 \cdot (THF)_2$-Bu^i_2AlH complex (where L – beta-ligand) protects the Eu^{2+} ion from oxidation by moisture and oxygen. However, with the complete oxidation of Bu^i_2AlH with oxygen, the Eu^{2+} ion is also oxidized to the Eu^{3+} ion, which is clearly seen by the disappearance of the green PL and the appearance of the structural spectrum of the red PL (Figure 39).

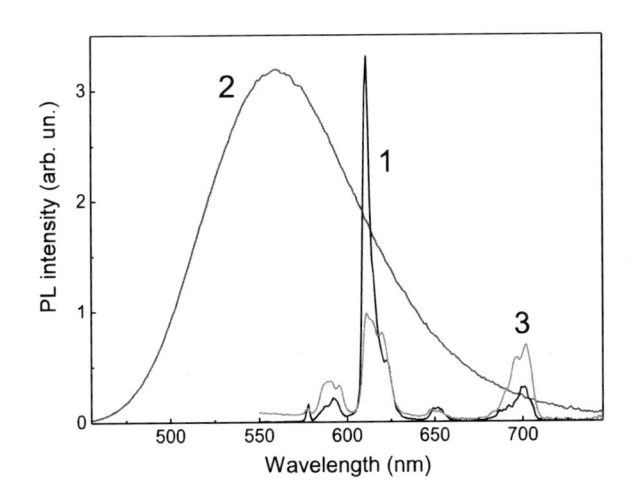

Figure 39. PL spectra before (1) and after (2) reaction of $Eu(acac)_3 \cdot H_2O$ and Bu^i_2AlH in THF. 3 – PL spectrum after air bubbling through the yellow-green reaction solution; $\lambda_{exc} = 394$ (1, 3), 443 (2) nm, $[Eu^{3+}]_0 = 10^{-2}$ M, $[Bu^i_2AlH]_0 = 4 \cdot 10^{-1}$ M.

The PL of the yellow-green solution described above is the first example of the liquid phase green PL of the Eu^{2+} ion due to $4f^65d^1 \rightarrow 4f^7$ transition. Indeed, a shorter-wavelength PL is more characteristic of the Eu^{2+} ion, which was first recorded as a very bright glow for frozen (77 K) aqueous solutions of $EuCl_2$ (450 nm) and $EuClO_4$ (463 nm) [49]. Then very weak PL was also detected (but at room temperature) for $EuCl_2$ in MeOH (489 nm [74]), complexes of Eu^{2+} with crown ethers, cryptands, and polyethylene glycols in MeOH (417-488 nm) [67, 75, 76], EuI_2 in THF (442 nm) [48], $NaEu(BH_4)_4 \cdot (DME)_n$ in THF (465 nm) [16], and also for the frozen solution (77 K) of Cp_2Eu in THF (430-450 nm) [15]. The examples of the red PL of the Eu^{2+} ion in solution are very seldom, although for solid matrices doped with Eu^{2+}, the red PL (612 nm) was recorded [77, 78].

Green CL is generated by a mechanism (Scheme 9), which is like the one for blue CL. In Scheme 9, P_2 denotes molecular oxidation products, among which $(Bu^iO)_2Al(OH)$ is the main one.

$$EuL_3 \cdot xH_2O + Bu^i_2AlH \xrightarrow{THF} Bu^i_2AlL + EuL_2 \cdot (THF)_2 + P_1 \tag{58}$$

$$Bu^i_2AlH + O_2 \longrightarrow RO_2^\cdot + P_2 \tag{59}$$

$$2RO_2^\cdot \longrightarrow O_2 + ROH + Me_2CHC(H)=O \ (or \ ^3Me_2CHC(H)=O^*) \tag{60}$$

$$^3Me_2CHC(H)=O^* \longrightarrow Me_2CHC(H)=O + BCL \ (430 \ nm) \tag{61}$$

$$^3Me_2CHC(H)=O^* + EuL_2 \cdot (THF)_2 \longrightarrow Me_2CHC(H)=O + {}^*EuL_2 \cdot (THF)_2 \tag{62}$$

$$^*EuL_2 \cdot (THF)_2 \longrightarrow {}^*EuL_2 \cdot (THF)_2 + CL \ (555 \pm 10 \ nm) \tag{63}$$

L = fod, dpm, acac, CH_3COO; x = 0, 1, 6; R = Bu^i, H.
P_1 = $(Bu^i_2Al)_2O$, Bu^iH, H_2; P_2 - molecular products.

Scheme 9.

The interaction starts with the reaction of Bu^i_2AlH with crystallization water, which is removed from coordination sphere of Eu^{3+} to afford a mixture of products (P_1), consisting of aluminoxane $(Bu^i_2Al)_2O$, isobutane, and hydrogen (58, scheme 9). Bu^i_2AlH taken in a big excess also reacts with Eu^{3+} (58) and oxygen (59-60). As result, Eu^{3+} is reduced to Eu^{2+}. One ligand L, removed from the coordination sphere of Eu^{3+}, is attached to the aluminum atom to form Bu^i_2AlL. The reaction $Bu^i_2AlH + O_2$ gives peroxyl radicals (59), disproportionation of which generates the triplet-excited state of isobutyric aldehide $^3Me_2CHC(H)=O^*$ (60). In Scheme 9, the oxidation of Bu^i_2AlH is reflected in its simplest form (59-60). The excited $^3Me_2CHC(H)=O^*$ molecules transfer energy to Eu^{2+} exciting it into the electronically exited state Eu^{2+*} (62), which is deactivated with emission of GR-CL (63). When the europium complex is absent in the reaction system (not loaded into the reactor), is observed only very weak BCL caused by emission of the triplet-excited aldehyde in a much shorter wavelength (430 nm) (63).

PL spectra of the isolated complexes $EuL_2 \cdot (THF)_2$ and of the bulky $EuL_2 \cdot (THF)_2$-Bu^i_2AlH complexes in the yellow-green solutions (Figure 36) are very similar. However, the PL quantum yields of Eu^{2+} in $EuL_2 \cdot (THF)_2$ are much lower than those in $EuL_2 \cdot (THF)_2$-Bu^i_2AlH. For example, the ratios of PL quantum yield of the pairs $EuL_2 \cdot (THF)_2$-$Bu^i_2AlH/EuL_2 \cdot (THF)_2$ are equal 28.0 (L = dpm), 7.0 (L =acac) and 3.2 (L = CH_3COO). The formation of bulky complexes is also confirmed by no precipitation of a solid even after complete removal of THF from yellow-green reaction solutions, though their concentration in this case becomes equal to 10^{-1} M and precipitating them requires the addition of heptane and heating. It is assumed that the bulky complexes have bridge-type four-center structure, in which the Eu^{2+} ion bonded to aluminum of Bu^i_2AlH through Bu^i radical (giving link

Eu^{2+}–Bu^i–Al) and the oxygen atom of the organic ligand (giving link Eu^{2+}–O–Al). Earlier such bridged-type four-centered structures have been detected in reactions of both trivalent [71] and divalent [72] lanthanide complexes with alkyl aluminum compounds and their hydrides, respectively.

An additional argument in favor of the mechanism of the indirect GR-CL is the enhancement of the GR-CL intensity with increasing in concentration of the energy acceptor Eu^{2+} ion. Enhanced CL can be described by the equation (Eq. I).

$$(I/I_0 - 1)^{-1} = [\alpha_1/\alpha(\phi_{En}/\phi_0 - 1)]^{-1} + [\alpha_1/\alpha(\phi_{En}/\phi_0 - 1)]^{-1}/K_{SV}[En] \quad (64)$$

where I and I_0 are CL intensities with and without enhancer Eu^{2+}, respectively; α and α_1 – photomultiplier sensitivity in the spectral region of emission of primary emitter $^3Me_2CHC(H)=O^*$ and the Eu^{2+} enhancer, respectively: ϕ_0 and ϕ_{En} are luminescence quantum yields of $^3Me_2CHC(H)=O^*$ and enhancer Eu^{2+}; $K_{SV} = k_{bim}\cdot\tau_0$ – Stern-Volmer constant.

The peculiarities of the GR-CL enhancement have been studied on the example of the $(Eu(acac)_3\cdot H_2O\text{-THF-}O_2)\text{-}Bu^i_2AlH$ system. On the Figure 40 (1) is shown that the GR-CL intensity increases with increasing in Eu^{2+} concentration, and in the coordinates $(I/I_0\text{-}1)^{-1}$ – $[Eu^{2+}]^{-1}$, the linear (R = 0.99) dependence is observed (Figure 40, line 2).

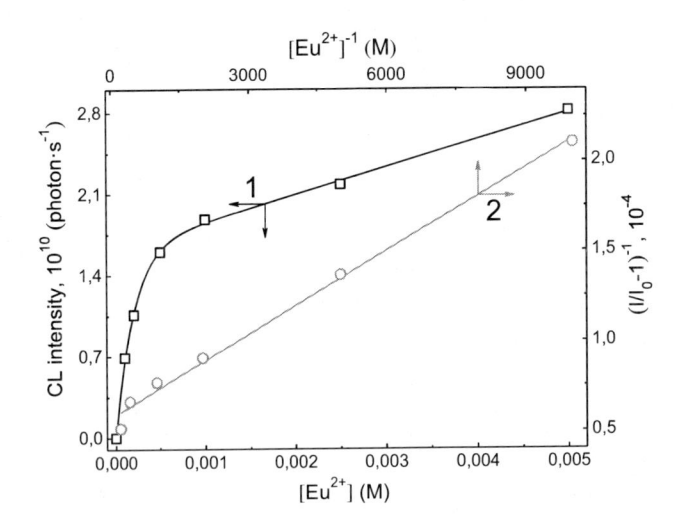

Figure 40. The dependence of the GR-CL intensity *vs* $[Eu^{2+}]_0$ which is equal to $[Eu(acac)_3\cdot 6H_2O]_0$ (1) and its linear anamorphosis in coordinates Eq. 1. $[Eu^{3+}]_0 = 10^{-4} - 5\cdot10^{-3}$ M, $[Bu^i_2AlH]_0 = 4\cdot10^{-1}$ M, V(THF) = 2 mL.

According to the Eq. 1, the intercept of ordinate by straight-line 2 A = $[\alpha_1/\alpha\cdot(\phi_{En}/\phi_0 - 1)]^{-1}$ and its slope B = $[\alpha_1/\alpha\cdot(\phi_{En}/\phi_0 - 1)]^{-1}\cdot[K_{SV}]^{-1}$ are equal to A = $5.765\cdot10^{-5}$ L·mol^{-1} and B = $1.543\cdot10^{-8}$ L·mol^{-1} for the CL enhancement by Eu^{2+}, respectively. The bimolecular rate constant calculation of energy transfer gives $k_{bim} = (2.7\pm1.4)\cdot10^{10}$ M^{-1}·s^{-1}.

Theoretically, a CL emitter ion Eu^{2+*} can be generated in two ways (Scheme 10). The first one is the direct energy transfer from the primary emitter $^3Me_2CHC(H)=O^*$ to the Eu^{2+} ion (2, scheme 10).

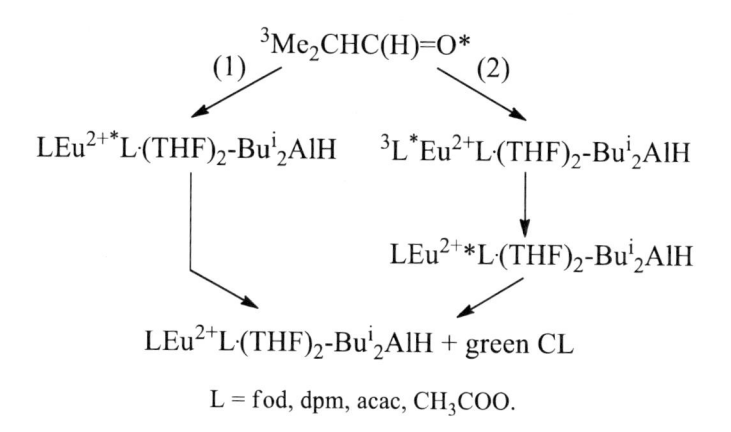

Scheme 10.

The second one includes the intermediate energy transfer from $^3Me_2CHC(H)=O^*$ to the ligand (L), and the subsequent energy transfer from the excited ligand (L*) to the Eu^{2+} ion (2). Analysis of the energy gaps between the excited levels of the donor $^3Me_2CHC(H)=O^*$ and the acceptor Eu^{2+}, as well as the donor $^3Me_2CHC(H)=O^*$ and acceptors of ligand allowed to conclude the following. The energy transfer from $^3Me_2CHC(H)=O^*$ to ligand coordinated with Eu^{2+} is possible only for the fod-ligand. The T_1-levels of other ligands are higher in energy than $^3Me_2CHC(H)=O^*$, i.e., energy transfer to these ligands are thermodynamically impossible. The excitation of Eu^{2+}, coordinated with acac- or dpm-ligand, most likely occurs by one-step mechanism according to (1), Scheme 10. In case of $Eu(fod)_3$, the excitation of Eu^{2+} is possible through both (1) and (2). In the first case, the energy gap between the levels of the T_1 ($^3Me_2CHC(H)=O^*$) and $4f^65d^1$ (Eu^{2+*}) is 5237.8 cm^{-1}. In the second case, the energy difference between the first and second energy transfer steps is 755.8 cm^{-1}, and the difference for the second and third steps is 4482 cm^{-1}. It turns out that the gain in resonance between the donor and acceptor (due to very small gap) in the first step is lost at the next one (due to the relatively large gap). Therefore, currently it is difficult to give priority to one of these variants of the energy transfer for $Eu(fod)_3$.

The high brightness and duration of CL, generated in systems ($EuCl_3 \cdot 6H_2O$-THF-O_2)-Bu^i_2AlH and ($Eu(fod)_3$-THF-O_2)-Bu^i_2AlH, opens the prospect of using them as chemical sources of blue and green light, respectively.

3.3. Divalent Eu²⁺ Ion – an Effective Inorganic Mediator of Energy Transfer from the Primary Emitter ³Me₂CHC(H)=O* on Tb³⁺ and Ru(bpy)₃²⁺ Ions

Original results were obtained in a comparative study of the effectiveness of Eu^{2+}, Tb^{3+} and $Ru(bpy)_3^{2+}$ ions as CL enhancers in the oxidation of Bu^i_2AlH, and their mutual influence on the chemiluminescent properties of each other [29, 31, 79].

As it turned out, by using Bu^i_2AlH, it is possible to successfully extract luminescent ions from crystalline hydrates of their salts into THF solution, not only for divalent lanthanides, but also for ions of other metals, for example, such as Tb^{3+} and $Ru(bpy)_3^{2+}$. Thus, the addition of Bu^i_2AlH to the slurry of $TbCl_3 \cdot 6H_2O$ or $Ru(bpy)_3Cl_2 \cdot 6H_2O$ in THF leads to transition of the corresponding metal ions Tb^{3+} or $Ru(bpy)_3^{2+}$ to solution. However, this ionic transfer occurs without reduction of these ions. The reason for this important nuance is that the ion reduction potentials for Tb^{3+} (-3.7 V [80]) and Ru^{2+} (-1.5 V [81]), it is much more negative than that of the Eu^{3+} (-0.35 V [80]). Therefore, Bu^i_2AlH cannot reduce these ions, and in this case, it only removes the crystallization water from the starting crystal hydrates of the salts of these ions.

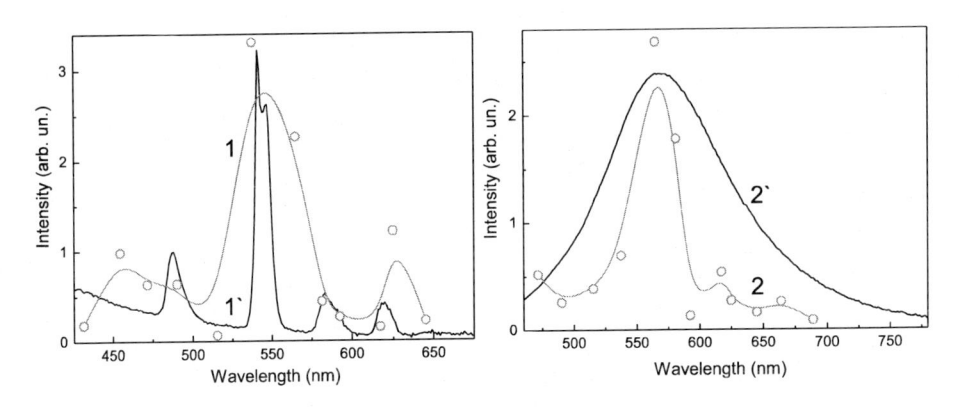

Figure 41. CL (1, 2) and PL (1`, 2`) spectra during reaction $TbCl_3 \cdot 6H_2O$ (1, 1`) or $Ru(bpy)_3Cl_2 \cdot 6H_2O$ (2, 2`) with Bu^i_2AlH in THF saturated by air at 298 K. $\lambda_{exc} = 365$ (1`), 450 (2`) nm. $N[Tb^{3+}]_0 = N[Ru(bpy)_3^{2+}]_0 = 0.02$ mmol; $N(Bu^i_2AlH) = 0.8$ mmol, V(THF) = 2 mL.

Immediately after the addition of Bu^i_2AlH, CL appears which for these ions is designated as CL-Tb and CL-Ru, respectively. In the first case, the emitter of glow is Tb^{3+*} ions ($\lambda_{max} = 485, 542, 625$ nm), and in the second – ion $Ru(bpy)_3^{2+*}$ ($\lambda_{max} = 570$ nm). According to [82] the luminescence in the case of ruthenium is caused by transfer from the metal-to-ligand charge-transfer (MLCT) excited state in the $Ru(bpy)_3^{2+}$ complex. The CL spectra in both cases correlate well with the PL spectra of the corresponding reaction solutions: for Tb, $\lambda_{max} = 494, 545, 586$ and 616 nm (Figure 41, 1), and for Ru, $\lambda_{max} = 570$ nm (Figure 41, 2), as well as with the known PL maxima of these ions in other compounds [82, 83]. The intensities of CL-Tb and CL-Ru are much higher than for BCL, and the corresponding ratios of CL-Tb/BCL = 100 and CL-Ru/BCL = 50 (Figure 42). However, compared with CL-Eu, the intensities of CL-Tb and CL-Ru are much smaller, and the ratios of the maximum intensities are CL-Eu/CL-Tb = ~10^5 and CL-Eu/CL-Ru = ~$2 \cdot 10^5$. It should be noted that this is the first example of comparing the intensity of CL, as well as the efficiency of enhancement of CL, divalent and trivalent lanthanide ions in identical conditions. From this comparison, it is obvious that the divalent europium ion is a much more efficient enhancer, even compared to a such brightly luminescing ion like Tb^{3+}.

It is important to note that the increase in the CL intensity for all three ions occurs without a change in the shape of the kinetic curve compare to BCL (Figure 42).

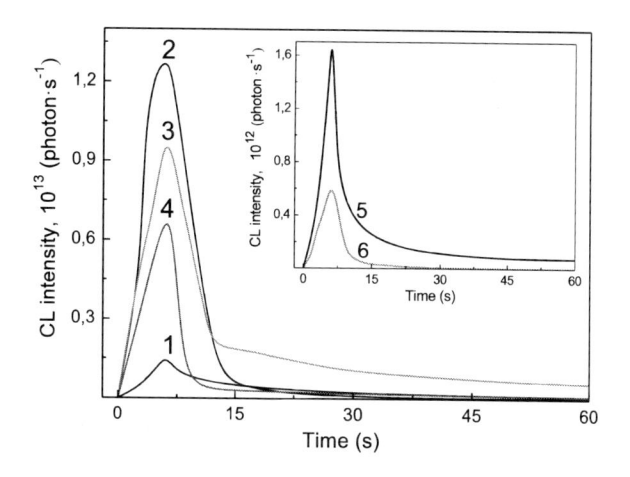

Figure 42. Kinetics of CL in THF at 298 K. 1 – BCL (intensity increase 10^6 times), 2 – CL-Eu, 3 – CL-Tb ($7 \cdot 10^4$), 4 – CL-Ru (10^5), 5 – CL-Eu-Tb, 6 – CL-Eu-Ru.

This means that these ions are effective CL enhancers and the enhancement mechanism can be characterized as the molecular-ion transfer of energy from the primary-excited $^3Me_2CHC(H)=O^*$ molecules to these ions to generate their electronically excited states, which are deactivated with emission of CL-Eu (1), CL-Tb (2), CL-Ru (3) (Scheme 11).

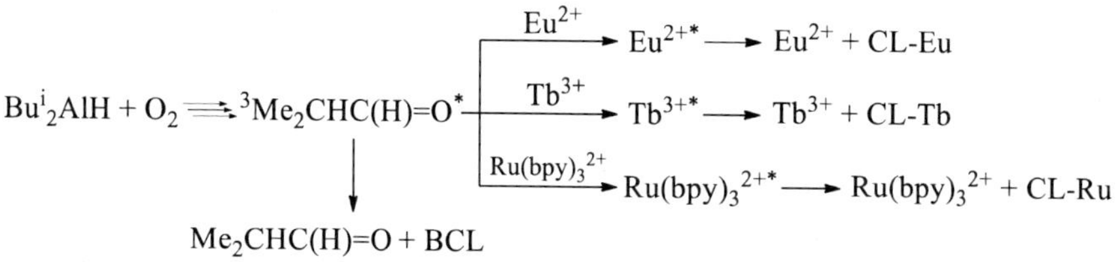

$$Eu^{2+} \longrightarrow Eu^{2+*} \longrightarrow Eu^{2+} + CL\text{-}Eu \qquad (65)$$

$$Bu^i_2AlH + O_2 \rightleftharpoons {}^3Me_2CHC(H)=O^* \quad Tb^{3+} \longrightarrow Tb^{3+*} \longrightarrow Tb^{3+} + CL\text{-}Tb \qquad (66)$$

$$Ru(bpy)_3^{2+} \longrightarrow Ru(bpy)_3^{2+*} \longrightarrow Ru(bpy)_3^{2+} + CL\text{-}Ru \qquad (67)$$

$$Me_2CHC(H)=O + BCL$$

Scheme 11.

An interesting finding was the discovery of a more significant effect of the enhancement of CL by ions of terbium and ruthenium in the presence of Eu^{2+} ions in the reaction solution. This effect take place at the simultaneous loading of two crystallohydrates $EuCl_3 \cdot 6H_2O$ and $TbCl_3 \cdot 6H_2O$ or $EuCl_3 \cdot 6H_2O$ and $Ru(bpy)_3Cl_2 \cdot 6H_2O$. In this case, the glow is designated as CL-Eu-Tb and CL-Eu-Ru, respectively (Figure 42, inset). The spectra of CL measured at the simultaneous loading of two crystallohydrates contain the luminescence of two ions. For example, the CL-Eu-Tb spectrum contains maxima of both Eu^{2+*} (465 nm) and Tb^{3+*} (485, 542, 584 nm) (Figure 43). For CL-Eu-Ru, the Eu^{2+*} emission maximum lies at 437 nm and the $Ru(bpy)_3^{2+*}$ emission take place only as a 'bend' at 560–570 nm (Figure 43). However, when CL-Eu-Ru was registered through a cut-off filter (GS-17) which is out the blue glow of Eu^{2+*} ion, a clear maximum at 570 nm characteristic of $Ru(bpy)_3^{2+*}$ emission is registered (Figure 43).

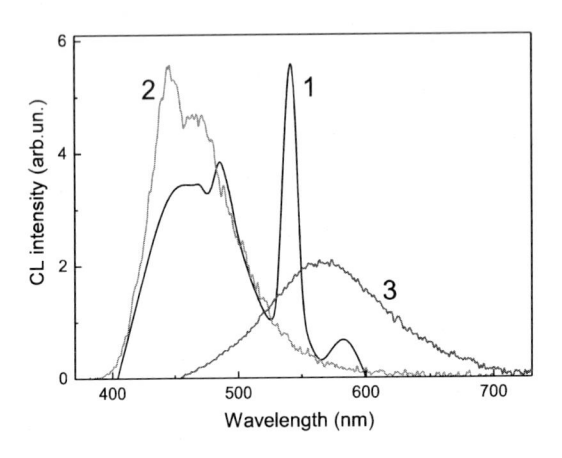

Figure 43. CL spectra measured at loading of two crystallohydrates together in THF at 298 K. 1 – CL-Eu-Tb, 2 and 3 – CL-Eu-Ru without and with cut off yellow filter GS-17, respectively.

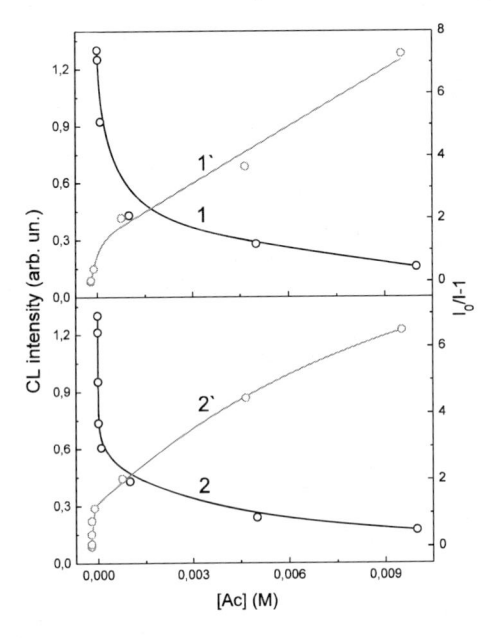

Figure 44. Dependence of the two-emitter CL intensity from the initial concentrations of acceptors (Ac), Ac = $[TbCl_3 \cdot 6H_2O]_0$ or $[Ru(bpy)_3Cl_2 \cdot 6H_2O]_0$ in THF at 298 K. 1 and 2 – CL-Eu-Tb and CL-Eu-Ru. 1` and 2` – anamorphosis of 1 and 2 dependences in the $(I_0/I - 1) - [Ac]$ coordinates. $[Tb^{3+}]_0 = 10^{-7}$ to 10^{-2} M; $[Ru(bpy)_3^{2+}]_0 = 10^{-5}$ to 10^{-2} M.

The two-emitter CL is also visible to the eye when it is observed in a darkened room. So, for CL-Eu–Tb without a filter, only the bright-blue emission of Eu^{2+*} is seen, but through the cut-off filter GS-17, the green luminescence of the Tb^{3+*} ions become clearly visible. For CL-Eu–Ru, both the intense blue emission of Eu^{2+*} ions and the red emission of $Ru(bpy)_3^{2+*}$ ions were clearly visible. In the case of two-emitter CL an increase in the initial concentration: $[TbCl_3 \cdot 6H_2O]_0$ in the range 10^{-5} to 10^{-2} M or $[Ru(bpy)_3Cl_2 \cdot 6H_2O]_0$ in the range 10^{-5} to 10^{-2} M, leads to a decrease in Eu^{2+*} emission intensity (Figure 44). This finding clearly indicates energy transfer from the Eu^{2+*} ion to the Tb^{3+} and $Ru(bpy)_3^{2+}$ ions. A liquid phase quenching of Eu^{2+} luminescence by any quenchers has not been described previously.

At low concentration, the Stern-Volmer dependencies are completely linear: for Ru in the range 10^{-7} to 10^{-6} M, and for Tb – 10^{-5} to 10^{-4} M (Figure 44). However, in the area of higher concentrations of quenchers, linear dependence is disturbed due to the loss of transparency of the reaction solution. Similar Stern-Volmer dependences were obtained to change the lifetime of the Eu^{2+*} ion (Figure 45).

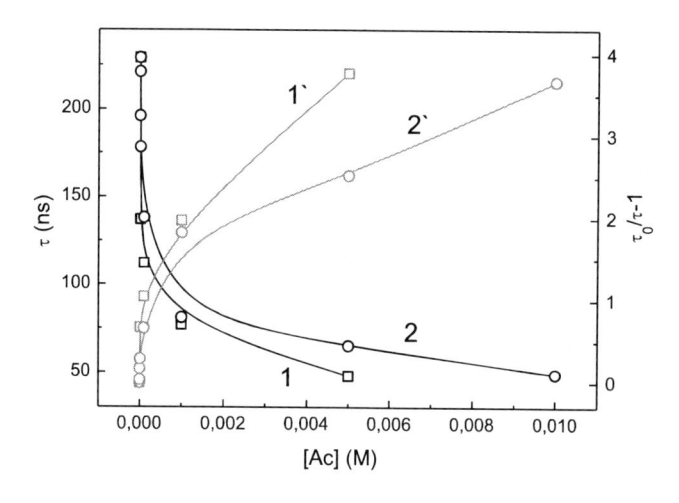

Figure 45. Dependence of the Eu^{2+*} ions lifetime for two-emitter CL from the initial concentrations of acceptors (Ac) $[TbCl_3 \cdot 6H_2O]_0$ or $[Ru(bpy)_3Cl_2 \cdot 6H_2O]_0$ in THF at 298 K. 1 and 2 – CL-Eu-Tb and CL-Eu-Ru. 1` and 2` – anamorphosis of 1 and 2 dependences in the $(\tau_0/\tau - 1) - [Ac]$ coordinates. $[Tb^{3+}]_0 = 10^{-7}$ to 10^{-2} M; $[Ru(bpy)_3^{2+}]_0 = 10^{-5}$ to 10^{-2} M.

In the case of two-emitter CL, the terbium and ruthenium ions partially quench the emission of Eu^{2+*} ions, and, on the contrary, the Eu^{2+} ion increases the enhancement effect CL by these ions during the oxidation of Bu^i_2AlH.

Thus, the Eu^{2+} ion is an efficient mediator in energy transfer from the primary $^3Me_2CHC(H)=O^*$ emitter to phosphors Tb^{3+} and $Ru(bpy)_3^{2+}$.

The efficiency of energy transfer along the chain $^3Me_2CHC(H)=O^* \rightarrow Eu^{2+} \rightarrow Tb^{3+}$ (or $Ru(bpy)_3^{2+}$) is provided by the following favorable factors. First of all, the energy of excited level of the donor is higher than that of the acceptor. That is clearly seen from a comparison of the energy of the radiating levels of the short-wave maxima for the three pairs under consideration: (1), (2), and (3).

$$2.9 \text{ eV}, 430 \text{ nm } (^3Me_2CHC(H)=O^*) > 2.7 \text{ eV}, 465 \text{ nm } (Eu^{2+}*) \tag{68}$$

$$2.7 \text{ eV}, 465 \text{ nm } (Eu^{2+}*) > 2.6 \text{ eV}, 485 \text{ nm } (Tb^{3+}*) \tag{69}$$

$$2.7 \text{ eV}, 465 \text{ nm } (Eu^{2+}*) > 2.2 \text{ eV}, 567 \text{ nm } (Ru(bpy)_3^{2+}*) \tag{70}$$

Comparison of the influence on the energy transfer efficiency of two other factors: the degree of overlap of the emission spectra of donors with the absorption spectra of acceptors and energy gaps (ΔE) between the radiative levels of donors and acceptors showed that the ΔE value is the determining factor. Indeed, the overlap of the $Eu^{2+}*$ emission spectrum with the $Ru(bpy)_3^{2+}$ absorption spectrum is greatest and characterized by the largest molar extinction coefficients (Figure 46).

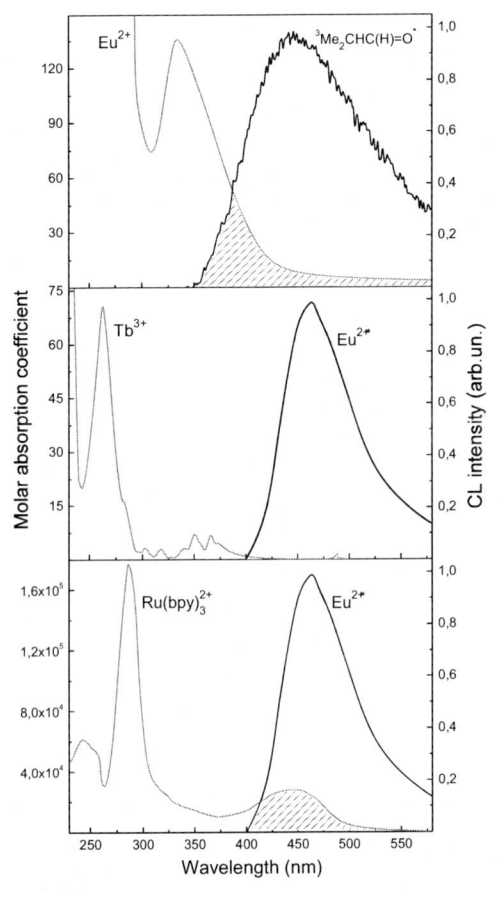

Figure 46. Overlap of the $^3Me_2CHC(H)=O^*$ (1) and $Eu^{2+}*$ (2) donors emission spectra with the Eu^{2+} (2), Tb^{3+} (4) and $Ru(bpy)_3^{2+}$ (5) acceptors absorption spectra. $[Eu^{2+}]_0 = 10^{-2}$ M, $[^3Me_2CHC(H)=O]_0 = 5 \cdot 10^{-4}$ M; $\lambda_{exc} = 310$ (1) and 400 (3) nm; $T = 77$ (1) and 298 (2-5) K.

Despite the large overlap of spectra for Ru, CL enhancement in this case was the weakest. The ΔE value between Eu^{2+*} (465 nm, 21505 cm^{-1}) and Tb^{3+*} (485 nm, 20619 cm^{-1}) emission levels is $\Delta E = 21505 - 20619 = 886$ cm^{-1}, and for Eu^{2+*} (465 nm, 21505 cm^{-1}) and $Ru(bpy)_3^{2+*}$ (567 nm 17637 cm^{-1}), $\Delta E = 21505 - 17637 = 3868$ cm^{-1}. So, the ΔE value is minimal for the Tb^{3+*} ion, and this ion is in greater resonance with the Eu^{2+*} ion compare to $Ru(bpy)_3^{2+*}$. For two-emitter CL with participant of ionic phosphors, therefore, the minimum energy gap between the radiation levels of the donor (Eu^{2+*}) and acceptor is the factor that determines the effectiveness of CL enhancement of the final emitter (Tb^{3+*} or $Ru(bpy)_3^{2+*}$).

In conclusion to this section, it should be noted that it was found for the first ime that the divalent europium ion is an effective mediator in energy transfer from the primary emitter $^3Me_2CHC(H)=O^*$ to ionic phosphors Tb^{3+}, $Ru(bpy)_3^{2+}$. Energy transfer *via* the mediator Eu^{2+} ion made possible emission enhancement of these ions to a much greater extent, compared with direct transfer of energy from $^3Me_2CHC(H)=O^*$.

CONCLUSION

The presented review shows that divalent lanthanide ions in chemiluminescent reactions act as both the initial reactants and luminescence emitters generated by the acceptance of energy from the initially excited reaction products. At the same time, there are still no examples of chemiluminescent reactions of the type (71), in which the excited ions of divalent lanthanides are formed as a result of electron acceptance by the trivalent ion Ln^{3+}.

$$Ln^{3+} + e^- \text{ (reductant)} \rightarrow Ln^{2+*} \rightarrow Ln^{2+} + h\nu_1 \qquad (71)$$

We also note that the detailed mechanism of excitation of the trivalent Ln^{3+} ion in the electron transfer reaction from the divalent Ln^{2+} ion in the already known reactions of the form (72) is also practically unexplored.

$$Ln^{2+} - e^- \text{ (oxidant)} \rightarrow Ln^{3+*} \rightarrow Ln^{3+} + h\nu_2 \qquad (72)$$

Point is that, it is quite possible to assume that the reaction mapped as (2) is not elementary, but a brutto reaction, i.e., the emitter Ln^{3+*} is formed due to the acceptance of energy from excited species, which in turn, are generated by the adoption of an electron from Eu^{2+}, and which are in close proximity to the formed Ln^{3+} ion. It seems that the details of such a mechanism can be revealed by a more detailed study of the elementary act of electron transfer from the Ln^{2+} ions (Ln = Pr, Nd) to water molecules and from the Eu^{2+} ions to the $^•OH$ radicals. Generally, the mechanisms of CL with participating Ln^{2+} ions are studied in less detail than that for Ln^{3+} ions and for many chemiluminescent reactions there is only a statement of the presence of a glow. The data presented in the review indicate that compounds of Ln^{2+} ions are stable enough to perform chemiluminescent experiments even with such chemically active ions as Pr^{2+} and Nd^{2+}, not to mention the much more stable ions in Eu^{2+}, Sm^{2+} and Yb^{2+}. The study of the CL of Ln^{2+} ions is promising in many aspects. For example, to create a CL-method for the selective determination of small quantities of europium (up to 10^{-12} M) in lanthanide compounds by reducing of Eu^{3+} to Eu^{2+} with the subsequent registration of blue CL during the oxidation of Bu^i_2AlH with oxygen. A more detailed study of the effect of low oxygen concentrations on blue CL during the oxidation of Bu^i_2AlH in the presence of Eu^{2+} will allow, using the CL, arising from the contact of gases with the $(Eu^{2+}\text{-}THF\text{-}O_2)\text{-}Bu^i_2AlH$ system, to determine the content of oxygen up to 10^{-13} M. This will make it possible to determine the oxygen content at significant distances from the Earth and possibly its presence on other planets. The study of CL in the hydrolysis of organometallic compounds Eu^{2+} is promising for creating signaling systems that generate red light upon contact with seawater. The use of CL in the oxidation of Eu^{2+} with hydrogen peroxide in the analysis of biologically active compounds is also promising. Bright blue CL of Eu^{2+*} ions, arising from the oxidation of $(C_2H_5)_3Al$ or $(C_4H_9)_3Al$ with oxygen in the presence of small amounts of Eu^{2+}, is promising for creating a chemiluminescent control of the oxidation of aluminum alkyls $(C_{2\text{-}22}H_{5\text{-}45})_3Al$ by oxygen, which is a key stage of a large-scale industrial process (the ALFOL Process) of producing higher fatty alcohols.

One should also expect interesting and practically important results on the detection and study of new chemiluminescent reactions with the participation of macromolecular complexes of Eu^{2+} ions, characterized by record high photoluminescence yields (ϕ_{PL}). Among them are the europium sandwich complexes: $[Eu(C_5Ph_5)_2]$ and $[Eu(C_5Ph_4H)_2(DME)]$ with $\phi_{PL} = 45\%$ and 41%, respectively [84]; $[Eu(Cp^{BIG})_2]$ were Cp^{BIG} = (4-nBu-C_6H_4)$_5$-cyclopentadienyl with $\phi_{PL} = 45\%$ [85]; $Eu(BH_4)_2\cdot(THF)_2$ ($\phi_{PL} = \sim70\%$) [86]; Eu^{2+}-containing aza-222 cryptate with $\phi_{PL} = 26\%$ [87].

REFERENCES

[1] Bulgakov, R. G. (1975). The formation of excited states of the uranyl and lanthanide ions in chemical and electrochemical reactions in solution. PhD diss. Institute of Chemistry, Ufa.

[2] Bulgakov, R. G., Kazakov, V. P., Parshin, G. S., Afonichev, D. D., Sharipov, G. L. (1975). Chemiluminescence reaction between UO^{2+}_2 and Eu^{2+} ions in solution. *High Energy Chem. 9:* 92-93.

[3] Elbanowski, M., Wierzchowski, J., Paetz, M., Slawinski, J. (1983). Chemiluminescent Oxidation Reaction of Eu^{2+} Ions with H_2O_2. *Z. Naturforsch. 38a:* 808-810.

[4]   Thomas, A. C., Ellis, A. B. (1984). Chemiluminescent reactions of bis(η^5-pentamethylcyclopentadienyl)ytterbium derivatives. *J. Chem. Soc., Chem. Commun.* 1270-1271.

[5]   Wildes, P. D., White, E. H. (1971). Dioxetane-sensitized chemiluminescence of lanthanide chelates. Chemical source of "monochromatic" light. *J. Am. Chem. Soc. 93:* 6286-6288.

[6]   McCapra, F., Watmore, D. (1982). Metal catalyzed light emission from a dioxetan. *Tetrahedron Lett. 23:* 5225-5228.

[7]   Benedict, B. L., Ellis, A. B. (1987). Zeolite-supported tetramethyl-1,2-dioxetane: new pathways to chemiluminescence. *Tetrahedron. 43:* 1625-1633.

[8]   Trofimov, A. V., Vasil`ev, R. F., Mielke, K., Adam, W. (1995). Energy transfer-enhanced chemiluminescence of adamantanone (n,π*) and ester (π,π*) singlet and triplet excited states in the thermolysis of silyloxyaryl-substituted spiroadamantyl dioxetanes. *Photochem. Photobiol. 62:* 35-43.

[9]   Voloshin, A. I., Shavaleev, N. M., Kazakov, V. P. (2000). Chemiluminescence of praseodymium (III), neodymium (III) and ytterbium (III) beta-diketonates in solution excited from 1,2-dioxetane decomposition and singlet–singlet energy transfer from ketone to rare-earth b-diketonates. *J. Lumin. 91:* 49-58.

[10] Bulgakov, R. G., Kazakov, V. P., Parshin, G. S., Dmitrieva, E. V. (1974). Chemiluminescence of rare earth ions (Dy^{3+}, Tb^{3+}) in concentrated H_2SO_4 under action of ozone. *High Energy Chem. 8:* 85-86.

[11] Belyakov, V. A., Vasil'ev, R. F. (1970). Molecular Photonics. Leningrad: Nauka.

[12] Sharov, V. S., Suslova, T. B., Deev, A. I., Vladimirov, Yu. A. (1980). Activation of chemiluminescence in lipid peroxide oxidation by tetracycline-europium compex. *Biofizika. 25:* 923-924.

[13] Kaczmarek, M. (2011). Chemiluminescence of the Reaction System Ce(IV) – Non-Steroidal Anti-Inflammatory Drugs Containing Europium(III) Ions and its Application to the Determination of Naproxen in Pharmaceutical Preparations and Urine. *J. Fluoresc. 21:* 2201-2205.

[14] Bulgakov, R. G., Kuleshov, S. P., Handozhko, V. N., Beletskaya, I. P., Tolstikov, G. A., Kazakov, V. P. (1988). Chemiluminescence in the interaction of (C_5H_5)$_3$Sm solutions and samarium-, europium-organic peroxides with H_2O. *Russ. Chem. Bull. 37:* 1689-1690.

[15] Bulgakov, R. G., Kuleshov, S. P., Handozhko, V. N., Beletskaya, I. P., Tolstikov, G. A., Kazakov, V. P. (1988). Chemiluminescence in the oxidation and photoluminescence of Sm and Eu bis- and tricyclopentadienyl compounds. *Russ. Chem. Bull. 37:* 1735-1735.

[16] Bulgakov, R. G., Tishin, B. A., Kuleshov, S. P., Makhaev, V. D., Borisov, A. P., Kazakov, V. P., Tolstikov, G. A. (1989). Photo- and chemiluminescence of

$NaLn(BH_4)\cdot nDME$ complexes upon reaction with O_2 in solution. *Russ. Chem. Bull. 38:* 434-435.

[17] Bulgakov, R. G., Kuleshov, S. P., Khandozhko, V. N., Beletskaya, I. P., Tolstikov, G. A., Kazakov, V. P. (1989). Chemi- and photoluminescence in autooxidation and hydrolysis of cyclopentadienides of Eu(II, III) and Sm(II, III) and interaction of their peroxides with H_2O. *Dokl. Akad. Nauk SSSR 304:* 114-118.

[18] Bulgakov, R. G., Kuleshov, S. P., Makhmutov, A. R. (2007). Lanthanide(III) ion as a luminescent and catalytically active reaction center of aniline condensation with butyraldehyde. *Russ. Chem. Bull. 56:* 443-445.

[19] Bünzli, J. C. G. (2010). Lanthanide Luminescence for Biomedical Analyses and Imaging. *Chem. Rev. 110:* 2729-2755.

[20] Bünzli, J. C. G., Piguet, C. (2005). Taking advantage of luminescent lanthanide ions. *Chem. Soc. Rev. 34:* 1048-1077.

[21] Bünzli, J. C. G., Eliseeva, S. V. (2011). Basics of Lanthanide Photophysics. In: Lanthanide Luminescence: Photophysical, Analytical and Biological Aspects, edited by P. Hänninen, H. Härmä, 1-45. Berlin: Springer Verlag.

[22] Bünzli, J. C. G. (2016). Lanthanide light for biology and medical diagnosis. *J. Lumin. 170:* 866-878.

[23] Lis, S. (2002). Luminescence spectroscopy of lanthanide(III) ions in solution. *J. Alloys and Comp. 341:* 45-50.

[24] Danja, B. A. (2016). A review of lanthanides as activators in luminescence. *IOSR-JAC 9:* 104-110.

[25] Zhang, X. R. Baeyens, W. R. G., Van Der Weken, G., Calokerinos, A. C., Nakashima, K. (1995). Chemiluminescence determination of captopril based on a Rhodamine B sensitized cerium(IV) method. *Anal. Chim. Acta 303:* 121-125.

[26] Wolfbeis, O. S., Durkop, A., Wu, M., Lin, Z. (2002). A europium-ion-based luminescent sensing probe for hydrogen peroxide. *Angew. Chem. Int. Ed. 41:* 4495-4498.

[27] Elbanowski, M., Makowska, B., Staninski, K., Kaczmarek, M. (2000). Chemiluminescence of systems containing lanthanide ions. J. *Photochem. Photobiol. A – Chem. 130:* 75-81.

[28] Bulgakov, R. G., Eliseeva, S. M., Galimov, D. I. (2013). The first observation of emission of electronically-excited states of the divalent Eu^{2+*} ion in the new chemiluminescent system $EuCl_3\cdot 6H_2O–Bu^i_2AlH–O_2$ and the energy transfer from Eu^{2+*} ion to the trivalent ion, Tb^{3+}. *J. Lumin. 136:* 95-99.

[29] Bulgakov, R. G., Eliseeva, S. M., Galimov, D. I. (2015). The first example of generation and emission of divalent Sm^{2+*} ion in a liquid-phase chemiluminescence in the system $SmCl_3\cdot 6H_2O–THF–Bu^i_2AlH–O_2$. *J. Photochem. Photobiol. A – Chem. 300:* 1-5.

[30] Bulgakov, R. G., Eliseeva, S. M., Galimov, D. I. (2015). The first registration of a green liquid-phase chemiluminescence of the divalent Eu^{2+*} ion in interaction of β-diketonate complexes $Eu(acac)_3 \cdot H_2O$, $Eu(dpm)_3$, $Eu(fod)_3$ and $Eu(CH_3COO)_3 \cdot 6H_2O$ with Bu^i_2AlH in THF with the participation of oxygen. *RSC Adv. 5:* 52132-52140.

[31] Bulgakov, R. G., Eliseeva, S. M., Galimov, D. I. (2016). Peculiarities of bright blue liquid-phase chemiluminescence of the Eu^{2+*} ion generated at interactions in the systems $EuX_3 \cdot 6H_2O$–THF–$R_{3-n}AlH_n$–O_2 (X=Cl, NO_3; R=Bu^i, Et and Me; n=0, 1). *J. Lumin. 172:* 71-82.

[32] Bulgakov, R. G., Kuleshov, S. P., Kinzyabaeva, Z. S., Fagin, A. A., Masalimov, I. R., Bochkarev, M. N. (2007). Chemiluminescence in the reaction of LnI_2 (Ln = Dy, Nd) with water. *Russ. Chem. Bull. 56:* 1956-1959.

[33] Bulgakov, R. G., Masyagutova, G. A., Mamykin, A. V., Ostakhov, S. S., Galimov, D. I., Khursan, S. L. (2017). Redox-chemiluminescence of the Eu^{3+*} ion and Xe^* atom generated in the solid-phase interaction of $EuSO_4$ and XeF_2. *J. Lumin. 192:* 801-807.

[34] Bochkarev, M. N., Fedushkin, I. L., Fagin, A. A., Petrovskaya, T. V., Ziller, J. W., Broomhall-Dillard, R. N. R., Evans, W. J. (1997). Der erste diskrete Thulium(II)-Komplex: [TmI$_2$(MeOCH$_2$CH$_2$OMe)$_3$]. *Angew. Chem. 109:* 12-124.

[35] Kazakov, V. P., Bulgakov, R. G. (1978). Chemiluminescence in reactions of disproportionation and uranium(5) oxidation by oxygen in solution. *Radiochemistry 20:* 102-108.

[36] Newton, T. W., Baker, F. B. (1965). A uranium(V)-uranium(VI) complex and its effect on the uranium(V) disproportionation rate. *Inorg. Chem. 4:* 1166-1170.

[37] Bulgakov, R. G., Kazakov, V. P., Lotnik, S. V. (1975). Pentavalent uranium, a quencher of excited-state of uranyl in solution. *High energy Chem.* 493-494.

[38] Parker, C. A. (1968). Photoluminescence of Solutions: With Applications to Photochemistry and Analytical Chemistry, 1st edition. Amsterdam, London, N.Y.: Elsevier.

[39] Hill, R. J., Kemp, T. J., Allen D. M., Cox A. (1974). Absorption spectrum, lifetime and photoreactivity towards alcohols of the excited state of the aqueous uranyl ion (UO^{2+}_2). *J. Chem. Soc., Faraday Trans. I. 70:* 847-857.

[40] Dobrolubskaya, T. S. (1968). Luminescence Methods for Determination of Uranium. Moscow: Nauka.

[41] Elbanowski, M., Staninski, K., Kaczmarek, M., Lis, S. (2001). Energy transfer in the chemiluminescent system: Eu(II)/(III)-N_3^--H_2O_2. *J. Alloys Compd. 323-324:* 670-672.

[42] Elbanowski, M., Staninski, K., Kaczmarek, M. (1998). The nature of the emitters in Eu(II)/(III)-coronand-H_2O_2 chemiluminescent systems. *Spectrochim. Acta A 54:* 2223-2228.

[43] Elbanowski, M., Staninski, K., Kaczmarek, M., Lis, S. (2003). A comparative study on chemiluminescence properties of some inorganic systems. *Int. J. Photoenergy 5:* 239-242.

[44] Elbanowski, M., Paetz, M., Slawinski, J., Ciesla, L. (1988). Chemiluminescence and fluorescence of the europium ions-adenine nucleotides system and its possible biological significance. *Photochem. Photobiol. 47:* 463-466.

[45] Elbanowski, M., Paetz, M., Slawinski, J., Ludwiczak, E. (1991). Chemiluminescence of nicotinamide adenine dinucleotide reduced disodium salt-europium ions-hydrogen peroxide system. *J. Photochem. Photobiol. A: Chem. 62:* 27-31.

[46] Paetz, M., Elbanowski, M. (1990). Chemiluminescence and mechanism of the reactions in the (Eu(II)-Eu(III)-deoxyribonucleic acid-H_2O_2 system. *J. Photochem. Photobiol. A: Chem. 55:* 63-69.

[47] Ellis, A. B., Thomas, A. C., Schlesener, C. J. (1984). Excited-state properties of cyclopentadienylytterbium complexes. Inorg. Chim. Acta 94: 20-21.

[48] Okaue, Y., Isobe, T. (1988). Characterizations of divalent lanthanoid iodides in tetrahydrofuran by UV-Vis, fluorescence and ESR spectroscopy. *Inorg. Chim. Acta 144:* 143-146.

[49] Bulgakov, R. G., Kazakov, V. P., Korobeinikova, V. N. (1973). Photoluminescence and electron transfer in acidic aqueous solutions of Eu(II). *Opt. Spectrosk. 35:* 497-501.

[50] Bulgakov, R. G., Kuleshov, S. P., Khandozhko, V. N., Beletskaya, I. P., Tolstikov, G. A., Kazakov, V. P. (1988). Chemiluminescence upon the reaction of solutions of $(C_5H_5)_3Sm$ and organosamarium and organoeuropium peroxides with H_2O. *Russ. Chem. Bull. 37:* 1504-1504.

[51] Evans, W. V., Dufford, R. T. (1923). Luminescence of compounds formed by the action of magnesium on para-dibromobenzene and related compounds. *J. Am. Chem. Soc.* 45: 278-285.

[52] Bulgakov, R. G., Minsker, S. K., Yakovlev, V. N., Tolstikov, G. A., Dzhemilev, U. M., Kazakov, V. P. (1983). Chemiluminescence during reaction of peroxide $(EtO)_2AlOOEt$ with H_2O. *Russ. Chem. Bull. 32:* 1271-1274.

[53] Bulgakov, R. G., Kuleshov, S. P., Yakovlev, V. N., Maistrenko, G. Ya., Tolstikov, G. A., Kazakov, V. P. (1986). Chemiluminescence testing of solutions to detect organometallic peroxides during oxidation of organometallic compounds by oxygen. *Russ. Chem. Bull. 35:* 2022-2024.

[54] Gundermann, K. D., McCapra, F. (1987) Chemiluminescence in organic Chemistry. Berlin: Springer-Verlag.

[55] Poluektov, N. S., Kononenko, L. I., Efryushina, N. P., Bel`tyukova, S. V. (1989). Spectrophotometric and Luminescent Methods for the Determination of Lanthanides. Kiev: Nauk. Dumka.

[56] Bulgakov, R. G., Kuleshov, S. P., Zuzlov, A. N., Mullagaleev, I. R., Khalilov, L. M. (2001). Dehydration of $LnCl_3 \cdot 6H_2O$ (Ln=Tb, Nd, Dy) in the reaction with i-Bu_3Al, Et_3Al, Et_2AlCl, $EtAlCl_2$ and formation of the complexes $LnCl_3 \cdot 3(BuO)_3PO$. *J. Organomet. Chem. 636:* 56-62.

[57] Galimov, D. I., Bulgakov, R. G. (2019). The first example of fluorescence of the solid individual compounds of Eu^{2+} ion: $EuCl_2$, EuI_2, $EuBr_2$. *Lumin. 34:* 127-129.

[58] Makhaev, V. D., Borisov, A. P., Tarasov, B. P., Semenenko, K. N. (1981). Synthesis and physicochemical properties of anionic borohydride complexes of cerium subgroup rare earths. *J. Inorg. Chem. 26:* 2645-2651.

[59] Bulgakov, R. G. (1990). Chemiluminescence of organometallic compounds in solution. Doctor diss. Institute of Chemistry, Ufa.

[60] Bulgakov, R. G., Maistrenko, G. Ya., Yakovlev, V. N., Kuleshov, S. P., Tolstikov, G. A., Kazakov, V. P. (1985). New property of organometallic compounds – redox-chemiluminescence during interaction with XeF_2 and O_2 in solution. *Dokl. Akad. Nauk SSSR 282:* 1385-1389.

[61] Tramsek, M., Zemva, B. (2006). Synthesis, properties and chemistry of xenon(II) fluoride. *Acta Chim. Slov. 53:* 105-116.

[62] Cottrell, T. L. (1958). The Strengths of Chemical Bonds, 2nd Edition. London: Butterworth.

[63] Johnston, H. S., Woolfolk, R. (1964). Reaction rates of xenon fluorides with oxides of nitrogen. *J. Chem. Phys. 41:* 269-273.

[64] Bulgakov, R. G., Eliseeva, S. M., Galimov, D. I. (2013). Reduction of Ln^{III} to Ln^{II} in reactions of $LnCl_3 \cdot 6H_2O$ (Ln = Eu, Yb, and Sm) with Bu^i_2AlH in THF. The formation of soluble luminescent complexes $LnCl_2 \cdot x$THF. *Russ. Chem. Bull. 62:* 2345-2348.

[65] Bulgakov, R. G., Kazakov, V. P., Tolstikov, G. A. (1990). Chemiluminescence of organometallics in solution. *J. Organomet. Chem. 387:* 11-64.

[66] Jiang, J., Higashiyama, N., Machida, K. I., Adachi, G. Y. (1998). The luminescent properties of divalent europium complexes of crown ethers and cryptands. *Coord. Chem. Rev. 170:* 1-29.

[67] Sabbatini, N., Ciano, M., Dellonte, S., Bonazzi, A., Balzani, V. (1982). Absorption and emission properties of a europium (II) cryptate in aqueous solution. *Chem. Phys. Lett. 90:* 265-268.

[68] Gusev, A. N., Hasegawa, M., Shimizu, T., Fukawa, T., Sakurai, S., Nishchymenko, G. A., Shul`gin, V. F., Meshkova, S. B., Linert, W. (2013). Synthesis, structure and luminescence studies of Eu(III), Tb(III), Sm(III), Dy(III) cationic complexes with acetylacetone and bis(5-(pyridine-2-yl)-1,2,4-triazol-3-yl)propane. *Inorg. Chim. Acta 406:* 279-284.

[69] Bulgakov, R. G., Kazakov, V. P., Tolstikov, G. A. (1989). Chemiluminescence of Organometallic Compounds, Moscow: Nauka.

[70] Vasil`ev, R. F., Tsaplev, Y. B. (2006). Light-created chemiluminescence. *Russ. Chem. Rev. 75:* 989-1002.

[71] Shen, Z., Ouyang, J., Wang, F., Hu, Z., Yu, F., Qian, B. (1980). The characteristics of lanthanide coordination catalysts and the cis-polydienes prepared therewith. *J. Polym. Sci.: Polym. Chem. 18:* 3345-3357.

[72] Evans, W. J., Champagne, T. M., Giarikos, D. G., Ziller, J. W. (2005). Lanthanide metallocene reactivity with dialkyl aluminum chlorides: Modeling reactions used to generate isoprene polymerization catalysts. *Organometallics 24:* 570-579.

[73] Sager, W. F., Filipescu, N., Serafin, F. A. (1965). Substituent Effects on Intramolecular Energy Transfer. I. Absorption and Phosphorescence Spectra of Rare Earth β-Diketone Chelates. *J. Phys. Chem. 69:* 1092-1100.

[74] Adachi, G. Y., Tomokiyo, K., Sorita, K., Shiokawa, J. (1980). Luminescence of divalent europium complexes with crown ethers and polyethylene glycols. *J. Chem. Soc., Chem. Commun.,* 914-915.

[75] Adachi, G. Y., Sorita, K., Kawata, K., Tomokiyo, K., Shiokawa, J. (1983). Luminescence of divalent europium complexes with crown ethers, cryptands and polyethylene glycols. *J. Less. Common Metals 93:* 81-87.

[76] Adachi, G. Y., Fujikawa, H., Tomokiyo, K., Sorita, K., Kawata, K., Shiokawa, J. (1986). Luminescence properties of divalent europium complexes with 15-crown-5 derivatives. *Inorg. Chim. Acta, 113:* 87-90.

[77] Swiatek, K., Godlewski, M., Ninisto, L., Leskela, M. (1992). SrS:Eu and CaS:Eu Thin Films: Influence of Host Lattice on Kinetics of Europium Emission. *Acta Phys. Pol. A, 82:* 769-772.

[78] Sastry, I. S. R., Bacalski, C. F., McKittrick, J. (1999). Preparation of Green-Emitting $Sr_{1-x}Eu_xGa_2S_4$ Phosphors by a Solid-State Rapid Metathesis Reaction. *J. Electrochem. Soc., 146:* 4316-4319.

[79] Galimov, D. I., Yakupova, S. M., Bulgakov, R. G. (2018). Divalent Eu^{2+} ion – an effective inorganic mediator of energy transfer from the primary chemiluminescence emitter $^3Me_2CHC(H)=O^*$ on Tb^{3+} and $Ru(bpy)_3^{2+}$ ions. *Luminescence 33:* 1365-1370.

[80] Arppe, R., Kofod, N., Junker, A. K. R., Nielsen, L. G., Dallerba, E., Sørensen, T. J. (2017). Modulation of the Photophysical Properties of 1-Azathioxanthones by Eu^{3+}, Gd^{3+}, Tb^{3+}, and Yb^{3+} Ions in Methanol. *Eur. J. Inorg. Chem. 2017:* 5246.

[81] Kalyanasundaram, K. (1982). Photophysics, photochemistry and solar energy conversion with tris(bipyridyl) ruthenium(II) and its analogues. *Coord. Chem. Rev. 46:* 159-244.

[82] Cui, X., Zhao, J., Karatay, A., Yaglioglu, H. G., Hayvali, M., Küçüköz, B. (2016). A $Ru(bipyridine)_3[PF_6]_2$ Complex with a Rhodamine Unit – Synthesis, Photophysical Properties, and Application in Acid-Controllable Triplet–Triplet Annihilation Upconversion. *Eur. J. Inorg. Chem. 32:* 5079-5088.

[83] Dong, Z. N., Wu, Y. S., Wang, Z., He, A., Li, M., Chen, M., Du, H., Ma, Q., Liu, T. (2013). Effect of temperature on the photoproperties of luminescent terbium sensors for homogeneous bioassays. *Lumin. 28:* 156-161.

[84] Kelly, R. P., Bell, T. D. M., Cox, R. P., Daniels, D. P., Deacon, G. B., Jaroschik, F., Junk, P. C., Le Goff, X. F., Lemercier, G., Martinez, A., Wang, J., Werner, D. (2015). Divalent tetra-and penta-phenylcyclopentadienyl europium and samarium sandwich and half-sandwich complexes: synthesis, characterization, and remarkable luminescence properties. *Organometallics 34:* 5624-5636.

[85] Harder, S., Naglav, D., Ruspic, C., Wickleder, C., Adlung, M., Hermes, W., Eul, M., Pöttgen, R., Rego, D. B., Poineau, F., Czerwinski, K. R., Herber, R. H., Nowik, I. (2013). Physical Properties of Superbulky Lanthanide Metallocenes: Synthesis and Extraordinary Luminescence of $[Eu^{II}(Cp^{BIG})_2]$ (Cp^{BIG}=(4-nBu-$C_6H_4)_5$-Cyclopentadienyl). *Chem. Eur. J. 19:* 12272-12280.

[86] Marks, S., Heck, J. G., Habicht, M. H., Ona-Burgos, P., Feldmann, C., Roesky, P. W. (2012). $[Ln(BH_4)_2(THF)_2]$ (Ln = Eu, Yb) – A Highly Luminescent Material. Synthesis, Properties, Reactivity, and NMR Studies. *J. Am. Chem. Soc. 134:* 16983-16986.

[87] Kuda-Wedagedara, A. N. W., Wang, C., Martin, P. D., Allen, M. J. (2015). Aqueous Eu^{II}-Containing Complex with Bright Yellow Luminescence, *J. Am. Chem. Soc. 137:* 4960-4963.

In: A Comprehensive Guide to Chemiluminescence ISBN: 978-1-53616-170-0
Editor: Luís Pinto da Silva © 2019 Nova Science Publishers, Inc.

Chapter 5

COMPUTATIONAL AND EXPERIMENTAL ASPECTS OF ACRIDINIUM ESTERS. CHEMILUMINESCENCE AND THEIR APPLICATIONS

Karol Krzymiński and Beata Zadykowicz*

Laboratory of Luminescence Research, Faculty of Chemistry,
University of Gdansk, Gdansk, Poland

ABSTRACT

Wide range of applications of chemiluminogenic acridinium salts in medical and chemical analytics oblige us to better understand the processes giving rise to the emission of light. The quantum-chemistry methods can be helpful to explain processes leading to chemiluminescence and physicochemical properties of investigated molecules. The density functional theory (DFT) and time-dependent density functional theory (TD DFT) calculations are performed to describe the mechanisms of reactions of acridinium esters in detail. The calculations indicate that the transformations of acridinium esters in an alkaline environment in the presence of oxidants give rise to electronically excited 10-substituted acridin-9-ones, but also products of hydrolysis of acridinium esters and concurrent dark reactions, not leading to the emission of light. The correct selection of computational methods are important issues for the rational design of new acridine-based chemiluminogens of potential utility. Since the late 1980s, the acridinium esters have had widespread success in modern ultra-sensitive luminescence diagnostics as markers and indicators - displacing to some extent the techniques based on radiolabels or luminol derivatives - in respect to which they present a number of advantages as emitters. In recent years, a number of new acridinium-based molecular systems have been developed, in order to obtain possibly high efficiency and dynamics of emission and, at the same time, sufficient chemical stability under assays and in storage. The structure of dozens of chemiluminogenic acridinium salts and relative compounds as well as their thermodynamic

* Corresponding Author's Email: karol.krzyminski@ug.edu.pl.

characteristics were assessed in detail - both experimentally and theoretically. Investigations has allowed to define the features of emitters and environment that exert a key influence on the development of light generation. Spectroscopic studies with the use of advanced techniques allowed to propose structure-property relationships, facilitating a rational design of new acridinium chemiluminogenic systems that can be useful in modern ultra-sensitive analytics.

1. THEORETICAL STUDIES ON ACRIDINIUM ESTERS

1.1. Computational Methods Used to Understand the Generation of Chemiluminescence of Acridinium Esters

Acridinium esters (AE) have been known as chemiluminogenic systems since the 1960s when McCapra's research group turned its attention to these derivatives. In the first scientific reports the synthesis and luminescence properties were described [1, 2]. About twenty years later, a practical importance was also presented and since that time AE have found numerous applications in medical, chemical and environmental analysis as chemiluminescent (CL) indicators and fragments of labels [3, 4]. The chemiluminogenic properties of AE and their application in analytics oblige us to better understand the processes giving rise to the emission of light. The quantum-chemistry methods have been helpful to explain processes leading to emission of light and physicochemical and spectroscopic properties of investigated molecules. The first use of computational chemistry methods in studies on acridine derivatives, described in the 1990s [5, 6] were concerned only with semi-empirical methods [7-9]. These methods are based on the Hartree–Fock formalism [10], but they employ many approximations and empirical corrections, which are omitted in *ab initio* methods. On the other hand, a parametrization can cause problems, especially when a structure of computed molecule is deviated from a structure used for parameterization. Also, the valance electrons are only included in the calculation and the core electrons are considered as pseudopotentials. The second computational method used to investigate acridinium derivatives is the density functional theory (DFT) [11], where the basic foundation is the Hohenberg-Kohn theorems [12], not the Schrödinger equation as in the case of *ab initio* methods. Therefore, the wave function in the Schrödinger equation has been replaced by the electron density function of the ground state. In practical terms, the modern DFT method is based on the Kohn–Sham equation (also called the one electron Schrödinger equation), in which a fictitious system (Kohn–Sham system) of non-interacting particles (typically electrons) generate the same density as any given system of interacting particles. Therefore, the electron density is represented through one-electron orbitals.

In the early 2000s, Błażejowski's group examined a sort of quantum-chemistry methods, such as semiempirical (MNDO, AM1, PM3, MNDO/d), *ab initio* (HF, MP2) and those based on density functional theory (DFT) with the 6-31G**, 6-31++G** and 6-311G** basis sets to predict structure and properties (enthalpies of formation, HOMO and LUMO energies and dipole moments) of various acridines [13]. It was noticed, that the standard enthalpies of formation, determined at the DFT level of theory using B3LYP functional [14-16], in which Becke's nonlocal exchange and the Lee-Yang-Parr correlation functionals were applied, and 6-31G** basis set [17, 18], correlate much better with the experimental values than those obtained at semiempirical or *ab initio* levels of theory. In DFT, the hybrid functionals (such as B3LYP [14-16] or M06-2X [19]) are considered generally quite reliable in predicting the thermochemical and physicochemical characteristics of gaseous molecules in the ground electronic state [20].

All mentioned quantum-chemistry methods (semiempirical, *ab initio* and DFT) can be employed to understand the mechanism of reactions or physicochemical properties of molecules in the gaseous phase. But the chemiluminescence of acridinium salts is a process, in which these derivatives may undergo oxidation in an alkaline environment, yielding electronically excited acridin-9-ones, that are able to emit visible light. Therefore, all the processes leading to CL occur in a solution, and it is necessary to use some method when taking into account the effect of the solvent. The PCM (polarizable continuum model) [21, 22] and the COSMO (conductor-like screening model) [23] are methods commonly used in computational chemistry to model solvation effects. The PCM solvation model uses the solvent as a polarizable continuum, rather than individual molecules, utilizing the exact dielectric boundary conditions. The COSMO solvation model approximates the solvent by a dielectric continuum, surrounding the solute molecules outside of a molecular cavity, using the rough scaling function of permittivity. The comparison of COSMO and PCM formalism showed that the differences between the models are negligible, but unfortunately deviations to experimental data might be significant.

The important thing in an investigation of systems based on acridine and their luminogenic properties or photochemical processes is an ability to study excited states. As mentioned above, the DFT method is a ground electronic state theory, but the extension, called the time-dependent DFT (TD DFT) [24], can be used that allows applying DFT to excited electronic states. The basic foundation of TD DFT is the Runge–Gross theorem [25] that connects the time-dependent density with the time-dependent external potential (such as electric or magnetic or both fields). This approach allows direct computation of excited states applying the perturbation theory. By using this method properties such as optical matrix elements and atomic forces can be calculated. The latter can be used to obtain a geometry optimisation or molecular dynamics run to be performed for a selected excitation. The TD DFT method allows us to obtain electronic excited states with rather low computational cost and can be applied for relatively large systems.

1.2. Reaction of Acridinium Esters with Oxidants in Alkaline Media

1.2.1. Origin of Chemiluminescence of Acridinium Esters

Acridinium esters may undergo oxidation by hydrogen peroxide or other oxidants in alkaline media relatively easily, yielding electronically excited molecules that are able to emit light in the visible range of the spectrum [26, 27]. The two premises that must be taken into account are whether oxidants dissociate in an alkaline media to the anionic form (environment of the reaction), and if the generated anions can react with acridinium cations.

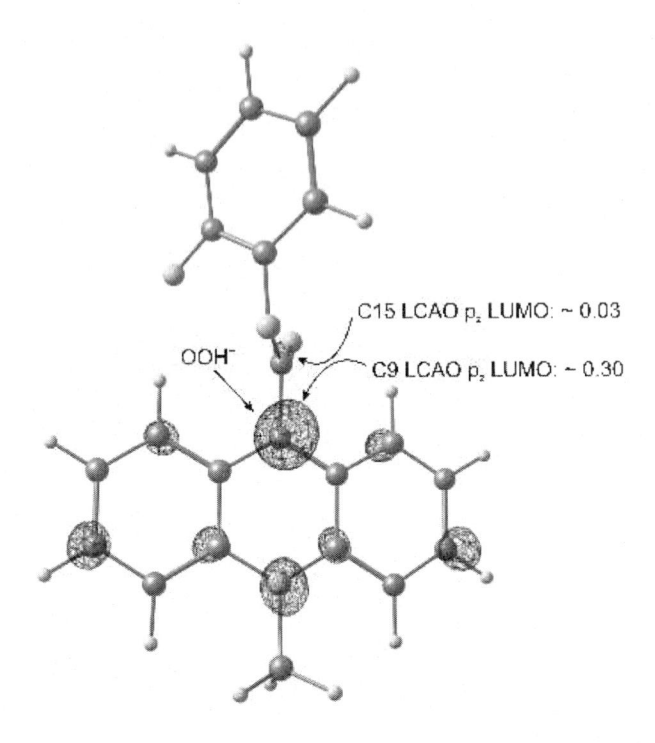

Figure 1. Computationally predicted Lowest Unoccupied Molecular Orbitals (LUMO) of acridinium ester with the values of the LCAO coefficient of the p_z atomic orbital in LUMO orbitals at the C9 and the C15 atom.

The first postulate was studied thoroughly applying different quantum-chemistry methods: semiempirical [6] and DFT [28-30]. The results suggest that hydrogen peroxide, which is the most commonly used and effective oxidant, reacts spontaneously with OH⁻ to yield OOH⁻. The reaction of CL of AE is initiated by nucleophilic attack of OOH⁻ at the electrophilic center in the cation. To better understand where the electrophilic center of the molecule is located, one can refer to the frontier molecular orbital theory [31], which describes the electronic structure of molecules. According to this theory, the bonding of molecules approximates the molecular orbitals as linear combinations of atomic orbitals (LCAO). The Lowest Unoccupied Molecular Orbital (LUMO) distribution of an electrophilic species determines the molecular centre sensitive to nucleophilic attack. The calculated value of the LCAO coefficient of the p_z atomic orbital in AEs' LUMO is ca. 10 times higher at the endocyclic C9 than at the carbonyl C15 atom (the ratio of the value of the LCAO coefficient of the p_z atomic LUMO orbital at the C9 and C15 among the investigated AEs bearing substituents in the phenyl group of different donor-acceptor character, is ranged from 8 to 16). This implies that C9 rather than C15 makes the site of the primary nucleophilic attack of the above-mentioned anions. On the other hand, the distribution of charges should also show which atom the largest deficit of the negative charge is on. Unfortunately, the Mulliken partial charge at C9 is lower than at the C15 atom of the carbonyl group (ca. 6 times) [28, 29]. This suggest that the population analysis does not indicate well the location of the electrophilic centre and using the frontier molecular orbital theory is the more appropriate approach.

1.2.2. Mechanism of Chemiluminescence of Acridinium Esters and Concurrent Reactions

The mechanism of oxidation of *N*-substituted-acridinium cations with hydrogen peroxide in alkaline media has been discussed by several authors [3, 26] and examined at the semiempirical and DFT level of theory [6, 28-30]. In the computational experiment, selected acridinium esters were optimized and then the trend of the energy values for the nucleophilic attack of anionic molecules at the endocyclic C9 atom was calculated. In the next step, all reactions leading to suggested reaction products (both "light" pathway and competitive dark ones) were proposed. In general, the mechanism of CL can be divided into three pathways, outlined in the Scheme 1 [6, 28-30]:

I. the processes leading to the electronically excited product and then the light emission,

II. the transformation of the acridinium cation to the non-emitting product,

III. the hydrolysis of the acridinium ester.

Only the first group of transformations are the processes leading to the chemiluminescent product – electronically excited 10-methylacridin-9-one; the second and the third group are the dark processes, not leading to the electronically excited products. The CL reaction starts from the competitive steps – nucleophilic attack of OOH$^-$ (leading to the emission of radiation) or OH$^-$ (creating of so-called 'pseudobase', leading to the dark processes (non-excited products) at the endocyclic carbon atom of acridine moiety (C9). As mentioned above, interest in nucleophilic attack at position 9 of acridine results from the frontier orbital theory, which states that the Lowest Unoccupied Molecular Orbital (LUMO) distribution of an electrophilic species determines the molecular centre sensitive to nucleophilic attack. Taking into account the competitive processes – leading to the CL (addition of the OOH$^-$) and the formation of the 'pseudobase' (addition of the OH$^-$) - the theoretically predicted data indicated that these steps are possible for all of the AEs. The Gibbs' free energy in aqueous phase, computed at standard temperature and pressure suggests that the addition of OH$^-$ to C9 is thermodynamically preferred over the addition of OOH$^-$ to the same carbon atom. In this case, the possible pathway of the reaction seems to be exclusively dependent on kinetic factors. On the other hand, the activation barriers characterizing the addition of OOH$^-$ or OH$^-$ to C9 of the AEs are negligible [28-30]. The value of the Gibbs' free energy of the addition of OH$^-$ at the endocyclic C9 is ca. 1.50 times lower than the Gibbs' free energy of the addition of OOH^- at the same carbon atom. The computational study shows that the substitution with an electron-donor group in position 2 and/or 6 and simultaneously electron-acceptor group in position 4 in the phenoxy fragment move the equilibrium towards the processes leading to the CL. Unfortunately, the additional substitution with an electron-donor group in position 2 of the acridine moiety leads to the formation of a 'pseudobase'.

The light pathway (pathway I in Scheme 1) represents the attack of OOH^- on carbon atom C9 of acridinium cation, followed by the reaction of the addition product with OH^-, the cyclization of the anion to cyclic intermediate, and the elimination of phenyl carbonate anion, leading to the energy-rich 10-methylacridin-9-one. In the case of selected AEs (substituted with NO_2 group in position 2 or Br atoms in positions 2 and 6 of phenoxy ring), the first pathway proceeds differently: after the attack of OOH^- onto carbon atom C9 of acridine moiety, followed by the reaction of the addition product with OH^-, the elimination of the adequate phenoxy anion and formed the cyclic intermediate – four-membered highly strained dioxetan structure, and the subsequent elimination of carbon dioxide leading to the energy-rich 10-methylacridin-9-one. The theoretically predicted thermodynamic data indicated that these two pathways are possible for most compounds. On the other hand, the enthalpy (gaseous phase), Gibbs' free energy (gaseous phase) and Gibbs' free energy (aqueous phase) of the second step at standard conditions is thermodynamically more preferred in the case of the compounds substituted with an electron-acceptor group in position 2 or 6 (or both) of the phenoxy moiety.

The non-chemiluminescent pathway (pathway II in Scheme 1) assumes the attack of OH^- on the C9 of the AEs, followed by the reaction of the addition product, the 'pseudobase', with OH^- and the elimination of phenoxy anion and carbon monoxide, leading to the formation of ground-state molecules of 10-methylacridin-9-one. Alternatively, the non-chemiluminescent pathway may include the attack of OOH^- on the carbonyl carbon atom C15 of the 'pseudobase', followed by the elimination of phenoxy anion, yielding complex intermediate (C9-OH/C15-OOH adduct), which decomposes to ground-state molecules of 10-methylacridin-9-one, carbon dioxide and water.

The theoretically predicted thermodynamic data indicated that these pathways are possible for all of the investigated molecular systems. The standard enthalpy (gaseous phase), Gibbs' free energy (gaseous phase) and Gibbs' free energy (aqueous phase) of the addition of the OH^- to the carbon atom in position 9 of acridine moiety, leading to the formation of the 'pseudobase', is thermodynamically more preferred than the addition of the OOH^- to the same carbon atom. In this case, the possible pathways of reaction are likely dependent on the kinetic factors. The activation barriers to the addition of OOH^- or OH^- to AEs and to the substitution at OOH^- of the carbonyl C15 atom of the 'pseudobase' and then the release of the phenoxy anion are negligible. However, the activation barriers exist only for the final steps of transformation of the 'pseudobase' to 10-methylacridin-9-one. Unfortunately, when the activation barriers are relatively small, the energy released is too small to electronically excite the molecules of final emitter (the calculated value is ca. 30–40 kcal mol^{-1}, while the literature value of electronically excite 10-methylacridin-9-one is 88.2 kcal mol^{-1} [32]) and contrarily if the thermodynamic functions are high enough to ensure electronic excitation of the 10-methylacridin-9-one formed (the calculated value is more than 100 kcal mol^{-1}), then the activation barrier characterizing this process is significantly higher than for the other processes. For these reasons, the pathway II, assuming the attack of OH$^-$ on the carbon atom 9 of AEs, could not lead to CL (pathway II represents the dark transformations of AEs).

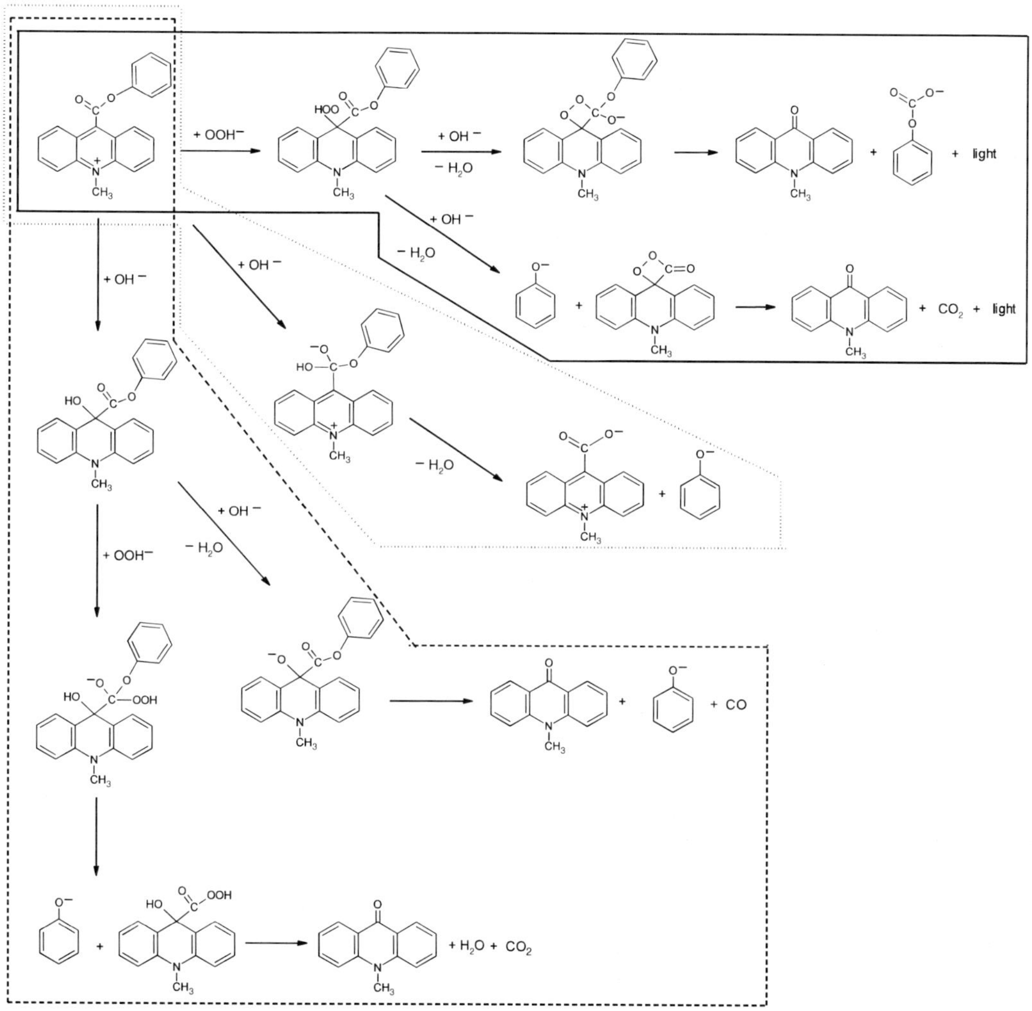

Scheme 1. Pathways of the reactions of acridinium ester with OOH$^-$ and OH$^-$: the processes leading to the electronically excited product and then the light emission (pathway I, marked with a solid line), the transformation of the acridinium cation to the non-emitting product (pathway II, marked with a dashed line), and the hydrolysis of the acridinium ester (pathway III, marked with a dotted line).

Pathway III (also a non-chemiluminescent one, outlined in Scheme 1) presents the occurrence of 10-methylacridinium-9-carboxylate anions, which are identified among the products of AEs' post-CL reaction mixtures [33]. 9-Carboxy-10-methylacridinium cation (or its zwitterion form – 10-methylacridinium-9-carboxylate) is most probably formed as a result of the hydrolysis of AEs, assuming the attack of OH^- onto the carbonyl carbon atom C15, followed by the reaction of the addition product with OH^-, and the elimination of phenoxy anion and water molecule leading to the product – 10-methylacridinium-9-carboxylate anions. The theoretically predicted data indicated, that this step is thermodynamically possible among the all of the investigated compounds and the above product was detected experimentally in easy measureable amounts.

The chromatographic (RP-HPLC) [33] and spectroscopic (MS) [30] analyses of post-CL reaction mixtures revealed that adequate 10-methylacridin-9-one, phenol and 10-methylacridinium-9-carboxylic acid derivatives make the final products after chemiluminogenic oxidation of AEs. It can be thus generally concluded, that the computationally predicted results remain in compliance with the experiments.

1.2.3. Formation and Decomposition of Cyclic Intermediates

On the light pathway of AEs' oxidation, which begins from the attack of OOH^- onto the C9 of acridinium cation, a four-member, highly strained dioxetanone cyclic structure is likely produced at certain stage of transformation (Scheme 1, pathway I). The similar intermediates are known – they are involved, for instance, in the mechanism of the firefly bioluminescence [34, 35] or 2-coumaranone [36] chemiluminescence. Regardless of whether the formation of a cyclic intermediate occurs with the elimination of the phenoxy anion or without it, this step makes the 'bottleneck' of the CL reaction and leads to the energy-rich molecules of excited product. The computationally predicted pathways of chemiluminogenic oxidation of AEs do not present clearly which pathway is the most probable and the obtained results suggest, that both are possible. The activation barriers predicted for both steps leading to a decomposition of cyclic intermediate are similar in the aqueous phase [29]. On the other hand, in the case of AEs substituted with an electron-acceptor group in position 2 or 6 (or both) of the benzene ring (eg. Br atom, NO_2 group) - where the elimination of R-phenoxy anion and the formation of the cyclic intermediate in neutral form occur - we disclosed only one thermodynamically possible pathway.

It is important to know what the nature of the decomposition of a cyclic intermediate is. In the four-membered highly strained dioxetanone entity there are two important bonds (O–O and C–C), which have an effect on the transformation in the energy-rich 10-methylacridin-9-one. In geometrical terms, the length of both bonds is in line with what is expected for these types of bonds and both ones appear to be covalent [28, 29]. A population analysis also presents a typical distribution of charges [28]. Both oxygen atoms have a negative charge and both carbon atoms have a positive charge. The negative charge of both oxygen atoms and the small charge difference indicates a repulsive electrostatic interaction between them. In the case of carbon atoms, the charge differences are significantly higher, which also indicates a repulsive effect of interaction, but attenuated by the polarization effect. The analysis of intramolecular interactions in the cyclic intermediate using the quantum theory of atoms in molecules (QTAIM) [37] and the topology of the electron localization function (data unpublished) indicates that the O–O bond character is charge-shifted, while the C–C makes a typical covalent bond. The analysis of the ellipticity, which measures π or σ bond character, demonstrates that both O–O and C–C bonds represent a sigma bond character. However, the value of ellipticity for O–O bond is very close to the value that indicates π bond character. On the other hand, the results of computational study on excited states of 1,2-dioxetane suggest that in the first step the O–O bond is broken and the molecule enters in a region of biradical character, and next the chemiexcitation occurs concomitantly with C–C bond breaking [38, 39]. The mentioned mechanism for the induced decomposition of 1,2-dioxetanes is an intramolecular version of the CIEEL (chemically initiated electron exchange luminescence) mechanism proposed in the literature [40].

1.2.4. Electronically Excited Product

The CL pathway leads to energy-rich 10-methylacridin-9-one, which when returning to the ground state emits light. The possibilities of formation of 10-methylacridin-9-one after the nucleophilic attack of OOH^- onto the carbon atom C9 in alkaline media are twofold; firstly, the cyclization of the anion to cyclic intermediate and then the elimination of phenyl carbonate anion, or secondly - the elimination of the phenoxy anion and formation of the cyclic intermediate and then the subsequent elimination of carbon dioxide. The energy released in the step leading to 10-methylacridin-9-one exceeds what is necessary to cause electronic excitation of the molecule (the literature value is 88.2 kcal mol^{-1} [32]). However, it is not clear yet, which pathway is more preferred. But the activation barriers predicted for processes leading to the formation of electronically excited molecules show, that for the second possibility they remain relatively high, which suggests that the formation of excited molecules proceeds via the elimination of phenyl carbonate anions rather than the consecutive release of phenoxy anions and carbon dioxide.

The formation of cyclic intermediate and then its decomposition is a rate determining step in acridinium CL reaction. To better understand the nature of an electronically excited product, an investigation of excited states has to be carried out. The preliminary TD DFT results indicate that the first triplet state (T_1) lay close to the ground state (S_0) at around the transition state of decomposition of cyclic intermediate [41]. The first singlet state (S_1) lay higher than T_1, but also close to S_0. According to this, the photoemission of the mixed state (from T_1 and S_1 states) of 10-methylacridin-9-one will be expected. This aspect is intensively studied in our group.

1.3. Perspectives of Application of Computational Methods in the Investigation of Acridinium Esters

The development of computer techniques and quantum chemistry methods allow us to obtain more and more accurate results, which are confirmed by the results of experiments. However, several aspects of computational investigation on AEs are still to be studied. Firstly, the formation of electronically excited product upon CL of acridinium derivatives – 10-methylacridin-9-one, is not well understood. In particular, the nonadiabatic transition through spin-orbit coupling between ground state and excited state or decomposition to reach acridin-9-one in excited state, should be examined. Using the CASSCF/CASPT2 method may be helpful. Secondly, there is a problem with the parameterization of the solvent. The COSMO and PCM solvation models utilize the solvent as a continuum, rather than as individual molecules. The interactions between a solvent and an investigated molecule are not taken into account. However, all the processes leading to CL occur in solution, so it is necessary to use some method which includes the effect of the solvent. Perhaps, a combined and sequential use of Monte Carlo simulations and quantum mechanical calculations (QM/MM) will allow to refine the description of the solvent in the investigation of acridinium chemiluminogenic salts. Also, it is worth paying attention to the studies on the mechanism of CL reactions in non-aqueous systems with the participation of organic oxidants and bases (discussed below), as they are, in general, more effective in terms of light emission than the aqueous systems.

2. EXPERIMENTAL INVESTIGATIONS ON ACRIDINIUM ESTERS

2.1. General Considerations

The experimental studies on physicochemical features of acridines capable of CL (acridinium esters and closely related compounds) were targeted at practical aspects. Simple AEs are treated here as model compounds, enabling us to gain information that can be useful in the designing and investigation of complex molecular systems, employed today as CL labels and indicators in medical diagnostics and environmental analysis. The study presents the capabilities of modern physicochemical methods in the investigation of light-emitting systems of potential usefulness. Examples of applications of chemilumino-genic acridines are also addressed.

2.1.1. Photophysics of Chemiluminescence

CL makes an attractive physicochemical process, sometimes observed in nature (then called bioluminescence), which relies on the conversion of chemical energy into the energy of electromagnetic radiation - usually occurring in the visible or near ultraviolet region. In order for the process to take place, electronically excited entities must appear in the reacting system - which is primarily limited by its energetics. As excited states are characterized by relatively short lifetimes, and the emission of radiation is quantized, the chemiluminogenic reaction should, in one stage, provide the amount of energy corresponding to at least the short-wave side of the observed spectrum. If one considers emission in the visible range, it should be roughly assessed as 170-290 kJ mol^{-1} [42]. Meeting energy requirements, however, does not guarantee the appearance of light emission, as the formation of entities emitting radiation is further limited by the need for effective permeation of the energy surfaces of the potential state of the ground state of substrate and the electronically excited states of the product. This requires similarities in the geometry of the transforming individuals as well as an effective overlap of the oscillation functions of the transition state and the electronically excited one (determined by the values of Franck-Condon factors). The efficiency of transitions among surfaces is further limited by the spin-spin rule and orbital symmetry. The above limitations decide that the CL - especially if it occurs in solutions (where the excess energy can be dissipated in collisions with solvent molecules) - is a rare phenomenon.

One of the basic parameters characterizing the processes of chemically triggered light emission is its efficiency. In quantitative terms, it can be expressed as the quantum yield of chemiluminescence (φ_{CL}), expressed by the following equation [42]:

$$\varphi_{CL} = \varphi_{CH} \times \varphi_{EX} \times \varphi_{F}$$

where φ_{CH} is the chemical yield of the formation of the product (in ground state and in excited states, P + P*), φ_{EX} is the excitation efficiency, i.e., the ratio of the number of molecules excited P* to the sum of P + P* and φ_{F} denotes the fluorescence quantum yield of the excited product P*. The above simple formula assumes the independence of the above components and does not take into account possible P* interactions with other individuals, that can be excited by energy transfer. It should be noted here that CL usually has the nature of chemically induced fluorescence, and chemo-phosphorescence processes are practically not observed.

The use of the CL in analytics requires, among others, a high efficiency of emission, which can be achieved by optimization of the above-mentioned parameters - in particular φ_{CH} and φ_{EX} - because they reflect structural modifications of the chemiluminogenic molecule (the substrate). Accordingly, knowledge concerning the mechanism of transformations and their physicochemical features (including the influence of the environment) make crucial aspects to be considered during the rational design of new CL systems that may find practical applications.

2.1.2. Classes of Chemiluminogenic Acridinum Salts

Among various classes of acridine-based molecular systems, investigated in order to be utilized in luminescence analytics, the quaternary salts of acridine-9-carboxylic acid derivatives are of great importance. Compounds of this type are presented in Figure 2.

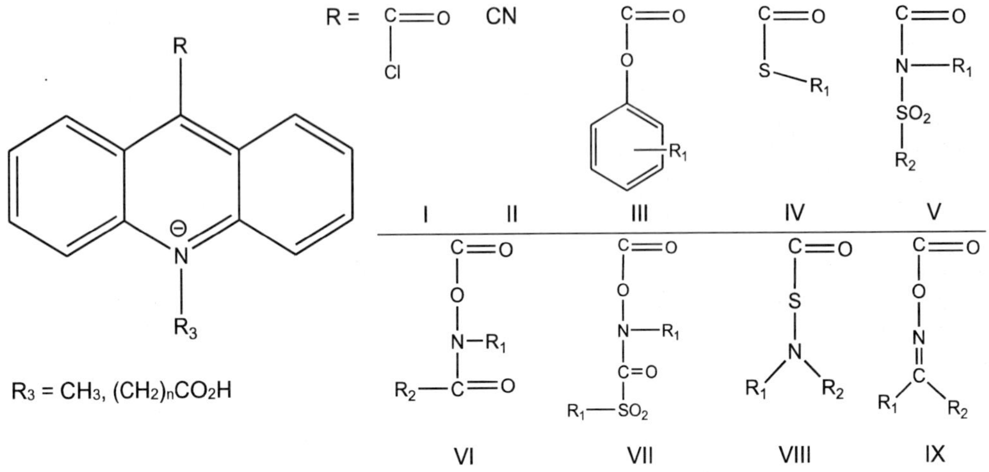

Figure 2. Chemiluminogenic 10-methylacridinium cations substituted in the 9-position by a group: chlorocarbonyl (I), cyano (II), phenoxycarbonyl (III), thioester (IV), carbonylsulfonamide (V), hydroxamic acid ester (VI), sulfohydroxamic acid ester (VII), carbonylthiolamine (VIII), carbonyloxime (IX). R_1 and R_2 denote aliphatic, aromatic groups, heteroatoms and other substitutents [43, 44].

The emission of such compounds typically occurs in alkaline hydrogen peroxide solutions. The CL of acridinium salts can also be generated indirectly - for example by introducing enzymes which produce hydrogen peroxide under the conditions of the experiment [27].

Unlike luminol derivatives, acridinium salts do not require catalysts to start the emission of light and are usually characterized by a high temporary emission intensity (up to 100 times higher than in the case of luminol). Acridinium salts are also better soluble in aqueous environments in relation to hydrazide derivatives - which is of great practical importance in biological sciences. The quantum yield of typical acridinium chemiluminogenic salts in aqueous solutions is about 2-3%, which can be considered a relatively high value if compared to the efficiency of luminol (ϕ_{CL} = 1.23%), used as a reference substance in this type of measurements [45].

In the past, the influence of the "leaving group" structure (R, Figure 2) on the CL features of the corresponding acridinium salts was observed. Sato [46] determined the relative efficiency of CL and stability in aqueous solutions of a series of aromatic AEs (group III, R_3 = CH_2CO_2H), differing in type and location of substituent in the benzene ring. The best emitters seemed to be those containing the methyl/methoxy group in the *ortho* position of benzene ring and a fluorine atom in the *para* position of one. If the stability of the compounds in solutions was considered, the 2-methylphenyl ester appeared as optimal in terms of maintaining its emissive parameters. Newer studies indicated that the built of the "leaving groups" primarily determines the kinetics of the CL process and has rather small impact on the efficiency of emissions. In turn, the substituents present in the acridine nucleus mainly determine the spectral range of the observed emission [33].

Further investigations on chemiluminogenic acridinium salts have shown that any acridine-9-carboxylic acid derivative containing a "leaving group" of $pK_a < 12$ (i.e., lower than pK_a of hydrogen peroxide) should emit radiation [47]. Among the AEs of group III, the main group of chemiluminescent acridines (Figure 2), it was stated that the emission efficiency increases linearly with the decrease in value of the pK_a of the phenoxy moiety - that is - when it becomes a "better leaving group" [48]. However, recently performed correlational studies on a large group of variously substituted 9-(phenoxycarbonyl)-10-methylacridinium salts suggest, that such a trend is actually observed - but within subgroups of closely related analogues [30].

On the other hand, acridinium salts containing less labile "leaving groups" (e.g., R = $CONH_2$, Figure 2) essentially are not chemiluminogenic, whereas the introduction in position 9 of an acridine nucleus an activating group (for example sulfonyl), makes the appropriate cations (V, Figure 2) chemiluminogenic [47]. Further studies indicated, that the acridinium sulfonamides are less susceptible to dark side reactions than AEs, as was suggested by mass spectrometry [49], but are characterized by generally lower than dynamics of temporary emission.

The effect of the alkyl group attached at the endocyclic acridine N atom (R_1 = methyl, ethyl, n-propyl, isopropyl and benzyl) on the CL efficiency of the corresponding AEs (Figure 2, group III) was investigated [50]. It turned out that the methyl derivative was characterized by the highest efficiency of emission (better by about 15-20% than other 10-substituted AEs); the chemical yield of the synthesis of the quaternary product was also the highest in this case. Thus, the attachment of a methyl group at the acridine N atom seems to be optimal among alkyl groups. Our own studies revealed that 10-methylated AEs are also significantly more effective in terms of CL efficiency than the corresponding salts containing hydrophilic substituents at the endocyclic N atom, such as carboxymethyl or acetyl (unpublished data).

As mentioned above, the target emitters in the chemiluminogenic oxidation of AEs and other salts presented in Figure 2 produce an electronically exited molecules 9-acridanones; 10-methyl-9-acridanone makes the simplest and attainable example and thus can be treated as a model compound. It seems advisable to take a closer look at such a fluorophore and its analogues. Analysis of fluorescence emission spectra of N-substituted acridin-9-ones (10-methyl-, 10-ethyl-, 10-phenyl-, 2-methyl-10-methyl- and 10-methyl-2-nitro-derivatives) in solvents of different character indicates that they fluoresce within a relatively narrow spectral range (ca. 395-455 nm). In protic environments, a slight shift towards longer wavelengths is observed, suggesting that their molecules may interact with the solvent molecules to form hydrogen-bound molecular systems. What's characteristic, in the fluorescence emission spectra of 10-substituted 9-acridanones several maxima appear (usually two), which may suggest the overlap of electronic and oscillatory transitions, but also - the emission from various excited states [32]. Theoretical studies seem to support the last thesis (see above).

The aromatic ring in N-substituted acridin-9-ones is not flat in the electron states S_0, S_1, and T_1, which underlines the fact that the central ring is mostly excluded from the coupling with other electrons. Polar solvents should flatten this fragment in the electronic ground state [32].

During the S_0-S_1 electronic excitation the symmetry of the 10-methylacridin-9-one decreases, the length of C=O bond increases and the valence angle determined by the central atoms changes (N10, C9, O15 and C9, N10, C16). Changes in the electron density accompanying the excitation of 10-methylacridin-9-ones occur mainly around the heterocyclic N atom (a decrease in density) and the adjacent aliphatic carbon atom (an increase in density). The values of dipole moment, reflecting the susceptibility of such compounds to non-specific interactions with other species, are moderate and increase in excited states in the order: $\mu(T_1) > \mu(S_1) > \mu(S_0)$. In the case of 10-methylacridin-9-one, they assume values of 4.04 D, 4.31 D and 5.46 D, respectively (gas phase) and somewhat increase in polar liquid environments.

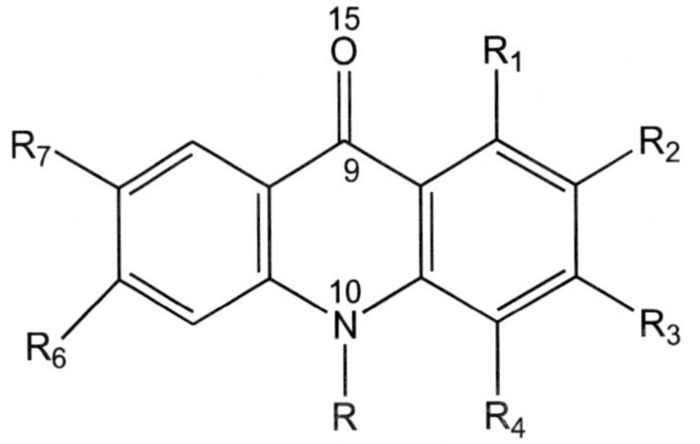

R = CH$_3$; -CH$_2$CH$_3$; C$_6$H$_5$, CH$_2$CO$_2$CH$_2$CH$_3$
R$_1$ - R$_7$ = H, OCH$_3$ in various combinations

Figure 3. Structural formula of acridin-9-one derivatives tested theoretically and experimentally for fluorescence emission properties ([32] and unpublished data).

Stationary fluorescence measurements and calculations (CNDO/S3) on the derivatives of 10-methylacridin-9-one, containing electron-donor substitutes located at various locations of the acridine ring were performed (Figure 3). Investigations were undertaken in order to select optimal fluorophores that efficiently emit radiation, bathochromic shifted in relation to the parent (unsubstituted molecule). Emission ranges above 500 nm are beneficial, avoiding difficulties associated with residual background signal and cross talks, as biological systems often fluoresce in the blue range of spectrum [47].

In general, the calculated fluorescence wavelengths maxima correlated well with those obtained experimentally. Incorporation of the substituent(s) such as OCH$_3$ to the acridine nucleus is associated with a bathochromic shift of the emission bands - but only if the substituents are introduced at specific locations. Positions 2, 4 and 7 of the acridine nucleus seem to be favorable in this respect. According to this, the 10-methylacridin-9-one of R = CH_3 and $R_2 = R_4 = R_7 = OCH_3$ expresses the longest emission among investigated compounds (maximum at > 525 nm). In turn, the attachment of OCH_3 groups to 1, 3 and/or C-6 positions does not lead to the desired spectral changes, and in some cases even causes a slight hipsochromic shift in the emission spectrum (e.g., $R_1 = OCH_3$ derivative, maximum at ca. 413 nm). Similarly to the previously observed facts [32], the fluorescence quantum yields of discussed acridan-9-ones get higher values in polar protic environments than in the aprotic ones, reaching values of 87-97% in ethanol or water.

2.2. Synthesis of Acridinium Esters and Relative Compounds

Over the last 30 years, the methods of synthesis of AEs and more complex molecular systems of practical importance have been developed. Recommended methods in newer literature are briefly presented below [51-54].

2..2.1. 9-(Phenoxycarbonyl)acridines

Compounds of this group (covalent organic bases, PCAs), being immediate precursors of AEs, are not capable for CL, although they include an aromatic ester function. Investigations of such compounds facilitate the identification of structure parameters and physicochemical analyses of the target substrates, that are chemiluminogenic acridinium salts.

The PCAs, containing various substituents in the benzene ring, were obtained in two steps, the first of which is the synthesis of the acid chloride of acridine-9-carboxylic. The next step involves the esterification of 9-(chlorocarbonyl)acridines with equimolar amount of appropriate phenol in the presence of an excess of tertiary base (*N,N,N*-triethylamine) and a catalytic amount of *N,N*-dimethyl-4-aminopyridine (DMAP), acting as catalyst in such a reaction (Figure 4) [55].

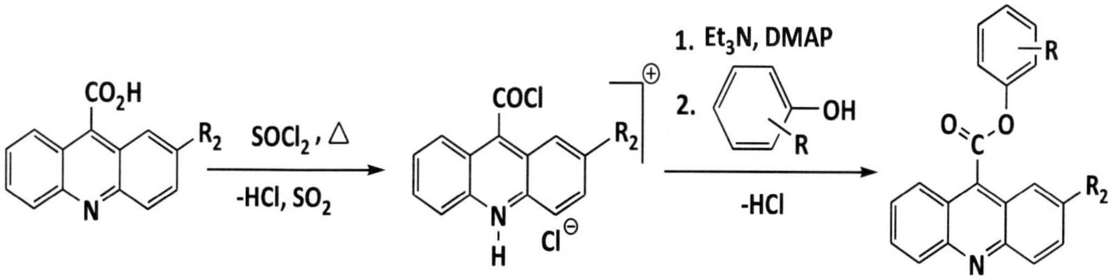

Figure 4. Synthesis scheme for 9-(phenoxycarbonyl)acridines (PCAs).

2.2.2. Acridinium Salts

9-Carboxy-10-methylacridinium cation makes the simplest derivative of acridine capable of weak CL [56]. It is also one of the products chemiluminogenic oxidation of AEs, which makes it interesting from a cognitive point of view [78]. The synthesis route for 9-carboxy-10-methylacridinium salts involves the protection of the carboxyl group (methyl ester function), the quaternization of the resulting ester with strong acid and deprotection of carboxyl group (alkaline hydrolysis of ester), followed by the separation of 9-carboxy-10-methylacridinium salt with trifluoromethanesulfonate (triflic) or hydrochloric acid in excess of dry ethyl ether (Figure 5).

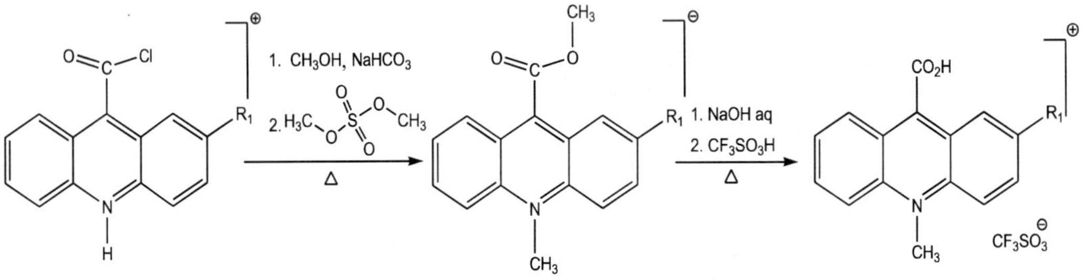

Figure 5. Synthesis of 9-carboxy-10-methylacridinium salts [56].

Chemiluminogenic acridinium esters can be obtained by the quaternization of 9-(phenoxycarbonyl)acridines with methyl triflate, serving as a highly reactive alkylation reagent. Its use is necessary, because the endocyclic acridine N atom in PCAs is deactivated towards electrophilic substitution. The methylation is carried out at RT under Ar in purified chloroform (obtained by passing by basic Al_2O_3 column immediately before the reaction), and in the presence of catalytic amounts of 2,6-di-*tert*-butylpyridine deposited on the polymeric support, acting as a proton scavenger (Figure 6).

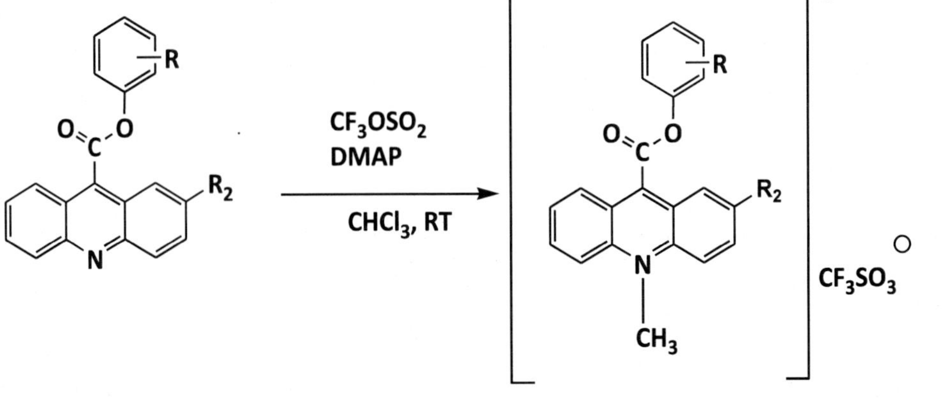

Figure 6. Synthesis route for 10-methylacridinium-9-(phenoxycarbonyl)acridine triflates (AEs) [28].

2.2.3. Reduced Forms of Acridinium Esters

Another important group of related compounds that can be classified as CL indicators are the acridan esters, being the reduced forms of AEs. Such acridans were tested in enzyme-linked immunoassays, where the emission of light is used for end-point detection instead of absorbance measurements taken in typical ELISA tests [42, 57]. The synthesis route for such molecular systems is presented in Figure 7.

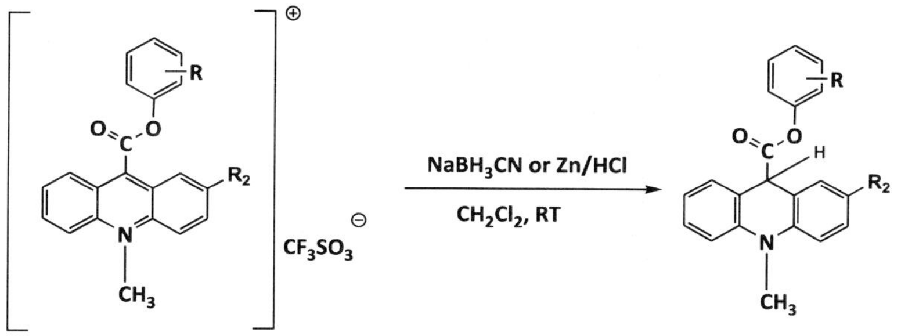

Figure 7. Synthesis route for reduced forms of AEs – acridane esters [57].

2.2.4. Products of Oxidation of Acridinium Esters

10-Substituted acridin-9-ones, in addition to their interesting photophysical properties and medicinal applications as drugs, make the main products of chemiluminogenic oxidation of AEs (see above). Such substances in an electronically excited state make reporter molecules, responsible for the light emission, occurring at the final stage of the chemiluminogenic oxidation of AEs.

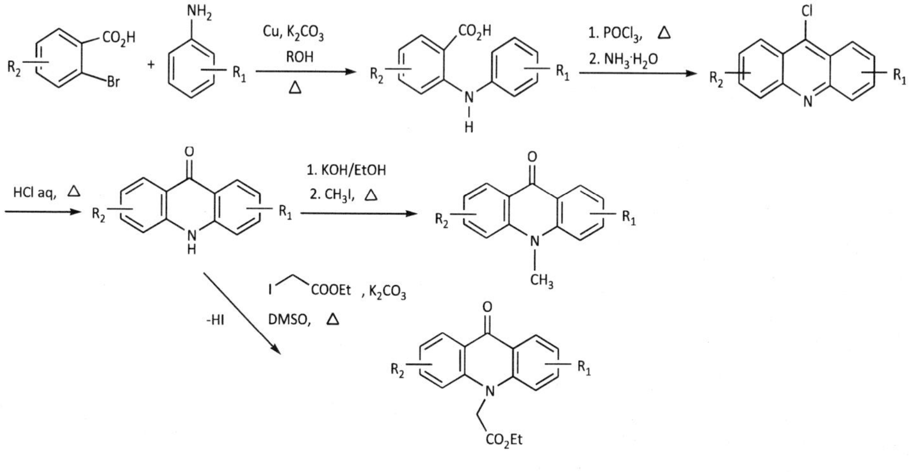

Figure 8. Synthesis route of 10-substituted acridin-9-ones – the main products of chemiluminogenic oxidation of AEs.

Acridin-9-ones can be obtained in four steps, including Jourdan-Ullmann condensation of benzoic acid and aniline derivatives, cyclization of resulting diphenylanthranilic acid derivative by the action of phosphoryl oxychloride and hydrolysis of the resulting 9-

chloroacridines. Alkylation of the endocyclic *N* atom in the resulting acridin-9-(10*H*)-ones with the appropriate reagents gives respectively 10-substituted acridin-9-ones (Figure 8).

2.2.5. Complex Acridinum Esters – Chemiluminogenic Labels

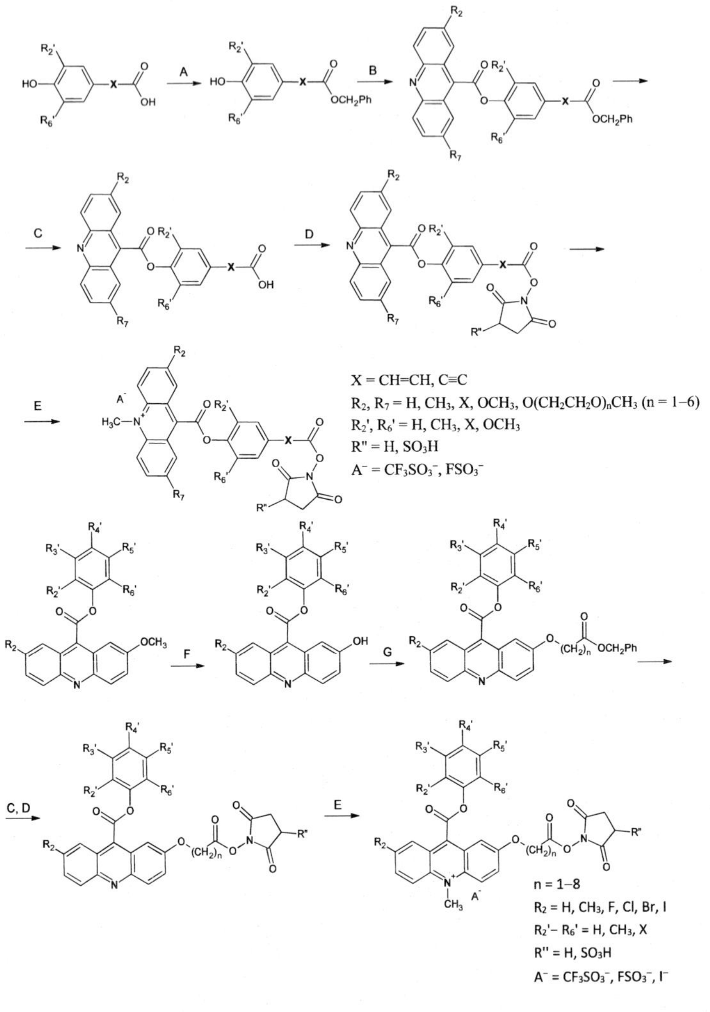

Figure 9. Synthesis routes for the synthesis of original acridinium labels of potential usability in immunodiagnostics [57].

Reagents and Conditions:

A = PhCH₂Cl, KI, K₂CO₃, DMF, RT; **B** = , Et₃N, DMAP, CH₂Cl₂, RT; **C** = HBr/CH₃CO₂H,

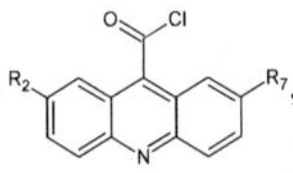

20-50°C; **D** = 1) DCC, DMF, 0-20°C; 2) 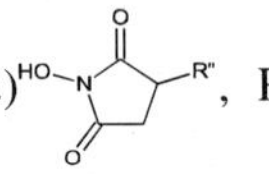 , RT; **E** = CF₃SO₃CH₃ or FSO₃CH₃, 2,6-di-*tert*-

butylpyridyne, CH₂Cl₂ or CHCl₃, RT; **F** = HBr/CH₃CO₂H, 80–150°C; **G** = 1) NaH, DMF, −20–0°C; 2) Hal(CH₂)ₙCO₂CH₂Ph (Hal = Cl, Br, I, OSO₂CF₃, OSO₂(4-CH₃Ph)), 20–70°C; **H** = 1) (COCl)₂, CH₂Cl₂, 40–60°C; 2) AlCl₃, CH₂Cl₂ or CHCl₃, 0–25°C; 3) OH⁻, H₂O, 80–100°C; 4) H₃O⁺, RT; **I** = 1) SOCl₂, 50–80°C; 2) RPhOH, Et₃N, DMAP, CH₂Cl₂, RT; **J** = NaBH₃R (R = H, CN), CH₂Cl₂, RT.

On the basis of simple AEs, more complicated structures can be obtained in order to be employed as CL labels in luminescent immunoassays. Examples of synthesis routes for original acridinium labels are presented in Figure 9. Detailed descriptions can be found in patent literature [52, 57].

2.3. Structure of Acridinium Esters in Crystalline Solid Phase

2.3.1. Short-Range Interactions among Acridinium Esters and Their Precursors

Detailed information on structures of acridines obtained in the crystalline solid phase is presented in the original literature [58, 59]. In addition to information on the interactions, occurring in the crystals, X-ray structural analysis enabled us to confirm unambiguously the identity of the new acridine-based molecular systems.

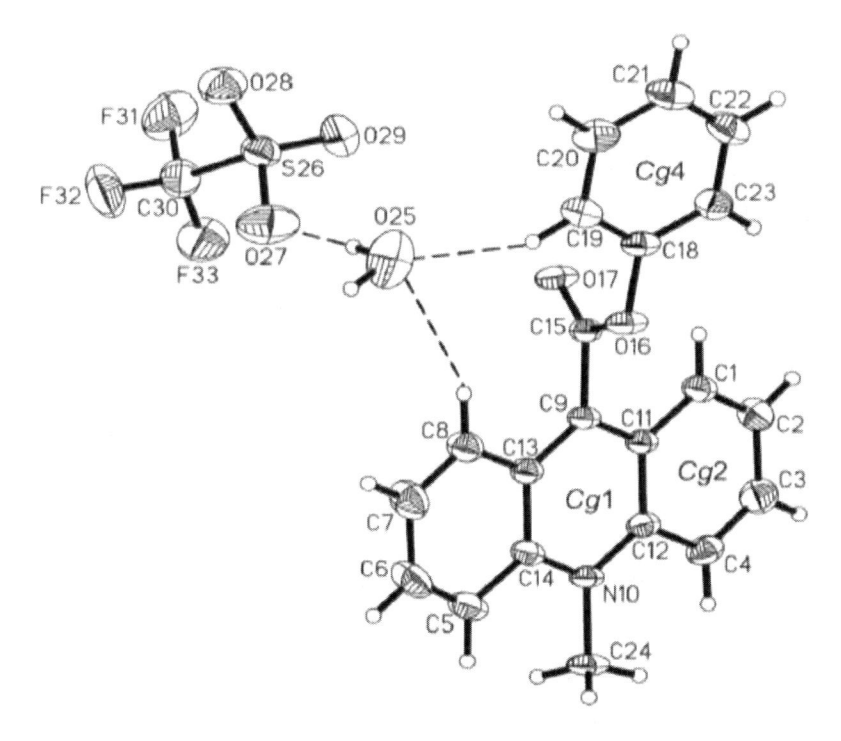

Figure 10. Crystalline solid phase structure of 9-phenoxycarbonyl-10-methylacridinium trifluoromethanesulfonate monohydrate along with atom numbering [59]. Reproduced with permission of the International Union of Crystallography (Crystallography Journals Online, https://journals.iucr.org/).

The representative structure, along with the numbering of atoms (common to all compounds), is shown in Figure 10.

An example of resolved crystal structure, representing AE involved in this article (compound 2,6-diFPhX) is presented below (Figure 11).

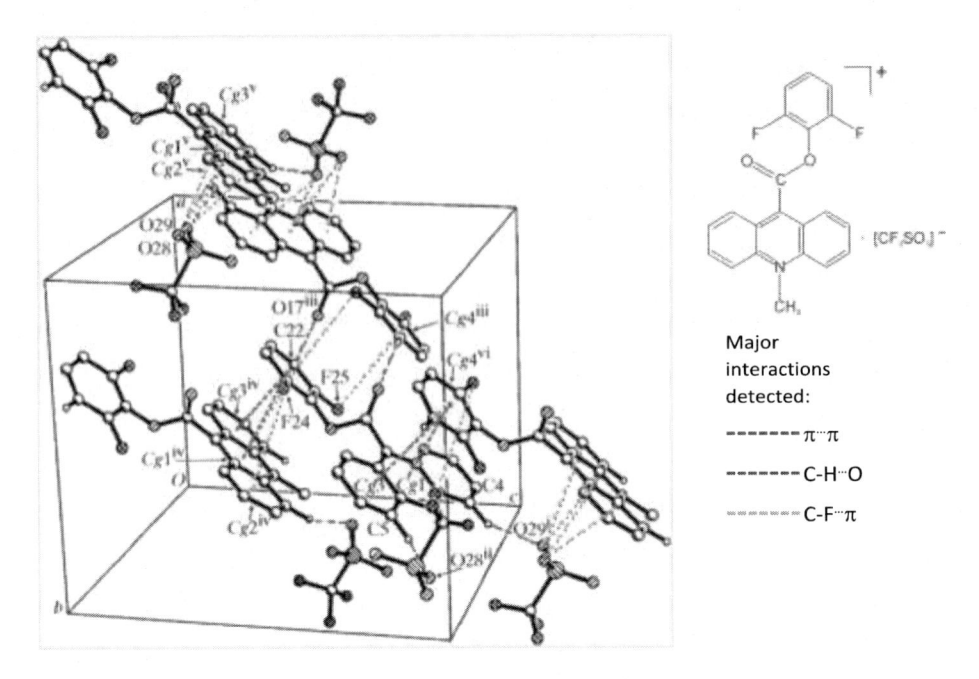

Figure 11. Crystalline solid phase structure of an exemplary AE (9-[(2,6-difluorophenoxy)carbonyl]-10-methylacridinium triflate, 2,6-diFX) together with major short-range interactions indicated. Reproduced with permission of the International Union of Crystallography (Crystallography Journals Online, https://journals.iucr.org/).

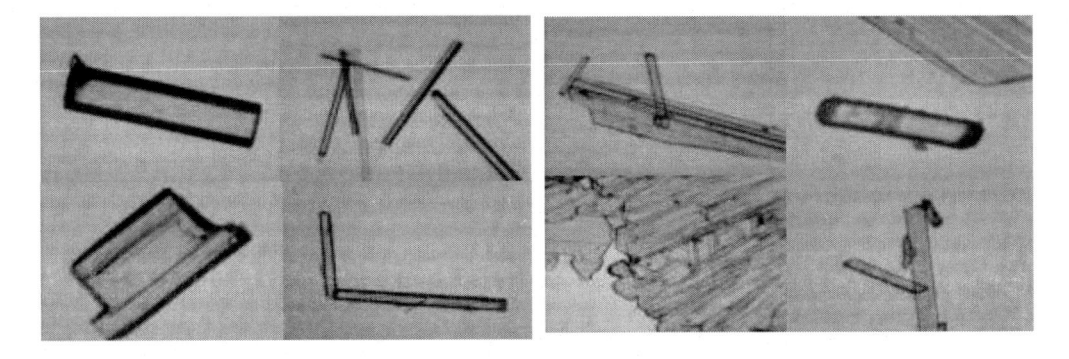

Figure 12. The photographs of monocrystals of exemplary acridinium esters subjected to crystallographic studies (left: 2,6-diFX, right: 4-MeX).

The ions and atoms involved among crystal structures, in addition to electrostatic interactions, take part in a number of short-range interactions, contributing to the cohesion forces and defining structural differences between them. They were selected on the basis of the criteria implemented in the PLATON program. The conditions for recognition of major types of interactions found among chemiluminogenic acridines in crystalline phase are specified in Table 1 [60, 61].

Table 1. The conditions for recognition of such interactions in crystalline solid phase among acridines [60, 61]

Hydrogen bonds X–H⋯A 	d(X⋯A) < R(X) + R(A) + 0.50 Å (X = C, N, O, halogen, S); d(H⋯A) < R(H) + R(A) − 0.12 Å (A = N, O, halogen, S) X–H⋯A angle > 100°; R − van der Waals radii of atoms (in Å): H = 1.20; C = 1,70; N = 1.55; O = 1,52; F = 1.47; Cl = 1.75; Br = 1.85; I = 1.98
X–H⋯π interactions 	X – the donor; H – hydrogen atom; Cg –the center of gravity of the aromatic ring; ω – the angle between H⋯C bond and the axis perpendicular to the plane of the ring.
π–π interactions 	$CgI⋯CgJ$ distance < 4.0 Å; β - angle between $CgI⋯CgJ$ a wector ⊥ do Cg(I) < 60°
C–X⋯π contacts (X = O, N, S, halogens).	(X⋯Cg) distance < R(X) + R(π) + 0,2 Å; C–X⋯π angle > 100°

- *Hydrogen bonding.* The parameters are shown in Table 1, where: D – is the distance between donor X and acceptor A; d – is the distance between the hydrogen atom and its acceptor A, Θ – is the angle between the donor, hydrogen atom and the acceptor, ϕ – the angle between a hydrogen atom, an acceptor and the acceptor atom linked to Y. Energies of classical hydrogen bonding (O-H⋯O, N-H⋯O) fall in the range of 30-35 kJ mol^{-1}. C-H⋯O interactions are classified as non-typical hydrogen bonding of energies of ca. 15 kJ mol^{-1}.

- *X-H⋯π interactions.* The non-classical hydrogen bonding, which, due to their population, play a significant role in the stabilization of molecular systems of investigated acridines. The X atom is typically C, sometimes N. Typical energies of such interactions fall in the range of 7-8 kJ mol^{-1}.

- *π⋯π Contacts.* The π⋯π contacts are characteristic for flat aromatic systems and thus are common among acridines. These weak, both positive and negative interactions, occur mainly between the layers created by flat aromatic fragments of acridine, although benzene rings are also engaged in this type of contacts. Their energies are on the level of 6 kJ mol^{-1}.

- *C-X⋯π contacts.* Among acridines investigated C-X⋯π type contacts are also common, usually involving fluorine atoms.

- *Halogen-halogen contacts (X−X)*. Such weak interactions can be either positive or negative, depending on the arrangement of C-X fragments.

The brief analysis of short-range interactions that occur among investigated crystals of luminogenic acridines (Figure 12) suggests that hydrogen bonds make the largest contribution to the energy of cohesive forces in such compounds.

At the opposite side are weak $\pi\cdots\pi$ interactions, with energies comparable to non-classical hydrogen bonds, such as C-H$\cdots\pi$. The hydrogen bonds type C-H$\cdots$O are present in all structures of crystalline acridines investigated (AEs, PCAs and 10-substituted acridin-9-ones). Interestingly, the typical O-H$\cdots$O hydrogen bond occurs only in one structure containing the incorporated molecules of water (unsubstituted acridinium ester, HX). Contacts of $\pi\cdots\pi$ type are as common as C-H$\cdots$O interactions among studied acridines. C-H$\cdots\pi$ interactions are observed in most of the structures, which involve both aromatic systems (Ph and Acr) and aliphatic or aromatic C atoms. Among them, the most numerous ones are those in which the H-donor is a methyl group attached to an acridine N atom; there are also numerous interactions, where the benzene ring takes an H-donor role. Interactions of C-F$\cdots\pi$ type, involving F from triflate anion ($CF_3OSO_2^-$) and both aromatic systems, exist in most of the crystal structures of AEs. Occasionally, there are also interactions containing other halogen atoms (e.g., Br) - if present in the structure. This group can also include the rare interactions S−O$\cdots\pi$, involving S atoms of anions, O and two aromatic rings, as well as N−O$\cdots\pi$ contacts. Other contacts are rare. Often, their occurrence is associated with the presence of a specific heteroatom, such as I (I$\cdots$I contacts).

2.3.2. Structural Features of Acridinium Salts

Chemiluminogenic acridinium salts make the largest group of compounds which has determined the structure in the crystalline solid phase. Most of them crystallize in the monoclinic system, characterized by four pairs of ions contained in the unit cell, and some AEs (2-methyl, 2-ethyl, 2,6-dichloro and 2,6-dibromo- derivatives substituted in the benzene ring) - in a triclinic system with two pairs of ions in the unit cell. Generally, a lower degree of order systems is preferred – probably due to the low degree of symmetry of the ions forming the crystal.

Among all 9-(phenoxycarbonyl)-10-methylacridinium cations the geometrical parameters are quite comparable (Table 2, Figure 13). Characteristic bond lengths, for example for C9−C15 and N10−CH$_3$, take typical values for this parameter (about 1.50 Å) and express little variability. The bond lengths of N10−CH$_3$ in the family of 10-methylacridinium salts are like the above, which suggests that the presence of anion in molecular system does not affect this parameter.

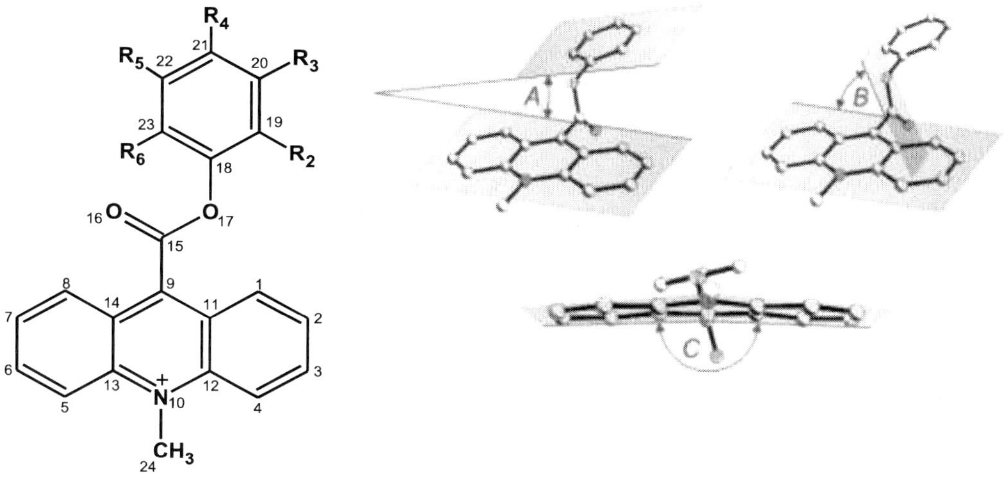

Figure 13. Selected structural parameters characterizing 9-phenoxycarbonyl-10-methylacridinium cations: *A* - orientation of benzene ring according to the acridine ring; *B* - angle of carboxyl group according to acridine ring, *C* - angle between left and right side of acridine ring [61].

Table 2. Selected structural parameters characterizing crystalline structures of represented acridinium esters and relative compounds. Codes denote the type and position of substituent in the benzene ring, X denote triflate anion ($CF_3OSO_2^-$)

Cpd (code)	A (°)	B (°)	C (°)	Bond length C9-C15 (Å)	Torsion angle C15-O16-C18-C19 (°)
9-(Phenoxycarbonyl)acridines (PCAs)					
2,6-diF	5.8	79		1.48	75
2,6-diI	46	54		1.49	89
9-(Phenoxycarbonyl)-10-methylacridinium triflates (AEs)					
4-MeX	3.0	83	0.01	1.50	83
2-tBuX	54	61	0.04	1.50	54
2,6-diFX	17	68	0.01	1.48	84
2,6-diBrX	36	61	0.01	1.50	81
Acridans					
10-methylacridin-9-one			0.03	1.24 (C15 = O)	
2,4,6-triMeAn [62]	80	89	0.10	1.48	80

A little more diverse are torsion angles C15–O16–C18–C19 (in the range of about 54–89°; the positive sign indicates the parallel arrangement of cations in crystals and vice-versa).

AEs differ significantly in the mutual arrangement of both the aromatic fragments (benzene and acridine rings), as well as in the degree of twisting the carboxyl groups in relation to acridine ring, as it is manifested by the dihedral angles *A*, which take values in the range of 3–54° among > 50 investigated acridinium salts up to date.

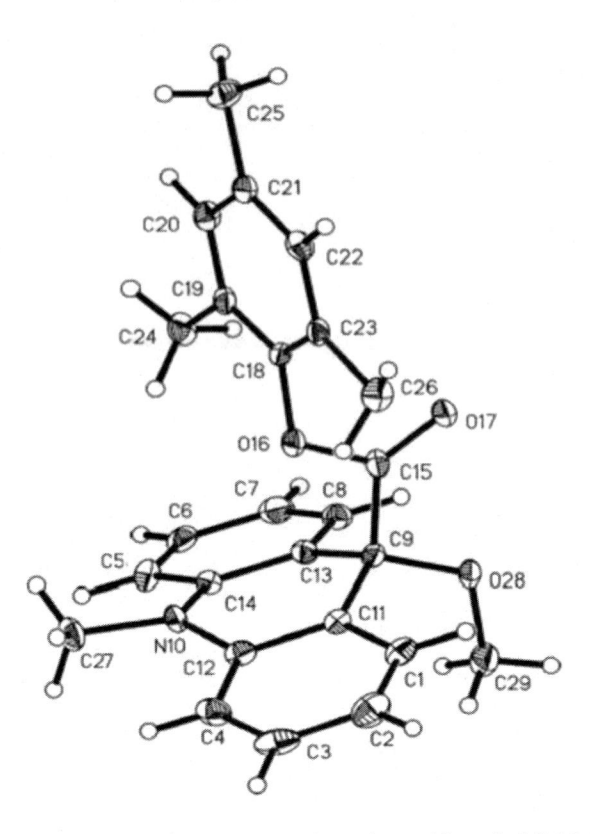

Figure 14. Crystal structure of the acridan ester, 9-methoxy-10-methyl-9,10-dihydroacridinyl mesityl 9-carboxylate (2,4,6-triMeAn, Table 2) [62]. Reprinted from Journal of Molecular Structure, Vol. 920, A. Niziolek, B. Zadykowicz, D. Trzybinski, A. Sikorski, K. Krzyminski and J. Blazejowski, 9-[(Mesityloxy)carbonyl]-10-methylacridinium trifluoromethanesulfonate and its derivative mesityl 9-methoxy-10-methyl-9,10-dihydroacridine-9-carboxylate: structural and physicochemical features, Pages 231-237, Copyright (2009), with permission from Elsevier.

By analyzing dozens of structures of AEs some trends can be noticed. Thus, the angle *A* assumes higher values in the case of sterically hindered systems – those containing volume substituents at the *ortho* positions of the benzene ring and lower in the case of systems bearing the substituents at the *para* position (examples are in Table 2). In the group of 2,6-dihalogen substituted cations involved in AEs, the angle *A* increases with increasing volume of the substituents in the benzene ring. By contrast, *C* angles close to zero indicate virtually planar acridinic aromatic system characterizing them. The structure of reduced form of exemplary acridinium ester (acridan ester) is unique in this group (Figure 14) [62].

This compound represents the first acridan ester-type adduct, for which the structure in the crystalline solid phase was resolved [62]. It makes the product of addition of OCH_3 moiety to the atom C9 of respective AE (2,4,6-triMeX) [62]. It can be associated with intermediate products, formed upon the attachment of nucleophilic ions to the 9-(phenoxycarbonyl)-10-methylacridinium cations in an alkaline medium. The possible mechanism of its formation is also proposed in [62], on the basis of DFT calculations. One makes a typical covalent compound that differs significantly in terms of the geometry from its cationic precursor (2,4,6-triMeX). The acridan ester 2,4,6-triMeAn is not flat – the distortion of the acridine nucleus from planarity is at the level of 0.1 Å (Table 2), resulting in its bending along the C9–N10 axis at an angle of about 20°. Thus, by the nucleophilic addition from the flat structure of acridinium moiety the "butterfly" structure is formed, which is presumably associated with the disappearance of much $\pi\cdots\pi$ contacts. Besides this, upon its formation the angle between both aromatic systems increases significantly ($A \sim 19°$ for 2,4,6-triMeX (acridinium ester) and ~80° for 2,4,6-triMeAn (acridan ester)).

2.4. Thermochemical Properties of Acridinium Esters and Related Compounds

Table 3. The formulas used in calculations based on thermochemical data for acridinium esters and relative compounds [63, 64]

Parameter	Formula
E_{cr} – crystal lattice energy; E_{el}, E_d i E_r – the energy of electrostatic, dispersive and repulsive interactions, respectively.	$E_{cr} = E_{el} + E_d + E_r$
Electrostatic energy	$E_{el} = \dfrac{1}{2}\sum_i\sum_{j\neq i}\dfrac{Ne^2}{4\pi\varepsilon_0}\dfrac{Q_iQ_j}{R_{ij}}$
Lennard – Jones eq.	$E_d + E_r = \dfrac{1}{2}\sum_i\sum_{j\neq i}\left(-\dfrac{D_iD_j}{R_{ij}^6} + \dfrac{A_iA_j}{R_{ij}^{12}}\right)$
Buckingham eq. N – Avogadro number; e – the elementary charge; ε_0 – vacuum permittivity; $Q_i(Q_j)$ – atomic partial charges; $D_i(D_j)$, $A_i(A_j)$ i $B_i(B_j)$ – atomic parameters; R_{ij} – the distances between interacting centers	$E_d + E_r = \dfrac{1}{2}\sum_i\sum_{j\neq i}\left[-\dfrac{D_iD_j}{R_{ij}^6} + A_iA_j\,\exp\left(-B_iB_jR_{ij}\right)\right]$
Enthalpy of volatilization α – degree of volatilization ($\alpha = p/p_0$, p – equilibrium vapor pressure at a given temperature, p^0 – standard pressure), Δ_vH^0 – enthalpy of volatilization, R – gas constant, T – temperature in K, T_v – temp. at which p attains p^0. Δ_vH^0 values needs standardization (anthracene)	$\ln\alpha = -\dfrac{\Delta_vH^0}{R}\times\dfrac{1}{T} + \dfrac{\Delta_vH^0}{R}\times\dfrac{1}{T_v}$

The crystal lattice (E_{kr}) energies of AEs and relative acridine derivatives were calculated as the sum of electrostatic (E_{el}) and dispersive (E_d) and repulsive (E_r) energies according to Buckingham or Lenard-Jones [63, 64]. Atomic parameters and other constants available in the literature were used for calculations. The sublimation enthalpy values were determined based on the analysis of TG and DSC curves. The Clausius-Clapeyron equation was utilized, using standardized values (anthracene) of volatilization enthalpy. Basic formulas used in calculations based on thermochemical data are summarized in Table 3.

Dozens of 9-(phenoxycarbonyl)-10-methylacridinium triflates and their precursors, 9-(phenoxycarbonyl)acridines, containing one or more alkyl groups (methyl, ethyl, isopropyl, *tert*-butyl), halogen atoms (F, Cl, Br, I) and other substituents (NO_2, CH_3O) located at various positions of the benzene ring, as well as relative compounds (acridane-9-ones and simple acridinium salts) were investigated by thermoanalytical methods including Differential Scanning Calorimetry (DSC) and Thermogravimetric Analysis (TGA) [65].

Investigated up-to-date PCAs (uncharged precursors of AE) melt in the range of 396-480 K, followed by their volatilization, separated from the melting signal typically by about 100 K. Simple shapes of their thermogravimetric curves (TG) indicate a one-stage courses of the volatilization, in which the molecules, maintaining their structure, are transferred from the solid to the gaseous phase. The only barrier in this phase transformation is the thermodynamic one (enthalpy of volatilization), thus such systems attain equilibrium immediately with the temperature increase. The temperature ranges, in which the volatilization of the compounds occur, express only slight variability. Thermal analyses (both DSC and TG methods) provide similar results, disclosing the temperatures of their sublimation oscillating near 600 K.

The DSC curves enabled a determination of the enthalpy of phase transformations occurring among PCAs, which fall in the range of about 23-40 kJ mol^{-1} for melting and about 79-110 kJ mol^{-1} for volatilization. The values of T_v and ΔH^0_v are usually slightly higher if the compounds contain polar groups or atoms that can additionally interact in the solid phase - for example through hydrogen bonds.

Thermal characteristics indicate that 9-(phenoxycarbonyl)acridines (and investigated acridin-9-ones) form molecular crystals, in which molecules participate in weak intermolecular interactions. In crystal lattices of such compounds a small number of short-range interactions were found, likely due to the lack of triflate ions in their structure. In order to gain a deeper insight into the nature of the cohesion forces present in these compounds, the crystal lattice energies (ΔH°_{cr}) and sublimation enthalpies (ΔH_s°) were determined. The first values fall in the range of 126-159 kJ mol^{-1} for PCAs and 105-143 kJ mol^{-1} for acridane-9-ones investigated [66]. Negative values of the enthalpy of formation (ΔH_f°) of most compounds indicate, that they are thermodynamically stable in the solid phase.

According to 9-(phenoxycarbonyl)-10-methylacridinium triflates, they form ionic compounds, thus melt at higher temperatures, expressing melting temperatures and enthalpies in the range of 466-542 K and 30-50 kJ mol^{-1}, respectively. Their complex decomposition begins at several dozen degrees above their melting points, which was not studied further.

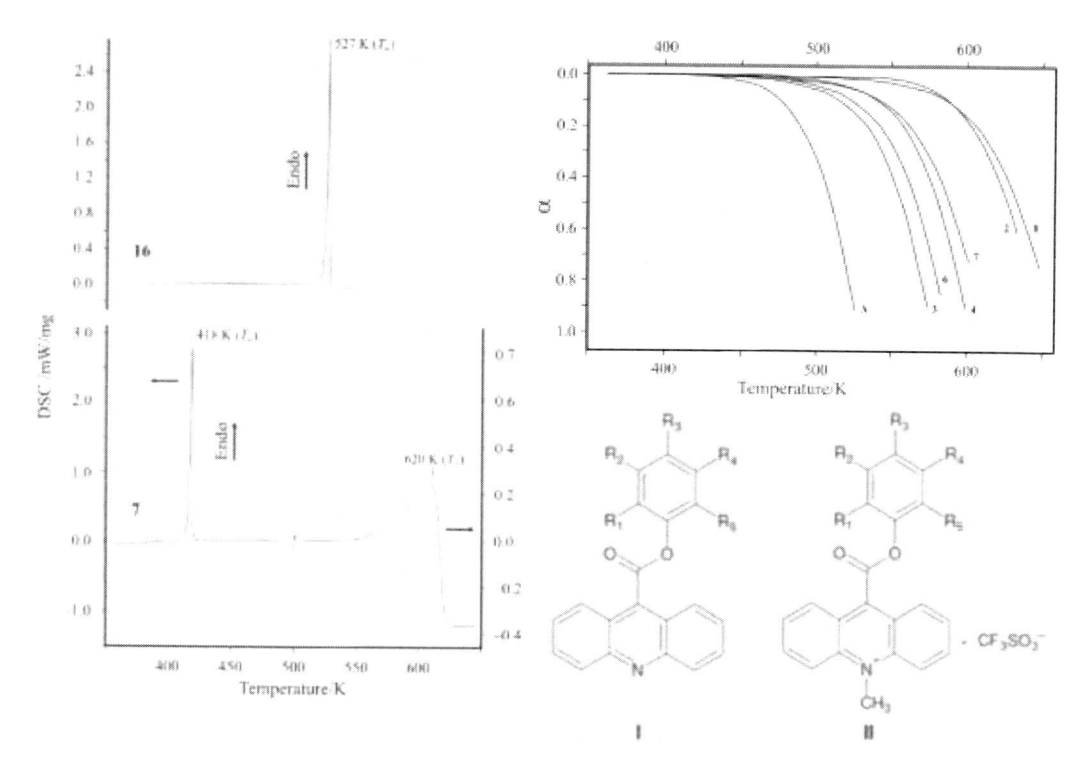

Figure 15. Left graphs: Exemplary DSC profile for an exemplary acridinium ester (2,6-diClX, left graph, upper) its precursor (left graph, lower). T_m – melting temperature; T_V – volatilization temperature (based on I derivative of DSC curve). Right graph: TGA profiles expressing volatilization degree (α) as a function temperature (T) for selected PCAs and anthracene (A). For experimental details and compounds numbering see [66]. Canonical structures of AEs (II) and their precursors, PCAs (I) subjected to DSC and TGA (R_1–R_6 = alkyl, halogens, NO_2, CH_3O at various positions) are presented in right down graph [65]. Reprinted by permission from Springer Nature: Journal of Thermal Analysis and Calorimetry, Thermochemistry and crystal lattice energetics of phenyl acridine-9-carboxylates and 9-phenoxycarbonyl-10-methylacridinium trifluoromethanesulphonates, K. Krzymiński, P. Malecha, P. Storoniak, B. Zadykowicz and J. Błażejowski, Copyright (2010) and Lattice energetics and thermochemistry of phenyl acridine-9-carboxylates and 9-phenoxycarbonyl-10-methylacridinium trifluoromethanesulphonates nitro, methoxy or halogen substituted in the phenyl fragment, B. Zadykowicz, K. Krzymiński, P. Storoniak and J. Błażejowski, Copyright (2010).

The crystal lattice energies calculated for tested AEs assume values in the range 560-605 kJ mol⁻¹, which can be considered as typical ones, characterizing ionic molecular systems containing complex single-negative ions. The dominant contribution to the crystal lattice energy of this type of compounds comes from the electrostatic interactions among oppositely charged monovalent ions. The AEs are characterized by similar and strongly negative values of formation enthalpy in solid phase at the level of -1000 kJ mol⁻¹. The results of thermoanalytical studies indicate that acridinium esters make thermodynamically stable substances below 500 K. They can therefore be stored for any time at room temperature if their use is planned.

Table 4. Thermochemical characteristics of exemplary acridinium esters and related compounds. T_m – melting temperature; T_v – volatilization temperature; $\Delta H_m°$ – standard melting enthalpy; $\Delta H_v°$ – standard volatilization enthalpy; $\Delta H_s°$ standard sublimation enthalpy (corrected for anthracene); $\Delta H_{kr}°$ – enthalpy of crystal lattice [65]

Cpd code	T_m (K)	T_v (K)	$\Delta H_m°$ (kJ mol^{-1})	$\Delta H_v°$ (kJ mol^{-1})	$\Delta H_s°$ (kJ mol^{-1})	$\Delta H_{cr}°$ (kJ mol^{-1})
9-(phenoxycarbonyl)acridines						
2-H	464	599	39.2	89.9	145	
2-*t*Bu	462	610	39.8	94.5	153	
2,6-diF	467	596	34.3	91.0	132	150
2,6-diBr	432	640	31.9	106.5	146	158
Simple acridinium salts						
Acr$^+$Cl$^-$·H$_2$O	468 (dec.)					725
Acr$^+$ I$^-$	490 (dec.)					628
9-(phenoxycarbonyl)-10-methylacridinium-triflates (AEs)						
2-*t*BuX	504		36.0			565
2,6-diFX	517		37.7			566
2,6-diBrX	520		44.1			583
Acridane-9-ones						
10-MeAon	479	565	29.7	98.0	105	130

2.5. Spectroscopic Features of Acridinium Esters

2.5.1. Electronic Absorption and Emission Spectra

Spectroscopic properties of representative acridinium esters and their uncharged precursors (PCAs) were assessed by analyzing their absorption (UV-Vis and fluorescence emission stationary electronic spectra, oscillation spectra (FT-IR) as well as nuclear magnetic resonance spectra (^{1}H and ^{13}C NMR). The experimental results were compared with their spectral characteristics, obtained by theoretical methods (DFT and/or semi-empirical levels of theory).

UV-Vis absorption spectra of 9-(phenoxycarbonyl)acridines are similar to the spectra of acridine itself and acridine-9-carboxylic acid. The efficiency of absorption of electromagnetic radiation by PCAs is generally small - as indicated by low values of their absorption coefficients at maximum (at the level of 10^{-4} M^{-1} cm^{-1}) [67]. Characteristic for acridines range of absorption above 300 nm, which makes a superposition of four bands, can be separated by deconvolution mathematical processing (Figure 16 left, upper graph).

Emission spectra of neutral esters (PCAs) are not very sensitive to the influence of the substituent(s) as well as the environment. The spectra express a single band with a maximum centered at 450-460 cm^{-1} if the acridine nucleus is unsubstituted. The compounds of this group are characterized by a low quantum yield of fluorescence (not exceeding a few percent), which is much lower than that of acridine itself (about 30%). The Stokes shifts at the level of 5600 cm^{-1} represent typical, moderate values expressed by organic fluorophores, indicating a lack of efficient relaxation processes that can eventually occur with the participation of solvent molecules in the excited states.

Most electron transitions occurring among PCAs are characteristic for aromatic systems $\pi \rightarrow \pi^*$ type. Some of the transitions to higher electronic levels ($S_0 \rightarrow S_2$) are of $n \rightarrow \pi^*$ type. A satisfactory agreement between the theory and the experiment in the position of the bands' maxima was obtained. The majority of absorption transitions occur within the acridine ring system, some occur in the benzene ring.

According to further calculations (semi-empirical AM1 and AM1/CI methods), the distance between the fluorescent S_1 states ($\pi \rightarrow \pi^*$) and non-fluorescent S_2 states ($n \rightarrow \pi^*$) among PCAs is small (from about 0.18 to 0.10 eV), which favors an effective relaxation processes, that may occur in the excited states - creating the possibility of strong vibrational coupling between them. On the other hand, the energy difference between the S_2 state and the nearest triplet state (T_4) is also small, which facilitates non-radiative internal conversion ($S_2 \rightarrow T_4$). The participation of the last process is likely significant, because of relatively high values of constant rates characterizing the ISC transitions. Thus, $S_2 \rightarrow T_i$ transitions can be the main pathway for non-radiative deactivation among 9-(phenoxycarbonyl) acridines [67]. These transitions are presented in simplified form using Jablonski's diagram (Figure 16 right).

The presence of acid in an environment is manifested by bathochromic shift of absorption and emission bands (by tens of nanometers) and weakening of fluorescence intensity, due to the formation of the corresponding 10*H*-acridinium cations (Figure 16 left, lower graph).

The long-wavelength ranges of absorption of 9-(phenoxycarbonyl)-10-methylacridinium cations fall in the region of 360-420 nm. The absorption bands of such molecular systems are quite clearly distinguished. The influence of the nature of the solvent on the position of the bands is manifested only slightly. The bathochromic shift of long-wavelength absorption bands derived from cation as well as the increase in Stokes shift values with the orientation polarization of the solvent indicates a greater variation in the charge distribution that occur in the excited state of acridinium cations. The fluorescence quantum yield of such compounds in protic solvents does not exceed 0.5%. The exception is cations containing the CF$_3$ group, where quantum efficiency is much higher, reaching up to several percent.

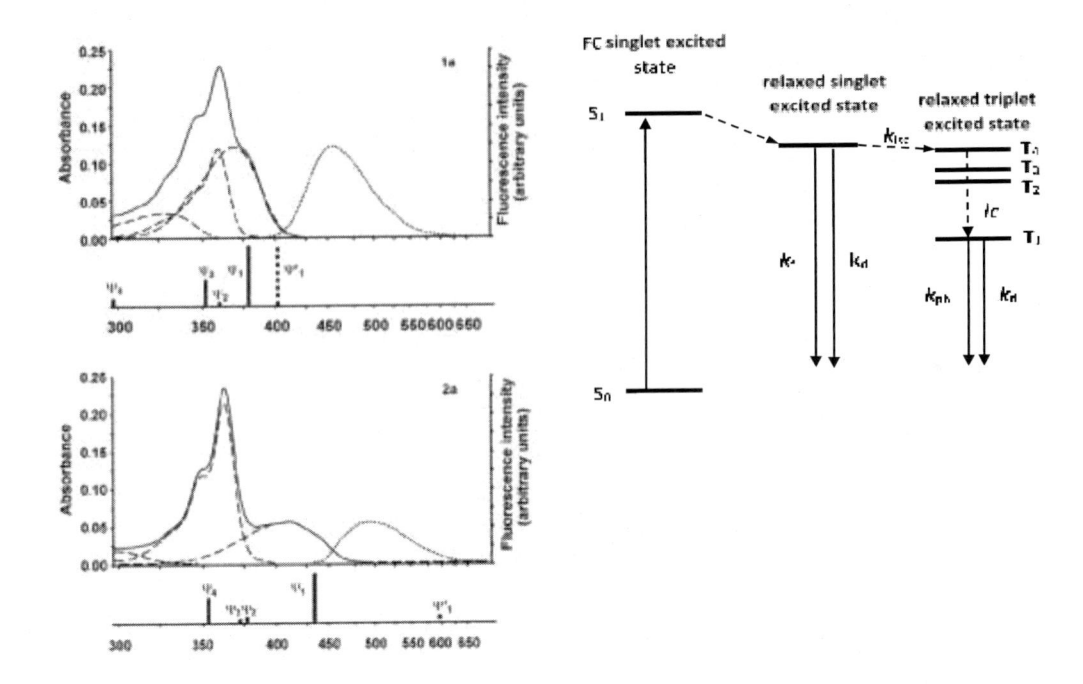

Figure 16. Left: Absorption (solid line: measured; dashed line: deconvoluted) and fluorescence (dotted line) spectra of 9-(phenoxycarbonyl)acridine and its 10-methylated cation together with positions of electronic absorption and fluorescence transitions predicted at the CNDO/S level of theory (ψ_1-ψ_4, lover graphs) [67]. Right: Simplified Jablonski's diagram presenting the behavior of electronically excited molecules of 9-(phenoxycarbonyl)acridines. Right: Simplified Jablonski's diagram presenting the behavior of electronically excited molecules of 9-(phenoxycarbonyl)acridines. Reprinted from Spectrochimica Acta Part A: Molecular and Biomolecular Spectroscopy, Vol. 70, K. Krzymiński, A.D. Roshal, A. Niziołek, Spectral features of substituted 9-(phenoxycarbonyl)-acridines and their protonated and methylated cation derivatives, Pages 394-402, Copyright (2008), with permission from Elsevier.

In the long-term range of the absorption spectrum of acridinium cations four bands appear; two of which (intensive ones) are linked to the acridine nucleus and the other two to the transitions occurring between the acridine and phenyl fragments (forbidden by quantum selection rules due to the mutual orientation of these molecular fragments).

The low quantum yield of 9-(phenoxycarbonyl)acridinium fluorescence is explained by semi-empirical calculations (AM1 and AM1/CI methods) [67]. According to them, the S_1 states (of π–π^* type) and S_2 (of n–π^* type) among cations involved in AEs are relaxed to the state at which efficient vibrational coupling takes place between them. The probable way of non-radiative deactivation is the $S_2 \rightarrow T_2$ ISC transition - due to the comparable energies of both states. The quantum yields predicted in the computations fall at the level of 0.03%, which clearly suggests that cations involved in AEs are very weak fluorophores themselves - as was also demonstrated in experiments. An electron-acceptor trifluoromethyl group, present in 2-CF_3 benzene ring-substituted cation, is likely responsible for the increase in the energy of transitions involving the acridine and benzene fragments to behave differently. In the latter case vibration relaxation is less effective, which is reflected in a much higher (up to several percent) quantum yield of fluorescence.

2.5.2. Structure - Property Relationships among Acridinium Esters

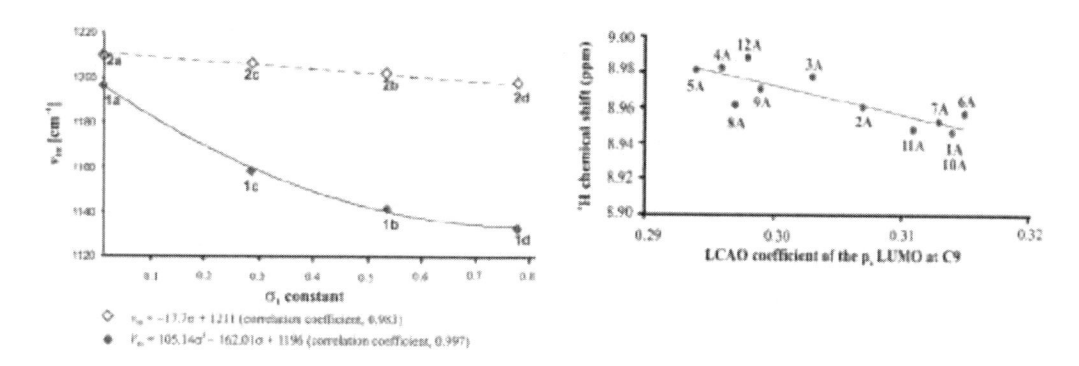

Figure 17. Left graph: Relationships among wavenumbers of C15-O17-C18 in-plane bending vibration (FT-IR) and the inductive sigma constants (σ_I) of the substituents in the 2-position of the benzene ring (H, F, OCH$_3$, NO$_2$ among AEs (dashed line) and their precursors, PCAs (solid line) [68]. Right graph: The ^{1}H4/^{1}H5 chemical shifts (DMSO-d6) *vs.* LCAO coefficients of the p$_z$ LUMO orbitals at C9 among a series of variously alkyl-substituted AEs in the benzene ring [69]. Reprinted from Spectrochimica Acta Part A: Molecular and Biomolecular Spectroscopy, Vol. 78, K. Krzymiński, P. Malecha, B. Zadykowicz, A. Wróblewska and J. Błażejowski, 1H and 13C NMR spectra, structure and physicochemical features of acridine-9 carboxylates and 9-phenoxycarbonyl-10-methylacridinium trifluoromethanesulphonates – alkyl substituted in the phenyl fragment, Pages 401-409, Copyright (2011), with permission from Elsevier.

For the FT-IR spectrophotometry studies, the compounds (AEs) were chosen to differ in the nature of the substituent in the 2-position of the benzene ring (H, F, OCH$_3$, NO$_2$) – the closest position near the C9 atom of acridine nucleus, being the site of attack of hydroperoxic ($^-$OOH) ions during generation of CL [68].

In Figure 17 (left), the relationship between the inductive constants of substituents (σ_I) and the wavenumbers of bending oscillations of the ester group (v_{III}) is presented. Both in the group of neutral esters (PCAs) and the relative salts (AEs) the trend is inversely proportional, but in the group of AEs - slow and linear, while in the group of PCAs – fast and non-linear.

Further trends were disclosed by analysis of ^{1}H and ^{13}C NMR spectra [69]. Among others, an interesting thing seems to be the linear dependence of the chemical shifts of ^{1}H4 and ^{1}H5 nuclei (in the phenoxy moiety – the "leaving group") on the values of LCAO coefficients of the LUMO p$_z$ of the C9 atom in the series of AEs investigated (Figure 17 right). As it was stated above (theoretical considerations), the values of the LCAO p$_z$ LUMO coefficients can be considered a measure of the compounds' susceptibility to a nucleophilic attack (e.g., by $^-$OOH). The data depicted in Figure 18 show, that the above coefficients take much higher values for C9 than C15 atoms in all tested acridines (PCAs and AEs). In general, they are also higher in the case of 9-phenoxycarbonyl-10-methylacridinium cations than in neutral esters. This underlines that the C9 atom will be the probable site of the nucleophilic attack of hydroperoxide ions in the case of AEs and also explains the low tendency to chemiluminescence among respective bases (PCAs).

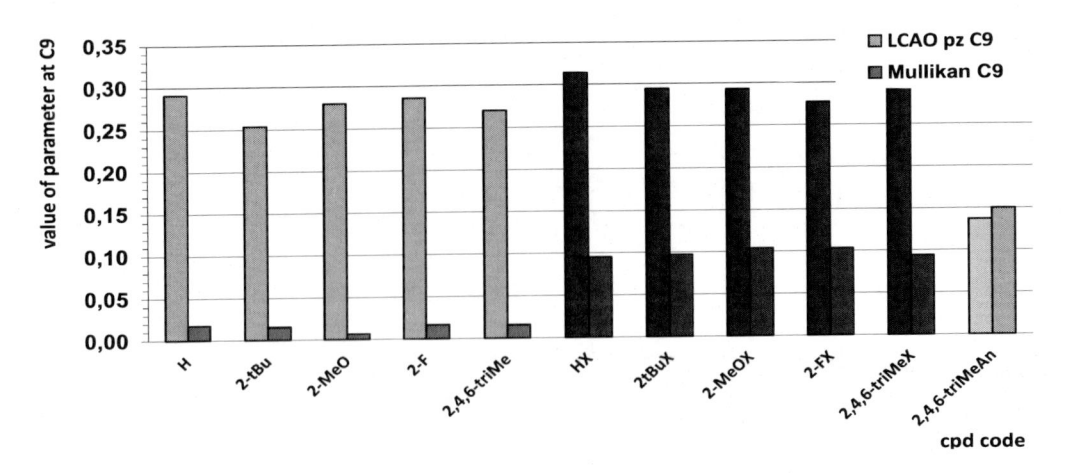

Figure 18. The values of LCAO p_z LUMO coefficients and Mullikan partial charges at C9 atom of exemplary AEs (darker colors) and their uncharged precursors (PCAs, lighter colors). The last two bars represent values obtained for the acridan ester derivative (Figure 14) [62].

The data obtained for acridane ester derivative (2,4,6-triMeAn) [Figure 14 [62]] seem to confirm the results of quantum-chemical calculations. Namely, in the latter molecular system, formed upon the attack of nucleophile onto the acridinium cation, a significant reduction in the value of the discussed parameter is observed.

2.6. Chemiluminogenic Systems Based on Acridinium Esters

2.6.1. Aqueous Environments

In works [28,29] the physicochemical and theoretical problems of variously substituted AEs' chemiluminescence in aqueous environments are discussed. Much attention was directed to compounds bearing alkyl and alkoxy groups, as such molecular systems are employed often as fragments of acridinium CL labels [51, 52].

The approximation of the CL decay curves with the kinetic equations describing the first order consecutive processes made possible to estimate the rate constants, characterizing the build-up and decay of emission (k_1 and k_2, respectively), reflecting the dynamics of the CL associated mainly with the structure of the "leaving group" in AEs. Among alkyl-substituted AEs, the dynamics of CL decay (triggered at optimal conditions for light generation) is diversified - from emissions lasting for a few seconds (flash type), through moderate to slow (glow type) (Figure 19). For sterically hindered cations (2,6-dimethyl, 2-*tert*-butyl), the k_2 constants are several times lower than those assessed for an unsubstituted derivative (HX).

The efficiency of emission for most AEs in aqueous environments can be considered as relatively high, especially when compared with the ϕ_{CL} of luminol (assuming the value of 1.23% in aqueous environment) [45]. The ϕ_{CL} for AEs reaches up to 2.4% in some cases (F-substituted AEs in benzene ring). The range of signal linearity as a function of concentration of luminescent substrate is six orders of magnitude (10^{-10} - 10^{-4} M), using typical PMT detector (the lowest LOD values characterize flash-emitting AEs). The range of linearity is much broader and LOD values are substantially lower in the case when a photon counting detector is in use. The CL parameters, characterizing the AEs, can be significantly improved in terms when selected cationic surfactants are added to the system (Figure 20); the feature is particularly pronounced in the family of AEs, contrary to 10-methylacridinium-9-sulfonamide derivatives [29].

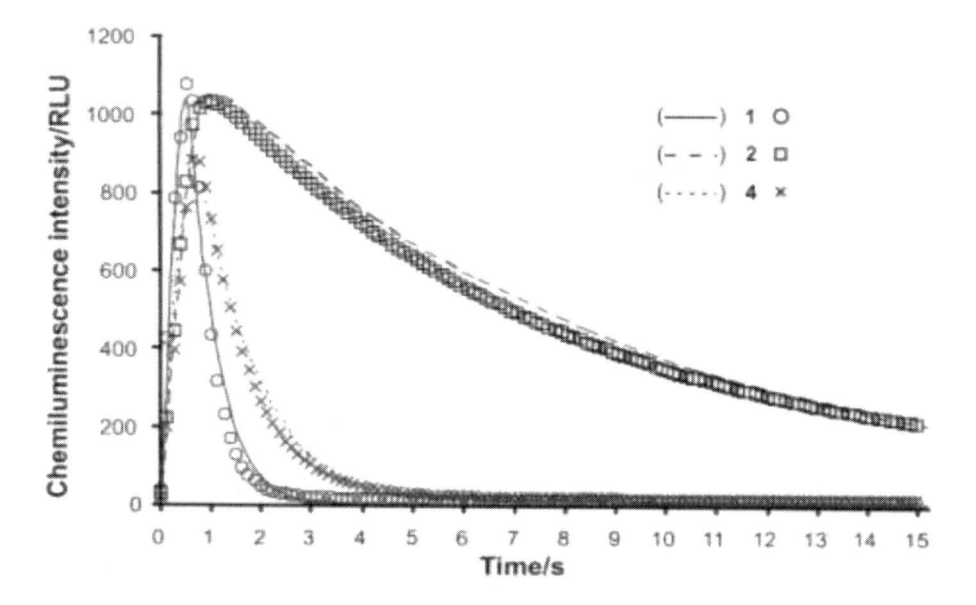

Figure 19. Experimental (points) and predicted (lines) profiles of CL emission for exemplary AEs in alkaline solution of H_2O_2. **1**: 9-(phenoxycarbonyl-10-methylacridnium triflate (HX); **2**: 9-[(2-methyl(phenoxycarbonyl)]-10-methylacridnium triflate (2-MeX) and **4**: 9-[(4-methyl(phenoxy carbonyl)]-10-methylacridinium triflate (4-MeX) [28]. Structural formulas of investigated compounds are given in Figure 15. Reprinted with permission from K. Krzymiński, A. Ożóg, P. Malecha, A.D. Roshal, A. Wróblewska, B. Zadykowicz and J. Błażejowski, Chemiluminogenic features of 10-methyl-9 (phenoxycarbonyl)acridinium trifluoromethanesulfonates alkyl substituted at the benzene ring in aqueous media, The Journal of Organic Chemistry, 76: 1072-1085. Copyright (2011) American Chemical Society.

To select systems of potential usefulness a parameter named utility (U) was introduced. The latter makes a product of the CL quantum yield and a factor determining the stability of AEs in solutions differentiating cations in terms of their suitability for analytics in aqueous systems. Among the alkylated derivatives, the 2-methyl (compound **2**) and 3,5-dimethyl derivative (compound 11) seem promising in that context (Figure 21, left). Data analysis enabled us to disclose dependencies and trends that could have a practical significance in designing new acridinium-based CL systems of potential utility. For example, utility varies non-linearly with the volume of the phenylcarbonate fragment, released upon chemiluminogenic transformations of AEs (Figure 21, right). The relationship reaches a maximum of around 220 Å^3, which suggests that the most favorable emissive properties are present by those AEs in which the volume of the phenylcarbonate moiety assumes moderate values [28].

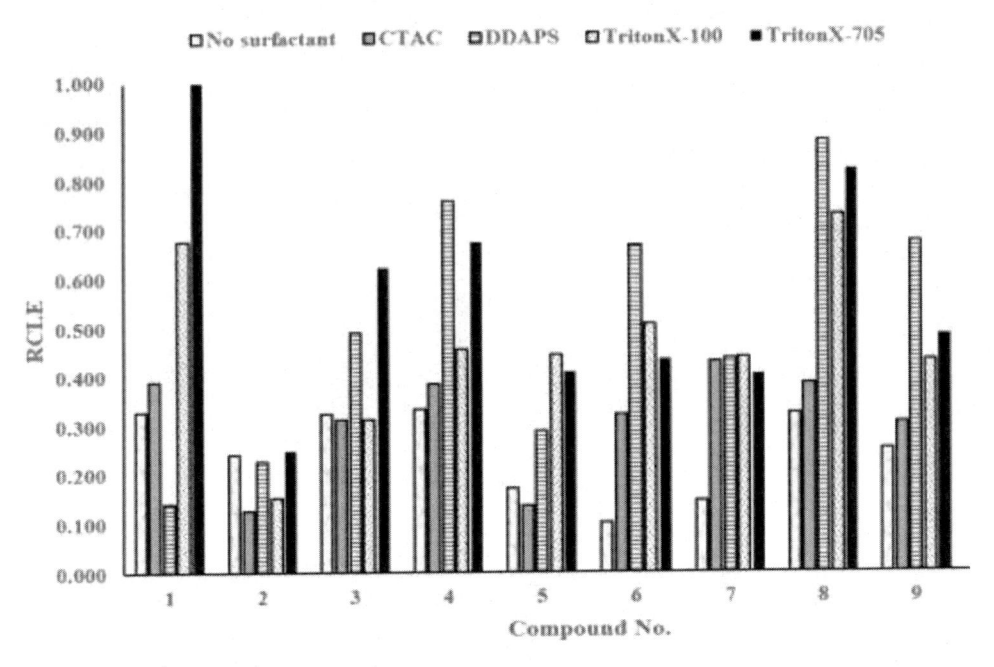

Figure 20. The changes in relative chemiluminescence efficiency (integral) (RCLE) among series of acridinium salts (1-9) derivatized in benzene and/or acridine rings, assessed in aqueous environments (H_2O_2/NaOH) containing surfactants [30]. Reprinted from Journal of Luminescence, Vol. 187, J. Czechowska, A. Kawecka, A. Romanowska, M. Marczak, P. Wityk, K. Krzymiński and B. Zadykowicz, Chemiluminogenic acridinium salts: A comparison study. Detection of intermediate entities appearing upon light generation, Pages 102-112, Copyright (2017), with permission from Elsevier.

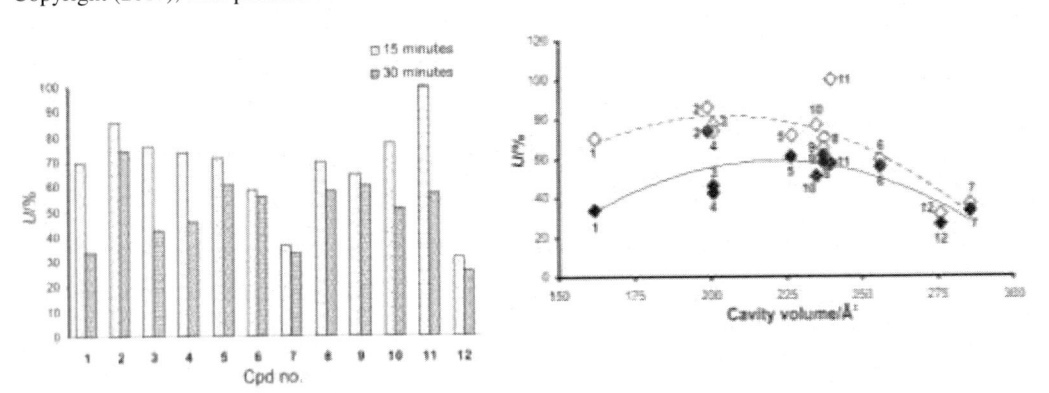

Figure 21. Left graph: Usefulness (*U*) of series of alkyl-substituted acridinium esters. Right graph: Usefulness against the volume of the hydration layer of the respective phenylcarbonate anions among alkyl-substituted AEs undergoing CL in aqueous environment (H_2O_2/NaOH). Upper graph (empty squares) − values corresponding to 15 minutes of incubation, lower graph (filled squares) − values corresponding to 30 minutes of incubation [28]. Reprinted with permission from K. Krzymiński, A. Ożóg, P. Malecha, A.D. Roshal, A. Wróblewska, B. Zadykowicz and J. Błażejowski, Chemiluminogenic features of 10-methyl-9 (phenoxycarbonyl)acridinium trifluoromethanesulfonates alkyl substituted at the benzene ring in aqueous media, The Journal of Organic Chemistry, 76: 1072-1085. Copyright (2011) American Chemical Society.

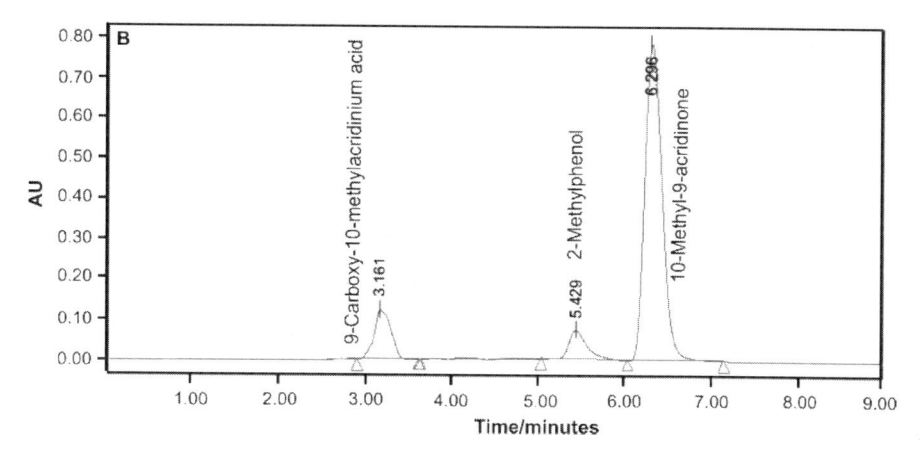

Figure 22. Products of chemiluminogenic oxidation (post-reaction mixture, H_2O_2/NaOH system) of 2-methyl substituted AE in benzene ring identified using RP-HPLC technique [28]. Reprinted with permission from K. Krzymiński, A.D. Roshal, B. Zadykowicz, A. Białk-Bielińska and A. Sieradzan, Chemiluminogenic properties of 10-methyl-9-(phenoxycarbonyl)acridinium cations in organic environments, Journal of Physical Chemistry, 114: 10550-10562. Copyright (2010) American Chemical Society.

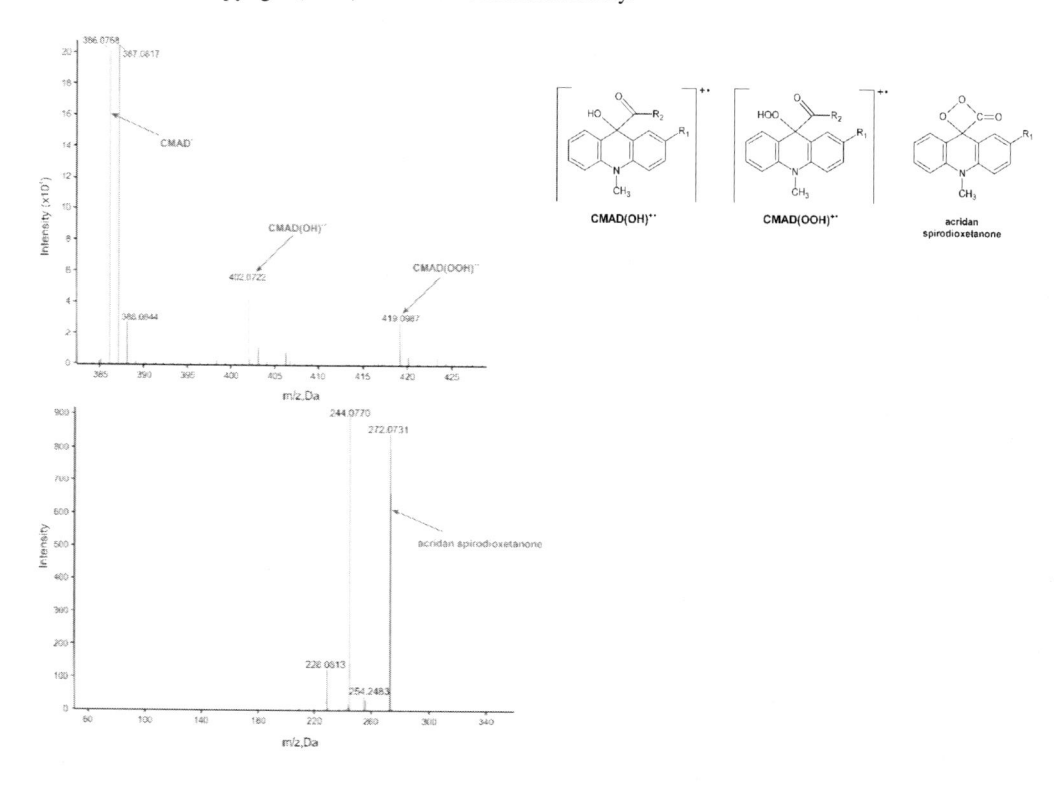

Figure 23. Upper graph: The signals of hydroxy- and hydroperoxy- radical cationic adducts (CMAD(OH)$^{•+}$ and CMAD(OOH)$^{•+}$, respectively) formed with the participation of an exemplary AE in basic H_2O_2 environment obtained by flow-injection ESI-QTOF MS analysis (positive mode). Lower graph: The positive mode MS signals of unstable acridan spirodioxetanone intermediate (as [M+1]$^+$ pseudocation) and its fragmentation ions, observed upon transformation of AE in basic H_2O_2 environment [71]. Reprinted from Journal of Luminescence, Vol. 187, J. Czechowska, A. Kawecka, A. Romanowska, M. Marczak, P. Wityk, K. Krzymiński and B. Zadykowicz, Chemiluminogenic acridinium salts: A comparison study. Detection of intermediate entities appearing upon light generation, Pages 102-112, Copyright (2017), with permission from Elsevier.

The phenylcarbonate anions mentioned above make chemical entities of low stability, as they undergo fast decomposition in slightly acidic or even neutral environments, transforming to respective phenols, that can be monitored by chromatographic techniques, such as RP-HPLC, LC-MS and others (Figure 22).

Interesting information flows from mass spectrometry (LC-MS) experiments performed for representative AEs in a flow-injection regime. High-class ESI-QTOF equipment enabled to observe instable intermediate entities, previously suggested and recently underlined by calculations [30]. Among them, the most interesting seems to be the acridan spirodioxetanone intermediate, for the first time observed experimentally [30]. Nucleophilic adducts derived from cations involved in AEs were also detected in the above-mentioned experiments (Figure 23).

2.6.2. Dark Transformations of Acridinium Esters

As it was suggested in the past, the loss of chemiluminogenic ability of AEs in aqueous environments is mainly related to the formation of "pseudo-base" forms [43, 44] and their further transformations (see theoretical considerations above). The changes observed in the UV-Vis spectra of AEs during the alkalinization of the environment indicate the coexistence of both - cationic and pseudo-basic forms (Figure 24). The determined equilibrium constants of pseudo-base formation take values at the level of 4.7 M^{-1} in the case of simplest (unsubstituted) AE (HX). The constants assume lower values if an electron-donor substituent is introduced into the acridine nucleus (e.g., methoxy group in position 2 and/or 7).

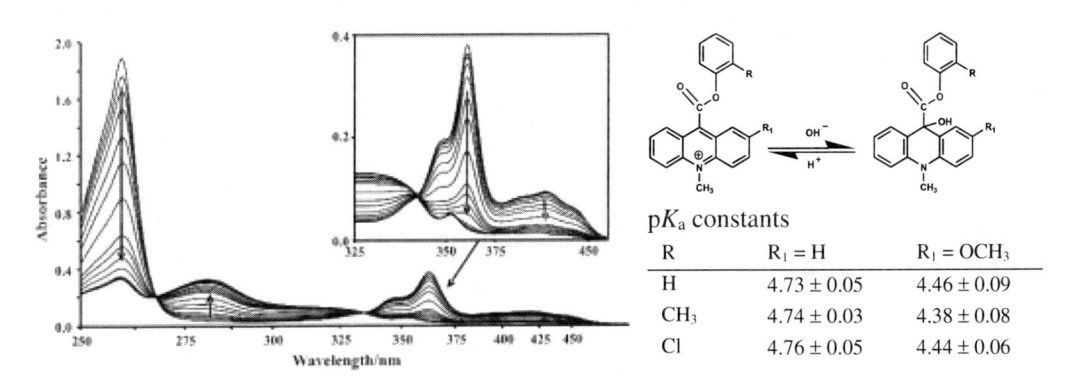

pK_a constants		
R	$R_1 = H$	$R_1 = OCH_3$
H	4.73 ± 0.05	4.46 ± 0.09
CH_3	4.74 ± 0.03	4.38 ± 0.08
Cl	4.76 ± 0.05	4.44 ± 0.06

Figure 24. The UV-Vis absorption spectra of unsubstituted acridinium ester (HX), recorded in aqueous environments of various acidity (pH range 2.0–12.0). Reprinted with permission from K. Krzymiński, A.D. Roshal, B. Zadykowicz, A. Białk-Bielińska and A. Sieradzan, Chemiluminogenic properties of 10-methyl-9-(phenoxycarbonyl)acridinium cations in organic environments, Journal of Physical Chemistry, 114: 10550-10562. Copyright (2010) American Chemical Society.

For representative AEs (R_1 = H; R = H, 2-CH_3 and 2,6-diCH_3-substituted in the benzene ring) the rate constants of alkaline hydrolysis were determined using the RP-HPLC technique [28] (Figure 25). The latter values enabled us to estimate the degree of conversion of the AEs to its respective hydroperoxide adduct (initiating the light path of transformations), applying kinetic analysis proposed in the literature for related systems [70].

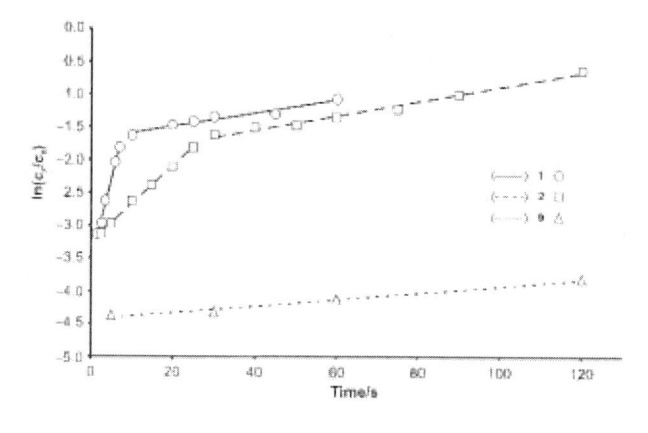

$$AE^+ + OH^- = ACO_2H + EOH$$

2-HX: $\ln(c_x/c_0) = 0.261(t) - 3.55$

2-MeX: $\ln(c_x/c_0) = 0.0545(t) - 3.18$

2,6-diMeX: $\ln(c_x/c_0) = 0.0046(t) - 4.41$

$k_{CL}/k_{OH} = 1.3{-}10$

Figure 25. Kinetics of alkaline hydrolysis (pH = 12.0) of AEs alkyl-substituted in the benzene ring (R = H, CH$_3$, 2,6-diCH$_3$); c_x and c_0 denote the actual concentrations of 9-carboxy-10-methylacridinium acid and the initial concentration of AE, respectively, assessed by RP-HPLC [28]. Reprinted with permission from K. Krzymiński, A. Ożóg, P. Malecha, A.D. Roshal, A. Wróblewska, B. Zadykowicz and J. Błażejowski, Chemiluminogenic features of 10-methyl-9-(phenoxycarbonyl)acridinium trifluoromethanesulfonates alkyl substituted at the benzene ring in aqueous media, The Journal of Organic Chemistry, 76: 1072-1085. Copyright (2011) American Chemical Society.

Determination of k_{OH} and k_{CL} constants enabled us to estimate the degree of conversion of AEX into luminogenic precursor, which decreases with the increase of sterical hindrance: 2-H > 2-Me > 2,6-diMe [28]. The latter finding is presumably responsible for the experimentally observed fact, that the emission efficiencies characterizing flash-type acridinium CL systems take generally higher values than the glow-emitting ones.

2.6.3. Chemiluminescence of Acridinium Esters in an Organic Environment

Chemiluminogenic properties of the AEs, differing in the type of substituent in both aromatic fragments (benzene and acridine rings) were investigated in organic environments of different properties, using various organic bases and concentrations of the triggering reagents (Figure 26) [33].

The main aim of the experiments, supported by statistical analysis, was to select the crucial parameters influencing acridinium chemiluminescence in organic environments and to characterize their emissive properties in such media. It was demonstrated that the efficiency of emission of the rate constants of CL decay (k_2) strongly depended both on the properties of the environment (solvent and base) as well as on the structure of the AE. A multiparametric analysis linked experimentally assessed parameters (such as kinetic constants of CL decay and integral emissions) to solvent empirical constants available in the literature. The Palm equation was applied: $k_{CL} = k_{CL}{}^0 + yY + pP + bB + eE$, where $k_{CL}/k_{CL}{}^0$ denote rate constants of CL decay in solvent/vacuum, Y - Kirkwood polarity constant, P - solvent polarizability, B - basicity constant and E - the electrophilicity index; y, p, b, and e denote specific coefficients of found equations.

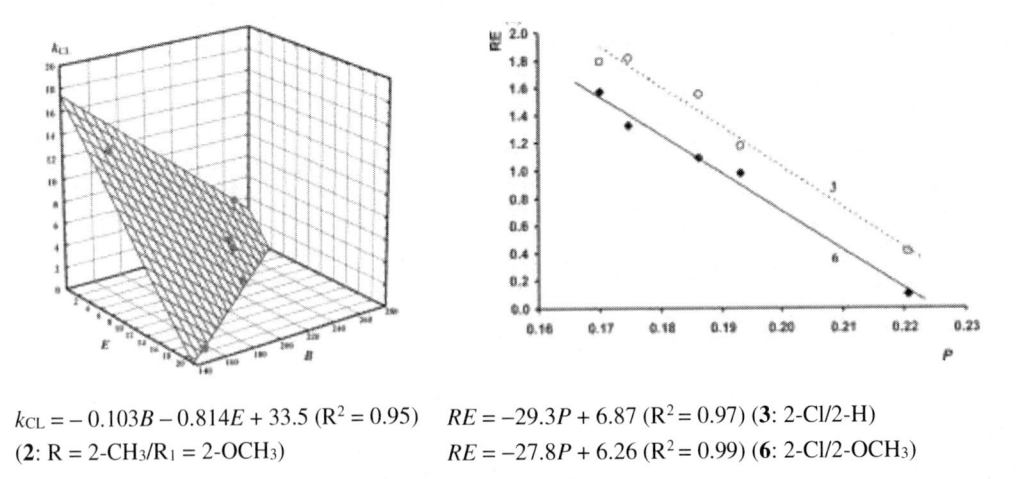

Oxidizer: H_2O_2

Bases: potassium hydroxide (KOH)
tetra-n-butyl ammonium hydroxide (TBAOH)
1,8-diazabicyclo[5.4.0]undec-7-ene (DBU)

Solvents: aliphatic alcohols, acetonitrile, dimethylsulfoxide

TBAOH   **DBU**

Figure 26. Acridinium esters, triggering reagents and solvents used in investigations of their chemiluminescence in organic environments [33]. Reprinted with permission from K. Krzymiński, A.D. Roshal, B. Zadykowicz, A. Białk-Bielińska and A. Sieradzan, Chemiluminogenic properties of 10-methyl-9-(phenoxycarbonyl)acridinium cations in organic environments, Journal of Physical Chemistry, 114: 10550-10562. Copyright (2010) American Chemical Society.

Organic acridinium CL systems express the tendency to slow down the emission rate with the increase in basicity and electrophilicity of the solvent; the last parameter seems to have pronounced influence (Figure 27 left).

$k_{CL} = -0.103B - 0.814E + 33.5$ (R^2 = 0.95) $RE = -29.3P + 6.87$ (R^2 = 0.97) (**3**: 2-Cl/2-H)
(**2**: R = 2-CH$_3$/R$_1$ = 2-OCH$_3$) $RE = -27.8P + 6.26$ (R^2 = 0.99) (**6**: 2-Cl/2-OCH$_3$)

Figure 27. The graphs presenting relationships of CL parameters of substituted AEs in organic environments on liquid phase properties. Left graph: Multiparametric relationship among kinetic constants of CL decay (k_{CL}) and solvent's basicity (*B*) and electrophilicity (*E*) in the case of **2** (base: TBAOH). Right graph: Relationships between relative integral efficiency (*RE*) and solvent polarizability (*P*) found for **3** and **6** in DBU/ethanol system (see Figure 26) [33]. Reprinted with permission from K. Krzymiński, A.D. Roshal, B. Zadykowicz, A. Białk-Bielińska and A. Sieradzan, Chemiluminogenic properties of 10-methyl-9-(phenoxycarbonyl)acridinium cations in organic environments, Journal of Physical Chemistry, 114: 10550-10562. Copyright (2010) American Chemical Society.

On the other hand, the relative efficiencies of emission (*RE*) seem to increase in the media of lower polarity and higher basicity; the last parameter appears to have weak or negligible influence (Figure 27 right). In Figure 28 the relative intensities (integrals) characterizing selected acridinium CL systems are compared. The appearance of this graph leads us to the conclusion that organic CL systems generally allow reaching more than two times higher (exciding 4%) ϕ_{CL}, than optimized, typical aqueous systems. The second conclusion is that the modification of the acridine nucleus with substituent of electron-donor character (OCH_3) has a beneficial effect on the emissive ability of respective AEs. In organic environments (acetonitrile, methanol) the range of AE emission slightly shifts towards shorter wavelengths (Figure 28, right).

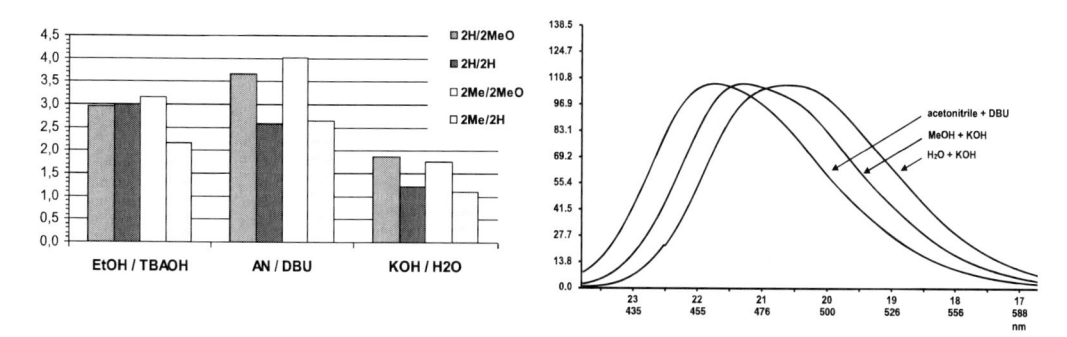

Figure 28. Left graph: Comparison of emitting ability (y axis: ϕ_{CL} in %) of selected AEs in various environments; 2H/2MeO; 2Me/2MeO; 2H/2H; 2Me/2H denote AEs substituted in benzene/acridine rings (compounds **1**, **2**, **4**, **5**, respectively, Figure 26. Right graph: Stationary CL spectra of **3** (Figure 26) recorded in various environments [33]. Reprinted with permission from K. Krzymiński, A.D. Roshal, B. Zadykowicz, A. Białk-Bielińska and A. Sieradzan, Chemiluminogenic properties of 10-methyl-9-(phenoxycarbonyl)acridinium cations in organic environments, Journal of Physical Chemistry, 114: 10550-10562. Copyright (2010) American Chemical Society.

2.7. Analytical Applications of Acridinium Indicators and Labels

2.7.1. Chemiluminescent Indicators

A novel method for determining the antioxidant activity of organic materials, based on the AEs chemiluminescence in an aqueous/alcoholic media was proposed [71]. The method involves assessment of the constant rates of CL decay (eventually intensity of integral emission), associated with the reaction of selected AEs with hydrogen peroxide in a pH-adjusted environment in the presence of an analyte of antioxidant properties.

Temporary emission intensities (I_{CL}) are presented as a function of time (*t*) according to the dependence: $\log I_{CL} = f(t)$ from which the pseudo-first order constants of CL decay (k_{CL}) can be determined. The last values remain in a certain range linearly dependent on the concentration of the tested analyte. The value of the linear coefficient (k^g_{CL}), derived from the relationship: $k_{CL} = f$(mass of analyte), makes a measure of the total antioxidant (positive values) or oxidant (negative values) capacity of the tested substance (or mixture of substances).

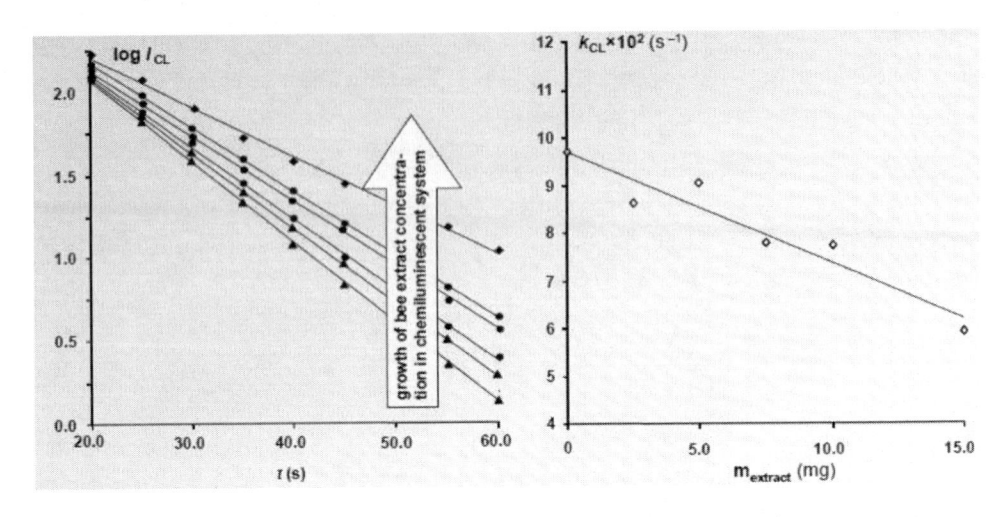

Figure 29. Left graph: Kinetics plots (log I_{CL} = $f(t)$) for different concentrations of bee extract (*Apiextract®*) in acridinium aqueous-alcoholic CL system (AE/H$_2$O$_2$/NaOH). Right graph: A plot of first order kinetic constants (k_{CL}) vs. mass of the extract. A slope coefficient of this relationship (k^g_{CL}) makes a quantitative parameter expressing an antioxidant activity of tested formulation [71]. Reprinted by permission from John Wiley and Sons: Luminescence, On the use of acridinium indicators for the chemiluminescent determination of the total antioxidant capacity of dietary supplements, K. Krzymiński, A.D. Roshal, P.B. Rudnicki-Velasquez, K. Żamojć, Copyright (2019).

Table 5. Comparison of analytical parameters assessed for the AE-based chemiluminometric method (ascorbic acid as reference antioxidant) and for the reference CL method (based on the use of luminol) [71]

CL parameters		Reference solution	Ascorbic acid	Ascorbic acid + muddling agent	Reference solution	Ascorbic acid	Ascorbic acid + muddling agent
			Reference method			Proposed method	
Solution optical density at wavelength of CL measurement	Unit	0.129	0.133	0.352	0.053	0.057	0.337
Total duration of CL emission	hrs	~ 28	~ 28	~ 11	-	-	-
	min	1.66×10^3	1.66×10^3	7.00×10^2	1.25	1.33	1.10
	sec	9.93×10^4	9.93×10^4	4.20×10^4	75	80	66
Time lag of CL rise	min	-	-	15	-	-	-
Time of maximal CL achieving	min	25	55	60	-	-	-
	sec	1.50×10^3	3.30×10^3	3.60×10^3	3	3	3
Maximal intensity of CL	a.u.	0.55	0.51	0.28	165.1	152.0	113.4
Light sum (integral CL)	a.u.×sec	4.88×10^4	4.42×10^4	1.20×10^3	1.74×10^3	1.81×10^3	1.13×10^3
Rate constant of CL quenching	sec^{-1}	1.83×10^{-5}	1.72×10^{-5}	4.95×10^{-5}	4.28×10^{-2}	3.25×10^{-2}	3.31×10^{-2}

The acridinium-based method expresses advantages over typical approaches utilizing luminol, because high temporary emission of (selected) AE enable work with highly colored and cloudy samples - which is often the case when testing biological extracts of complex composition. The usefulness of the method was demonstrated in the study of some reference antioxidants (quercetin, ascorbic acid) as well as exemplary complex formulations, including plant extracts and food additives. Among the tested analytes was natural bee product, *Apiextract®*, owing its antioxidant properties to the flavonoid and carotenoid fractions, disclosed in the 3-D fluorescence emission spectrum. The comparison between the acridinium-based test and the reference test based on reference substrate, luminol, is presented in Table 5. Numerical values, characterizing studied analytical parameters indicate significant advantages of the acridinium-based method in terms of its sensitivity, repeatability of results, versatility and average time of analysis.

2.7.2. Acridinium Chemiluminescent Labels

The current state of knowledge about AEs can be considered as advanced. However, it mainly concerns analytical aspects. Medical immunodiagnostics, utilizing the CL in the antigen-antibody assays (Chemiluminescence Immunoassays (CLIA), Enzymatic Chemiluminescence Immunoassays (ECLIA), Figure 30) make dynamically developing branch of modern analytics [47].

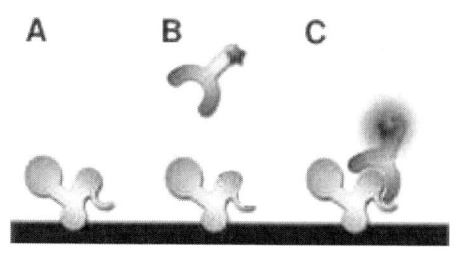 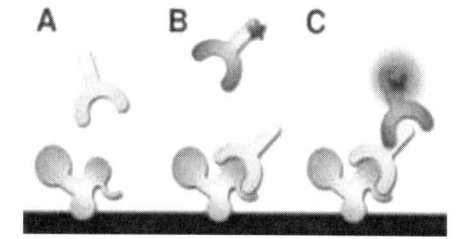

A: an antigen coated on microplate (analyte, e.g., IgG)
B: AE-labeled antibodies (anti-IgG)
C: CL reaction enabling to detect the antigen

A: binding of the antibody (analyte) to the antigen immobilized on plate
B: binding of the secondary antibody directed against the first antibody
C: a chemiluminescent reaction that detects specific antibodies

Figure 30. The scheme of modified direct CLIA test utilizing immobilized antigen (left graph) and modified indirect (sandwich) CLIA tests utilizing immobilized antigen (right graph) with the use of chemiluminogenic labels.

Luminescent methods constitute an attractive alternative to typical methodologies, utilizing UV-Vis absorption (e.g., ELISA tests) - due to their high sensitivity, small amounts of reagents needed for assays and simplicity of measurements (no complicated optics needed) - which speak for their versatility and relatively low cost. The detection limits of analytes in the femtomol range of concentrations (10^{-15} M) are easily accessible in most cases; however, detection limits of analytes at the attomole or even lower levels were reported [72]. No other assay types are characterized by such high sensitivities. Moreover, the methods based on CL allow the elimination of problems associated with the use of radioactive isotopes (such as special laboratories, waste storage, radioactive decay, etc.), while maintaining comparable or higher sensitivity.

The acridinium salts presented in this article have gained wide interest from researchers and clinicians as light probes in modern trace analytics [3]. Among them, an aromatic esters and *N*-sulfonamides of 10-substituted acridine-9-carboxylic acid make the most important CL emitters of practical importance [52, 73]. Examples of newer acridinium labels, tested for high emission efficiency are presented in Figure 31.

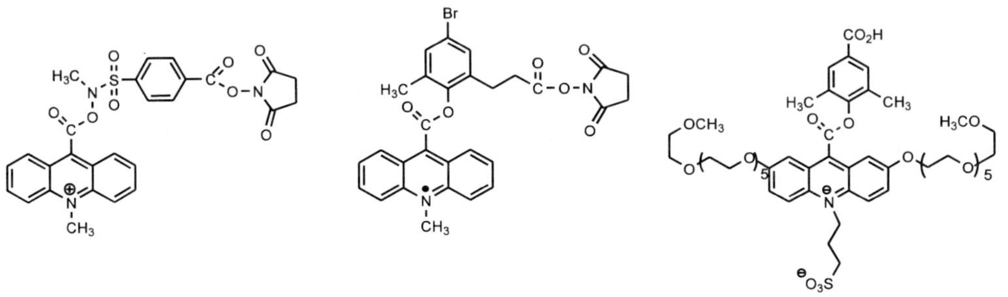

Figure 31. Chemiluminogenic acridinium labels characterized by high emission efficiency [76, 77].

Acridinium derivatives, being employed in luminescence bioanalysis, express a number of advantages over luminol derivatives - they are characterized by general high efficiency and dynamics of emission, excellent solubility in aqueous systems and no need for catalyst to trigger the CL. The latter feature increases signal-to-noise ratio and reduces background signal, contrary to typical substrates [43]. The above advantages, as well as relative ease of their modification and simplicity of measurements with their participation causing a dynamic development of analytical methods with the use of acridnium salts [47, 74].

Applications of chemiluminogenic acridinium salts cover a wide area of medical and environmental analytics where they are employed as CL labels (Figure 32) - in a chemically bound form with an analyte, as well as CL indicators - in an unbound state - whose spectral changes after calibration allow a determination of trace amounts of tested substance in a given system.

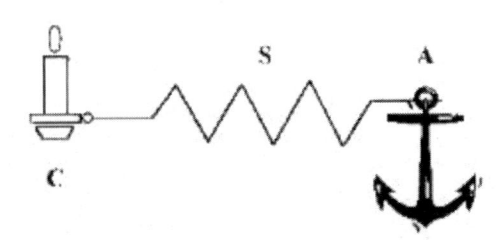

Figure 32. A general scheme of chemiluminogenic labels. C denotes a fragment capable for CL, S - a separating fragment, A - an active (binding) group [4].

CLIA or ECLIA tests with the participation of acridinium salts have been utilized in many biomedical assays, including human α-fetoprotein, gonadotropins, TSH hormones, viruses, nucleic acids, dioxigenin, hydrogen peroxide and others [44, 47].

Generally, in luminescence diagnostics the two types of acridinium esters as CL labels are in use. In most assays the labels where the electronically excited product (reporter molecule) is detached from the analyte (e.g., immune complex) upon light emission are in use. This variant commonly makes a case in antigen-antibody luminogenic immunoassays. However, in certain assays the labels where the excited product remains attached to the analyte upon CL are employed. The compounds representing the second group of acridinium labels were used, for example, in so-called Hybridization Protection Assays (HPA) [74].

The reactivity of original immunodiagnostic reagents obtained with the participation of new acridinium CL labels (Figure 9) in indirect CLIA tests with *Toxoplasma gondii* antigens is illustrated in Figure 33 [57]. Their reactivity, understood as serospecificity, clearly demonstrate practical utility of the new CL substrates as well as their superior performance in comparison to commercial acridinium CL label, supplied in marketed luminescence tests [75].

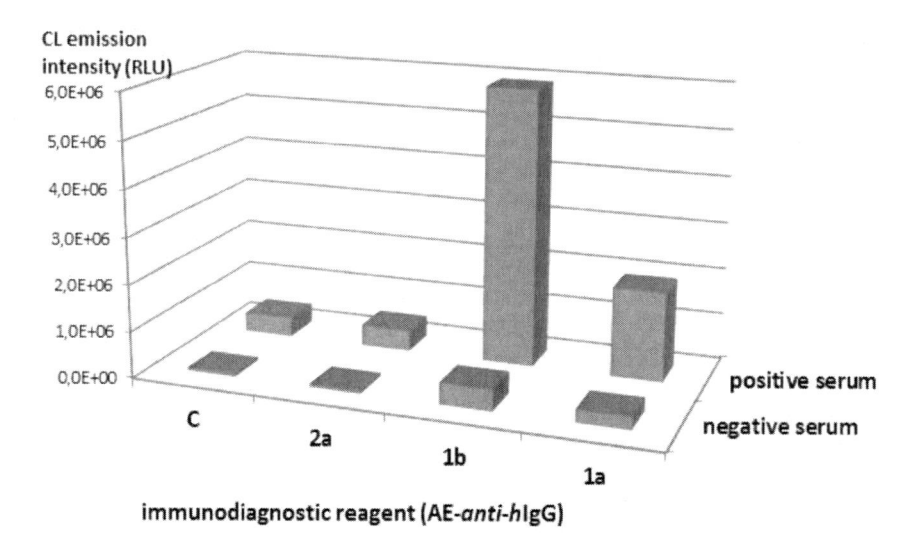

Figure 33. Comparison of the emission intensity of exemplary immunodiagnostic reagents in indirect CLIA test, obtained with the participation of new acridinium CL labels (**1a**, **1b**, **2a**, Figure 9) and the control (reference) label (**C**) in the presence or absence of specific anti-*Toxoplasma gondii* antibodies, assessed in the serum of patients (serospecificity). **C** label was used according to the producer's description [75] and original labels **1a** - **2a** were applied according to the optimized procedures described in [57].

A known disadvantage of chemiluminogenic acridinium compounds is their susceptibility to dark side reactions (discussed above) - that are hydrolysis and the ease of formation of 'pseudobases' in an aqueous environment of neutral or alkaline pHs [28, 33]

- which complicate, to some extent, their practical uses. Investigations are still being sought to improve emission properties and the stability of acridinium chemiluminogenic substrates in aqueous systems. However, despite the above drawbacks, AEs seem to be extremely beneficial and promising reagents to be employed in modern luminescence bioanalysis.

REFERENCES

[1]   McCapra, F. and Richardson, D. G. (1964). The mechanism of chemiluminescence: A new chemiluminescent reaction. *Tetrahedron Lett.* 5: 3167-3172.

[2]   McCapra, F., Richardson, D. G. and Chang, Y. C. (1965). Chemiluminescence involving peroxide decompositions. *Photochem. Photobiol.* 4: 1111-1121.

[3]   Weeks, I., Beheshti, I., McCapra, F., Campbell, A. K. and Woodhead, J. S. (1983). Acridinium esters as high-specific-activity labels in immunoassay. *Clin. Chem.* 9: 1474-1479.

[4]   Zomer, G. and Stavenuiter, J. F. C. (1989). Chemiluminogenic labels, old and new. *Anal. Chim. Acta* 227: 11-19.

[5]   Rak, J., Krzymiński, K., Skurski, P., Jóźwiak, L., Konitz, A., Dokurno, P. and Błażejowski, J. (1998). X-Ray, quantum mechanics and density functional methods in the examination of structure and tautomerism of N-methyl-substituted acridin-9-amine derivatives. *Aust. J. Chem.* 51: 643-651.

[6]   Rak, J., Skurski, P. and Błażejowski, J. (1999). Toward an understanding of the chemiluminescence accompanying the reaction of 9-carboxy-10-methylacridinium phenyl ester with hydrogen peroxide. *J. Org. Chem.* 64: 3002-3008.

[7]   Dewar, M. J. S. and Thiel, W. (1977). Ground states of molecules. The MNDO Method. Approximations and Parameters. *J. Am. Chem. Soc.* 99: 4899-4907.

[8]   Dewar, M. J. S., Zoebisch, E. G., Healy, E. F. and Stewart, J. J. P. (1985). AM1: A new general purpose quantum mechanical molecular model. *J. Am. Chem. Soc.* 107: 3902-3909.

[9]   Stewart, J. J. P. (1989). Optimization of parameters for semiempirical methods I. Method. *J. Comput. Chem.* 10: 209-220.

[10]   Roothaan, C. C. J. (1951). New developments in molecular orbital theory. *Rev. Mod. Phys.* 23: 69-89.

[11]   Labanowski, J. K. (1991). *Density Functional Methods in Chemistry.* ed. Andzelm, J. W., Springer-Verlag, New York.

[12]   Hohenberg, P. and Kohn, W. (1964). Inhomogeneous electron gas. *Phys. Rev.* 136: B864-B871.

[13]   Boużyk, A., Jóźwiak, L., Wróblewska, A., Rak, J. and Błażejowski, J. (2002). Structure, properties, thermodynamics, and isomerization ability of 9-acridinones. *J. Phys. Chem. A* 106: 3957-3963.

[14] Becke, A. D. (1993). A new mixing of Hartree–Fock and local density-functional theories. *J. Chem. Phys.* 98: 1372-1377.

[15] Becke, A. D. (1993). Density-functional thermochemistry. The role of exact exchange. *J. Chem. Phys.* 98: 5648-5652.

[16] Lee, C., Yang, W. and Parr, R. G. (1988). Development of the Colle-Salvetti correlation-energy formula into a functional of the electron density. *Phys. Rev. B* 37: 785-796.

[17] Hariharan, P. C. and Pople, J. A. (1973). The influence of polarization functions on molecular orbital hydrogenation energies. *Theor. Chim. Acta* 28: 213-222.

[18] Hehre, W. J., Radom, L., Schleyer, P. v. R. and Pople, J. A. (1986). *Ab Initio Molecular Orbital Theory,* Wiley, New York.

[19] Zhao, Y. and Truhlar, D. G. (2008). The M06 suite of density functionals for main group thermochemistry, thermochemical kinetics, noncovalent interactions, excited states, and transition elements: two new functionals and systematic testing of four M06-class functionals and 12 other functionals. *Theor. Chem. Acc.* 120: 215-241.

[20] Jursic, B. S. (1996). *Theoretical and Computational Chemistry.* ed. Seminario, J. M., Elsevier, Amsterdam.

[21] Tomasi, J. and Persico, M. (1994). Molecular interactions in solution: an overview of methods based on continuous distributions of the solvent. *Chem. Rev.* 94: 2027-2094.

[22] Barone, V., Cossi, M. and Tomasi, J. (1997). A new definition of cavities for the computation of solvation free energies by the polarizable continuum model. *J. Chem. Phys.* 107: 3210-3221.

[23] Klamt, A. and Schurmann, G.Z. (1993). COSMO: a new approach to dielectric screening in solvents with explicit expressions for the screening energy and its gradient. *J. Chem. Soc. Perkin* 2: 799-805.

[24] Ullrich, C. A. and Yang, Z. (2014). A brief compendium of time-dependent density functional theory. *Braz. J. Phys.* 44: 154-188.

[25] Runge, E. and Gross, E. K. U. (1984). Density-functional theory for time-dependent systems. *Phys. Rev. Lett.* 52: 997-1000.

[26] McCapra, F. (1970). Chemiluminescence of organic compounds. *Pure Appl. Chem.* 24: 611-629.

[27] Dodeigne, C., Thunus, L. and Lejeune, R. (2000) Chemiluminescence as diagnostic tool, A review. *Talanta* 51: 415-439.

[28] Krzymiński, K., Ożóg, A., Malecha, P., Roshal, A. D., Wróblewska, A., Zadykowicz, B. and Błażejowski, J. (2011). Chemiluminogenic features of 10-methyl-9-(phenoxycarbonyl)acridinium trifluoromethanesulfonates alkyl substituted at the benzene ring in aqueous media. *J. Org. Chem.* 76: 1072-1085.

[29] Zadykowicz, B., Czechowska, J., Ożóg, A., Renkevich, A. and Krzymiński, K. (2016). Effective chemiluminogenic systems based on acridinium esters bearing

substituents of various electronic and steric properties. *Org. Biomol. Chem.* 14: 652-668.

[30] Czechowska, J., Kawecka, A., Romanowska, A., Marczak, M., Wityk, P., Krzymiński, K. and Zadykowicz, B. (2017). Chemiluminogenic acridinium salts: A comparison study. Detection of intermediate entities appearing upon light generation. *J. Lumin.* 187: 102-112.

[31] Fleming, I. (1976). *Frontier Orbitals and Organic Chemical Reactions,* Wiley, London, New York.

[32] Boużyk, A., Jóźwiak, L., Kolendo, A. Y. and Błażejowski, J. (2003). Theoretical interpretation of electronic absorption and emission transitions in 9-acridinones. *Spectrochim. Acta A* 59: 543-558.

[33] Krzymiński, K., Roshal, A. D., Zadykowicz, B., Białk-Bielińska, A. and Sieradzan, A. (2010). Chemiluminogenic properties of 10-methyl-9-(phenoxycarbonyl) acridinium cations in organic environments. *J. Phys. Chem.* 114: 10550-10562.

[34] Roca-Sanjuán, D., Delcey, M. G., Navizet, I., Ferré, N., Liu, Y.-J. and Lindh, R. (2011). Chemiluminescence and fluorescence states of a small model for coelenteramide and cypridina oxyluciferin: A CASSCF/CASPT2 study. *J. Chem. Theory Comput.* 7: 4060-4069.

[35] Navizet, I., Roca-Sanjuán, D., Yue, L., Liu, Y.-J., Ferré, N. and Lindh, R. (2013). Are the bio- and chemiluminescence states of the firefly oxyluciferin the same as the fluorescence state? *Photochem. Photobiol.* 89: 319-325.

[36] Schramm, S., Navizet, I., Karothu, D. P., Oesau, P., Bensmann, V., Weiss, D., Beckert, R. and Naumov, P. (2017). Mechanistic investigations of the 2-coumaranone chemiluminescence. *Phys. Chem. Chem. Phys.* 19: 22852-22859.

[37] Silvi, B. and Savin. A. (1994). Classification of chemical bonds based on topological analysis of electron localization functions. *Nature* 371: 683-686.

[38] Farahani, P., Roca-Sanjuan, D., Zapata, F. and Lindh, R. (2013). Revisiting the nonadiabatic process in 1,2-dioxetane. *J. Chem. Theory Comput.* 9: 5404-5411.

[39] Augusto, F. A., Francés-Monerris, A., Galván, I. F., Roca-Sanjuán, D., Bastos, E. L., Baader, W. J. and Lindh, R. (2017). Mechanism of activated chemiluminescence of cyclic peroxides: 1,2-dioxetanes and 1,2-dioxetanones. *Phys. Chem. Chem. Phys.* 19: 3955-3962.

[40] Ciscato, L. F. M. L., Bartoloni, F. H., Weiss, D., Beckert, R. and Baader, W. J. (2010). Experimental evidence of the occurrence of intramolecular electron transfer in catalyzed 1,2-dioxetane decomposition. *J. Org. Chem.* 75: 6574-6580.

[41] Nakazono, M., Oshikawa, Y., Nakamura, M., Kubota, H. and Nanbu, S. (2017). Strongly chemiluminescent acridinium esters under neutral conditions: synthesis, properties, determination, and theoretical study. *J. Org. Chem.* 82: 2450-2461.

[42] Schulman, S. G., Schulman, J. and Rakicioğlu, Y. (2001). *Chemiluminescence in Analytical Chemistry*. eds Garcia-Campaña, A. M. and Bayenes, W. R. G., Marcel Dekker, New York-Basel. pp 67–81.

[43] Zomer, G., Stavenuiter, J. F. C., Van den Berg, R. H. and Jansen, E. H. J. M. (1991). Synthesis, chemiluminescence, and stability of acridinium ester labeled compounds. *Pract. Spectrosc*. 12: 505-521.

[44] Krićka, L. J. (2003). Clinical applications of chemiluminescence. *Analitica Chimica Acta*, 500: 279-286.

[45] Ando, Y., Niwa, K., Yamada, N., Irie, T., Enomoto, T., Kubota, H., Ohmiya, Y. and Akiyama, H. (2007). Development of a quantitative bio/chemiluminescence spectrometer determining quantum yields: re-examination of the aqueous luminol chemiluminescence standard. *Photochem. Photobiol*. 83: 1205-1210.

[46] Sato, N. (1996). Synthesis and properties of new luminescent 10-carboxymethyl-acridinium derivatives. *Tetrahedron Lett*. 37: 8519-8522.

[47] Van Dyke, K., Van Dyke, C. and Woodfork, K. (2002). *Luminescence biotechnology instruments and applications*. CRC Press, Boca Raton, London-New York-Washington.

[48] McCapra, F. (1976). Chemical Mechanisms in bioluminescence. *Acc. Chem. Res*. 9: 200-208.

[49] Adamczyk, M., Fishpaugh, J. R., Gebler, J. C., Mattingly, P. G. and Shreder, K. (1998). Detection of reaction intermediates by flow injection electrospray ionization mass spectrometry: reaction of chemiluminescent N-sulfonylacridinium-9-carboxamides with hydrogen peroxide. *Eur. J. Mass Spectrom*. 4: 121-125.

[50] Batmanghelich, S., Woodhead, J. S., Smith, K. and Weeks, I. (1991). Synthesis and chemiluminescent evaluation of a series of phenyl N-alkylacridinium 9-carboxylates. *J. Photochem. Photobiol. A* 56: 249-254.

[51] Smith, K., Yang, J.-J., Li, Z., Weeks, I. and Woodhead, J. S. (2009). Synthesis and properties of novel chemiluminescent biological probes: 2- and 3-(2-succinimidyl-oxycarbonylethyl)phenyl acridinium esters. *J. Photochem. Photobiol. A* 203: 72-79.

[52] Natrajan, A., Jiang, Q. and Sharpe, D. (2012). Stable acridinium esters with fast light emission. United States Patent, US8119422B2.

[53] Natrajan, A., Jiang, Q., Sharpe, D. and Costello, J. (2007). *High quantum yield acridinium compounds and their uses in improving assay sensitivity*, United States Patent US7309615B2.

[54] Renotte, R., Sarlet, G., Thunus, L. and Lejeune, R. (2000). High stability and high efficiency chemiluminescent acridinium compounds obtained from 9-acridine carboxylic esters of hydroxamic and sulphohydroxamic acids. *Luminescence* 15: 311-320.

[55] Carey, F. A. and Sundberg, R. J. (1990). *Advanced Organic Chemistry*, B: Reactions and Synthesis. Plenum Press, New York.

[56] Rauhut, M. M., Sheehan, D., Clarke, R. A., Roberts, B. G. and Semsel, A. M. (1972). Chemiluminescence from the reaction of 9-chlorocarbonyl-10-methylacridinium chloride with aqueous hydrogen peroxide. *J. Am. Chem. Soc.* 30: 3587-3592.

[57] Krzymiński, K. J., Czechowska, J., Serdiuk, I. E., Karska, N., Kur, J. W., Holec-Gąsior, L., Ferra, B. and Kasprzykowski, F. (2016). *Salts of aryl esters of N-substituted acridine-9-carboxylic acids, aryl esters of N-substituted acridane-9-carboxylic acids, method of their preparation and their use.* Patent Application No. P.416813, Republic of Poland.

[58] Sikorski, A., Krzymiński, K., Niziołek, A. and Błażejowski, J. (2005). 9-(2,6-Difluorophenoxycarbonyl)-10-methylacridinium trifuoromethanesulfonate and its precursor 2,6-difluorophenylacridine-9-carboxylate. *Acta Crystallogr., Sect. C: Cryst. Struct. Commun.* 61: o690-o694.

[59] Trzybiński, D., Krzymiński, K., Sikorski, A. and Błażejowski, J. (2010). 10-Methyl-9-phenoxycarbonylacridinium trifluoromethanesulfonate monohydrate. *Acta Crystallogr. Sect. E: Struct. Rep. Online* 66: o906-o907.

[60] Spek, A. L. (2003). Single-crystal structure validation with the program PLATON. *J. Appl. Crystallogr.* 36: 7-13.

[61] Trzybiński, D. (2012). *The crystalline structure of acridine derivatives of chemiluminogenic properties.* Thesis, University of Gdansk, Poland.

[62] Niziołek, A., Zadykowicz, B., Trzybiński, D., Sikorski, A., Krzymiński, K. and Błażejowski, J. (2009). 9-[(Mesityloxy)carbonyl]-10-methylacridinium trifluoromethanesulfonate and its derivative mesityl 9-methoxy-10-methyl-9,10-dihydroacridine-9-carboxylate: structural and physicochemical features. *J. Mol. Struct.* 920: 231-237.

[63] Jones, J. E. (1924). On the determination of molecular fields.—II. From the equation of state of a gas. *Proc. R. Soc. London, Ser. A* 106: 463-477.

[64] Buckingham, R. A. (1938). The classical equation of state of gaseous helium, neon and argon. *Proc. R. Soc. London, Ser. A* 168: 264-283.

[65] (a) Zadykowicz, B., Krzymiński, K., Storoniak, P. and Błażejowski, J. (2010). Lattice energetics and thermochemistry of phenyl acridine-9-carboxylates and 9-phenoxycarbonyl-10-methylacridinium trifluoromethanesulphonates nitro, methoxy or halogen substituted in the phenyl fragment. *J. Therm. Anal. Calorim.* 101: 429-437. (b) Krzymiński, K., Malecha, P., Storoniak, P., Zadykowicz, B. and Błażejowski, J. (2010). Thermochemistry and crystal lattice energetics of phenyl acridine-9-carboxylates and 9-phenoxycarbonyl-10-methylacridinium trifluoromethanesulphonates. *J. Therm. Anal. Calorim.* 100: 207-214.

[66] Storoniak, P., Krzymiński, K., Boużyk, A., Kovalchuk, E.P. and Błażejowski, J. (2003). Melting, volatilization and crystal lattice enthalpies of acridin-9(10*H*)-ones. *J. Therm. Anal. Calorim.* 74: 443–450.

[67] Krzymiński, K., Roshal, A. D. and Niziołek, A. (2008). Spectral features of substituted 9-(phenoxycarbonyl)-acridines and their protonated and methylated cation derivatives. *Spectrochim. Acta A* 70: 394-402.

[68] Zadykowicz, B., Ożóg, A. and Krzymiński, K. (2010). Vibrational spectra of phenyl acridine-9-carboxylates and their 10-methylated cations: A theoretical and experimental study. *Spectrochim. Acta A* 75: 1546-1551.

[69] Krzymiński, K., Malecha, P., Zadykowicz, B., Wróblewska, A. and Błażejowski, J. (2011). ^{1}H and ^{13}C NMR spectra, structure and physicochemical features of acridine 9-carboxylates and 9-phenoxycarbonyl-10-methylacridinium trifluoro-methanesulphonates – alkyl substituted in the phenyl fragment. *Spectrochim. Acta A* 78: 401-409.

[70] Miller, G. W., Morgan, C. A., Kieber, D. J., King, D. W., Snow, J. A., Heikes, B. G., Mopper, K. and Kiddle, J. J. (2005). Hydrogen peroxide method intercomparision study in seawater. *Mar. Chem.* 97: 4-13.

[71] Krzymiński, K. J., Roshal, A. D., Rudnicki-Velasquez P. B. and Żamojć, K. (2019). On the use of acridinium indicators for the chemiluminescent determination of the total antioxidant capacity of dietary supplements. *Luminescence*, In Press, doi.org/10.1002/bio.3629

[72] Roda, A., Guardigli, M., Michelini, E., Mirasoli, M. and Pasini, P. (2003). Analytical bioluminescence and chemiluminescence. *Anal. Chem.* 462-470.

[73] Adamczyk, M., Fino, J. R., Mattingly, P. G., Moore, J. A. and Pan, Y. (2004). Chemiluminescence quenching of pteroic acid–N-sulfonyl-acridinium-9-carboxa-mide conjugates by folate binding protein. *Bioorg. Med. Chem. Lett.* 14: 2313-2317.

[74] Browne, K., Deheyn, D. D., El-Hiti, G. A., Smith, K. and Weeks, I. (2011). Simultaneous quantification of multiple nucleic acid targets using chemiluminescent probes. *J. Am. Chem. Soc.* 133: 14637-14648.

[75] (a) http://www.enzolifesciences.com/ADI-907-001/chemiluminescent-labeling-kit/ (b) https://www.ibl-international.com/en/acridinium-protein-labeling-kit (assessed on April 12, 2019).

[76] Natrajan, A., Jiang, Q., Sharpe, D., Costello, J. (2007). *High quantum yield acridinium compounds and their uses in improving assay sensitivity,* United States Patent No. US 7,309,615B2, publication date: Dec. 12.

[77] Renotte, R. Sarlet, G., Thunus, L., Lejeune, R. (2000). High stability and high efficiency chemiluminescent acridinium compounds obtained from 9-acridine carboxylic esters of hydroxamic and sulphohydroxamic acids. *Luminescence* 15: 311–320.

[78] Trzybiński D., Zadykowicz, B., Wera, M., Serdiuk, I.E., Sieradzan, A., Sikorski, A., Storoniak, P. and Krzymiński, K. Structure, formation, thermodynamics and interactions in 9-carboxy-10-methylacridinium-based molecular systems (2016). *New J. Chem.* 40: 7359-7372.

In: A Comprehensive Guide to Chemiluminescence   ISBN: 978-1-53616-170-0
Editor: Luís Pinto da Silva   © 2019 Nova Science Publishers, Inc.

Chapter 6

Phenothiazine Derivatives as Enhancers of Peroxidase-Catalyzed Chemiluminescence and Their Application in Bioanalysis

Ivan Yu. Sakharov[*]

Department of Chemistry, Lomonosov Moscow State University, Moscow, Russia

Abstract

Some *N*-alkyl phenothiazines with different ionic groups were studied as enhancers of chemiluminescence formed upon luminol oxidation catalyzed by plant peroxidases. It was shown that the phenothiazines with positively charged groups do not possess an enhancing ability, while phenothiazines with negatively charged groups significantly increase the chemiluminescence intensity. 3-(10′-Phenothiazinyl)propane-1-sulfonate (SPTZ) and 3-(10'-phenothiazinyl) propionic acid (PPA) showed the highest activity in enhanced chemiluminescence reaction (ECR) catalyzed both horseradish (HRP) and soybean (SbP) peroxidases. The addition of 4-dialkylaminopyridines such as 4-morpholinopyridine (MORPH), 4-pyrrolidinopyridine and 4-dimethylaminopyridine to a substrate solution also increased chemiluminescence; this increase was maximum in the presence of MORPH. Kinetic studies of the enzyme oxidation of SPTZ by hydrogen peroxide demonstrated that the addition of MORPH (secondary enhancer) significantly increased the rate of production of SPTZ cation radical and did not affect the rate of decomposition of this radical. Detection limit values of HRP in ECR with PPA/ MORPH and SPTZ/ MORPH were lower than that with *p*-iodophenol. To demonstrate feasibility of phenothiazine enhancers, ultra-sensitive chemiluminescent assays for determination of small molecules, proteins and DNA were developed. These enhancers allow new opportunities for increasing the sensitivity of determination of analytes by chemiluminescent assays.

[*] Corresponding Author Email: sakharovivan@gmail.com.

1. INTRODUCTION

Enzyme immunoassay, based on the specific binding of the detected compound with the corresponding antibodies, is currently one of the most used analytical methods and is intensively applied in various fields of medicine, agriculture, microbiological and food industry, and environmental monitoring. This is due to the high affinity and unique specificity of the reaction between the antigen and the antibody, as well as the low detection limit of the enzyme label.

In practice its heterogeneous variant, namely enzyme-linked immunosorbent assay (ELISA) is widely used. There are several ELISA formats, for example, direct and indirect competitive assays, sandwich assay, etc. Schemes of these ELISAs are different, but in all the assays the determination of the catalytic activity of the enzyme label is required at their last stage.

Cationic isoenzyme of horseradish peroxidase (HRP) is widely used as a label in ELISA [1-3]. Peroxidases from other sources, for example, soybean and tobacco anionic peroxidases are also used in bioanalysis [4-6], although less often. The wide application of HRP results from this enzyme being commercially available in highly purified state, inexpensive, and having high specific activity and stability. Calf alkaline phosphatase may be also applied in ELISA [7, 8]. However, the latter enzyme is used only in such cases where the use of HRP is impossible.

Usually the catalytic activity of peroxidase is determined by a colorimetric method towards chromogenic substrates [9-11]. Among peroxidase substrates, 3,3',5,5'-tetramethylbenzidine is currently the most popular. However, in those assays where a decrease in the detection limit and an extension of the working range are required, the chemiluminescent method for determining peroxidase activity is used. In this case luminol (5-Amino-2,3-dihydrophthalazine-1,4-dione) and hydrogen peroxide are employed as substrates. It is well known that plant peroxidases can catalyze luminol oxidation [12-15]. However, they are poor catalysts in this reaction. This problem was overcome by using enhanced chemiluminescence reaction (ECR).

2. ENHANCED CHEMILUMINESCENCE REACTION

This reaction was discovered by Larry Kricka and coworkers [16-19] and is based on the fact that some compounds (enhancers), after their introduction in the substrate solution, can accelerate peroxidase-catalyzed oxidation of luminol. The enhancer and luminol exhibit synergistic properties, which is expressed in the non-additivity of the observed chemiluminescent signals for the individual oxidation of the substrates.

The first paper where D-luciferin was used as an enhancer in ECR was published in 1983 [20]. The use of this compound led to an increase in the signal/background ratio by 80 times. Later it was shown that compounds whose reduction potential is close to or greater than the reduction potential of luminol can exhibit the enhancing ability.

Later, a mechanism of ECR was studied in detail which is based on a "ping-pong" mechanism of peroxidase catalysis. In the first step the enzyme reacts with hydrogen peroxide producing Compound I (eq.1). Then, the enhancer molecule reacts with EI. The products of this reaction are Compound II and a radical enhancer (eq. 2). In the next step, in which Compound II reacts with a second molecule of enhancer, ferric peroxidase and another radical enhancer are produced (eq. 3). These three reactions proceed under enzymatic control.

$$E + H_2O_2 => EI + H_2O \tag{1}$$

$$EI + SH => EII + S^{\cdot} + H_2O \tag{2}$$

$$EII + SH => E + S^{\cdot} + H_2O \tag{3}$$

Here, E is the ferric enzyme (resting state), EI and EII are Compound I and Compound II, the oxidized intermediates of peroxidase which are by two and one oxidation equivalents above the resting state, respectively, and SH and $S^{\cdot}$ are an electron donor substrate (enhancer) and its radical product of its one-electron oxidation, respectively [21].

After carrying out the enzymatic reactions, a number of non-enzymatic reactions take place, the main ones being: oxidation of luminol anion molecules by the radicals of the enhancer; formation of luminol diazaquinone due to interaction of luminol radicals with oxygen or in the process of disproportionation of luminol radicals; formation of luminol peroxide due to the interaction of luminol radicals with superoxide radicals or of the luminol diazaquinone with hydrogen peroxide anion; the decomposition of luminol peroxide into nitrogen and 3-aminophthalate, which is produced in an excited state. The transition of 3-aminophthalate to the ground state is accompanied by the emission of a quantum of light. These reactions are presented below:

$$S^{\cdot} + AH^- => SH + A^{\cdot -} \tag{4}$$

$$A^{\cdot -} + O_2 => A + O_2^{\cdot -} \tag{5}$$

$$A^{\cdot -} + O_2^{\cdot -} + H^+ => AO_2H^- \tag{6}$$

$$2\ A^{\cdot -} => A + AH^- \tag{7}$$

$$A + HO_2^- => AO_2H^- \tag{8}$$

$$AO_2H^- \Rightarrow 3\text{-}AP + N_2 + H^+ + h\upsilon \tag{9}$$

where LH^- - luminol anion (under alkaline conditions), $L^{\cdot-}$ - luminol radical, L – luminol diazaquinone, AO_2H^- - luminol peroxide, 3-AP – 3-aminophthalate.

Thus, the enhancer plays a role of electron "shuttle" between the active site of peroxidase and a target molecule (luminol). Mediators of peroxidases are used in order to increase a rate of oxidation of poor substrates [22, 23]. The emission spectrum produced by luminol oxidation shows a maximum at 425 nm [24]. Moreover, some studies demonstrated that emission spectrum is the same in the presence and absence of enhancers.

3. ENHANCERS

Enhancement of chemiluminescence reaches several hundred times and depends on the nature of the enhancer, the concentrations of reagents and of reaction conditions [25]. For the last 40 years it was found that some derivatives of phenol, 6-hydroxybenzothiazole, naphthol, aniline, lophine and phenylboric acid are good enhancers [25, 26]. The most popular enhancer for HRP is 4-iodophenol (PIP). Here it should be noted that the compounds mentioned above are enhancers only in HRP-catalyzed ECR. Their enhancing ability is low, if anionic peroxidases were used instead of cationic HRP [12].

Although HRP-catalyzed ECR with PIP was successfully applied in the development of ultrasensitive immunochemical kits for the determination of various compounds, the chemiluminescent signal formed in this reaction is not stable in time and is quickly quenched [27, 28]. This drawback is a reason for higher errors in chemiluminescent enzyme immunoassay. The detection limit of HRP in ECR with PIP was 1.1 pM [27]. Although this value is low and lower that that in colorimetric method, sometimes for development ultrasensitive assays it is required higher sensitivity. Due to this, the search for new enhancers with improved properties is constantly ongoing. These studies allowed the discovery of phenothiazine enhancers, which have the highest enhancing ability so far.

4. PHENOTHIAZINE ENHANCERS

The screening of phenothiazine derivatives containing different N-substituents (Figure 1) as potential enhancers of peroxidase-dependent chemiluminescence showed that phenothiazines with positively charged ionic groups do not possess the required enhancing ability [29]. Moreover, some of them are even weak quenchers. On the other hand, phenothiazines with negatively charged groups increased the chemiluminescence intensity. The maximum enhancing effect was found for 3-(10'-phenothiazinyl)propane-1-sulfonate (SPTZ) and 3-(10'-phenothiazinyl) propionic acid (PPA) [29].

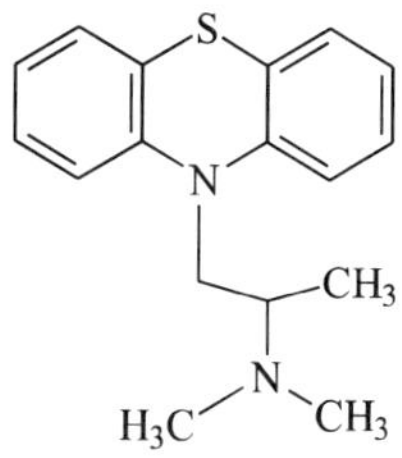

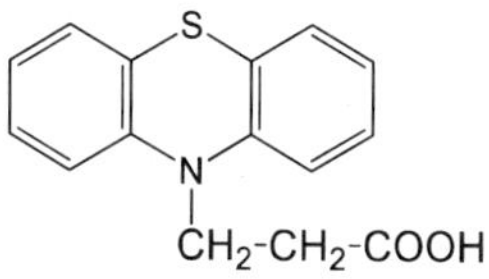

Diprazine 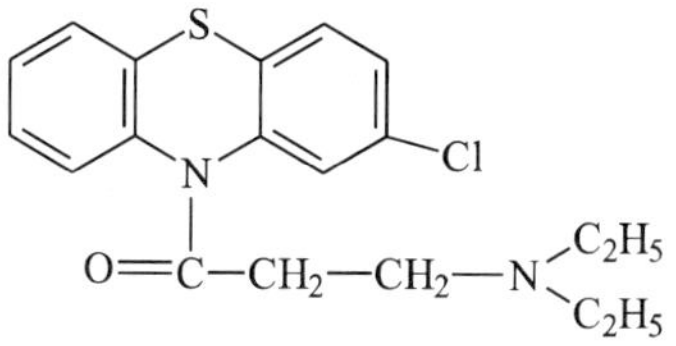   **Chlorocyzine** 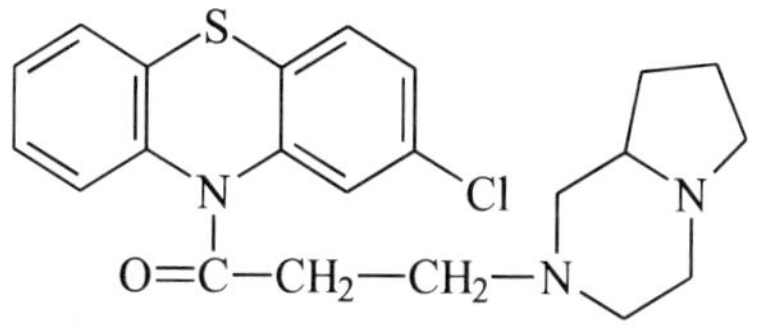   **Nonachlazine**

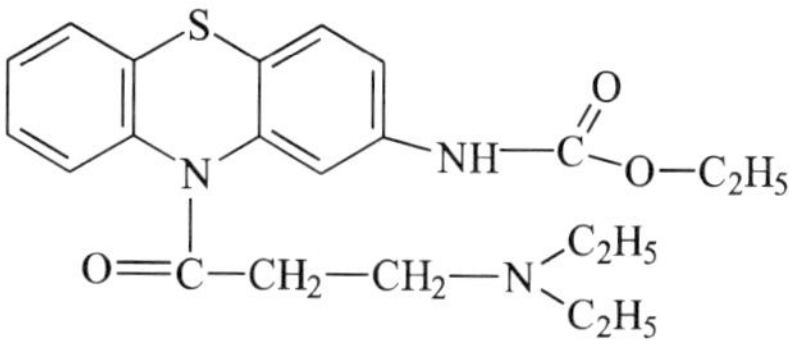

Ethacizine 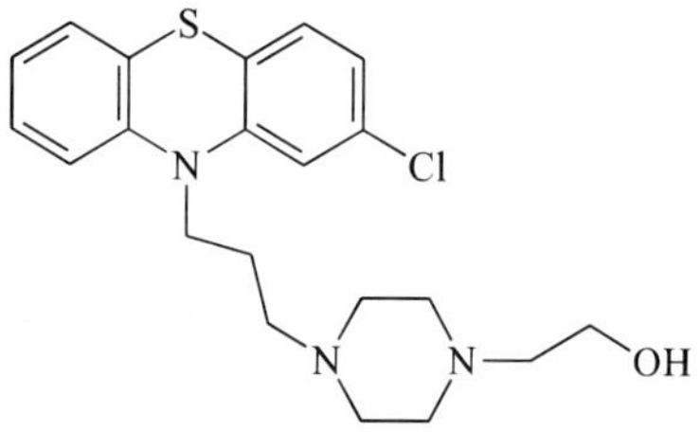   **Perphenazine**

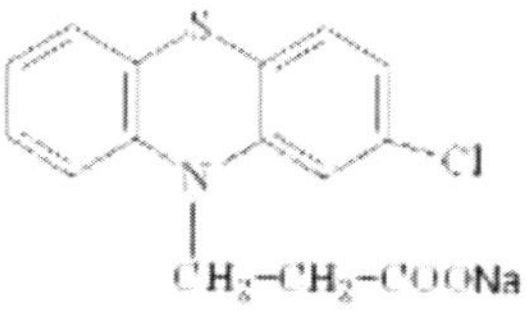

3-(2-Chlorophenothiazi-   **3-(10'-phenothiazinyl)-**   **3-(10'-phenothiazinyl)-**
ne-10-yl)propionate   **propane-1-sulfonate**   **propionic acid**
sodium (CPTP)   **(SPTZ)**   **(PPA)**

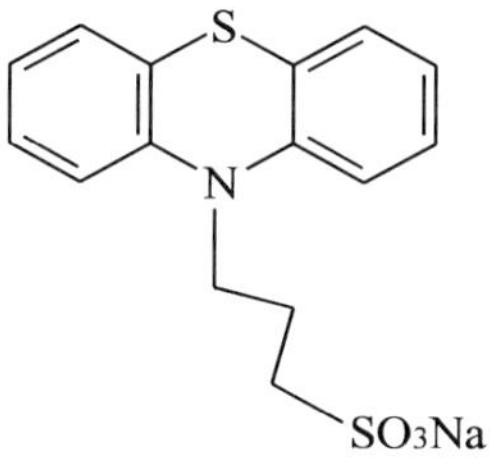

Figure 1. Chemical structures of studies phenothiazines.

It should be noted that 2,2'-azino-bis(3-ethylbenzothiazoline-6-sulfonic acid), one of the most efficient peroxidase substrates [39, 31], also bears a negative charge (sulfonic group). The presence of negatively charged groups is likely to be essential for productive binding of the bulky substrates to the active site of peroxidases.

As mentioned above, most enhancers can only increase HRP-catalyzed chemiluminescence. In contrast, SPTZ and PPA are efficient enhancers in ECR catalyzed by both cationic (HRP) [32-34] and anionic peroxidases (SbP, sweet potato peroxidase) [35-38].

Another feature of phenothiazine enhancers is that there are some compounds which are able to improve their enhancing ability [30]. It was shown that the introduction of some 4-dialkylaminopyridines such as 4-morpholinopyridine (MORPH), 4-dimethylaminopyridine (DMAP), and 4-pyrrolidinopyridine (PPY) to a reaction mixture containing luminol, H_2O_2 and SPTZ resulted in an increase of chemiluminescence intensity. Since 4-dialkylaminopyridines only enhanced chemiluminescence in the presence of a primary enhancer (SPTZ) and did not act as enhancers in the absence of SPTZ, these compounds were named as secondary enhancers.

5. MECHANISM OF PHENOTHIAZINE ENHANCERS

A mechanism of the action of MORPH in ECR with SPTZ was studied previously. It was proved that, in the presence of MORPH, the rate of the peroxidase-catalyzed oxidation of SPTZ with hydrogen peroxide is significantly higher [39]. Therefore, the production of SPTZ cation radical ($SPTZ^{.+}$) in higher concentrations results in an increase of chemiluminescence intensity.

MORPH increases the SPTZ-dependent chemiluminescence intensity more than 3 times. Other 4-dialkylaminopyridines such as DMAP and PPY also act as secondary enhancers, albeit with less efficiency than MORPH. Interestingly, pyridine itself was not a secondary enhancer. Note also that a replacement of the pyridyl fragment in DMAP with phenyl resulted in dimethylaminobenzene not possessing any enhancing ability. Likewise, it was an effective quencher of SPTZ-dependent CL catalyzed by peroxidase.

Given this, the secondary enhancers are active in ECR only if their chemical structure includes both pyridine and dialkylamine fragments. MORPH increases the rate of $SPTZ^{.+}$ production only in the presence of peroxidase; in the absence of the enzyme, SPTZ is not oxidized regardless of whether MORPH was in the reaction solution.

As a catalyst of the SPTZ oxidation, hemin (a prosthetic group in plant peroxidases) was also studied. It is very interesting that while using hemin, the addition of MORPH to the substrate solution did not affect the rate of SPTZ oxidation. Thus, these results allowed for the conclusion that for an implementation of the enhancing ability, 4-dialkylaminopyridines should get bound to a protein fragment of peroxidase located near the entrance in the canal of the active site, where adsorption of peroxidase substrates commonly occurs.

6. FAVORABLE CONDITIONS FOR PHENOTHIAZINE ENHANCERS

In the case of the use of phenothiazine enhancers, the substrate solution composed by luminol, hydrogen peroxide, and primary and secondary enhancers dissolved in a buffer. To optimize this system, an "one-variable-a time" method is inefficient. The use of factorial design allowed to find the most favorable conditions for SPTZ and PPA, which are: 80 mM Tris, pH 8.3, 0.17 mM luminol, 2.1 mM SPTZ, 8.75 mM MORPH, and 1.75 mM H_2O_2; 100 mM Tris, pH 8.3, 1.0 mM luminol, 5.2 mM PPA, 9.3 mM MORPH and 3.0 mM H_2O_2; respectively [33, 40]. It should be noted that under these conditions, the background signals were very low.

The comparison between SPTZ/MORPH, PPA/MORPH and PIP performed under optimized conditions showed that the detection limit for HRP SPTZ/MORPH and PPA/MORPH was equal to 0.09 pM, which was lower than that of PIP (1.1pM). It should be noted that the sensitivity of HRP determination with both phenothiazine enhancers was similar, but significantly higher than that with PIP (Figure 2) [40]. The obtained results demonstrated that the above phenothiazine derivatives in combination with MORPH are potent enhancers of peroxidase-induced chemiluminescence.

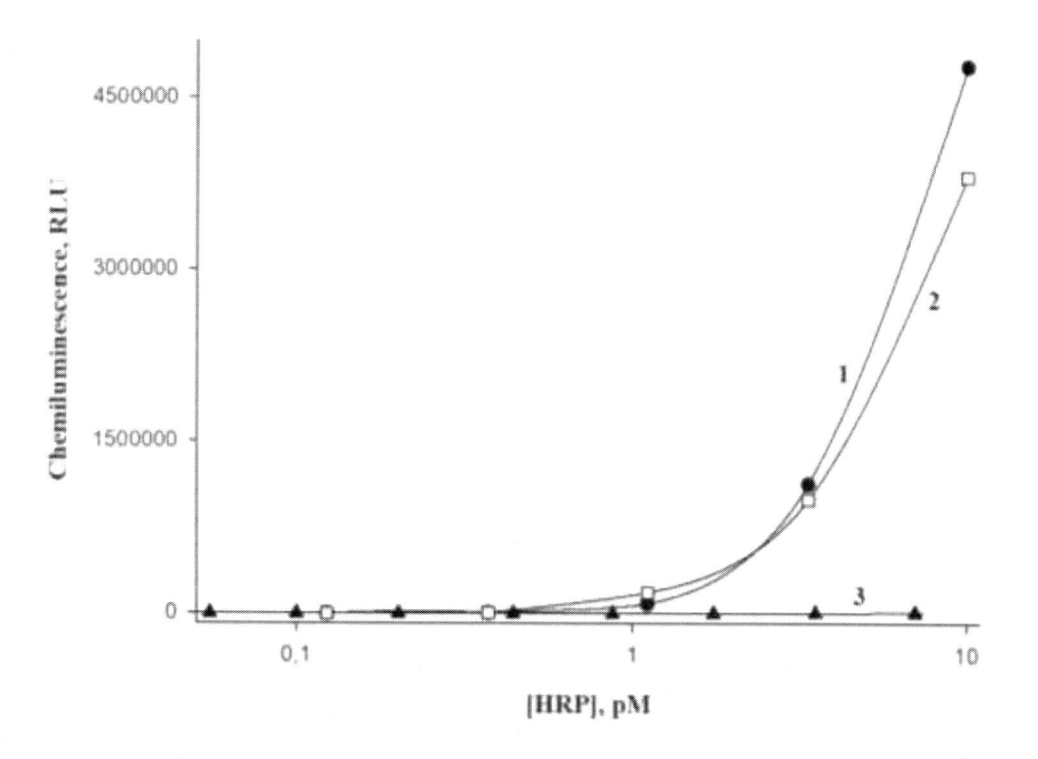

Figure 2. Dependence of chemiluminescence intensity produced upon the HRP-catalyzed oxidation of luminol in the presence of (*1*) PPA, (*2*) SPTZ and (*3*) PIP as primary enhancers as a function of HRP concentration. Reprinted with permission from ref 40.

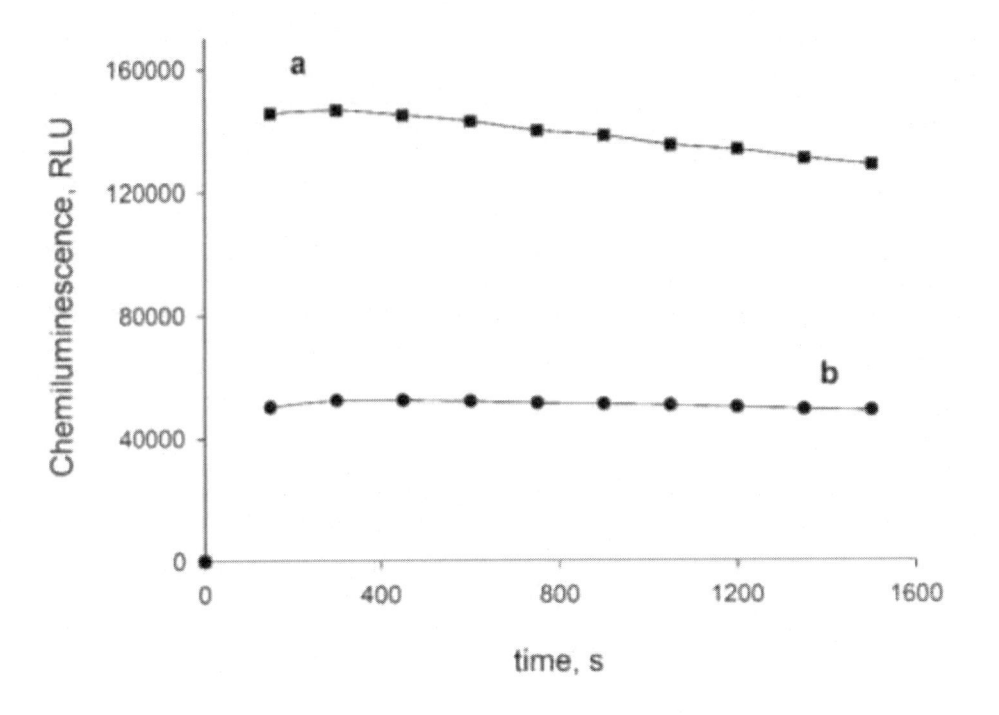

Figure 3. Kinetic curves of peroxidase-induced chemiluminescence produced in the presence of SPTZ/ MORPH (*a*) and only SPTZ (*b*). Reprinted with permission from ref 35.

7. STABILITY OF PHENOTHIAZINE CHEMILUMINESCENCE

As mentioned above, one of the drawbacks of PIP is a quick quenching of its chemiluminescence. In contrast, phenothiazines/MORPH enhancers produce long-term chemiluminescent signal (Figure 3) [33]. Moreover, long-term chemiluminescent signal is produced in the presence of only phenothiazines (without of MORPH). However, in this case the light intensity is lower.

8. APPLICATION OF PHENOTHIAZINE ENHANCERS IN BIOANALYSIS

The high enhancing ability of the studied phenothiazines and long-term chemiluminescence allowed applying these enhancers in the development of ultrasensitive enzyme immunoassays. So, by using SbP as an enzyme label and SPTZ/ MORPH as enhancers, it was constructed a chemiluminescent sandwich ELISA for the determination of human thyroglobulin with a detection limit of 0.2 ng/mL [41]. It should be noted that with the use of HRP and PIP as an enzyme label and enhancer, respectively, the detection limit increased ten times (2 ng/mL) [41]. These results demonstrated the advantages of phenothiazine enhancers over PIP.

Later, have been developed sensitive enzyme immunoassays for the determination of ochratoxin A in 21 various agricultural commodities, aflatoxin M1 in milk, aflatoxin B1 in rice and mung beans, okadaic acid in shellfish, dexamethasone in milk, 2,4-dichlorophenoxyacetic acid (2,4-D) in oranges and mandarins, chloramphenicol in milk, phenol in wastewater, methylglyoxal-modified low density lipoproteins and human troponin I, serum biomarkers related to Down's syndrome, dihydroartemisinin in human plasma, by using phenothiazine enhancers in combination with MORPH as primary and secondary enhancers, respectively [34, 37, 38, 40-48]. These enhancers were also successfully applied to a fiber-optic immunosensor for the detection of Crimean-Congo hemorrhagic fever IgG antibodies in patients [49]. Moreover, some sensitive chemiluminescent assays with phenothiazines use for DNA detection were also developed [50-52].

CONCLUSION

This review describes some phenothiazine derivatives which can be used as potent enhancers in peroxidase-catalyzed ECR. Only phenothiazines having negatively charged groups (SPTZ, PPA) present enhancing activity, whereas phenothiazines with positively charged groups are inactive in ECR. Contrary to other enhancers, phenothiazines are active both in the presence of cationic peroxidases (HRP) and in the presence of anionic peroxidases. A feature of phenothiazine enhancers (primary enhancers) is that their enhancing activity can be improved by secondary enhancers. The mechanism of the action of secondary enhancers was studied. The most active secondary enhancer is MORPH. It should be noted that phenothiazines/MORPH enhancers produce long-term chemiluminescent signal. Phenothiazines/MORPH enhancers were successfully applied to improve the sensitivity of peroxidase-linked assays for determination of small molecules, proteins and nucleic acids. These developments open up very promising perspectives for the development of ultrasensitive analytical methods with high signal strength and low background.

ACKNOWLEDGMENTS

This work was supported by the Russian Science Foundation (Grant No. 17-14-01042).

REFERENCES

[1] Hendrickson, O. D., Smirnova, N. I, Zherdev, A. V., Sveshnikov, P. G., Dzantiev, B. B. (2016). Competitive photometric enzyme immunoassay for fullerene C-60 and

its derivatives using a fullerene conjugated to horseradish peroxidase, *Microchim. Acta,* 183: 211-217.

[2]   Li, D., Ying, Y., Wu, J., Niessner, R., Knopp, D. (2013). Comparison of monomeric and polymeric horseradish peroxidase as labels in competitive ELISA for small molecule detection, *Microchim. Acta,* 180: 711-717.

[3]   Qu, Z.Y., Xu, H., Xu, P., Chen, K.M., Mu, R., Fu, J.P., Gu, H.C. (2014). Ultra-sensitive ELISA Using Enzyme-Loaded Nanospherical Brushes as Labels, *Anal. Chem,* 86: 9367-9371.

[4]   Sakharov, I.Y., Efremov, E.E. (2006). Use of Soybean Peroxidase in Chemilumine-scent Enzyme-Linked Immunosorbent Assay, *J. Agric. Food Chem.* 54: 1584-1587.

[5]   Sakharov, I.Y., Berlina, A.N., Zherdev, A.V., Dzantiev, B.B. (2010). Advantages of Soybean Peroxidase over Horseradish Peroxidase as the Enzyme Label in Chemiluminescent Enzyme-Linked Immunosorbent Assay of Sulfamethoxy-pyridazine, *J. Agric. Food Chem.* 58: 3284–3289.

[6]   Gazaryan, I.G., Yu, M., Rubtsova, Kapeliuch, Y.L., Rodriguez-Lopez, J.N., Lagrimini, L.M., Thorneley, R.N.F. (1998). Luminol Oxidation by Hydrogen Peroxide Catalyzed by Tobacco Anionic Peroxidase: Steady-State Luminescent and Transient Kinetic Studies, *Photochem. Photobiol.* 67: 106-110.

[7]   Abraham, A., Albrechtsen, S.E. (2001). Comparison of penicillinase, urease and alkaline phosphatase as labels in enzyme-linked immunosorbent assay (ELISA) for the detection of plant viruses, *J. Plant Diseases Propection,* 108: 49-57.

[8]   Harper, K., Kerschbaumer, R.J., Ziegler, A., Macintosh, S.M., Cowan, G.H., Himmler, G., Mayo, M.A., Torrance, L. (1997). A scFv-alkaline phosphatase fusion protein which detects potato leafroll luteovirus in plant extracts by ELISA, *J. Virol. Meth.,* 63: 237-242.

[9]   Sakharov, I.Y., Vesga Blanco, M.K., Sakharova, I.V. (2002). Substrate Specificity of African Oil Palm Tree Peroxidase, *Biochemistry (Moscow)* 67: 1043-1047.

[10]   Conyers, S.M., Kidwell, D.A. (1991). Chromogenic substrates for horseradish peroxidase, *Anal. Biochem.* 192: 207-211.

[11]   Josephy, P.D., Eling, T., Mason, R.P. (1982). The horseradish peroxidase-catalyzed oxidation of 3,5,3',5'-tetramethylbenzidine – free-radical and charge-transfer complex intermediates, *J. Biol. Chem.,* 257: 3669-3675.

[12]   Alpeeva, I.S., Sakharov, I.Y. (2005). Soybean peroxidase-catalyzed oxidation of luminol by hydrogen peroxide, *J. Agric. Food Chem.* 53: 5784–5788.

[13]   Alpeeva, I.S., Sakharov, I.Y. (2007). Luminol oxidation catalyzed by royal palm leaf peroxidase, *Appl. Biochem. Microbiol.* 43: 25–28.

[14]   Alpeeva, I.S., Sakharov, I.Y. (2007). Luminol–hydrogen peroxide chemilumine-scence produced by sweet potato peroxidase, *Luminescence* 22: 92–96.

[15]   Hushpulian, D.M., Poloznikov, A.A., Savitski, P.A., Rozhkova, A.M., Chubar, T.A., Fechina, V.A., Lagrimini, L.M., Tishkov, V.I., Gazaryan, I.G. (2007). Biocatalytic

properties of recombinant tobacco peroxidase in chemiluminescent reaction, *Biocatalysis Biotransformation* 25: 163-170.

[16] Thorpe, G. H. G., Kricka, L.J., Gillespie, E., Moseley, R., Amess, R., Buggett, N., Whitehead, T.P. (1985). *Anal. Biochem.* 145: 96–100.

[17] Thorpe, G.H.G., Kricka, L.J., Moseley, S.B., Whitehead, T.P. (1985). *Clin. Chem.* 31: 1335–1341.

[18] Kricka, L. J., Ji, X.J., Thorpe, G.H.G., Edwards, B., Voyta, J., Bronstein, I. (1996). *J. Immunoassay* 17: 67-83.

[19] Kricka, L. J., Cooper, M., Ji, X.P. (1996). Synthesis and characterization of 4-iodophenylboronic acid: A new enhancer for the horseradish peroxidase-catalyzed chemiluminescent oxidation of luminal, *Anal. Biochem.* 240: 119-125.

[20] Whitehead, T.P., Thorpe, G.H.G., Carter, T.N.J., Groucutt, C., Kricka, L.J. (1983). Enhanced luminescence procedure for sensitive determination of peroxidase-labeled conjugates in immunoassay, *Nature* 305: 158-159.

[21] Roswell, D.F., White, E.H. (1978). The chemiluminescence of luminal and related hydrazides, *Methods Enzymol.* 57: 409-423.

[22] Alpeeva, I.S., Soukharev, V.S., Alexandrova, L., Shilova, N.V., Bovin, N.V., Csoregi, E., Ryabov, A.D., Sakharov, I.V. (2003). Cyclometalated ruthenium(II) complexes as efficient redox mediators in peroxidase catalysis, *J. Biol. Inorg. Chem.* 8: 683–688.

[23] Hushpulian, D.M., Fechina, V.A., Kazakov, S.V., Sakharov, I.Y., Gazaryan, I.G. (2003). Non-Enzymatic Interaction of Reaction Products and Substrates in Peroxidase Catalysis, *Biochemistry (Moscow)* 68; 1006-1011.

[24] Marquette, C.A., Blum, L.J. (2009). Chemiluminescent enzyme immunoassays: a review of bioanalytical applications. *Bioanalysis*, 1: 1259-1269.

[25] Yang, L., Jin, M., Du, P., Chen, G., Zhang, C., Wang, J., Jin, F., Shao, H., She, Y., Wang, S., Zheng, L., Wang, J. (2015). Study on Enhancement Principle and Stabilization for the Luminol-H_2O_2-HRP Chemiluminescence System, *PLoS ONE* 10: e0131193.

[26] Chen, G., Jin, M., Du, P., Zhang, C., Cui, X., Zhang, Y., Wang, J., Jin, F., She, Y., Shao, H., Wang, S., Zheng. L. (2017). A review of enhancers for chemiluminescence enzyme immunoassay, *Food Agric. Immunology* 28: 315–327.

[27] Sakharov, I.Y. (2001). African Oil Palm Tree Leaf Peroxidase Catalyzed Peroxidation of Luminol Produces a Stable Chemiluminescence Signal, *Biochemistry (Moscow)* 66: 515-519.

[28] Xu, K., Sun, Y., Li, W., Xu, J., Xu, B., Cao, B., Jiang, Y., Zheng, T., Li, J., Pan, D. (2014). Multiplex chemiluminescent immunoassay for screening of mycotoxins using photonic crystal microsphere suspension array, *Analyst* 139: 771-777.

[29] Vdovenko, M.M., Vorobiev, A.K., Sakharov, I.Y. (2013). Phenothiazine Derivatives as Enhancers of Peroxidase–Dependent Chemiluminescence, *Russ. J. Bioorgan. Chem.* 39: 176–180.

[30] Childs, R.E., Bardsley, W.G. (1975). The steady-state kinetics of peroxidase with 2,2'-azino-di-(3-ethyl-benzthiazoline-6-sulphonic acid) as chromogen, *Biochem. J.* 145: 93–103.

[31] Arnao, M.B., Acosta, M., Rio, J.A., Garcia-Canovas, F. (1990). Inactivation of peroxidase by hydrogen peroxide and its protection by a reductant agent, *Biochim. Biophys. Acta,* 1038: 85–89.

[32] Marzocchi, E., Grilli, S., della Ciana, L., Prodi, L., Mirasoli, M., Roda, A. (2008). Chemiluminescent detection systems of horseradish peroxidase employing nucleophilic acylation catalysts, *Anal. Biochem.* 377: 189–194.

[33] Vdovenko, M.M., Demiyanova, A.S., Chemleva, T.A., Sakharov, I.Y. (2012). Optimization of horseradish peroxidase-catalyzed enhanced chemiluminescence reaction by full factorial design, *Talanta* 94: 223–226.

[34] Tao, X., Wang, W., Wang, Z., Cao, X., Zhu, J., Niu, L., Wu, X., Jiang, H., Shen, J. (2014). Development of a highly sensitive chemiluminescence enzyme immunoassay using enhanced luminol as substrate, *Luminescence* 29: 301–306.

[35] Vdovenko, M.M., Della Ciana, L., Sakharov, I.Y. (2009). 3-(10'-Phenothiazinyl) propane-1-sulfonate is a potent enhancer of soybean peroxidase-induced chemiluminescence, *Anal. Biochem.* 392: 54–58.

[36] Vdovenko, M.M., Della Ciana, L., Sakharov, I.Y. (2010). Enhanced chemiluminescence: A sensitive analytical system for detection of sweet potato peroxidase, *Biotechnol. J.* 5: 886–890.

[37] Zhao, F., Chai, D., Lu, J., Yu, J., Liu, S. (2015). Novel chemiluminescent imaging microtiter plates for high-throughput detection of multiple serum biomarkers related to Down's syndrome via soybean peroxidase as label enzyme, *Anal. Bioanal. Chem.* 407: 6117–6126.

[38] Zehnacker, L., Nevers, M.C., Sinou, V., Parzy, D., Créminon, C., Parzy, D., Azoulay, S. (2015). Development of sensitive direct chemiluminescent enzyme immunoassay for the determination of dihydroartemisinin in plasma, *Anal. Bioanal. Chem.* 407: 7823–7830.

[39] Sakharov, I.Y., Vdovenko, M.M. (2013). Mechanism of action of 4-dialkylaminopyridines as secondary enhancers in enhanced chemiluminescence reaction, *Anal. Biochem.* 434: 12–14.

[40] Sakharov, I.Y., Demiyanova, A.S., Gribas, A.V., Uskova, N.A., Efremov, E.E., Vdovenko, M.M. (2013). 3-(10'-Phenothiazinyl)propionic acid is a potent primary enhancer of peroxidase-induced chemiluminescence and its application in sensitive ELISA of methylglyoxal-modified low density lipoprotein, *Talanta* 115: 414–417.

[41] Vdovenko, M.M., Zubkov, A.V., Kuznetsov, G.I., Della Ciana, L., Kuzmina, N.S., Sakharov, I.Y. (2010). Development of ultra-sensitive soybean peroxidase-based CL-ELISA for the determination of human thyroglobulin, *J. Immunol. Meth.* 362: 127–130.

[42] Yu, F.Y., Vdovenko, M.M., Wang, J.J., Sakharov, I.Y. (2011). Comparison of Enzyme-Linked Immunosorbent Assays with Chemiluminescent and Colorimetric Detection for the Determination of Ochratoxin A in Food, *J. Agric. Food Chem.* 59: 809–813.

[43] Vdovenko, M.M., Stepanova, A.S., Eremin, S.A., Van Cuong, N., Uskova, N.A., Sakharov, I.Y. (2013). Quantification of 2,4-dichlorophenoxyacetic acid in oranges and mandarins by chemiluminescent ELISA, *Food Chem.* 141: 865–868.

[44] Yu, F.Y., Gribas, A.V., Vdovenko, M.M., Sakharov, I.Y. (2013). Development of ultrasensitive direct chemiluminescent enzyme immunoassay for determination of aflatoxin B1 in food products, *Talanta* 107: 25–29.

[45] Vdovenko, M.M., Lu, C.C., Yu, F.Y., Sakharov, I.Y. (2014). Development of ultrasensitive direct chemiluminescent enzyme immunoassay for determination of aflatoxin M1 in milk, *Food Chem.* 158: 310–314.

[46] Vdovenko, M.M., Hung, C.T., Sakharov, I.Y., Yu, F.Y. (2013). Determination of okadaic acid in shellfish by using a novel chemiluminescent enzyme-linked immunosorbent assay method, *Talanta* 116: 343–346.

[47] Vdovenko, M.M., Papper, V., Marks, R.S., Sakharov, I.Y. (2014). Chemiluminescent assay of phenol in wastewater using HRP-catalysed luminol oxidation with and without enhancers, *Anal. Methods* 6: 8654-8659.

[48] Vdovenko, M.M., Byzova, N.A., Zherdev, A.V., Dzantiev, B.B., Sakharov, I.Y. (2016). Ternary covalent conjugate (antibody–gold nanoparticle–peroxidase) for signal enhancement in enzyme immunoassay, *RSC Adv.* 6: 48827- 48833.

[49] Algaar, F., Eltzov, E., Vdovenko, M.M., Sakharov, I.Y., Fajs, L., Weidmann, M., Mirazimi, A., Marks, R.S. (2015). *Anal. Chem.* 87: 8394−8398.

[50] Bodulev, O.L., Gribas, A.V., Sakharov, I.Y. (2018). Microplate chemiluminescent assay for HBV DNA detection using 3-(10′-phenothiazinyl)propionic acid/N-morpholinopyridine pair as enhancer of HRP-catalyzed chemiluminescence, *Anal. Biochem.* 543: 33–36.

[51] Sakharov, I.Y. (2018). Microplate Chemiluminescent Assay for DNA Detection Using Apoperoxidase-Oligonucleotide as Capture Conjugate and HRP-Streptavidin Signaling System, *Sensors* 18: 1289-1300.

[52] Kolosova, A.Y., Sakharov, I.Y. (2019). Triple Amplification Strategy for Improved Efficiency of Microplate-Based Assay for Chemiluminescent DNA Detection, *Anal. Letters* 52: 1352-1362.

In: A Comprehensive Guide to Chemiluminescence
ISBN: 978-1-53616-170-0
Editor: Luís Pinto da Silva
© 2019 Nova Science Publishers, Inc.

Chapter 7

BIOLUMINESCENCE IN SQUIDS

Jeremy D. Mirza[1,2] and Anderson G. Oliveira[2,]*
[1]Institute of Environmental, Chemical and Pharmaceutical Sciences,
Federal University of Sao Paulo, Diadema, Brazil
[2]Institute of Oceanography, University of Sao Paulo, Sao Paulo, Brazil

ABSTRACT

Bioluminescence is a chemical reaction involving the release of chemical energy in the form of light following the oxidation of a substrate (luciferin), in the presence of an enzyme (luciferase). In some bioluminescent organisms, this process involves the binding of the luciferin to a "photoprotein" in order for light to be produced. This phenomenon exists in a wide variety of taxa and is predominantly found in deep dwelling marine organisms. Bioluminescence is an example of convergent evolution, and numerous organisms have developed their own unique mechanisms and reaction pathways. Among oceanic squid, bioluminescence has a variety of functions within different species. However, until now only two systems have been looked at in any detail in terms of the chemistry of the reaction. These are *Watasenia scintillans* (Firefly squid) and *Sthenoteuthis oualaniensis* (formerly *Symplectoteuthis oualaniensis*). As of now, there is still a lot to be learned about these commercially important and well-studied species, in addition to other deep-dwelling bioluminescent squid. This chapter aims to collate the available information on each of these systems as well as provide a brief overview of some lesser-known examples of squid bioluminescence.

* Corresponding Author's Email: anderson.garbuglio@usp.br.

1. INTRODUCTION

Bioluminescence is a chemical process many organisms utilise to produce light and has been studied in a wide range of taxa, in terms of its chemistry, evolutionary history and purpose in ecology [1]. This ability of emitting light via a chemical reaction can be found in a diverse range of phyla, from unicellular bacteria and protists to more complex organisms such as cephalopods and elasmobranchs [1]. A diverse range of organisms in marine environments are able to produce light, in contrast to the few terrestrial and freshwater examples [2]. Examples of non-marine bioluminescent organisms include fireflies, glow worms, the millipede *Luminodesmus sequoia,* and several species of fungi [1, 3].

Marine bioluminescence is far more diverse, and is found in over 700 genera belonging to 16 phyla [4, 5]. This phenomenon has evolved independently at least 30 times, across both marine and terrestrial genera, and around 80% of bioluminescent organisms occur in the oceans [6]. They range from polar to tropical latitudes, and throughout the water column from the surface down to abyssal depths. The majority of marine taxa have been observed between 200 and 800 m depth, in the disphotic zone [6]. It is estimated that up to 95% of organisms in this depth range have the ability to emit light [1, 7]. This high value implies that bioluminescence provides multiple forms of intraspecific and interspecific communication in the deep ocean [5].

Bioluminescent occurs in a variety of morphological forms in different species, leading to unique mechanisms of light emission. Some animals have highly developed complex organs, which are controlled by the nervous system, to house their bioluminescent components, meaning that they can control the emission of luminescence as a response to stress or stimulation [1]. Other organisms, such as bacteria and fungi can emit light continuously as a low intensity glow and show no response to external stimuli [2]. Some these organisms also have multiple forms of bioluminescence (e.g., Anglerfish) that have alternate ecological functions [8].

Chemically, all bioluminescent reactions involve the oxidation of a substrate (a small organic molecule generically called luciferin) in the presence of an enzyme (luciferase). This reaction produces an unstable complex (usually a cyclic peroxide) that breaks down to produce a compound called oxyluciferin, and gives off a large amount of energy as light. The notable feature of this reaction is that more energy is released as light instead of as heat energy. In some bioluminescent systems, e.g., *Aequorea victoria*, this entire reaction can be contained within a structure known as a photoprotein, and is triggered by a specific ion, in this case calcium [2]. The reaction rate for bioluminescence is controlled by the luciferase enzyme or photoproteins that bind to either ions or other cofactors to produce light. This gives to some organisms the ability to control their light emission, normally through their nervous system [2, 9].

Luciferases along with photoproteins are unique, and are derived from multiple evolutionary lineages. In marine environments, the luciferase enzyme is thought to have evolved from less specific oxygenase enzymes [10]. The oxygenase enzymes mutated as a result of animals migrating to deeper waters to either evade visual predators, or to predate on organisms that have migrated to deeper water [6]. Mutations in oxygenase enzymes resulted in external luminescence being exhibited. Species that exhibited luminescence using luciferase enzymes would have an evolutionary advantage over those that did not, resulting in multiple evolutions of this enzyme, across a wide range of phyla [1].

Unlike the enzymatic component of the reaction, luciferins are highly conserved across phyla, and four major luciferin compounds account for the majority of light production in the ocean. The four main marine groups of luciferins are FMN (flavin mononucleotide), used by bacteria, tetrapyrrole used by dinoflagellates, Vargulin used by some fish and ostracods and coelenterazine used by at least eight different phyla, but was first discovered in coelenterates. Groups in which luminescent organisms use coelenterazine as there source of luciferin include protozoa, radiolarians, ctenophores, cnidarians, multiple arthropods, and some fish. A large proportion of these organisms are assumed to have taken up this luciferin through their diet [11-13]. The coelenterazine was originally given the name due to its presence in coelenterates *Aequorea victoria* and *Renilla sp.* [14]. The coelenterazine luciferin (Figure 1) itself is an imidazolopyrazine, occurring in a range of both luminous and non-luminous marine organisms [15-17].

Some organisms use modified versions of coelenterazine as their substrate and have evolved to utilise unique mechanisms for producing light. Such examples can be found within the cephalopoda class of molluscs in at least two different genus of squid in *Watasenia scintillans* and *Sthenoteuthis oualaniensis* (formerly *Symplectoteuthis oualaniensis*). *W. scintillans* uses a disulphated version of coelenterazine, whilst *S. oualaniensis* uses an oxidised form called dehydrocoelenterazine (Figure 1). These species along with other luminescent squid have been observed and studied for their ability to produce light for over one hundred years. However, as of now the only *W. scintillans* and *S. oualaniensis* have been researched in detail in terms of the biochemical pathway for light emission as well as the components involved in the reaction.

As of now, there is still a lot to be learned about these commercially important and well-studied species, in addition to other deep-dwelling bioluminescent squid that have not been looked at in any detail in terms of the chemistry of the reaction, despite numerous ecological studies of luminescence. This chapter aims to collate the available information on both of these well characterised systems as well as provide a brief overview of some lesser known examples of squid bioluminescence, that warrant further investigation, at least in terms of the chemistry of their bioluminescence.

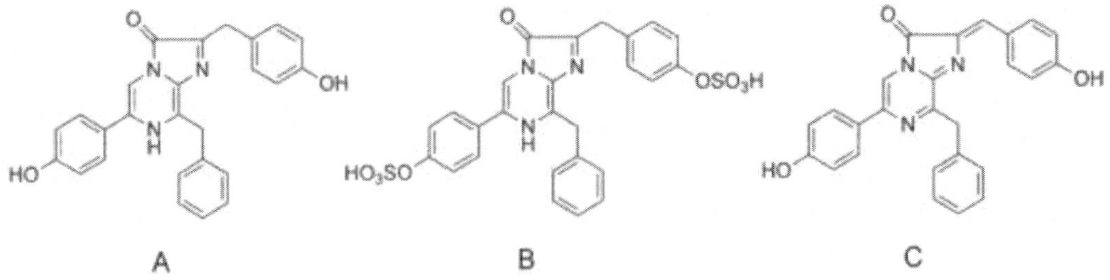

Figure 1. Chemical structure of coelenterazine (A), and its derivatives, coelenterazine disulphate (B) and dehydrocoelenterazine (C).

2. *WATASENIA SCINTILLANS* BIOLUMINESCENCE

2.1. Introduction

The deep-sea squid *Watasenia scintillans* is an example of unique bioluminescence amongst species of squid and has been studied for over 100 years, first by Watase in 1904 [18], who it is named after. It is commonly known as the firefly squid as the bioluminescent flashes produced resemble those seen in fireflies, only the light produced is blue in color [19].

W. scintillans is found in the Sea of Japan and between March and May, during which time females of the species migrate inshore to spawn, and can be collected in large numbers [20]. They are incredibly abundant as it is estimated that hundreds of millions of tons of *W. scintillans* are caught annually. During the migration period, almost all of the squids caught by net are females bearing fertilised eggs, with the most observable migration site being toward the southern part of Toyama Bay near to the coastal city of Namerikawa, Japan. In the bay, the females swim into shallow water, release their eggs and die which ends their one year life cycle. Unlike the females, the male life cycle is unknown due to their relative rarity and because they spend their entire adult life away from the coast in open water. *W. scintillans* is relatively small in size compared with other cephalopods and has a mantle length around 4-6 cm when mature [20, 21].

The bioluminescence of *W. scintillans* differs from other squids such as *Sthenoteuthis oualaniensis*. It produces flashes of blue light, following a series of reactions involving ATP, Mg^{2+} and molecular oxygen [21]. Studies of the light emission reaction, along with the physiology and behaviour of *W. scintillans* are summarised in this section, with particular emphasis on the biochemistry of its luminescence.

2.2. Physiology and Behaviour

The bioluminescence of *W. scintillans*, like many cephalopods is restricted to dermal light organs, photophores, that are scattered along its ventral surface. Approximately 800 of these dermal light organs, which are less than 1 mm in diameter, cover the mantle and head of the squid [19]. Additionally, *W. scintillans* has 5 photophores around each eye and 3 black-pigmented light organs that occur on the tips of the fourth pair of ventral arms. The photophores along the body emit a continuous dim blue light in the dark, whilst the light organs on the arm tips emit an intense flash of blue light after a mechanical or electrical stimulus [21]. In addition to these photophores, *W. scintillans* has 3 visual pigments, the most prominent of which is based on retinal and is distributed across the entire retina [22]. A second pigment is based on 3-dehydroretinal and a retinal derivative was identified as the chromophore for the third visual pigment, which was based on 4-hydroxyretinal [22, 23].

Given that there are few observations outside of spawning sites around Toyama Bay and that *W. scintillans* appears to spend the majority of its life in the deep ocean, the potential ecological functions of luminescence remain unclear. The ventral photophores, as in many deep-sea cephalopods, possibly functions as a means of counter-illumination, allowing the squid to hide its silhouette by blending in with the color of the surrounding water, allowing it to avoid detection from deeper dwelling predators [24]. The flashing ability may be used as a means of intra-specific communication and recognition, as seen in many bioluminescent organisms like the fireflies it is named after [24]. However, until it is possible to observe these deep-sea squid during the majority of their life-cycle, a defined role for bioluminescence in *W. scintillans* cannot be determined.

2.3. Biochemistry of *W. scintillans* Luminescence

W. scintillans was first described by Watase [18] over a century ago, and is named after him. Early studies, such as those of Harvey [25] stated that the reaction showed no sign of a luciferin-luciferase reaction for bioluminescence and [26] identified that molecular oxygen was necessary for the reaction to occur. The bioluminescent light produced, was initially proposed to be caused by bacterial symbionts occurring within the light organs [27]. However, this was not supported by a later study using scanning electron microscopy that showed rod-like proteinaceous structures occurring within the light organs [28, 29].

The first attempts to elucidate the structure of components within the system were carried out over a series of experiments by Goto and colleagues. Using the light organs located at the arm tips of over 10,000 specimens, they isolated a blue fluorescent compound coelenteramide disulphate. This sulphated compound was presumed to be the oxidized product of *W. scintillans'* luciferin, and was referred to as *Watasenia* oxyluciferin [19, 30].

Following this, Inoue et al., [31] were able to isolate and synthesise a compound obtained from the liver of *W. scintillans*, which was determined to be the pre-cursor to *Watasenia* luciferin. The researchers identified that the oxyluciferin compound previously determined was analogous to the oxyluciferin of a marine ostracod *Vargula hilgendorfii* [30]. Using this information, they isolated and identified a compound with the structure of coelenterazine disulphate, which was found in photophores located on the arms, head, mantle and around the eyes [34] This compound was also found in liver, and was determined to be the *Watasenia* luciferin [32]. However, they were unable to test this compound for light activity, as it was not possible to prepare a light-emitting extract from *W. scintillans'* light organs.

Multiple studies by [21, 33] allowed for a better understanding of the chemistry of the bioluminescent reaction. First, it was shown that the reaction was dependent on adenosine triphosphate (ATP) similar to the bioluminescent reaction in fireflies [33]. The reaction had the highest yield of light activity at a pH of 8.8. Light emission also required $MgCl_2$, a soluble component (luciferin) and an insoluble membrane-bound luciferase [29, 33]. Years later the reaction was confirmed to be a luciferin-luciferase reaction that requires molecular oxygen, with the luciferin identified as coelenterazine disulphate [21]. This study showed that the spectrum for the blue bioluminescent light had a peak at 470 nm and was enhanced by a high alkaline pH, with the highest light activity recorded at a pH of 8.26. Furthermore, addition of synthetic coelenterazine disulphate to a luminescing mixture containing ATP, Mg^{2+} and a homogenate of the light organs, resulted in a sharp increase in light activity [21]. Finally the author proposed a potential reaction pathway for *W. scintillans'* bioluminescence that was further built on after the role of molecular oxygen throughout the reaction was better understood [19]. An illustration of the proposed reaction pathway for *W. scintillans* bioluminescence is shown in Figure 2.

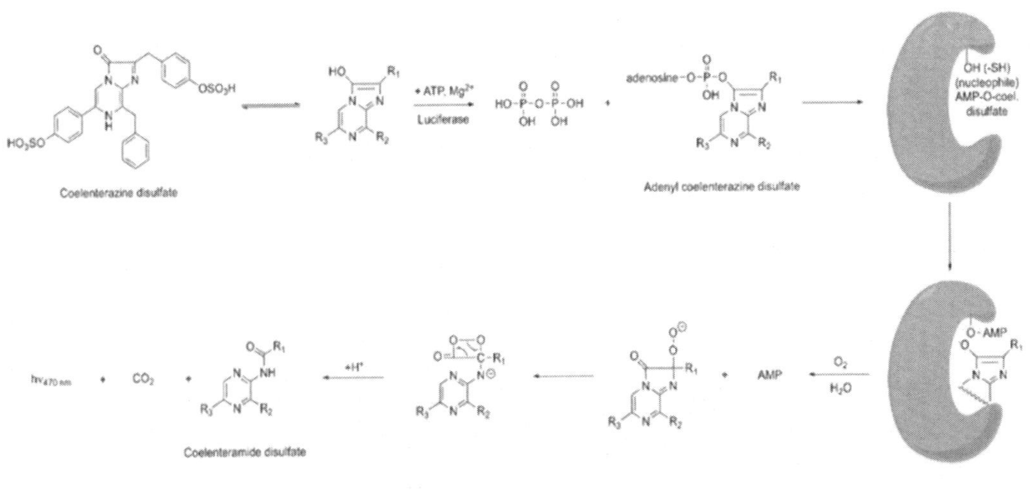

Figure 2. Reaction scheme proposed by Tsuji [21] for *Watasenia* bioluminescence (redrawn).

First, an enzyme-catalysed enolization at the keto oxygen at the C-3 carbon of the luciferin occurs. The enol group is adenylated by ATP forming adenyl coelenterazine disulphate. The AMP group is removed and molecular oxygen is added to the C-2 carbon, forming an unstable intermediate which spontaneously decomposes producing CO_2 and the oxyluciferin along with the emission of blue light (470 nm) [19, 34]. This scheme proposed that the adenosine monophosphate (AMP) group allowed for specific interaction between the luciferin and membrane-bound luciferase, however no experimental evidence was presented due to the difficulty in isolating and purifying the luciferase in a stable form [19, 29].

A later study aimed to address this by partial purification of the luciferase to maintain its activity, and then apply it to study the ATP dependent reaction of with luciferin [29]. They were able to stabilise and partially purify the luciferase using a high concentration of sucrose (2M), and found the optimal temperature for luminescence was 5°C [29]. The luciferin was found to be specific to the enzyme in this reaction, as any modification to its structure lead to a great loss in activity. Furthermore, ATP could be replaced by a modified analogue adenosine 5'-[γ-thio] triphosphate (ATP-γ-S) that had a backbone structure configurationally identical to ATP and the light emission reaction was as effective. Other analogues of ATP that did not have this same backbone structure were shown to be ineffective in propagating the reacting, thus showing the importance of the structure of ATP in this reaction. The quantum yield obtained using coelenterazine disulphate (luciferin) was 0.36 [29]. Finally, it was argued that unlike the proposed reaction mechanism [19], this process did not involve the formation of adenyl luciferin as an intermediate step as the quantity of AMP produced by the reaction was substantially less than the amount of coelenteramide disulphate (oxyluciferin) [29].

A previous study by Okada [28] had used scanning electron microscopy and observed rod-like proteinaceous structures occurring within the light organs. Recently, studies have attempted to analyse these structures within the light organs in the arms of *W. scintillans* [35, 36]. The former group of researchers identified rod-like bodies within the light organs and were able to isolate and use these to successfully achieve light emission *in vitro*. Furthermore, using x-ray diffraction, they showed that these rod-like bodies were well-ordered microcrystals [35]. Gimenez and colleagues [36] used mass spectrometry to analyse protein microcrystals from the light organs on the arm tips of *W. scintillans*. They found that the proteins analysed shared 19-21% sequence identity with firefly luciferases. Finally, they propose that this family of enzymes showed convergent evolution in phylogenetically distant species, which can use these enzymes to produce bioluminescence by differing methods and a variety of substrates [36].

The most recent study looking at *W. scintillans*, reviewed the reaction pathway proposed by Tsuji [19], highlighting its limitations as at the time it was not possible to provide strong theoretical or experimental evidence for two steps of the reaction [37]. The study aimed to find theoretical justification for these two key steps: the addition of molecular oxygen to the luciferin and the process leading to the formation of the 'light emitter' [37]. Substantial support for Tsuji's proposed mechanism for *W. scintillans'* bioluminescence was demonstrated, specifically that the oxygenation reaction involving the luciferin could occur via a single-electron transfer mechanism. In addition, Ding and Liu [37] proposed that the 'light emitter' is produced via the mechanism of gradually reversible charge-transfer-induced luminescence. Finally, using their hypothetical model for the reaction they proposed a quantum yield for the light emitter to be 43%, which was higher than the previous estimate [29, 37].

2.4. Summary

In summation, *W. scintillans* is a squid native to the Sea of Japan that produces a dim continuous blue bioluminescence from ventral photophores, as well as a bright blue flash of luminescence (470 nm) from light organs on its arm tips after being mechanically stimulated. The reaction involves the oxidation of coelenterazine disulphate (luciferin) in the presence of Mg^{2+}, ATP and a membrane-bound luciferase. This system still warrants a large amount of investigation both from an ecological and biochemical perspective. As the luciferase enzyme is very unstable, it is still yet to be completely isolated, purified and characterised preventing lab-based investigations of the individual steps of the reaction process. Additionally, little is known of the life cycle of male firefly squid along with the ecological function of bioluminescence for *W. scintillans*. By observing these animals in the deep ocean and reattempting to purify the membrane-bound luciferase, more can be learnt about this unique bioluminescent system. This will provide a better understanding of related systems such as fireflies along with the origins of bioluminescence as a phenomenon.

3. *STHENOTEUTHIS OUALANIENSIS* BIOLUMINESCENCE

3.1. Introduction

Along with *Watasenia scintillans*, the other example of squid bioluminescence that has been studied in detail is found in the Purpleback Flying Squid *Sthenoteuthis oualaniensis* (formerly *Symplectoteuthis oualaniensis*) [38]. This species has been studied for over three decades in terms of its bioluminescent reaction, luciferin substrate and the enzyme involved. *S. oualaniensis* belongs to the order Teuthoidea and the family Ommastrephidae, which is from the same family as two other luminescent squids, its sister species the Luminous Flying Squid (*S. luminosa*) and the Humboldt Squid, *Dosidicus gigas* [39].

Species within this family include some of the largest and most abundant fast-swimming oceanic squid to be described [40]. The biochemistry of *S. oualaniensis* luminescence will be discussed, along with the areas of research that still require more investigation. In addition to this, the other lesser studied species in this genus, *S. luminosa*, will be briefly characterised in terms of what is known in regards to its bioluminescence.

3.2. Biochemistry of *Sthenoteuthis oualaniensis* Luminescence

This species is distributed throughout both the Indian and Pacific Oceans, though most observations by fishermen have occurred at latitudes between 30° north and 20° south [41]. However, despite being an abundant species, there is little information on the biology of *S. oualaniensis* and nothing regarding the function of bioluminescence in this animal [40].

S. oualaniensis has a maximum length of approximately 40 cm [42], with a mantle length of 15-20 cm [2]. There is little information on their vertical distribution in the water column, though they can be seen in surface waters at night where they are attracted either to the lights from ships or to fish that are also attracted to these lights [40]. First-hand observations along with stomach analysis from ten individual squid showed they predate on bioluminescent lantern fish (Myctophidae) that migrate to the surface at night [40], so it is possible that luminescence is used as a means to aid predation as seen in other species of squid [43]. However, as of now, there have been no studies that investigate the ecological role of bioluminescence has for *S. oualaniensis*. It is possible that this assists in some form of predation strategy or intra-specific recognition however there is no substantial information on *in situ* observations of bioluminescence in this species.

Like *W. scintillans*, *S. oualaniensis* produces an intense blue (peak wavelength between 456-480 nm) flash of bioluminescent light from luminogenic organs called photophores that are distributed along their body. Specifically, it has a large, oval-shaped cluster of photophores on the anterodorsal surface of its mantle [2]. Like *W. scintillans*, *S. oualaniensis* uses a modified form of coelenterazine as its luciferin substrate, dehydrocoelenterazine (Figure 1).

Since 1981, several studies attempting to understand the biochemistry of *S. oualaniensis* bioluminescence have occurred [44]. One of the main limitations of studying this system along with bioluminescence in other cephalopods was that it has been incredibly difficult to extract luminescent tissue whilst maintaining stable light activity. Individual components of luminescent reactions can be unstable *in vitro*, either through denaturing of proteins involved, or via the oxidation or degradation of the luciferin and other cofactors that may be involved [2].

Tsuji and Leisman [44] were the first to be able to study the biochemistry of light emission in *S. oualaniensis*. They observed that the bioluminescence originated from oval-shaped granules within their photophores light organs. Individual photophores were shown to emit a yellow fluorescent light when stimulated with near-UV wavelengths of light. The insoluble granules within the photophores were 2-3 µm in diameter and 100 µm long, emitting light as an intense flash of blue light that decayed rapidly *in vitro*. The durability of the light emission reaction could be extended by the addition of reducing agents such as EDTA and ethylene glycol bistetraacetic acid that could extend the time the suspension remained capable of producing light from 3 to 6 hours. The components of the bioluminescent reaction were identified as being fixed and membrane bound, similar to *W. scintillans* [30] and another luminescent squid more closely related to *S. oualaniensis*, *Ommastrephes pteropus* [45].

Moreover, light emission could be triggered by the addition of monovalent cations such as K^+, Rb^+, Na^+, Cs^+, NH^+ and Li^+ (in order of decreasing effect), always in the presence of molecular oxygen [44]. A 0.6M concentration of KCl or NaCl, and a pH of 7.8 were found to be the optimal conditions for bioluminescent activity, though light activity was still observable in a pH range of 6.5-9.0 [44]. Under a variety of conditions, the emission spectrum of *S. oualaniensis* was shown to be a blue light between 455-458 nm, strongly suggesting that this was a unique system in comparison to other luminescent squid such as *O. pteropus*, which also emits blue light though at a higher wavelength of 474 nm [45].

Takahashi and Isobe [46, 47] studied the components of the reaction in more detail and were able to partially purify the enzyme involved in luminescence. This system was identified to use a photoprotein in its luminescent reaction. The photoprotein was isolated from the granular light organs in the photophores and was named symplectin [47, 48]. For photoprotein-based systems, the luciferin must be bound to the uncharged form of the photoprotein (apoprotein) with all required present in order to form the active photoprotein that can emit luminescence.

In order to identify the luciferin molecule in the reaction, the photophores containing this molecule were first extracted using methanol. Following this, the luciferin was separated from other small molecules via chromatography using a Sephadex LH-20 column [46]. An eluted fraction contained dehydrocoelenterazine, which had a characteristic deep-red coloration, and while it did not produce light emission following chromatography, it was shown to have a small amount of luminescent activity when extracted using methanol and acetone as solvents. Therefore, the lack of light activity observed in the chromatographic fraction may have been caused by instability of the molecule, or because an additional factor necessary for light emission was removed at this step. With a new potential luciferin identified, a suggested mechanism was proposed to explain how this modified luciferin could act as the chromophore during light emission [46].

Dehydrocoelenterazine, an oxidised form of the more common luciferin coelenterazine (Figure 1), was identified as the luciferin required in this reaction to produce light. The majority of marine systems, including coelenterates, cnidarians. Ctenophores and some molluscs utilise unmodified coelenterazine to produce bioluminescence, and the reasons as to why some species like *W. scintillans* and *S. oualaniensis* use a modified form is still not clear.

Following partial purification of symplectin, dehydrocoelenterazine, along with dithiothreitol (DTT) and glutathione (GSH) adducts were added to the photoprotein which resulted in immediate light emission, showing that the luminescent system of *S. oualaniensis* could be reconstructed *in vitro* [47]. Further experiments suggested that coelenterazine was not directly involved in this luminescent system as the addition of this compound to the photoprotein extract did not initially produce light, and a relatively low amount of light was produced in comparison to the experiments using dehydrocoelenterazine [47]. Thus, a new and unique luminescent system was proposed for *S. oualaniensis* that used dehydrocoelenterazine as its substrate, which could bind readily to a photoprotein called symplectin to produce light [47].

Researchers then aimed to prove both the identity of the luciferin and the proposed reaction pathway for *S. oualaniensis* [49]. Additionally, the reaction pathway was confirmed as addition of DTT and GSH to the dehydrocoelenterazine produced a luminous adduct of the luciferin which emitted light in the presence of an extract of the photoprotein [50]. Further research by the same group was conducted in order to separate then clone the photoprotein symplectin, and obtain its structure [49].

To purify symplectin, the photophores were extracted using pH 7.6, 50 mM phosphate buffer containing 0.25 M sucrose, 1 mM EDTA, 1 mM DTT. Impurities were then removed by further extraction using the same buffer containing 0.4 M KCl [48]. Symplectin was isolated using 0.6 M KCl solution at pH 6, and purified using a size-exclusion chromatography column coupled to a HPLC system. The photoprotein was eluted as a 200 kDa oligomer, with trace amounts of a 60 kDa monomer. A limited tryptic digestion of the extract resulted in cleavage into two fragments of 40 kDa and 16 kDa [48]. The 60 kDa and 40 kDa fractions both emitted light when warmed and were shown to be fluorescent in SDS PAGE analysis [2, 48]. All monomeric fractions had amino acid sequences that did not resemble any previously studied photoproteins, and were similar to carbon-nitrogen domains present in mammalian biotinidase and vanin [48].

With dehydrocoelenterazine established as the luciferin, and symplectin identified as the photoprotein, a mechanism (Figure 3) for this luminescent reaction was proposed [51]. Cloning and sequencing of the total amino acid sequence for symplectin showed it consisted of 501 amino acids with 11 available cysteine residues that could potentially bind to the dehydrocoelenterazine to form the complete photoprotein [52]. Thus from this it was proposed that dehydrocoelenterazine binds covalently to the amino acid residue Cysteine-390 in symplectin (Cys390), in order for light emission to occur [52].

Further studies by Isobe and colleagues attempted to confirm that Cys390 was the active site for covalent bond formation in this reaction [53-55]. Light emission experiments using fluorinated analogues of dehydrocoelenterazine showed the importance of dynamic chirality in the *S. oualaniensis* bioluminescence [54]. This was argued to show support for the proposed hypothesis that Cys390 was the active site for the reaction on the symplectin photoprotein. Additionally, they argued that as dehydrocoelenterazine could be replaced by certain analogues based the relative binding affinity they had to the cysteine residue, that this showed that Cys390 is where covalent binding in the reaction occurs [55].

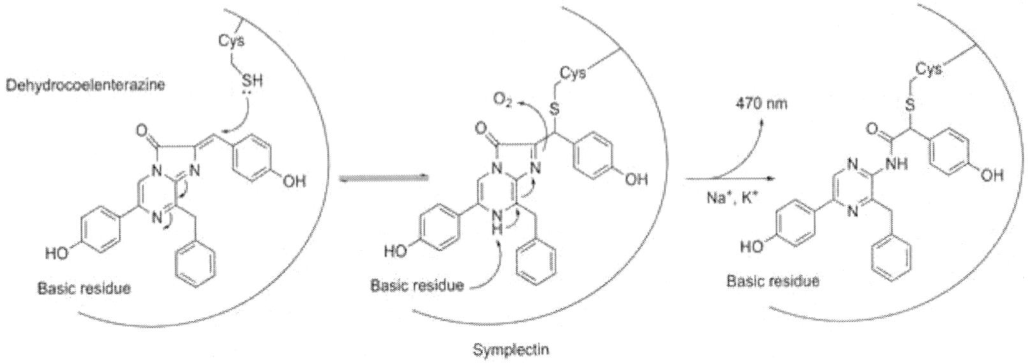

Figure 3. Proposed mechanism for reconstitution and luminescence reaction of symplectin for *S. oualaniensis* light emission. (Adapted from Isobe et al., [51]).

Whilst their research, strongly suggests that a cysteine amino acid residue is the site for covalent bonding, modifying the luciferin alone, would not be sufficient to show that Cys390 specifically is the active site for luminescence. It is feasible that other cysteine residues may be the active site of the symplectin photoprotein [56]. It will only be possible to confirm whether Cys390 is the active site for symplectin through site-directed mutagenesis of this amino acid residue and subsequent observations of a change in light activity.

3.3. A Brief Description of Bioluminescence in *S. luminosa*

The other example of bioluminescence in this genus is found in the Luminous Flying Squid, *S. luminosa*, although the only descriptions of this bioluminescent reaction are from Shimomura´s book [2] on general luminescence, with no other published works on this organism. Commonly caught off the Pacific coast northeast of Tokyo, this squid is thought to have similar mechanisms of light emission to *S. oualaniensis*. The luminescence in *S. luminosa* is found within two narrow strips of photophores spanning the length of its ventral surface as well as two small patches of photophores around its eyes. It´s liver contains a large amount of the same luciferin (dehydrocoelenterazine) as *S. oualaniensis* in addition to a small amount of coelenterazine [2].

The photoprotein of *S. luminosa* is similar to that of *S. oualaniensis* as bioluminescence is pH dependent, and was soluble in buffers containing between 0.6-1.0 M salt. To purify this photoprotein, an extract prepared from the photophores around the eyes was partially purified using size exclusion chromatography. The photoprotein was eluted at molecular masses of 400 kDa and 50 kDa, with the large fraction probably being an aggregate of the 50 kDa fraction [2]. Whilst this fraction produced light in the presence of dehydrocoelenterazine, further purification steps will be required to characterize the bioluminescence of *S. luminosa*.

3.4. Summary

S. oualaniensis bioluminescence has been studied for over 30 years, and a great deal has been identified in regards to its reaction and the individual components. Dehydrocoelenterazine binds covalently to the apo-form of the photoprotein symplectin to produce a blue bioluminescent light. The symplectin photoprotein is 60 kDa, comprising of 501 amino acids, 11 of which are cysteine residues that could potentially bind covalently to the luciferin dehydrocoelenterazine. Further studies proposed that the active site of the reaction is at the amino acid Cys390 on symplectin and that dynamic chiral properties of the interaction between symplectin and dehydrocoelenterazine are significant in allowing this reaction to occur effectively. Recently Francis and colleagues (2017) have proposed a structural model of symplectin based on the protein vanin-1 in which they argue that it is not clear if Cys390 would be available for bonding to dehydrocoelenterazine and may instead form a disulphide bridge with a Cys385 amino acid residue. This research emphasises the need for further study of the active site for this photoprotein, most feasibly through site-directed mutagenesis.

Whilst the luminescent system of *S. oualaniensis* has been well characterised, its sister species *S. luminosa* still requires a lot of research in order to identify the protein involved in its luminescence, and demonstratively confirm that dehydrocoelenterazine is the luciferin. Finally, in addition to further investigating the chemical components of this reaction, there is potential to further study why these squid emit light, as there is little to no information on the ecological function of bioluminescence in both species.

4. BIOLUMINESCENCE OF OTHER SQUID

4.1. Species with Unknown Luminescent Reactions

The two systems, already discussed have had a lot of investigation into the biochemistry of their bioluminescent systems. This is in part because both species can be fished in large numbers at particular times of year. For other species, this is simply not possible although their large biomass and high density of photophores can potentially allow for future studies using only a few individual specimens. Such opportunities exist for the future study of large cephalopods such as *Dosidicus gigas* commonly known as the Humboldt squid [57], found in the mesopelagic zone of the southeastern Pacific Ocean [58, 59]. This is one of the largest species known to emit bioluminescence, growing up to 2 m in length and weighing 65 kg [60, 61]. The Humboldt squid is the largest member of the Ommastrephidae family, and emits light from oval shaped photophores found in the muscle on its mantle, fins, head, arms and tentacles [60, 62, 64]. The squid has been observed to emit a bright blue light from these photophores [63], and emission could also be chemically induced to separated regions of its arms and tentacles by the addition of 2% hydrogen peroxide diluted with seawater [64].

Several other examples of unique systems of luminescence exist among squid, though given the paucity of available specimens for biochemical research, there is little to no information on their chemistry, however studies have instead focused on their physiology and the possible ecological functions of bioluminescence in these species.

One such squid is *Abralia veranyi*, found off the coast of the Bahamas [65]. This species produces bioluminescence from ventral photophores at a peak wavelength of blue light at 490 nm. Given the distribution of photophores along its body, it is feasible that *A. veranyi* uses bioluminescence as a form of counter-illumination to evade detection in the water column [65]. This is a strategy where the photophores continuously emit light at an intensity that matches the ambient light levels of the surrounding water. If a predator would look up through the water column, instead of seeing the silhouette of the animal they would just see the lighter coloration of the upper part of the water column [1]. In addition to squids having their own systems of bioluminescence, some species exhibit symbiotic relationships with bacteria. One such example is seen in the Hawaiian Bobtail Squid *Eupryma scolopes*, which harbours the bioluminescent bacteria *Vibrio fischeri* within complex light organs on its ventral surface to act as a means of camouflage [66].

Squid do not just utilise bioluminescence as a means of camouflage, some like *Taningia danae* have evolved to utilise it as a means of communication during both predation and courtship. Several observations made by Kubodera et al., [43] noted that bioluminescent emissions occurred while attacking its prey. They proposed that this could act as a method of blinding the prey, or possibly as a means of illumination so that *T. danae* could judge the distance to its target in an otherwise dark environment [43].

A final example that demonstrates the diverse array of bioluminescent behaviours amongst squid can be seen in *Vampyroteuthis infernalis* [67]. This squid is found in tropical and temperate oceanic waters at depths between 600 and 1200 m. *V. infernalis* has evolved to produce bioluminescent ink instead of regular ink, thus allowing it to startle predators at depth, who would be adapted to low light conditions. Additionally, Robinson et al., [67] were able to show that coelenterazine and a luciferase are the main components for this reaction.

The species mentioned above highlight the diversity in squid bioluminescence, both in terms of chemical mechanisms and in terms of the functional role of light emission. A number of other species of luminescent squid exist, and considering our lack of knowledge regarding the ocean, it is plausible that there are many more that we are yet to discover.

4.2. Phylogenetic Similarities of Luminescent Cephalopods

Up until now, only two species have been well characterised in terms of their luminescence, primarily due to biochemical experiments requiring large amounts of material and because of the inaccessibility of the deep ocean up until recently. For oceanic species that cannot be collected in large quantities there needs to be an alternate strategy for analysis that does not require as great a quantity of biomass. One possible option is to use a transcriptomic approach to see if other known luciferases or photoproteins are present in the transcriptome of some luminescent species.

Such a study was carried out using transcriptomes for six luminescent species of squid with unidentified luminescent systems (*Phylliroe bucephalum, Chiroteuthis calyx, Pterygioteuthis hoylei, V. infernalis, D. gigas,* and *Octopoteuthis deletron*) [68]. These transcriptomes were compared to the known sequence of the photoprotein symplectin from *S. oualaniensis,* using BLAST analysis. The results showed that *P. hoylei* has a photoprotein that strongly resembled symplectin, whilst *D. gigas, P. bucephalum,* and *V. infernalis* shared similarities with the photoprotein [68].

Whilst this approach requires the organism being studied to have a light emission system that has already been characterised, it allows for a more rapid means to identify similar light emission systems, as well as providing information on the evolution of particular luminescent reactions. Such analysis has been performed using symplectin in order to study its evolutionary history in relation to bioluminescent squid [56]. This study utilised the transcriptomes of several luminescent squid species in order to identify symplectin-like proteins within them. Homologs of symplectin were found in *P. hoylei* and *D. gigas,* though this alone is not sufficient to identify the type of bioluminescent system used, as *W. scintillans* was also shown to have a symplectin group in its transcriptome [56]. As more transcriptomes become available, the ability to identify and characterise new luminescent species becomes easier. Which will hopefully lead to identifying lesser known luminescent systems such as those of *T. danae, V. infernalis* and *D. gigas.*

CONCLUSION

Bioluminescence occurs in a wide variety of squid, and has a diverse range of mechanisms and functions. As of now two species, *W. scintillans* and *S. oualaniensis* have been characterised in terms of their biochemistry. In addition to these, several other squid species have been observed utilising bioluminescence in a number of strategies. The current collective knowledge on squid bioluminescence has been summarised in this chapter. However, this has highlighted the lack of information from both a biochemical and ecological perspective on bioluminescent light emission among these organisms. As the depths of the ocean become more accessible, there will be more possibilities to study many of these bioluminescent systems, in addition to potentially discovering new ones.

REFERENCES

[1] Haddock, S. H., Moline, M. A., and Case, J. F. (2010). Bioluminescence in the sea. *Ann. Rev. Mar. Sci.*, *2*: 443-493.

[2] Shimomura, O. (2006). *Bioluminescence: chemical principles and methods*. World Scientific.

[3] Hastings, J. W., and Davenport, D. (1957). The luminescence of the millipede, *Luminodesmus sequoiae. Biol. Bull.*, *113:* 120-128.

[4] Herring, P. J. (1987). Systematic distribution of bioluminescence in living organisms. *J. Biolumin. Chemilumin.*, 1: 147-163.

[5] Moline, M. A., Oliver, M. J., Mobley, C. D., Sundman, L., Bensky, T., Bergmann, T., Bissett, W. P., Case, J., Raymond, E. H. and Schofield, O. M. (2007). Bioluminescence in a complex coastal environment: 1. Temporal dynamics of nighttime water-leaving radiance. *J. Geophys. Res. Oceans*, *112*.

[6] Widder, E. A. (2010). Bioluminescence in the ocean: origins of biological, chemical, and ecological diversity. *Science*, *328*: 704-708.

[7] Pieribone, V., and Gruber, D. F. (2005). *Aglow in the dark: the revolutionary science of biofluorescence*. Harvard University Press.

[8] Pietsch TW. 2009. *Oceanic Anglerfishes: Extraordinary Diversity in the Deep Sea*. Berkeley: Univ. of Calif. Press. pp. 576.

[9] Wilson, T., and Hastings, J. W. (1998). Bioluminescence. *Annu. Rev. Cell Dev. Biol.*, *14*: 197-230.

[10] Widder, E. A. (1999). Bioluminescence. In *Adaptive mechanisms in the ecology of vision* (pp. 555-581). Springer Netherlands.

[11] Tsuji, F. I., Barnes, A. T., and Case, J. F. (1972). Bioluminescence in the marine teleost, *Porichthys notatus*, and its induction in a non-luminous form by *Cypridina* luciferin. *Nature* 237: 515-516.

[12] Frank, T. M., Widder, E. A., Latz, M. I., and Case, J. F. (1984). Dietary maintenance of bioluminescence in a deep-sea mysid. *J. Exp. Biol.*, *109*: 385-389.

[13] Harper, R. D., and Case, J. F. (1999). Disruptive counterillumination and its anti-predatory value in the plainfish midshipman Porichthys notatus. *Mar. Biol.*, *134*: 529-540.

[14] Shimomura, O., and Johnson, F. H. (1975). Regeneration of the photoprotein aequorin. *Nature* 256: 236-238.

[15] Shimomura, O., Inoue, S., Johnson, F. H., and Haneda, Y. (1980). Widespread occurrence of coelenterazine in marine bioluminescence. *Comp. Biochem. Physiol. B, 65:* 435-437.

[16] Shimomura, O. (1987). Presence of coelenterazine in non-bioluminescent marine organisms. *Comp. Biochem. Physiol.* 86B: 361-363.

[17] Thompson, J. F., Geoghegan, K. F., Lloyd, D. B., Lanzetti, A. J., Magyar, R. A., Anderson, S. M., and Branchini, B. R. (1997). Mutation of a protease-sensitive region in firefly luciferase alters light emission properties. *J. Biol. Chem., 272:* 18766-18771.

[18] Watase, S. (1904). The luminous organ of the firefly squid. *Dobutsugaku Zasshi, 17:*119-123.

[19] Tsuji, F. I. (2005). Role of molecular oxygen in the bioluminescence of the firefly squid, *Watasenia scintillans*. *Biochem. Biophys. Res. Commun, 338:* 250-253.

[20] Hayashi, S. (1995). Fishery biological studies of the firefly squid, *Watasenia scintillans* (Berry). *Toyama Bay. Bull. Toyama Pref. Fish. Res. Inst,* 7: 128.

[21] Tsuji, F. I. (2002). Bioluminescence reaction catalyzed by membrane-bound luciferase in the "firefly squid," *Watasenia scintillans. Biochim. Biophys. Acta Biomembr., 1564:* 189-197.

[22] Matsui, S., Seidou, M., Uchiyama, I., Sekiya, N., Hiraki, K., Yoshihara, K., and Kito, Y. (1988). 4-Hydroxyretinal, a new visual pigment chromophore found in the bioluminescent squid, *Watasenia scintillans. Biochmim. Biophys. Acta, 966:* 370-374.

[23] Seidou, M., Sugahara, M., Uchiyama, H., Hiraki, K., Hamanaka, T., Michinomae, M., ... and Kito, Y. (1990). On the three visual pigments in the retina of the firefly squid, *Watasenia scintillans. J. Comp. Physiol. A, 166:* 769-773.

[24] Bush, S. L., Robison, B. H., and Caldwell, R. L. (2009). Behaving in the dark: locomotor, chromatic, postural, and bioluminescent behaviours of the deep-sea squid *Octopoteuthis deletron* Young 1972. *Biol. Bull., 216:* 7-22.

[25] Harvey, E. N. (1922). Studies on bioluminescence. XIV. The specificity of luciferin and luciferase. *Gen. Physiol.* 4: 285-295.

[26] Shoji, R. (1919). A physiological study on the luminescence of *Watasenia scintillans* (Berry). *Am. J. Physiol.* 47: 534-557.

[27] Harvey, E. N. (1952) Bioluminescence, Academic Press, New York.

[28] Okada, Y. K. 1966. Observations on rod-like contents in the photogenic tissue of *Watasenia scintillans* through the electron microscope. *Bioluminescence in progress, 611,* p. 625.

[29] Teranishi, K. and Shimomura, O. (2008). Bioluminescence of the arm light organs of the luminous squid *Watasenia scintillans. Biochim. Biophys. Acta, 1780:* 784-92.

[30] Goto, T., Iio, H., Inoue, S., and Kakoi, H. (1974). Squid bioluminescence I. Structure of Watasenia oxyluciferin, a possible light-emitter in the bioluminescence of Watasenia scintillans. *Tetrahedron Lett.*, *15*: 2321-2324.

[31] Inoue, S., Sugiura, S., Kakoi, H., Hasizume, K., Goto, T., and Iio, H. (1975). Squid bioluminescence II. Isolation from *Watasenia scintillans* and synthesis of 2-(p-hydroxybenzyl)-6-(p-hydroxyphenyl)-3, 7-dihydroimidazo [1, 2-a] pyrazin-3-one. *Chem. Lett.*, *4*: 141-144.

[32] Inoue, S., Taguchi, H., Murata, M., Kakoi, H., and Goto, T. (1977). Squid bioluminescence IV. Isolation and structural elucidation of Watasenia dehydropreluciferin. *Chem. Lett.*, *6*: 259-262.

[33] Tsuji, F. I. (1985). ATP-dependent bioluminescence in the firefly squid, *Watasenia scintillans*. *Proc. Natl. Acad. Sci. USA.*, *82*: 4629-4632.

[34] Inoue, S., Kakoi, H., and Goto, T. (1976). Squid bioluminescence III. Isolation and structure of Watasenia luciferin. *Tetrahedron Lett.*, *17*: 2971-2974.

[35] Hamanaka, T., Michinomae, M., Seidou, M., Miura, K., Inoue, K., and Kito, Y. (2011). Luciferase activity of the intracellular microcrystal of the firefly squid, *Watasenia scintillans*. *FEBS Lett.*, *585*: 2735-2738.

[36] Gimenez, G., Metcalf, P., Paterson, N. G., and Sharpe, M. L. (2016). Mass spectrometry analysis and transcriptome sequencing reveal glowing squid crystal proteins are in the same superfamily as firefly luciferase. *Sci. Rep.*, *6*: 27638.

[37] Ding, B. W., and Liu, Y. J. (2016). Bioluminescence of Firefly Squid via Mechanism of Single Electron-Transfer Oxygenation and Charge-Transfer-Induced Luminescence. *J. Am. Chem. Soc.*, *139*: 1106-1119.

[38] Lesson R. P. (1830) Mollusques. Voyage dans la Coquille, 2, 239-246. [Molluscs. Voyage in the Coquille, 2, 239-246].

[39] Duncan, D. (1941) Fishing giants of the Humboldt. *Natn. Geogr. Mag. 79*: 373-400.

[40] Young, R. E. (1975). A brief review of the biology of the oceanic squid, *Symplectoteuthis oualaniensis* (Lesson). *Comp. Biochem. Physiol. B*, *52*: 141-143.

[41] Burgess L. A. (1970) A report on the Cephalopoda of the Hawaii area. Report of the director, National Marine Fisheries Service, Hawaii branch.

[42] Okutani, T., and Tung, I. H. (1978). Reviews of biology of commercially important squids in Japanese and adjacent waters. I. Symplectoteuthis oualaniensis (Lesson). *The Veliger*, *21*: 87-94.

[43] Kubodera, T., Koyama, Y., and Mori, K. (2007). Observations of wild hunting behaviour and bioluminescence of a large deep-sea, eight-armed squid, Taningia danae. *Proc. Royal Soc. Lond.*, *274*: 1029-1034.

[44] Tsuji, F. I., and Leisman, G. B. (1981). K+/Na+-triggered bioluminescence in the oceanic squid *Symplectoteuthis oualaniensis*. *Proc. Natl. Acad. Sci. USA* 78: 6719-6723.

[45] Girsch, S. J., Herring, P. J., and McCapra, F. (1976). Structure and preliminary biochemical characterization of the bioluminescent system of *Ommastrephes pteropus* (Steenstrup) (Mollusca: Cephalopoda). *J. Mar. Biol. Assoc. U.K.*, *56*: 707-722.

[46] Takahashi, H., and Isobe, M. (1993). *Symplectoteuthis* bioluminescence. (1). Structure and binding form of chromophore in photoprotein of a luminous squid. *Bioorg. Med. Cbem. Lett.* 3: 2647-2652.

[47] Takahashi, H., and Isobe, M. (1994). Photoprotein of luminous squid, *Symplectoteuthis oualaniensis* and reconstruction of the luminous system. *Cbem. Lett.* 5: 843-846.

[48] Fujii, T., Ahn, J. Y., Kuse, M., Mori, H., Matsuda, T., and Isobe, M. (2002). A novel photoprotein from oceanic squid (*Symplectoteuthis oualaniensis*) with sequence similarity to mammalian carbon–nitrogen hydrolase domains. *Biochem. Biophys. Res. Commun.*, *293*: 874-879.

[49] Isobe, M., Kuse, M., Yasuda, Y., and Takahashi, H. (1998). Synthesis of 13C-dehydrocoelenterazine and model studies on *Symplectoteuthis* squid bioluminescence. *Bioorg. Med. Chem. Lett.*, 8: 2919-2924.

[50] Kuse, M., and Isobe, M. (2000). Synthesis of 13C-dehydrocoelenterazine and NMR studies on the bioluminescence of a Symplectoteuthis model. *Tetrahedron*, *56*: 2629-2639.

[51] Isobe, M., Fujii, T., Kuse, M., Miyamoto, K., and Koga, K. (2002). 19F-Dehydrocoelenterazine as probe to investigate the active site of symplectin. *Tetrahedron*, *58*: 2117-2126.

[52] Isobe, M., Kuse, M., Tani, N., Fujii, T., and Matsuda, T. (2008). Cysteine-390 is the binding site of luminous substance with symplectin, a photoprotein from Okinawan squid, *Symplectoteuthis oualaniensis*. *Proc. Jpn. Acad. Ser. B*, *84*: 386-392.

[53] Phakhodee, W., Toyoda, M., Chou, C. M., Khunnawutmanotham, N., and Isobe, M. (2011). Suzuki–Miyaura coupling for general synthesis of dehydrocoelenterazine applicable for 6-position analogs directing toward bioluminescence studies. *Tetrahedron*, *67*: 1150-1157.

[54] Kongjinda, V., Nakashima, Y., Tani, N., Kuse, M., Nishikawa, T., Yu, C. H., ... and Isobe, M. (2011). Dynamic chirality determines critical roles for bioluminescence in symplectin–dehydrocoelenterazine system. *Chem. Asian J.*, *6*: 2080-2091.

[55] Chou, C. M., Tung, Y. W., and Isobe, M. (2014). Molecular mechanism of Symplectoteuthis bioluminescence—Part 4: Chromophore exchange and oxidation of the cysteine residue. *Bioorg. Med. Chem.*, *22*: 4177-4188.

[56] Francis, W. R., Christianson, L. M., and Haddock, S. H. (2017). Symplectin evolved from multiple duplications in bioluminescent squid. PeerJ 5: e3633.

[57] d'Orbigny, A. D. (1835). *Voyage dans l'Amérique méridionale: le Brésil, la république orientale de l'Uruguay, la république Argentine, la Patagonie, la*

république du Chili, la republ. de Bolivia, la républ. du Pérou (Vol. 1). Pitois-Levrault. [Voyage in South America: Brazil, the Eastern Republic of Uruguay, the Argentine Republic, Patagonia, the Republic of Chile, the Republic of Bolivia, the Republic of Peru (Vol 1). Pitois Levrault].

[58] Roper, C. F., Sweeney, M. J., and Nauen, C. (1984). Cephalopods of the world. An annotated and illustrated catalogue of species of interest to fisheries.

[59] Schmiede, P., and Acuña, E. (1992). Regreso de las jibias (*Dosidicus gigas*) a Coquimbo. *Revista Chilena de Historia Natural, 65*, 389-390. [Return of the squid (*D. gigas*) to Coquimbo. *Chilean Magazine of Natural History, 65*: 389-390].

[60] Wormuth, J. H. (1976). The biogeography and numerical taxonomy of the Oegopsid squid family Ommastrephidae in the Pacific Ocean (Vol. 23). Univ of California Press.

[61] Nigmatullin, C. M., Nesis, K. N., and Arkhipkin, A. I. (2001). A review of the biology of the jumbo squid *Dosidicus gigas* (Cephalopoda: Ommastrephidae). *Fish. Res., 54*: 9-19.

[62] Nesis, K. N. (1970). The biology of the giant squid of Peru and Chile, *Dosidicus gigas. Oceanology, 10*: 108-118.

[63] García-Tello, P. (1964). Nota preliminar sobre una observación de bioluminiscencia en *Dosidicus gigas* (D'Orb) Cephalopoda. *Bol. Univ. Chile, 46*: 27-28. [Preliminary notes on an observation of bioluminescence in *Dosidicus gigas* (D'Orb) Cephalopoda. *Bol. Univ. Chile*, 46, 27-28].

[64] Lohrmann, K. B. (2008). Subcutaneous photophores in the jumbo squid *Dosidicus gigas* (d'Orbigny, 1835) (Cephalopoda: Ommastrephidae). *Magazine of marine biology and oceanography, 43*: 275-284.

[65] Herring, P. J., Widder, E. A., and Haddock, S. H. (1992). Correlation of bioluminescence emissions with ventral photophores in the mesopelagic squid *Abralia veranyi* (Cephalopoda: Enoploteuthidae). *Mar. Biol., 112*: 293-298.

[66] Jones, B. W., and Nishiguchi, M. K. (2004). Counterillumination in the Hawaiian bobtail squid, *Euprymna scolopes* Berry (Mollusca: Cephalopoda). *Mar. Biol., 144*: 1151-1155.

[67] Robinson, B. H., Reisenbichler, J. C, Hunt, J. C, and Haddock, S. H. (2003). Light production by the arm tips of the deep-sea cephalopod *Vampyroteuthis infernalis. Biol. Bull.* 205: 102-109.

[68] Hersh, T. (2014). A Transcriptomics-Based Approach to Novel Photoprotein Analysis in Pelagic Molluscs.

In: A Comprehensive Guide to Chemiluminescence   ISBN: 978-1-53616-170-0
Editor: Luís Pinto da Silva   © 2019 Nova Science Publishers, Inc.

Chapter 8

METAL-ENHANCED CHEMILUMINESCENCE

Joshua Moskowitz and Chris D. Geddes[*]

Institute of Fluorescence and Department of Chemistry and Biochemistry,
University of Maryland, Baltimore County, Baltimore, Maryland, US

ABSTRACT

This chapter discusses the enhancement of chemiluminescent emission utilizing metal-nanoparticles in close proximity to a luminescent molecule, known as Metal-Enhanced Chemiluminescence (MEC). This phenomenon occurs due to non-radiative energy transfer from a chemiluminescent emitter to a metal nanoparticle and is closely related to Metal-Enhanced Fluorescence (MEF) which is also discussed in this chapter. The initial discovery of MEC and Surface Plasmon Coupled Chemiluminescence (SPCC) are also detailed. SPCC involves the excitation of plasmons in continuous metal films as opposed to nanoparticles, producing highly directional emission. The use of microwaves to increase chemiluminescent reaction rates, known as Microwave Triggered Metal Enhanced Chemiluminescence (MT-MEC) is also detailed in this chapter. Techniques based on MEC provide for increased sensitivity over traditional chemiluminescence assays, and the development of highly sensitive MT-MEC protein assays will be discussed. Finally, substrate geometry and multiplexed MT-MEC assays are also included in this chapter.

1. INTRODUCTION

Chemiluminescence involves the generation of chemically-induced electronic excited states in molecular species, which relax via radiative emission [1-5]. This process has the advantage of requiring relatively simple instrumentation compared to traditional

---

[*] Corresponding Author's Email: Geddes@umbc.edu.

fluorescence, as excitation sources and other optics associated with traditional fluorescence are not required. However, chemiluminescent emitters often suffer from low quantum efficiency, making them less attractive in probe and assay development [6, 7]. Enhanced chemiluminescent emission holds potential to increase sensitivity in a variety of detection assays such as protein, immunoassays, and small molecule detection. In this chapter we discuss the enhancement of chemiluminescent emission utilizing plasmonic nanoparticles in close proximity to a luminescent molecule, known as Metal-Enhanced Chemiluminescence (MEC). This phenomenon is due to energy transfer from a donor molecular species to an acceptor metal nanoparticle, as opposed to metal catalysis of a chemiluminescent reaction which is also found in the literature. MEC is closely related to Metal-Enhanced Florescence (MEF), which has been widely cited in the literature and will be introduced in this chapter [8-12]. This chapter also includes the initial discoveries of MEC and surface plasmon coupled chemiluminescence (SPCC) by our laboratory. In addition, microwave-triggered metal enhanced chemiluminescence (MT-MEC) which utilizes microwaves to increase chemiluminescent reaction rates, and the development of MT-MEC protein assays will be discussed. This chapter will also include the dependence of MT-MEC on substrate geometry, and multiplexed MT-MEC assays.

2. METAL-ENHANCED FLUORESCENCE (MEF)

In Metal-Enhanced Fluorescence (MEF) an excited fluorophore in the near electric field of a metal nanoparticle induces a mirror dipole in the metal, leading to radiative emission from the nanoparticle, characteristic of the fluorophore [8-10]. The mechanism of this enhancement involves both an absorption and emission component. First, the metal can use its relatively large size to act as a nanoantenna for the fluorophore, effectively increasing the extinction cross section of the fluorophore [8, 9, 11, 12, 13]. This absorption enhancement mechanism does not apply to metal-enhanced chemiluminescence (MEC), as there is no excitation source leading to molecular absorption. Second, energy transfer may take place from the fluorophore to metal, increasing the emission rate relative to the free-space fluorescence [8]. Experimental observation of a decrease in the fluorescence lifetime in the presence of metal nanoparticles compared to the free space fluorophores has been made by multiple groups [13-17]. This decrease is observed simultaneously with an increase in emission intensity [13-17], distinguishing MEF from quenching, which displays a drop in intensity. The rate of plasmon decay is well-known to be much faster than the free space fluorophore emission rate [13], supporting the notion that the metal is radiating the light. Here, a strong energy transfer is expected when emission from the fluorophore overlaps with absorption of the metal [13]. It is the transfer of energy from fluorophore to the metal causing plasmon excitation, and the subsequent fast emission by the metal, which is thought to underpin both the MEF and MEC mechanisms (Figure 1) [6, 18].

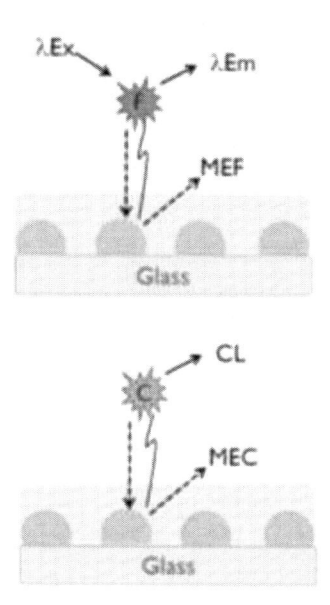

Figure 1. Graphical representations of MEF (top), and for MEC (Bottom). F—Fluorophore, C—Chemiluminescence species, and CL—Chemiluminescence. [Redrawn from Ref. 18].

3. METAL-ENHANCED CHEMILUMINESCENCE (MEC)

In MEC, a chemically excited molecule can couple with and non-radiatively transfer energy to a plasmonic nanoparticle. The interaction of a chemiluminescent species with silver nanoparticles has been shown to produce large increases in emission intensity compared to the free-space emission, and it has been shown that surface plasmons can be directly excited by chemically-induced electronically excited molecules [6, 7, 18]. It is important to note that the wavelength of emission must overlap with the nanoparticle absorbance for energy transfer to occur.

3.1. Metal Nanoparticle Film Synthesis and Characterization

MEC was initially discovered on Silver-Island films (SiFs), with specific nanoparticle properties allowing for energy transfer from the molecule to the plasmonic particle [18]. Synthesis and characterization of these metal nanoparticle films is therefore critical to MEC. In the literature, the deposition of silver nanoparticles on a substrate is commonly carried out using both chemical and physical methodologies. One liquid chemical method uses a Tollens reaction to reduce diamine silver (I) to solid silver particles on a glass substrate [19-21]. The physical method uses thermal vapor deposition to evaporate solid metal with a controlled rate and under high pressure, forming a film composed of metal nanoparticles on the substrate [22, 23].

A variety of characterization methods may be used to determine the metal nanoparticle film properties. Absorption spectroscopy is commonly used to provide an extinction spectrum of the nanoparticle film, with peaks corresponding to discrete plasmon resonances [13]. Figure 2 shows extinction spectra from various metal films. The extinction spectra can be broken down into both absorption and scattering components experimentally with scattering measurements. For particles comparable in size to the incident radiation, the phase of the external field varies over the particle, and Mie Theory calculations are used to solve for the scattering component of extinction [9, 13]. Atomic force microscopy (AFM) and Scanning Electron Microscopy (SEM) may also be used to image the metal films with nanometer resolution, providing information regarding particle size and shape in the film.

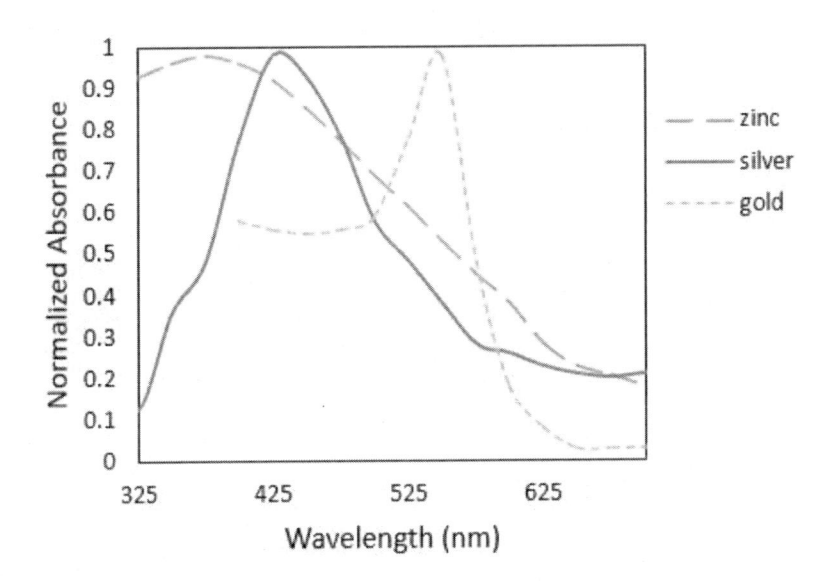

Figure 2. Normalized extinction of zinc, gold and silver nanostructured particles on a glass substrate [Modified from Ref. 6].

3.2. MEC Measurements

The MEC phenomenon was first observed by placing a drop of chemiluminescent solution between two silver-island film glass slides (Figure 3). Chemiluminescent spectra from the reaction between hydrogen peroxide and 9,10-diphenylanthracene provided blue emission, while spectra from the reaction between hydrogen peroxide and 9,10-bis(phenylethynyl)-anthracene provided green emission. Both of these reactions produced emission that overlaps with the absorbance of the silver island film. Emission enhancements compared to an un-silvered glass control substrate were shown to be 4x and 10x for the blue and green chemiluminescence, respectively (Figure 3) [6].

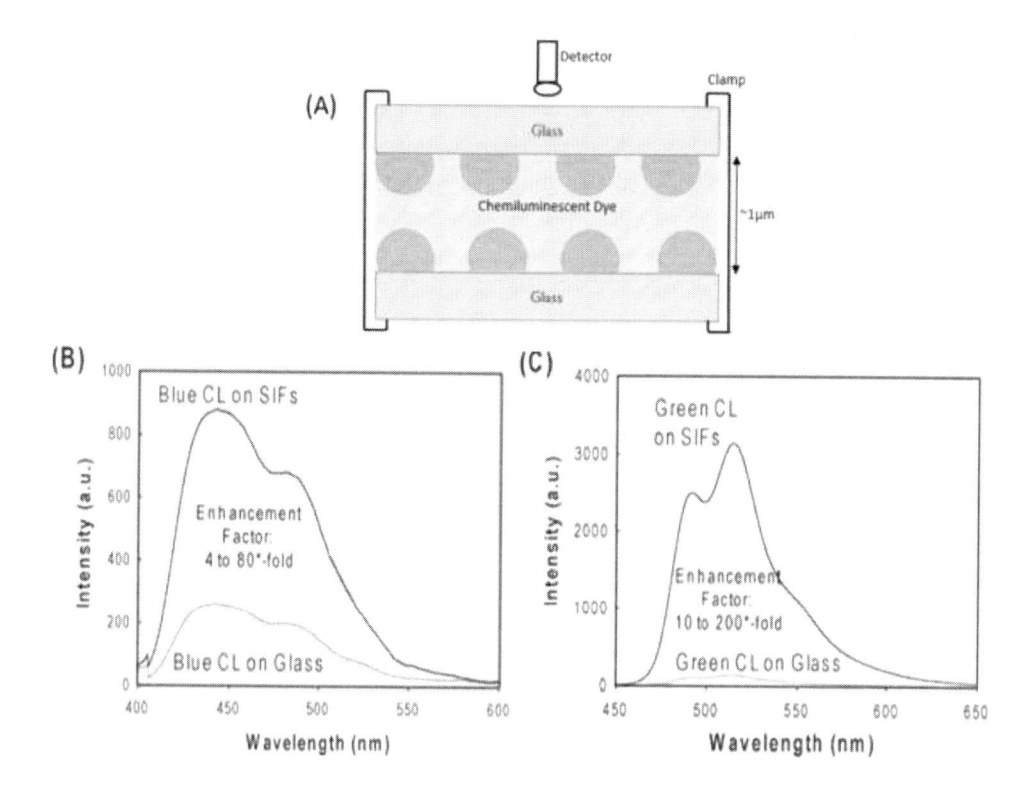

Figure 3. (A) Schematic depiction of the experimental geometry for metal-enhanced chemiluminescence (MEC) studies. Chemiluminescence spectra of (B) blue and (C) green chemiluminescence dyes on glass and SIFs [Modified from Ref. 6].

In addition to serving as energy acceptors, metal nanoparticles may act as reaction catalysts as well [24-26]. The possibility of the silver catalyzing the chemiluminescent reaction described above was also investigated. Emission intensity as a function of time on glass, silver islands, and a continuous silver sheet were measured. The silver islands produced an emission depletion rate approximately 1.7x faster than on glass, $0.034s^{-1}$ for the SiF vs $0.019 \, s^{-1}$ on glass. This observation may be attributed to either the non-radiative energy transfer from molecule to plasmon or the silver catalysis of the chemiluminescent reaction. The decay profile of the silver island film was then compared to that of a solid silver sheet. If the silver was causing catalysis then the solid silver sheet would be expected to induce more catalysis compared to the metal nanoparticle island film. Interestingly, the rate of decay was found larger on the silver-island film compared to the solid metal sheet, eliminating silver catalysis as an explanation for the observed emission increase (Figure 4) [6]. These observations suggest that chemically–induced excited molecular species couple with and induce plasmon resonance in silver island particles. In addition, the reduced emission lifetime in the presence of silver nanoparticles compared to the glass control is similar to lifetime decreases observed in MEF, suggesting MEC to share a similar mechanism as MEF [6].

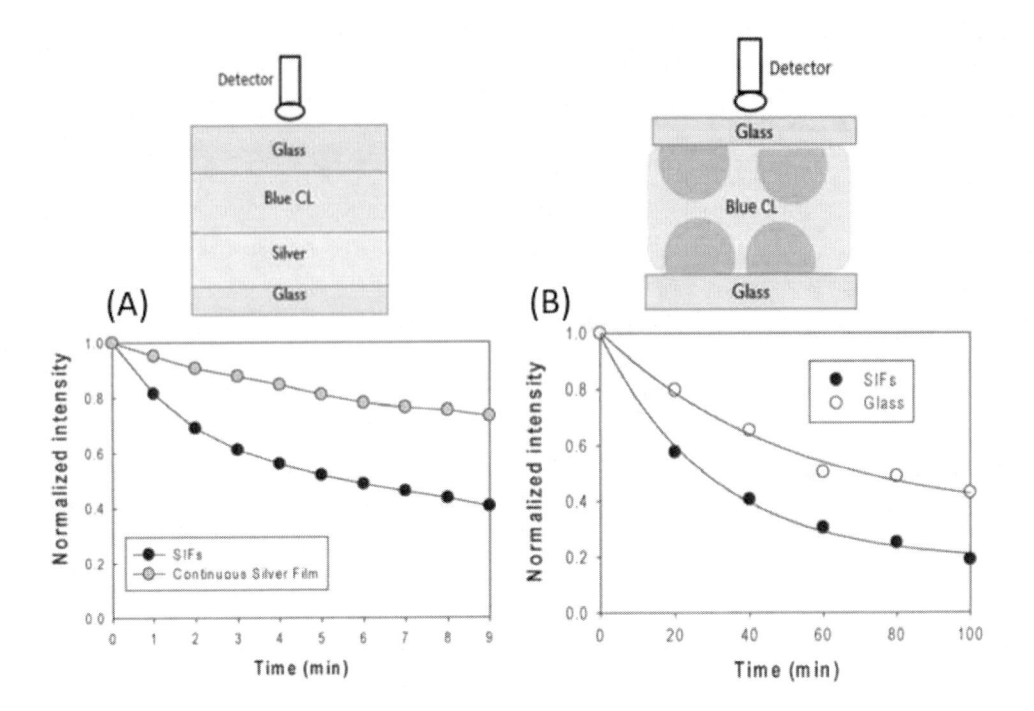

Figure 4. Decay of emission intensity of a blue chemiluminescence dye placed on (A) a continuous strip of silver film and (B) SIFs and glass, versus time. The decay rates were calculated by fitting the data to a first order exponential curve. Experimental schematics are also shown (Top) [Modified from Ref. 6].

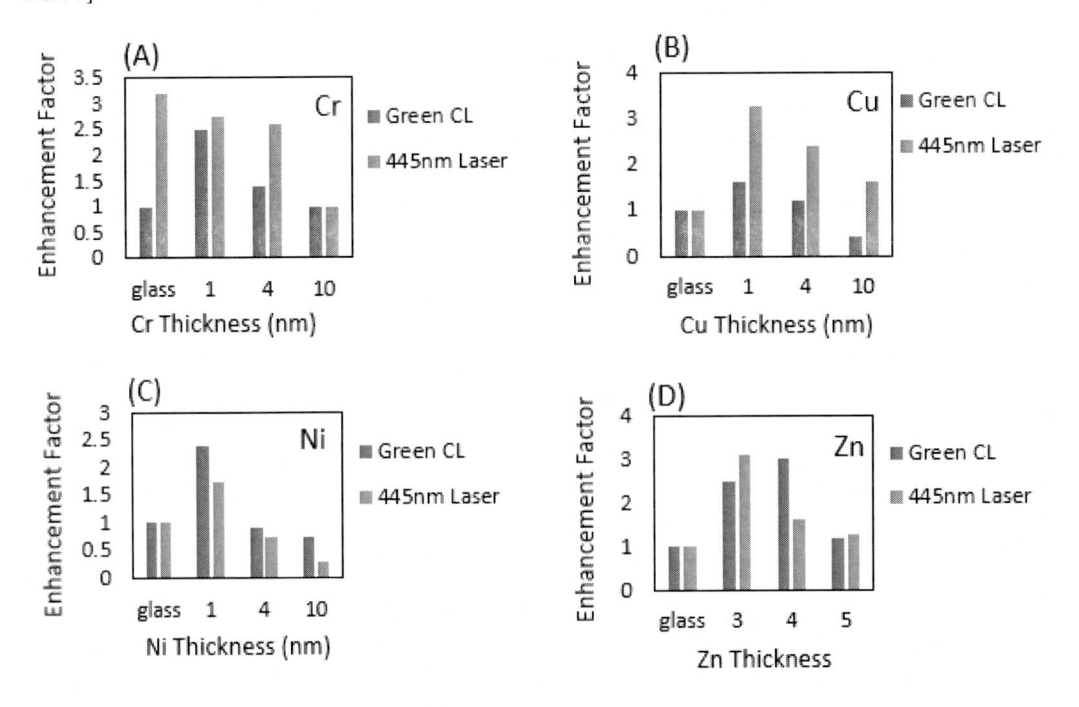

Figure 5. Enhancement factor versus thickness of the metal film for (A) chromium (Cr), (B) copper (Cu), (C) nickel (Ni), (D) zinc (Zn) for both a green chemiluminescence solution (green CL) and a 445 nm laser [Modified from Ref. 6].

There are also metals other than silver that support plasmon resonance in the visible spectrum, which include chromium, copper, nickel and zinc. MEC has been demonstrated with these metals over a range of metal nanoparticle sizes (Figure 5) [6]. It is observed that as the particle sizes grow into each other, approaching a continuous metal sheet, emission enhancement is decreased in a similar manner to MEF. This is explained by the transformation of discreet plasmonic nanoparticles into the bulk metal, which no longer couples strongly with molecular emitters [6].

4. SURFACE PLASMON COUPLED CHEMILUMINESCENCE (SPCC)

In addition to metal nanoparticle islands, chemically-induced excited electronic states can couple with and excite plasmons in continuous metal films [27-32]. In the nanoparticles, discreet plasmon resonance frequencies exist, which couple directly with the excited molecular species. As the metal nanoparticle is elongated, the restoring force of the dipole weakens and the plasmon resonance frequency is red-shifted. In a continuous metal sheet (bulk metal) plasmon resonances can still be excited, and these resonances propagate through the metal, as opposed to the discreet resonances in nanoparticles which do not propagate [13]. Surface Plasmon Resonance (SPR) is a common technique in laboratories today, which involves excitation of these propagating surface plasmons in continuous metal films, where the plasmons must be excited at a critical angle [33-36]. This critical angle is dependent on the refractive index of the metal surface, allowing for a wide range of detection assays, which often function based on a biochemical reaction bringing species into proximity of the surface causing a refractive index change [33]. In addition to critical angle excitation, propagating surface plasmons can also be excited locally by fluorescent species in close proximity of the metal, a technique known as surface plasmon coupled emission (SPCE) [37]. The subject of this section, Surface Plasmon Coupled Chemiluminescence (SPCC) involves the excitation of plasmons in continuous metal films with chemiluminescent species in close proximity to the metal. Following this excitation, emission is detected from the backside of the metal film. This emission is characteristic of the free-space chemiluminescence, except it is highly directional emission as compared to free-space emission which is isotropic [27-32].

The experimental setup for SPCC is shown in Figure 6 [27]. The chemiluminescent solution is placed on a metal film with a hemi-cylindrical prism on the back side of the metal film. A rotating stage is then used for directional emission measurements. In one experiment, blue chemiluminescent dye was placed on a 47 nm thick continuous silver film. Figure 7a shows the free-space emission spectrum compared to the SPCC spectrum. The spectra appear similar, indicating the metal is emitting at a frequency characteristic of the chemiluminescent species, and that the metal is not significantly altering the emission wavelength [27]. The top half of Figure 7b shows the free space emission compared to the coupled emission on the bottom half. Emission from the bottom half is clearly directional and *p*-polarized (perpendicular to the metal surface), compared to the top half which displays equal emission at all measurement angles [27]. This p-polarized emission is due to coupling of molecular dipoles oriented perpendicular to the metal surface allowing for enhanced emission, while *s*-oriented dipoles (parallel to the metal surface) cancel out with their own mirror image on the surface, preventing *s*-emission from the coupled system [30]. The experimental setup is shown in Figure 7c. Here, chemiluminescence emission from the metal coupled system is observed from both the sample side of the metal and the back side of the prism, while isotropic free-space emission is measured from the front side of the metal.

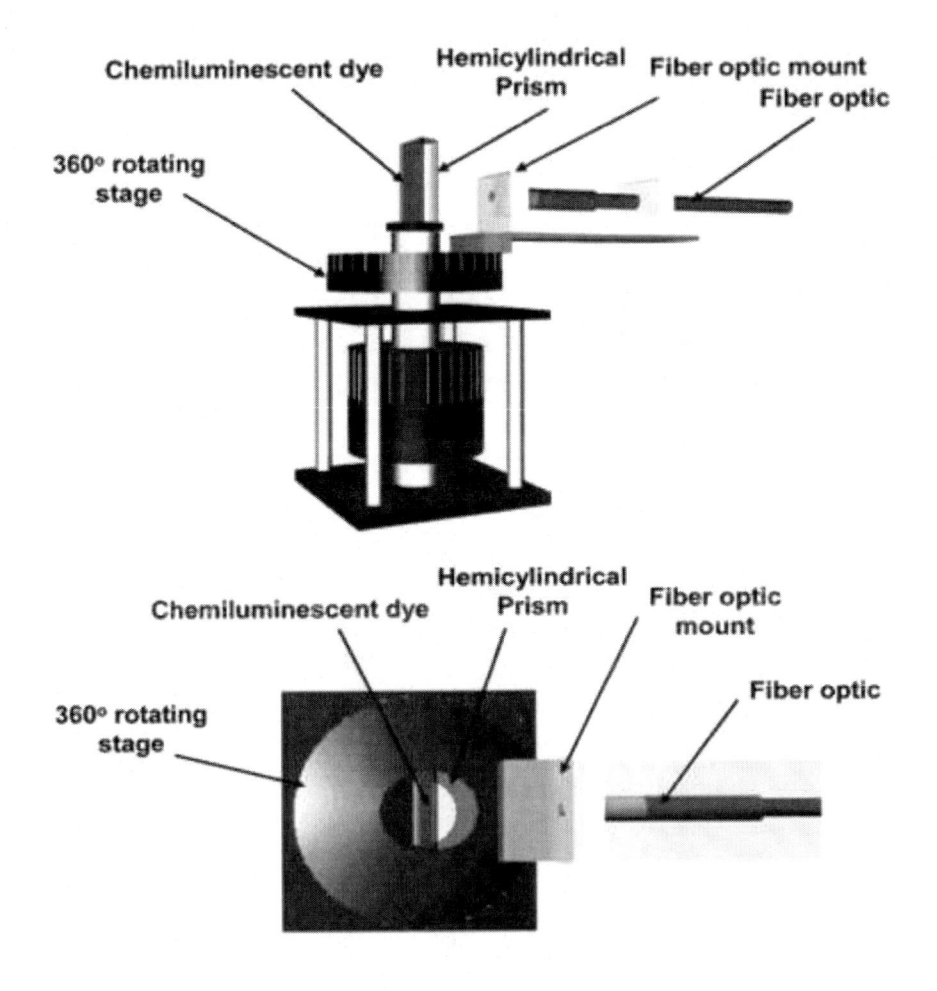

Figure 6. Experimental geometry used for surface plasmon-coupled chemiluminescence (SPCC). (Top) side view; (bottom) top view [Modified from Ref. 27].

SPCC is observed on a variety of metals which support plasmon resonances in the visible spectrum. While silver tends to have the strongest plasmon resonances, other metals such as gold, aluminum, zinc, and nickel have been demonstrated to display SPCC as well [28, 31, 32]. Coupled emission from the back side of a prism at p and s polarizations as well as with no polarization for silver, gold, and aluminum films was measured [28]. The metals display directional emission, with p-polarized emission clearly dominating over s-polarized emission, suggesting surface plasmons to be responsible for the SPCC emission [28]. Figure 8 shows emission polarization, demonstrating SPCC from a thin zinc film [32].

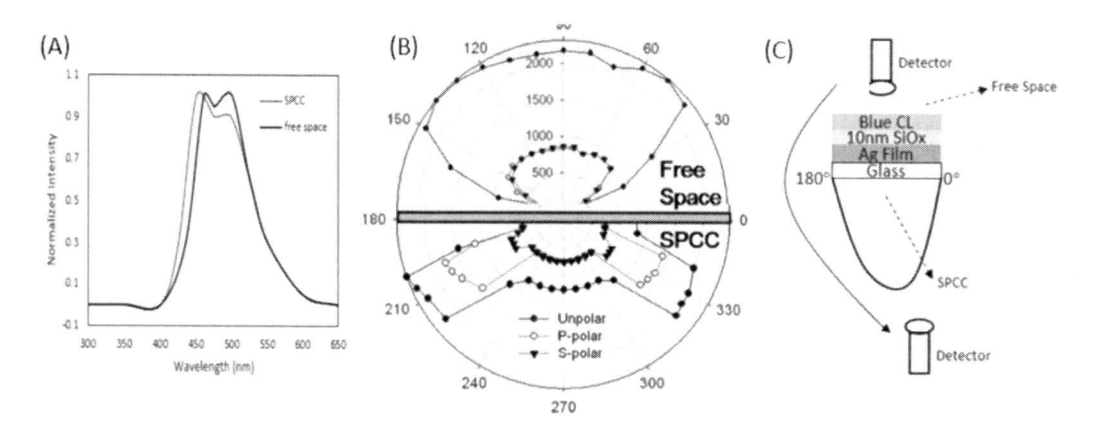

Figure 7. (A) Emission spectra of both the free-space emission and SPCC (B) Free-space emission and SPCC (C) Schematic for surface plasmon coupled chemiluminescence (SPCC) from 47 nm thick silver films. [Modified from Ref. 6].

Fresnel curves plot the metal reflectivity as a function of the emission angle and may be used to predict the optimal thickness of the metal film in SPCC [29]. It is worth noting the application of Fresnel curves to SPCC has been adapted from use in SPR predictions. Figure 9 shows Fresnel curves for iron films with thicknesses ranging from 15-50 nm. The reflectivity minimum is shown to occur in the 15nm films, demonstrating that SPCC emission is theoretically decreased in the thicker metal films [30]. These curves are also used to predict p-polarized emission to be dominate over s-polarization in SPCC [30].

Emission intensity decay rates were also measured for blue chemiluminescent dyes for both free space and SPCC emission [28]. For the SPCC measurements, only p-polarized emission was measured, because SPCC only consists of p-polarized light. This decay is due to depletion of the reagents involved in the chemiluminescence reaction, and was found to follow first-order decay kinetics. Figure 10 compares the free space to SPCC emission for the blue dye on an aluminum film substrate and shows decay intensities normalized to their respective starting values. The rate of decay is found to be similar for SPCC compared to free space emission [28]. This finding indicates that the aluminum is not serving as a catalyst for the chemiluminescent reaction, which would be expected to speed up the depletion of the reactants [28].

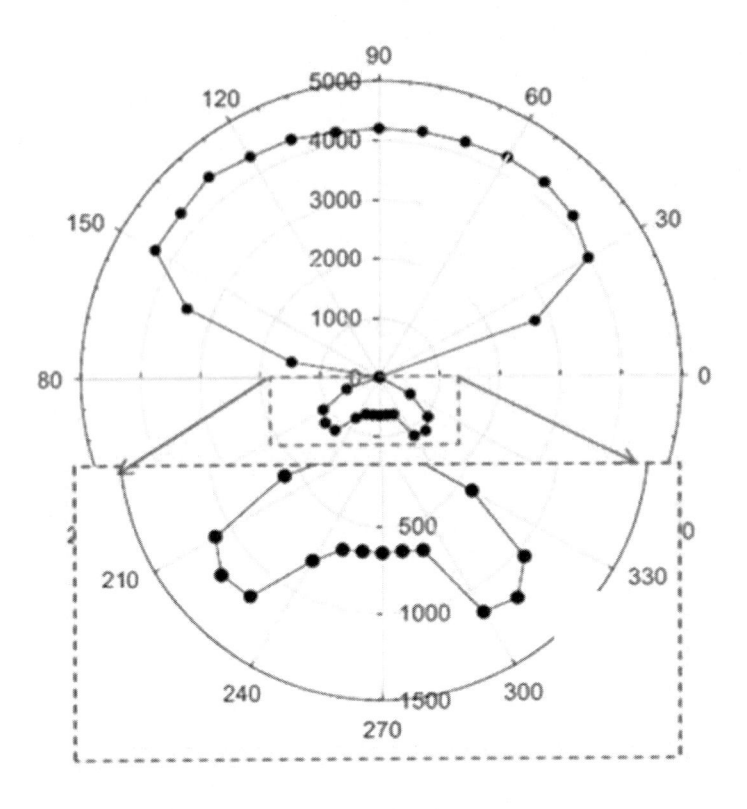

Figure 8. Polar plot of free-space chemiluminescence emission (a.u.) 0°-180° and SPCC 180°-360° for a zinc film [Modified from Ref. 32].

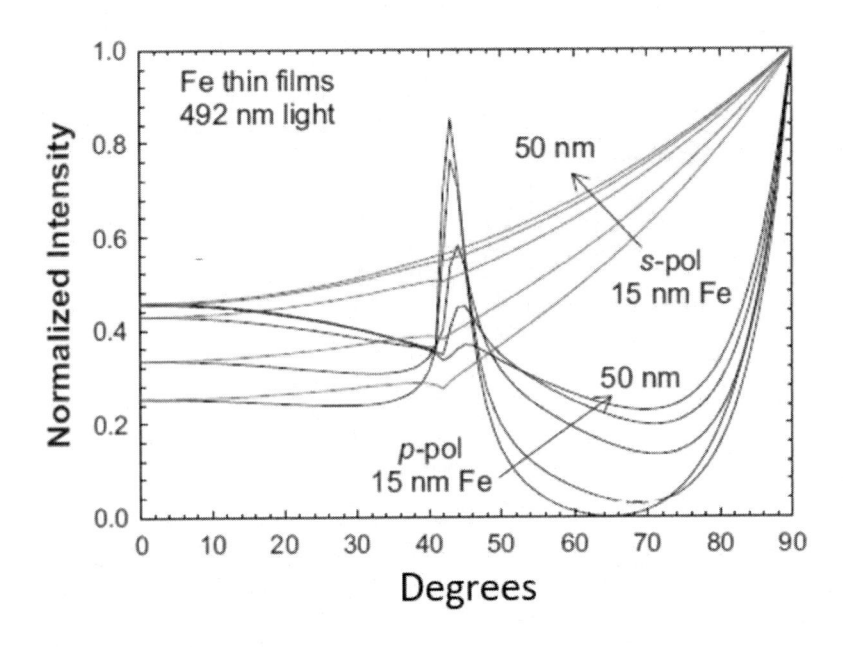

Figure 9. Four-phase Fresnel reflectivity curves for *p*- and *s*-polarized lights at 492 nm for iron thin film thicknesses of 15, 20, 30, 40 and 50 nm [Modified from Ref. 30].

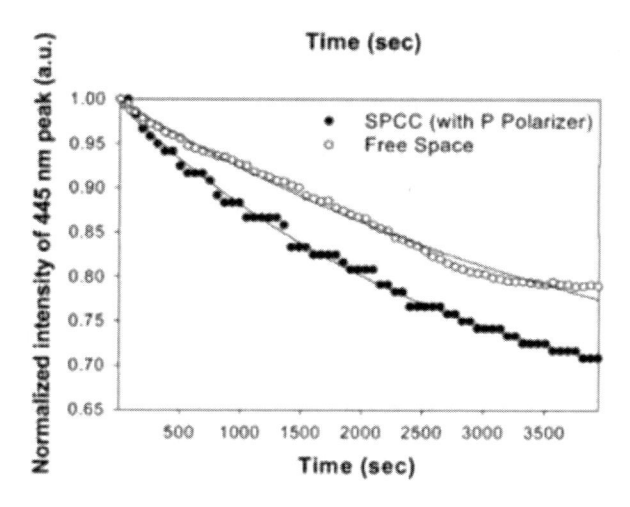

Figure 10. Chemiluminescence intensity decays from aluminum films for both free-space and coupled emission, normalized to the same initial intensity [Modified from Ref. 28].

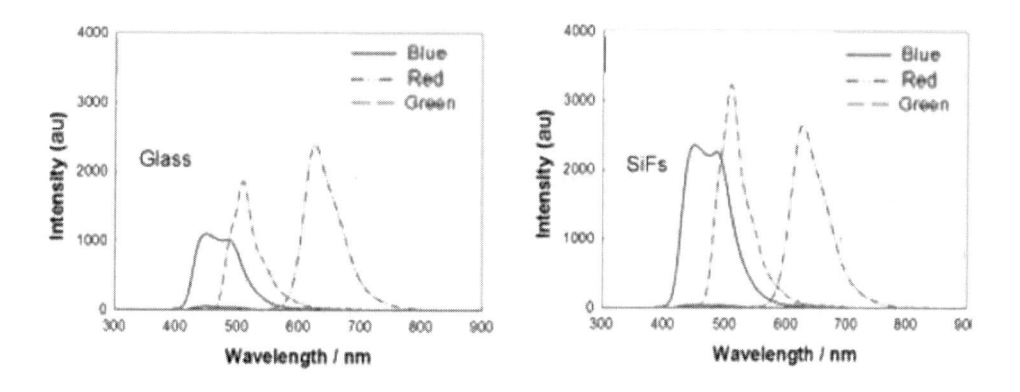

Figure 11. Chemiluminescence emission spectra of blue, green, and red chemiluminescent reagents on glass surfaces (Left) and on silver island films (SiFs) (Right), after 10 s microwave (Mw) exposure [Modified from Ref. 38].

5. MICROWAVE-TRIGGERED METAL-ENHANCED CHEMILUMINESCENCE (MT-MEC)

The acceleration of MEC reactions can be achieved using low-power microwaves. This phenomenon, named Microwave-Triggered Metal-Enhanced Chemiluminescence (MT-MEC), increases both sensitivity and speed over other MEC assays, by increasing chemiluminescent reaction rates using kinetic energy from the microwaves [38-42]. This opens up the possibility for both rapid and sensitive chemiluminescent assays.

Aslan et. al. measured the time dependent chemiluminescence emission from a blue acriden based reagent on both silver-island films and on glass, with and without microwave application [38]. Here, chemiluminescence emission was measured as total photon counts, and the emission was demonstrated to spike with successive microwave pulses. With each microwave pulse chemiluminescent species were depleted, resulting in emission decay over 2000 seconds. It is worth noting the microwave power was held at a low power (20% power) preventing surface drying of the solution on the substrate. The solution heating due to the microwave was found to be ~8°C for 30µl of sample. Total photon counts on both silver-island films and glass substrates were found to be significantly higher with microwave application compared to without microwaves, yielding approximately a 2.45-fold increase in photon flux with application of the microwave. In addition to several microwave pulses, it is also possible to apply one relatively long microwave pulse, which is expected to yield similar results compared to the successive microwave pulses [38]. Clearly, MT-MEC provides a rapid approach to sensitive and rapid chemiluminescent assays, with potential for reducing detection times from hours down to seconds.

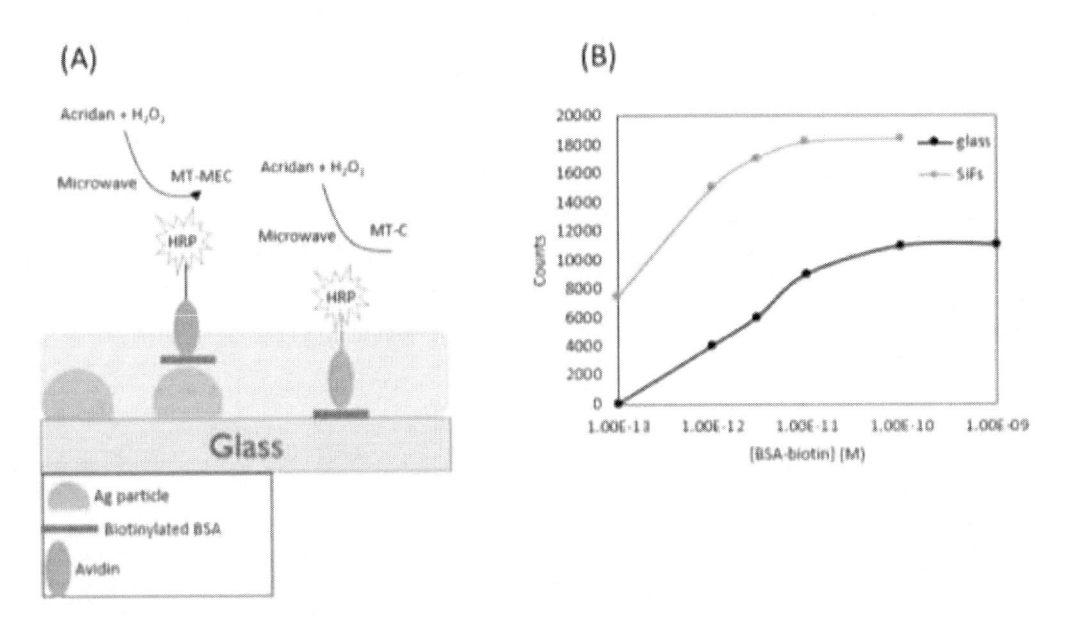

Figure 12. Microwave-triggered metal-enhanced chemiluminescence-based detection of proteins on SiFs and glass microscope slides. Horseradish peroxidase (HRP) chemiluminescence assay on both glass and SiFs. (B) Integrated photon flux of chemiluminescence emission for different concentrations of biotinylated-BSA from both SiFs and glass [Redrawn from Ref. 6].

MT-MEC also allows for multicolor detection [38]. Emission from blue, red, and green chemiluminescent species was detected before and after 10 second microwave exposures (Figure 11). Emission enhancement factors following microwave application were calculated as the ratio of emission with microwave exposure to emission without microwave exposure. For MT-MEC, emission enhancement was calculated as the ratio of microwave exposure on Sifs to no microwave exposure on a glass substrate, and was quite large, up to 125-fold for green emission (Table 1) [38]. These results open up potential for assay panels capable of simultaneous emission detection from several reactions.

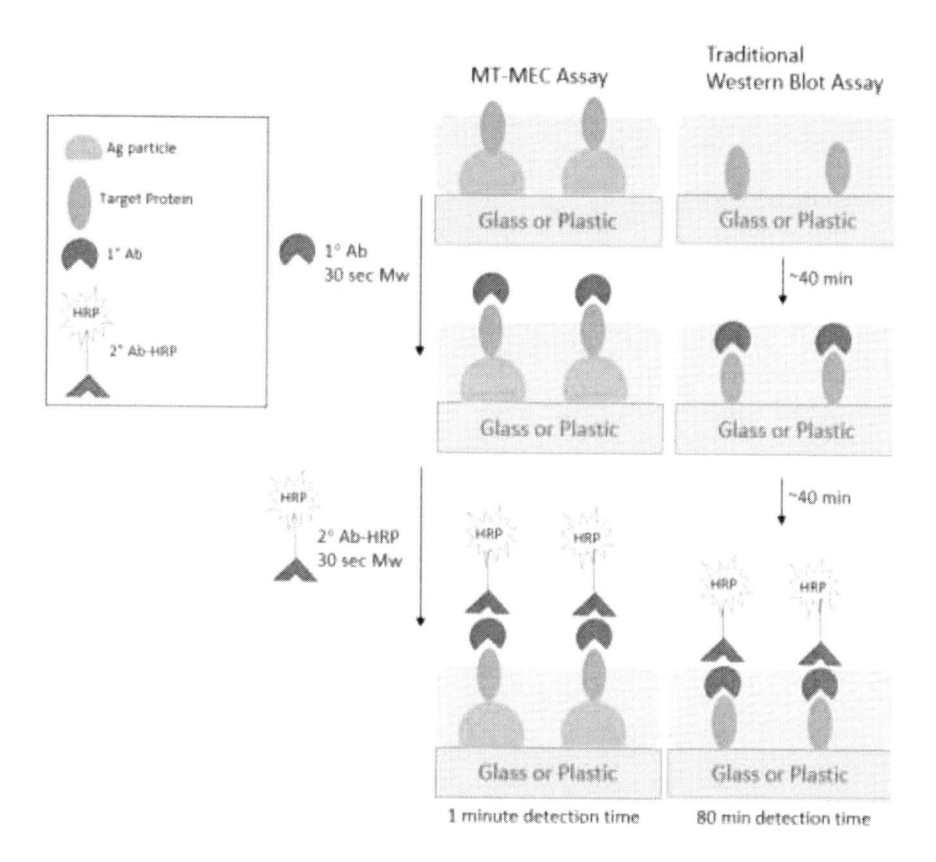

Figure 13. Procedure for the MT-MEC immunoassay (Mw, low-power microwave heating) [Redrawn from Ref. 40].

Table 1. Calculated Enhancement Values for Chemiluminescent Reagents Due to MT and MT-MEC (Mw-Microwave) [Redrawn from Ref. 38]

	Blue	Green	Red
SiFs: SiFs(Mw)/SiFs (no Mw)	54.0	85.0	69.0
Glass: glass(Mw)/glass (no Mw)	22.0	75.0	63.0
MT-MEC: SiFs(Mw)/glass (no Mw)	54.0	125	70.0

5.1. Application of MT-MEC to Biological Detection Assays

MT-MEC may also be utilized for the development of highly sensitive and rapid protein assays [39-41]. Metal nanoparticles act as energy acceptors from a chemically induced electronic excited state in a molecular species. Therefore, if a protein interaction brings the excited molecular species into close proximity of the metal, the protein can be detected and quantitated. In addition, microwaves can be used to simultaneously accelerate both the chemiluminescent reaction and the protein-ligand binding process, leading to a rapid protein assay [39, 40]. A model protein assay was constructed utilizing the strong interaction between biotin and streptavidin. In this experiment biotinylated-BSA was first incubated on a SiF substrate. Here, BSA is known to adsorb spontaneously in metal films and colloids, with electrostatic and hydrophobic interactions being implicated in the binding mechanism [43-45]. HRP-labelled streptavidin was then added, allowing for binding between biotin and streptavidin, and bringing the HRP label into close proximity of the metal. HRP serves as the catalysis for the chemiluminescent reaction, providing for chemiluminescent emission in close proximity to the metal [40]. Acridan (lumophore) and peroxide chemiluminescent reagents are then added, which strongly luminesce in the presence of the HRP catalysis (Figure 12) [40].

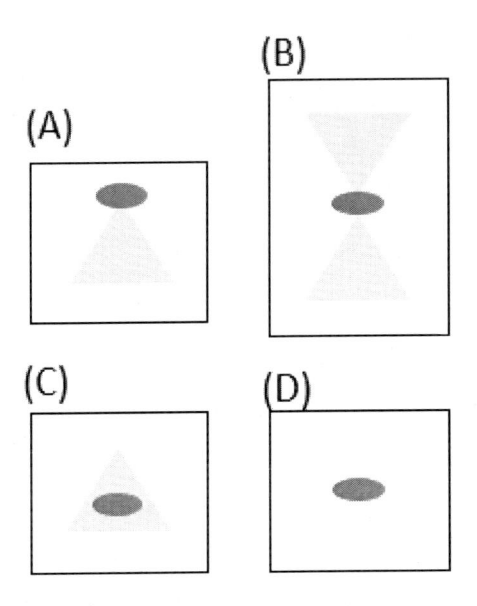

Figure 14. (A-D) Sample geometry depicting the chemiluminescence sample (blue circle), glass substrate (white square), and aluminum triangle geometries (12.3-mm length; 1-mm gap size for disjointed bow-tie geometry) [Modified from Ref. 39].

Results in Figure 12b demonstrate enhanced emission upon microwave acceleration of the assay. The microwaves in this process are thought to produce a localized heating around the metal nanoparticles, accelerating the chemiluminescent reaction, and providing for a rapid protein assay [40]. A control experiment is also performed on a glass slide without the silver nanoparticles (Figure 12b) with results indicating emission enhancement in the presence of the metal. Quantitation is achieved through the incubation of the biotinylated-BSA on the metal substrate over a range of concentrations. Here, a larger concentration of biotinylated-BSA on the surface is expected to bring more HRP labelled streptavidin into close proximity of the metal. The HRP is then expected to increase the chemiluminescent reaction, leading to quantitative determination of the biotin-streptavidin interaction. Results from the assay demonstrate femtomolar sensitivity, which was achieved in a 2-minute assay time [40]. These results demonstrate a sensitive and rapid model MT-MEC based assay, opening up potential for a variety of protein assays based on MT-MEC. MT-MEC protein assays were also compared to traditional western blot assays. MT-MEC was found to decrease detection time to <2 minutes compared with traditional western blot analysis times of approximately 80min. The time breakdown of various steps in each assay is depicted in Figure 13 [40].

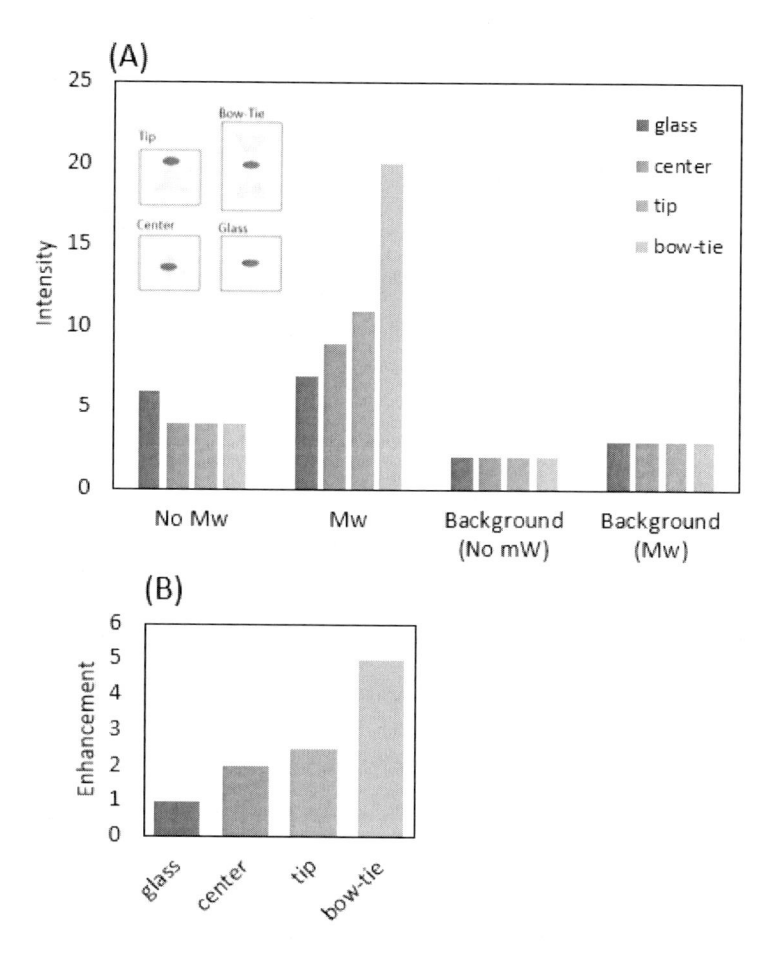

Figure 15. (A) Chemiluminescence emission before (No Mw) and after (Mw) exposure to low-power microwave (Mw) pulses from glass coverslips incubated with 1 μM BSA-biotin and 1 μM HRP-streptavidin positioned on glass substrates modified with and without 12.3-cl Al triangle 75 nm thick. (Background) Chemiluminescence emission before and after exposure to low-power microwave (Mw) pulses from control glass coverslips incubated with 1.5% BSA and 1 μM HRP-streptavidin. (B) Chemiluminescent microwave (Mw) enhancement rations (Mw/no Mw) upon application of low-power microwaves pulses (Mw) for different sample geometries [Modified from Ref. 39].

5.2. Dependence of MT-MEC on Substrate Geometry

MT-MEC in previous sections of this chapter has been conducted on a silver-island film substrate, allowing for emission enhancement. However, solid metal films shaped in a variety of geometries may be utilized for microwave acceleration of chemiluminescent reactions. The metals here work by focusing microwaves onto the reaction material, increasing reaction speed [39, 42]. FDTD (Finite Difference Time Domain) simulations are commonly used to visualize electric fields resulting from various metal geometries, and the interaction of microwaves with the metal can also be imaged with this technique [39, 42]. FDTD can then be utilized to optimize metal substrate geometries, allowing for optimal electric fields being directed at the chemiluminescent reaction, and leading to optimized reaction kinetics. FDTD was used to simulate the electric field resulting from 2.45GHz microwave application to both single aluminum triangles and double triangles pointed inward in bow-tie geometry. Maximum electric field enhancements were found at the corners or tips of the metal structures compared to the center, and in the gap between the two triangles in bow-tie geometry [39]. The various geometries are depicted in Figure 14.

A chemiluminescent reaction was subsequently performed in the gap between the two triangles in the bow-tie geometry, utilizing focused microwaves to increase the reaction rate [39]. To experimentally verify the dependence of substrate geometry on MT-MEC predicted by FDTD simulations, 75nm thick aluminum films were first deposited on a glass substrate using thermal vapor deposition. A chemiluminescent solution was then deposited onto either a single triangle or into the gap between two triangles in the bow-tie geometry. Results demonstrated 100x and 500x chemiluminescent emission enhancements from the single and bow-tie geometry, respectively. It is worth noting that emission measured from the center of a single aluminum triangle was found to produce negligible emission enhancement, which is consistent with the FDTD simulations of a diminished electric field in the triangle center. Reaction rates were determined by fitting emission measurements as a function of time to a single exponential decay curve. 45x and 95x reaction rate enhancements were found for the single and bow-tie triangle geometries, respectively [39].

A model protein assay was developed to demonstrate the applicability of optimized substrate geometries to biological detection assays [39]. The biotin-streptavidin interaction was utilized to bring a chemiluminescent reaction catalyst into proximity of the metal substrate, and emission was monitored. This protein assay was performed for several substrate geometries, and maximum emission enhancements were found for the triangular bow-tie geometry [Figure 15] [39]. These emission enhancements clearly demonstrate the potential for the development of a wide variety of biological assays which utilize optimized substrate geometries for increased sensitivity and speed.

5.3. Multiplexed MT-MEC Assay

A multiplexed MT-MEC assay was performed in order to simultaneously detect multiple chemiluminescent species with high sensitivity [39]. In this assay, a square aluminum film was used as a substrate, and different color chemiluminescent species were placed at each corner of the aluminum square. This is beneficial because maximum electric field intensities are present at corners of the metallic structures, providing for the greatest emission enhancement with microwave irradiation. 300-fold emission enhancements were observed at the corners of the square aluminum film, while negligible enhancements were observed in the center of the metal [39]. These results clearly show the applicability of MT-MEC to the development of high throughput and multiplexed chemiluminescent detection assays.

CONCLUSION

In summary, this chapter has reviewed the discovery and development of the MEC, SPCC, and MT-MEC phenomena. The mechanism behind MEC involves energy transfer from a chemically-induced electronic excited state in a molecular species to a metal nanoparticle island film. This interaction leads to an enhanced emission, which radiates from the metal at a frequency characteristic of the chemiluminescent species. Plasmons in the metal are thought to relax via radiative emission at an enhanced rate compared to the molecular emitter, leading to more intense and brighter chemiluminescent emission in the presence of metallic nanoparticles. In continuous metal films, SPCC involves the excitation of propagating plasmons via chemiluminescent species. These propagating plasmons relax via highly directional emission, compared to the isotropic emission from the free space emission. It has also been demonstrated that microwave irradiation can be used to accelerate the MEC process, by providing energy to kinetically accelerate chemiluminescent reactions. MT-MEC has also been applied to a model protein assay which utilizes the strong interaction between biotin and streptavidin to demonstrate both a sensitive and rapid biological assay. Optimization of the metal substrate geometry has also been demonstrated using FDTD simulations to map out microwave induced electric field intensities along a metal substrate. Metallic triangles in the bow-tie configuration have been shown to provide for the greatest emission enhancement compared to a variety of substrate geometries. It is the gap between the two triangles in the bow-tie geometry which provides the greatest field enhancement, providing for optimized MT-MEC assays with enhanced sensitivity and speed. Finally, a multiplexed assay has been developed allowing for the simultaneous detection of four chemiluminescent solutions, opening up potential for a variety of high-throughput chemiluminescent assays. In conclusion, MEC represents a novel analytical platform capable of both sensitive and rapid detection assays, with potential for a wide variety of chemical and biological target species.

REFERENCES

[1] Siraj, N., El-Zahab, B., Hamdan, S., Karam, T. E., Haber, L. H., Li, M., Fakayode, S. O., Das, S., Valle, B., Strongin, R. M. (2016). Fluorescence, Phosphorescence, and Chemiluminescence. *Anal. Chem.* 88: 170-202.

[2] Das, S., Powe, A. M., Baker, G. A., Valle, B., El-Zahab, B., Sintim, H. O., Lowry, M., Fakayode, S. O., McCarroll, M. E., Patonay, G. (2012). Molecular Fluorescence, Phosphorescence, and Chemiluminescence Spectrometry. *Anal. Chem.* 84: 597-625.

[3] Vacher, M., Fdez Galván, I., Ding, B.-W., Schramm, S., Berraud-Pache, R., Naumov, P., Ferré, N., Liu, Y.-J., Navizet, I., Roca-Sanjuán, D., Baader, W. J., Lindh, R. (2018). Chemi- and Bioluminescence of Cyclic Peroxides. *Chem. Rev.* 118: 6927-6974.

[4] Kuntzleman, T. S., Rohrer, K., Schultz, E. (2012). The Chemistry of Lightsticks: Demonstrations to Illustrate Chemical Processes. *J. Chem. Educ.* 89: 910-916.

[5] Timofeeva, I. I., Vakh, C. S., Bulatov, A. V., Worsfold, P. J. (2018). Flow analysis with chemiluminescence detection: Recent advances and applications. *Talanta* 179: 246-270.

[6] Aslan, K., Geddes, C. (2009). Metal-enhanced chemiluminescence: advanced chemiluminescence concepts for the 21st century. *Chem. Soc. Rev.* 38: 2556-2564.

[7] Chowdhury, M. H., Aslan, K., Malyn, S. N., Lakowicz, J. R., Geddes, C. D. (2006). Metal-enhanced chemiluminescence. *J. Fluoresc.* 16: 295-299.

[8] Geddes, C. D., Metal-Enhanced Fluorescence: Progress towards a Unified Plasmon-Fluorophore Description. *Metal-Enhanced Fluorescence* 2010, 1.

[9] Yongxia, Z., Oragan, A., Geddes, C. D. (2009). Wavelength dependence of metal-enhanced fluorescence. *J. Phys. Chem. C* 113: 12095-12100.

[10] Dragan, A. I., Mali, B., Geddes, C. D. (2013). Wavelength-dependent metal-enhanced fluorescence using synchronous spectral analysis. *Chem. Phys. Lett.* 556: 168-172.

[11] Yongxia, Z., Aslan, K., Previte, M. J. R., Geddes, C. D. (2008). Metal-enhanced e-type fluorescence. *Appl. Phys. Lett.* 92: 013905-1-3.

[12] Aslan, K., Lakowicz, J. R., Geddes, C. D. (2005). Metal-Enhanced Fluorescence Using Anisotropic Silver Nanostructures: Critical Progress to Date. *Anal. Bioanal. Chem.* 382: 926-933.

[13] Pelton, M., Bryant, G. W., *Introduction to metal-nanoparticle plasmonics*. Hoboken, New Jersey: John Wiley & Sons, Inc., [2013]: 2013.

[14] Novotny, L., van Hulst, N. (2011). Antennas for light. *Nat. Photonics* 5: 83-90.

[15] Li, J., Krasavin, A. V., Webster, L., Segovia, P., Zayats, A. V., Richards, D. (2016). Spectral variation of fluorescence lifetime near single metal nanoparticles. *Sci. Rep.* 6: 21349.

[16] Weitz, D. A., Garoff, S., Hanson, C. D., Gramila, T. J., Gersten, J. I. (1982). Fluorescent lifetimes of molecules on silver-island films. *Opt. Lett.* 7: 89-91.

[17] Kuhn, S., Hakanson, U., Rogobete, L., Sandoghdar, V. (2006). Enhancement of single-molecule fluorescence using a gold nanoparticle as an optical nanoantenna. *Phys. Rev. Lett.* 97: 017402/1-4.

[18] Chowdhury, M. H., Aslan, K., Malyn, S. N., Lakowicz, J. R., Geddes, C. D. (2006). Metal-enhanced chemiluminescence: radiating plasmons generated from chemically induced electronic excited states. *Appl. Phys. Lett.* 88: 173104-1-3.

[19] Saito, Y., Wang, J. J., Smith, D. A., Batchelder, D. N. (2002). A Simple Chemical Method for the Preparation of Silver Surfaces for Efficient SERS. *Langmuir* 18: 2959-2961.

[20] Rycenga, M., Cobley, C. M., Jie, Z., Weiyang, L., Moran, C. H., Qiang, Z., Dong, Q., Younan, X. (2011). Controlling the Synthesis and Assembly of Silver Nanostructures for Plasmonic Applications. *Chem. Rev.* 111: 3669-3712.

[21] Pribik, R., Dragan, A. I., Zhang, Y., Gaydos, C., Geddes, C. D. (2009). Metal-enhanced fluorescence (MEF): physical characterization of Silver-island films and exploring sample geometries. *Chem. Phys. Lett.* 478: 70-74.

[22] McPeak, K. M., Jayanti, S. V., Kress, S. J. P., Meyer, S., Iotti, S., Rossinelli, A., Norris, D. J. (2015). Plasmonic Films Can Easily Be Better: Rules and Recipes. *ACS Photonics* 2: 326-333.

[23] Hulteen, J. C., Van Duyne, R. P (1995). In Nanosphere lithography: a materials general fabrication process for periodic particle array surfaces. *J. Vac. Sci. Technol.* 13: 1553-1558.

[24] Li, N., Liu, D., Cui, H. (2014). Metal-nanoparticle-involved chemiluminescence and its applications in bioassays. *Anal. Bioanal. Chem.* 406: 5561-5571.

[25] Zhang Z. F., Cui, H., Lai, C. Z., Liu, L. J. (2005). Gold nanoparticle-catalyzed luminol chemiluminescence and its analytical applications. *Anal. Chem.* 77: 3324–3329.

[26] Li Q. Q., Liu, F., Lu, C., Lin, J. M. (2011). Aminothiols sensing based on fluoro-surfactant-mediated triangular gold nanoparticle-catalyzed luminol chemilumine-scence. *J. Phys. Chem. C* 115: 10964–10970.

[27] Chowdhury, M. H., Malyn, S. N., Aslan, K., Lakowicz, J. R., Geddes, C. D. (2007). First observation of surface plasmon-coupled chemiluminescence (SPCC). *Chem. Phys. Lett.* 435: 114-118.

[28] Chowdhury, M. H., Malyn, S. N., Asian, K., Lakowicz, J. R., Geddes, C. D. (2006). Multicolor directional surface plasmon-coupled chemiluminescence. *J. Phys. Chem. B* 110: 22644-22651.

[29] Aslan, K., Weisenberg, M., Hortle, E., Geddes, C. D. (2009). Fixed-angle observation of surface plasmon coupled chemiluminescence from palladium thin films. *Appl. Phys. Lett.* 95: 123117.

[30] Aslan, K., Weisenberg, M., Hortle, E., Geddes, C. D. (2009). Surface plasmon coupled chemiluminescence from iron thin films: Directional and approaching fixed angle observation. *J. Appl. Phys.* 106: 14313-14318.

[31] Weisenberg, M., Aslan, K., Hortle, E., Geddes, C. D. (2009). Directional surface plasmon coupled chemiluminescence from nickel thin films: Fixed angle observation. *Chem. Phys. Lett.* 473: 120-125.

[32] Aslan, K., Geddes, C. D. (2009). Surface plasmon coupled chemiluminescence from zinc substrates: Directional chemiluminescence. *Appl. Phys. Lett.* 94: 073104.

[33] Lakowicz, J. R., *Principles of fluorescence spectroscopy.* New York: Springer, c2006. 3rd ed.: 2006.

[34] Gupta, R., Dyer, M. J., Weimer, W. A. (2002). Preparation and characterization of surface plasmon resonance tunable gold and silver films. *J. Appl. Phys.* 92: 5264-5271.

[35] Homola, J., *Surface plasmon resonance based sensors.* Berlin: Springer, 2006.: 2006.

[36] Wei, W., Zhang, X., Ren, X. (2015). Plasmonic circular resonators for refractive index sensors and filters. *Nanoscale Res. Lett.* 10: 211-211.

[37] Lakowicz, J. R. (2004). Radiative decay engineering 3. Surface plasmon-coupled directional emission. *Anal. Biochem.* 324: 153–169.

[38] Aslan, K., Malyn, S. N., Geddes, C. D. (2006). Multicolor microwave-triggered metal-enhanced chemiluminescence. *J. Am. Chem. Soc.* 128: 13372-13373.

[39] Previte, M. J. R., Aslan, K., Geddes, C. D. (2007). Spatial and Temporal Control of Microwave Triggered Chemiluminescence: A Protein Detection Platform. *Anal. Chem.* 79: 7042-7052.

[40] Previte, M. J. R., Aslan, K., Malyn, S. N., Geddes, C. D. (2006). Microwave Triggered Metal Enhanced Chemiluminescence: Quantitative Protein Determination. *Anal. Chem.* 78: 8020-8027.

[41] Previte, M. J., Aslan, K., Malyn, S., Geddes, C. D. (2006). Microwave-Triggered Metal-Enhanced Chemiluminescence (MT-MEC): Application to Ultra-fast and Ultra-sensitive Clinical Assays. *J. Fluoresc.* 2006, 16: 641-647.

[42] Previte, M. J., Geddes, C. D. (2007). Microwave-Triggered Chemiluminescence with Planar Geometrical Aluminum Substrates: Theory, Simulation and Experiment. *J. Fluoresc.* 17: 279-287.

[43] Maleki, M. S., Moradi, O., Tahmasebi, S. (2017). Adsorption of albumin by gold nanoparticles: Equilibrium and thermodynamics studies. *Arab. J. Chem.* 10: S491-S502.

[44] Wang, X., Herting, G., Wallinder, I. O., Blomberg, E. (2015). Adsorption of bovine serum albumin on silver surfaces enhances the release of silver at pH neutral conditions. *Phys. Chem. Chem. Phys.* 17: 18524-18534.

[45] Williams, R. L., Williams, D. F. (1988). Albumin adsorption on metal surfaces. *Biomaterials* 9: 206-212.

In: A Comprehensive Guide to Chemiluminescence
Editor: Luís Pinto da Silva

ISBN: 978-1-53616-170-0
© 2019 Nova Science Publishers, Inc.

Chapter 9

CRYSTALLINE FIREFLIES: FROM MODELS FOR BIOLUMINESCENCE TO SOLID STATE CHEMILUMINESCENCE REACTIONS

*Stefan Schramm**

Friedrich-Schiller University Jena, Germany,
New York University Abu Dhabi, UAE

ABSTRACT

The impressive bioluminescence reaction of the firefly marks the start of this chapter. By discussing the firefly's light emission mechanism, the synthesis of the luciferin is explained, photochemical properties of its amino-derivatives are described and the need for model compounds is disclosed. With the help of these molecules, fundamental but hard-to-study steps in complex bioluminescence reaction mechanisms can be made accessible and examined in closer detail. A system that suits well this purpose are chemiluminescent 2-coumaranones. Their efficient and easy synthesis together with studies on their chemiluminescence mechanism are reviewed and put into perspective with the increasing demand of innovative model compounds for established bioluminescent systems. The high energy intermediate of the 2-coumaranone chemiluminescence is a 1,2-dioxetanone. From understanding its electronic structure, the question is asked whether these or similar peroxides can also show light emission in macroscopic molecular crystals. Such molecules – the crystalline fireflies – are examined for their chemiluminescence behavior and potential for applications.

* Corresponding Author's Email: Dr.Stefan.Schramm@gmail.com, Stefan.Schramm@uni-jena.de, Stefan.Schramm@nyu.edu.

"Not everything that is visible depends upon light for its visibility. This is only true of the proper color of things. Some objects of sight which in light are invisible, in darkness stimulate the sense; that is, things that appear fiery or shining. This class of objects has no simple common name, but instances of it are fungi, flesh, heads, scales, and eyes of fish. In none of these is what is seen their own proper color. Why we see these at all is another question."

Aristotle – De anima, Book 2, D6r

1. INTRODUCTION

Hardly any person does not marvel at the sight of a shiny and flashing glow of a firefly. Bioluminescence and chemiluminescence are truly spectacular phenomena and are widely recognized in society. Everyone, who once had a chance to observe the combination of two colorless liquids in a dark room during an experimental chemistry lecture immediately illuminating the entire auditorium, will deeply remember this moment. Moreover, chemiluminescence is well known to us in everyday life in the form of glow sticks, which are used in emergency rescues on sea as well as for entertainment purposes.

A chemical reaction that results in the emission of light is defined as chemiluminescence. When this reaction occurs in a biological organism and in most cases catalyzed by an enzyme called luciferase it is commonly referred as bioluminescence. The best-known example for a bioluminescent organism is certainly the firefly, which numerous species can be found all around the globe and even in such remote locations with low temperatures like Tibet [1]. Nevertheless, a multiplicity of other organisms are also bioluminescent. According to a recent study, more than 3500 species show this phenomenon [2]. Among them are some worms, many marine organisms such as jellyfish, crustaceans, mollusks, as well as fungi and even bacteria. Nature has produced a particularly wide variety of organisms and biochemical systems that are able to emit light. Today, it is commonly accepted that bioluminescence has developed more than 40 times independently within evolution [3] a clear testimony to the evolutionary advantage of this luminous phenomenon.

A variety of photochemical reactions in living organisms are known, in which light is converted into chemical energy such as the photochemical processes in the retina of our eyes or the production of vitamin D [4]. Bioluminescence and chemiluminescence, on the other hand, are unique processes in which the energy that stored with chemical bonds leads to the emission of "cold" light (in comparison to light that is emitted during combustion reaction). Thus, they represent a kind of reversal of the before mentioned photochemical reactions.

2. UNDERSTANDING THE GLOWING OF FIREFLIES

Historically, one of the first explored bioluminescent systems was the firefly. Already in the 1950s, authors started to investigate its luminescence mechanism [5, 6]. In the firefly, the oxidation of a molecule called firefly luciferin (**1**) to oxyluciferin (**2**) takes place in an enzymatically catalyzed process. The enzyme involved herein is referred to as firefly luciferase - a kinase that activates the luciferin (**1**) by phosphorylation in advance of the actual light emitting process. This activation allows the oxidation of the resulting conjugate (**3**) by triplet oxygen forming a 1,2-dioxetanone (**4**) as the high-energy intermediate. 1,2-Dioxetanones are highly strained four-ring heterocycles containing two oxygen atoms in the form of a peroxy group. Thus, compound (**4**) is thermally highly unstable and readily decomposes with CO_2 cleavage. This leads to the formation of the oxyluciferin (**2**)* in its excited state. During its relaxation to the electronic ground state, it emits light, of the wavelength that corresponds to the energy difference of the two states (Figure 1) [7].

Although such mechanistic studies started very early, many questions regarding the firefly mechanism but also in the in the field of bioluminescence in general remain unanswered to the present day. These include but are not limited to the exact color coding mechanism of different organisms that utilize the same light emitting molecule (firefly luciferin / oxyluciferin), the tautomerization and protonation state of the oxyluciferin during the light emission and also the addition of oxygen to the phosphorylated luciferin (**3**) forming the peroxy anion species (**5**).

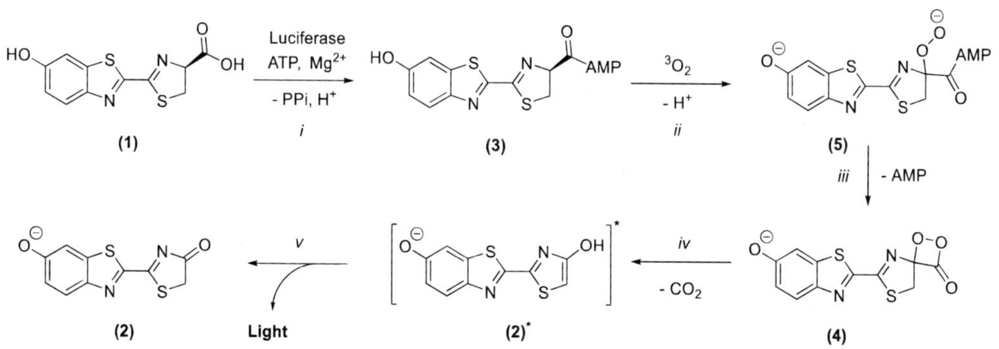

Figure 1. Chemical mechanism of firefly bioluminescence, i) Phosphorylation of firefly luciferin in the active site of the luciferase with the help of ATP and Magnesium ions, ii) Single electron transfer with molecular oxygen and formation of the peroxy species (**5**), iii) nucleophilic attack of the peroxygroup on the alpha carbon and cleavage of AMP leading to the formation of the dioxetanone (**4**), iv) decomposition of the high energy intermediate under the cleavage of CO2 resulting in the oxyluciferin in its excited state (**2**)*, v) light emission from the oxyluciferin during its relaxation into the ground state. Whether the oxyluciferin is emitting light from a keto or an enol form is still questionable.

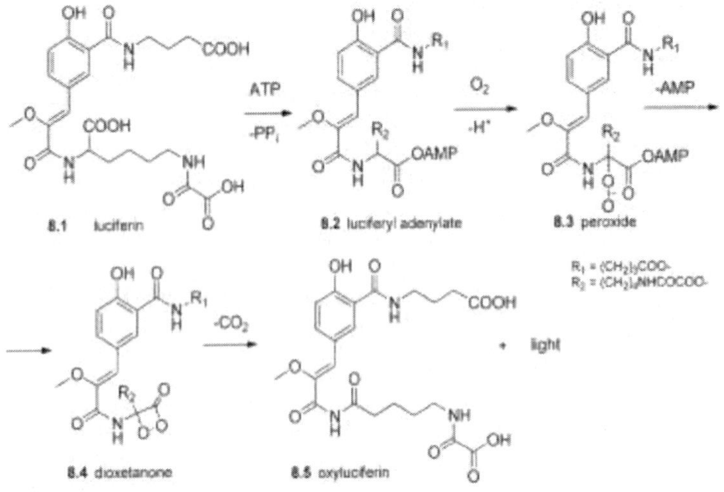

Figure 2. Fridericia heliota luciferin and their proposed bioluminescence mechanism. Reprinted with permission of the Royal Society of Chemistry from reference [7].

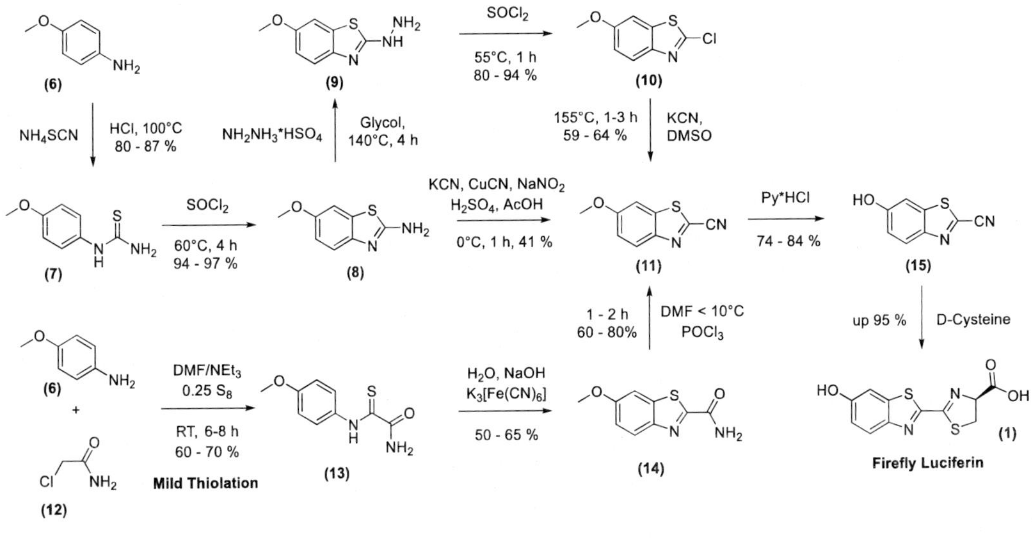

Figure 3. Different synthetic approaches for the synthesis of firefly luciferin (**1**). The key intermediate in all routes is -hydroxybenzothiazole-2-carbonitrile (**15**) or respectively its methyl ether (**11**).

Newly discovered luciferins and emitters of other living organisms such as those of the Siberian earthworm *Fridericia heliota* or the fungus luciferin, firstly discovered in 2015, make bioluminescence research a current and intensively investigated research area with many unanswered questions [7].

There are several ways for the synthesis of firefly luciferin described in recent literature [8-11]. (Figure 3) All of them bear the key intermediate 6-hydroxybenzothiazole-2-carbonitrile (15) or respectively its methyl ether (11). Starting from compound (15) firefly luciferin (1) can be synthesized with almost quantitative yield via the condensation with D-cysteine in a phosphate buffer [12]. Therefore, an efficient firefly luciferin synthesis has to aim for an efficient and also convenient way of producing 6-hydroxybenzothiazole-2-carbonitrile or its derivatives. The most systematic route starts from p-anisidine (6) by its reaction with ammonium thiocyanate forming the thiourea derivative (7) [8]. After this, the benzothiazole moiety of the molecule can be build up via a cyclisation reaction that is induced by the treatment of compound (7) with thionyl chloride. The amine (8) can be directly transformed into 6-methoxybenzothiazole-2-carbonitrile (11) via a Sandmeyer reaction [9, 13]. But this yields a compound (11) with a fairly low yield. Therefore, compound (8) can be reacted to its corresponding hydrazide (9) and reacted with thionyl chloride to for the corresponding chloride (10). The reaction of compound (10) with potassium cyanide in DMSO leads to 6-methoxybenzothiazole-2-carbonitrile (11) in moderate to good yield. Alternative approach to this complex procedure was developed by Beckert and co-workers by applying the method of the mild thiolation [11, 10]. In this approach, p-anisidine (6) is reacted with chloro acetamide (12) and sulfur yielding molecule (13). This intermediate can be conveniently oxidized with potassium hexacyanoferrate(III) to the amide (14). The treatment of compound (14) with phosphor oxychloride results 6-methoxybenzothiazole-2-carbonitrile (11). Lastly, the ether bond of compound (11) must be cleaved with pyidinium hydrochloride to form 6-hydroxybenzothiazole-2-carbonitrile (15) that can be used to produce firefly luciferin.

Considering the complex molecular structure of firefly luciferin (1), it is easy to see that such difficult and costly manufacturing process (Figure 3) makes its exploration, and the exploration of other luciferins, which tend to be even more complex, very difficult. A rather specific substrate structure of the luciferins in evolutionary adaptation to the luciferases is a further barrier for a fundamental chemical modification of the system and thus far-reaching research on novel bioluminescence substrates. This limits synthetic modifications to small variation of the luciferin structure like the introduction of amino group on the benzothiazole moiety.

Aminoluciferins (Figure 4) that can be deduced by such methodology found broad interest due to their light emission in the red to near infrared region of the electromagnetic spectrum [14-16]. Since the amount of scattering in tissues is anti-proportional to the wavelength of light emission, these molecules found useful application as agents for deep tissue imaging. A recent study by Schramm et al. gave a very interesting insight into the solid-state photochemistry of N,N-Dimethylaminoluciferin (17) [17]. When crystals of this compound are exposed to blue LED light, they crack parallel to the orientation of the carboxyl groups in the crystal and release carbon dioxide (Figure 5).

This phenomenon can be visually experienced as "jumping" of the crystals when they are put under blue light. While it resembles the motility of thermosalient [18] and photosalient crystals, [19, 20] it cannot be classified as such since during this photo-induced decarboxylation the crystal loses its entire crystallinity and turns into an amorphous agglomerate. Besides these impressive photomechanical effects, the fluorescence of the material is also drastically increased by a factor of 81 during this reaction (Figure 5 F-M).

This can be explained by the formation of the reaction product: decarboxy-N,N-dimethylaminoluciferin (20) (Figure 6).

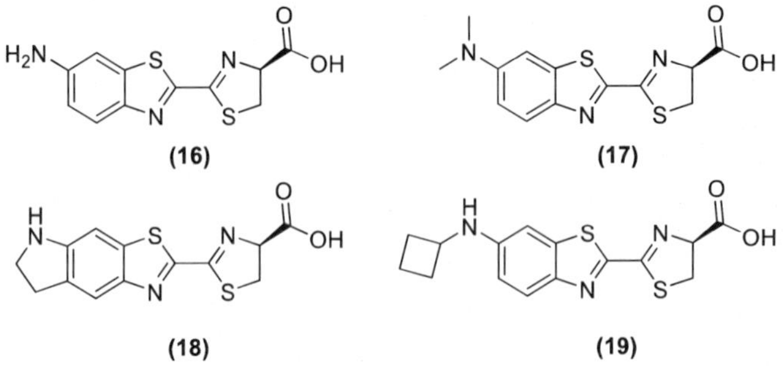

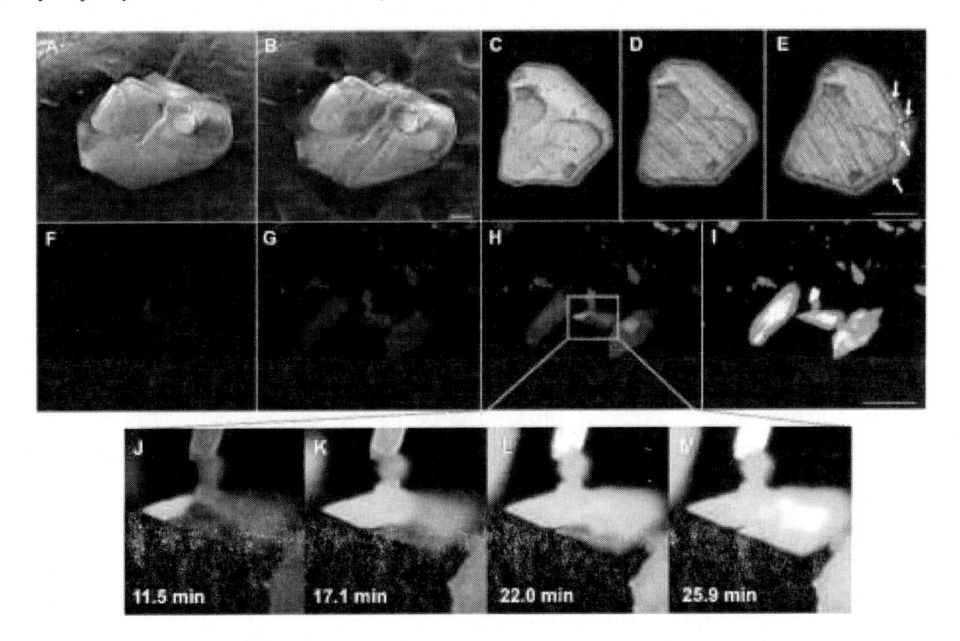

Figure 4. Aminoluciferin (16) and some of ist commonly used derivatives N,N-Dimethylaminoluciferin (17), Cyclopentylaminluciferin (18) and cybLuc (19).

Figure 5. N,N-Dimethylaminoluciferin (17) and its solid state photochemistry. When crystals of it are exposed to blue LED light, they show strongly disintegrative effects ("jumping") and release carbon dioxide. During this photo-induced solid-state decarboxylation, the solid-state fluorescence of the reactions product strongly increases. Reprinted with permission of Wiley-VCH from [17].

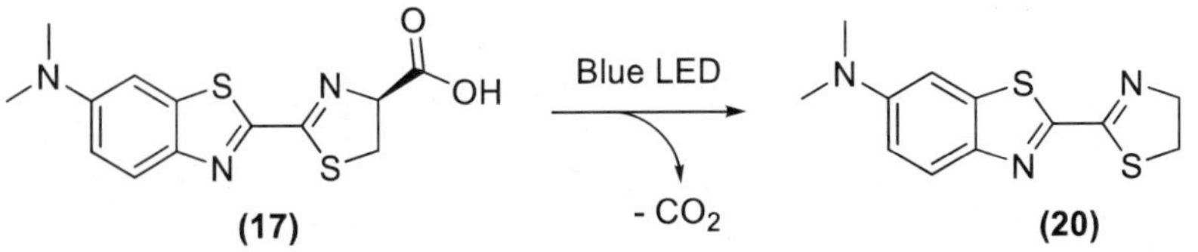

Figure 6. The formation of decarboxy-N,N-dimethylaminoluciferin (20) during the photo-induced solid-state decarboxylation of N,N-dimethylaminoluciferin (17).

Decarboxy-N,N-dimethylaminoluciferin (20) is strongly fluorescent in the solid-state due to the efficient stacking of the π-conjugated system. This results in bright fluorescence emission from an excimer and a monomer species of the molecule.

In summary, while firefly luciferin and its derivatives found important bio-technological applications, their synthesis remains lengthy and complicated. Thus, for the individual chemist, it is both difficult to prepare and expensive to buy this compound, which, therefore, limits its applicational and basic research potential. A novel luciferin may only be varied to a structurally narrow scale so that it still fits into the active site of the catalyzing enzyme - luciferase.

3. 2-COUMARANONES AS A MODEL SYSTEM FOR FIREFLY BIOLUMINESCENCE

From these numerous problems, a list of requirements can be deducted for a suitable model system that should have a similar luminescence mechanism as established bioluminescence systems. Moreover, the luminescence of such a model system should not necessarily be coupled to an enzyme and it should be synthetically easily accessible. Ideally, its emission quantum yields should also be in a range that suits well for its use as an analytical reagent. In recent publications [2, 21-26] Schramm et al. were able to show that the chemiluminescent 2-coumaranones have favorable properties for this purpose and allow conclusions to be drawn about known bioluminescence reactions. 2-Coumaranones are a class of heterocyclic compounds structurally derived from benzofuran-2 (3H) -one. They can be synthesized in a wide structural range using the highly efficient one-pot Tscherniac-Einhorn 3-Component Reaction (TsE-3CR) from commercially available chemicals (Figure 7) [21, 22, 26].

Glyoxylic acid monohydrate (21) is combined with a para substituted phenol (23) derivative and an amide or carbamate (22) in a mixture of acetic acid and sulfuric acid. This reaction mixture is typically stirred at room temperature for 48 – 72 hours. During this period of time the pure reaction product either precipitates and has to be filtered and washed with water or has to be precipitated after completion of the reaction by pouring the reaction mixture into water. With this easy and convenient method, a wide variety of 2-coumaranones (25 - 40) can be synthesized in moderate to good yields. (Figure 8) [2, 21, 22, 24, 25, 26, 27, 28, 29, 30, 31].

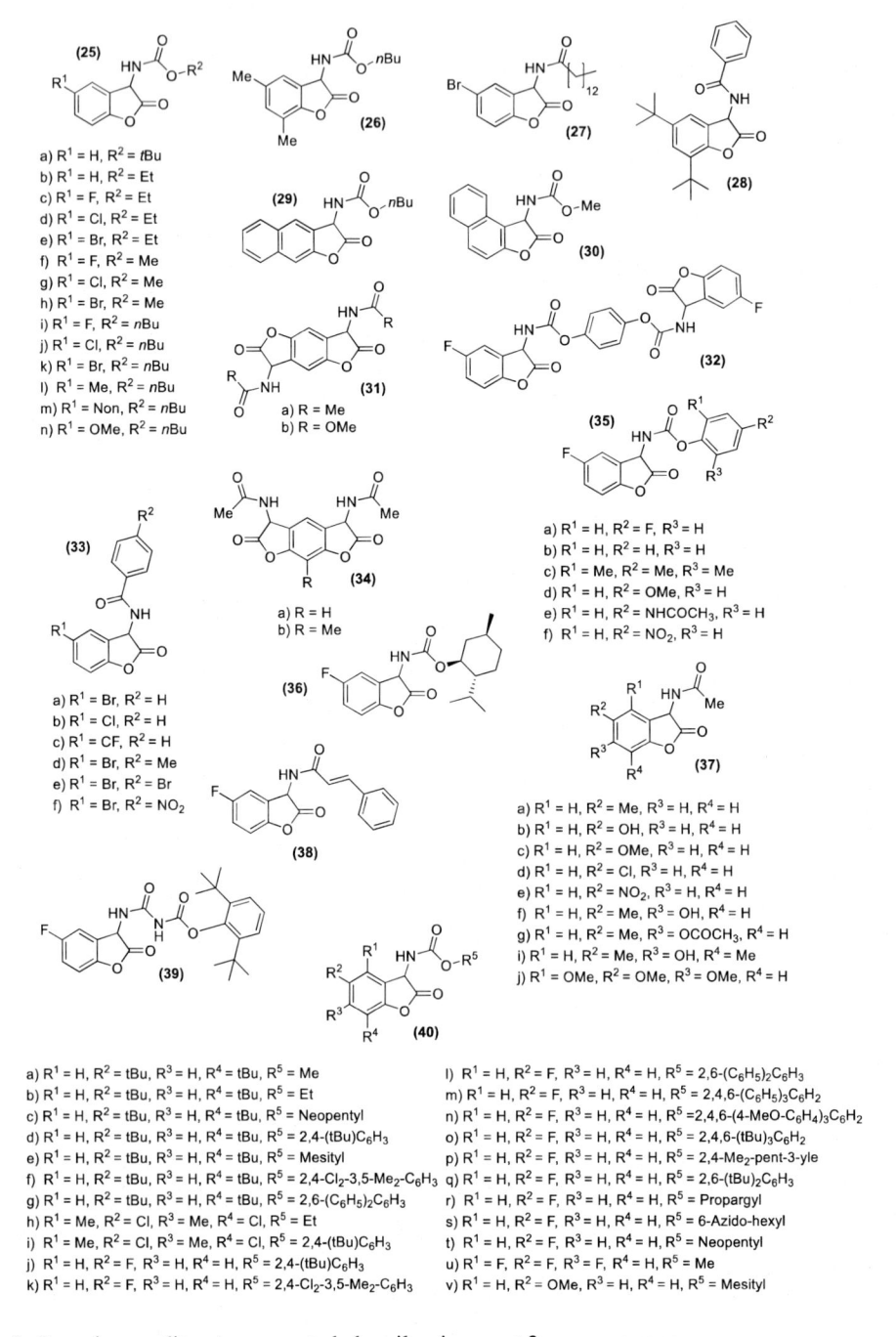

Figure 7. Synthesis of chemiluminescent 2-coumaranones via a one-pot Tscherniac-Einhorn 3-Component Reaction (TsE-3CR).

Figure 8. Overview on literature reported chemiluminescent 2-coumaranones.

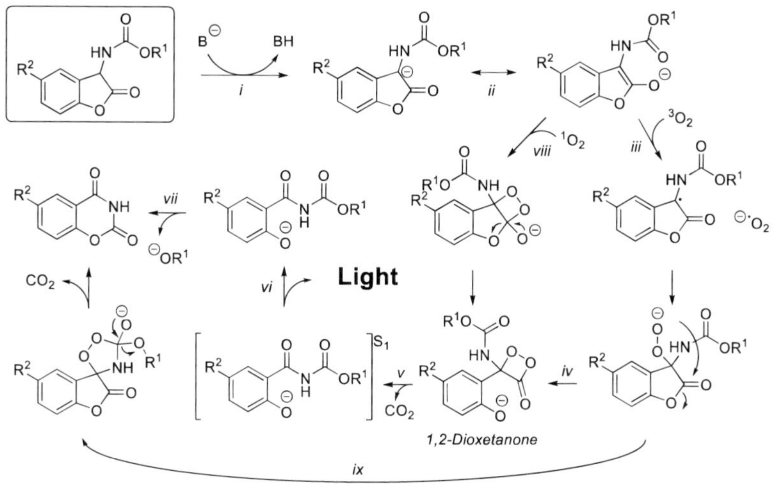

Figure 9. Chemiluminescence mechanism of the 2-coumaranones: i) Deprotonation with a base in a polar aprotic solvent, ii) Stabilization as an electron rich enolate, iii) Single electron transfer with triplet oxygen leading to the formation of an peroxy anion, iv) cyclisation oft the peroxy anion resulting in a 1,2-dioxetanone, v) Decomposition of the 1,2-dioxetanone resulting in an open-chain salicylamide like emitter structure in its excited state, vi) relaxation of the emitter to the ground state involving the release of light vii) its cyclisation to the final product of the reaction, viii) direct reaction with singlet oxygen leading directly to the 1,2-dioxetanone, ix) parallel side reaction without the emission of light that results in the same final reaction product as in the case of the chemiluminescence reaction.

When added to a base in polar aprotic solvents 2-coumaranones show intense chemiluminescence, which is in many cases even visible to the naked eye at daylight. The emission color strongly depends on the chosen phenol and amide/carbamate component. While the combination of medium deactivated phenols with carbamates (e.g., 25b-h) typically results in very bright blue emission, the color of emission changes from blue for alkyl amides (e.g., 27) to green for benzamide (e.g., 33a) and orange/red for p-bromo benzamide (e.g., 33e). Generally, the chemiluminescence quantum yield for carbamate derivatives is larger than those for amides. The decay kinetics of the chemiluminescence reaction can be tuned into either a glow kinetic regime by choosing sterically hindered carbamates in combination with medium deactivated phenol components (e.g., 40j, l, m) or flash kinetics by using medium activated phenol components with small carbamates (e.g., 40a, b).

The chemiluminescence mechanism was recently studied in detail by Schramm et. al. [32, 24, 23] These studies indicate the following reaction sequence. A 2-coumaranone is deprotonated by a base in a polar aprotic solvent on the alpha carbon of its lactone (*i*). The newly formed deprotonated species is stabilized as an electron rich enolate (*ii*). A single electron transfer with triplet oxygen induces the production of super oxide radical anions and leads to the formation of a peroxy anion (*iii*). The cyclisation of the peroxy anion leads to the generation of a 1,2-dioxetanone as a high energy intermediate (HEI) (*iv*). This 1,2-dioxetanone decomposes under the cleavage of CO_2 and produces the emitter of the chemiluminescence reaction, an acylated open-chain salicylamide derivative in its excited state (*v*). Its relaxation to the grounds state leads to the emission of light (*vii*). The primary reaction product is highly unstable under strongly alkaline conditions and reacts to the secondary reaction product, which can be isolated (*vii*). The direct reaction of the electron rich enolate with singlet oxygen can also lead to the formation of the HEI (*viii*). A parallel side reaction without the emission of light that results in the same final reaction product as in the case of the chemiluminescence reaction lowers the quantum yield in case of the usage of small carbamates (*ix*).

Comparison of the chemiluminescence mechanism of the 2-coumaranone chemiluminescence with the firefly bioluminescence reveals a multitude of interesting similarities [32]. It is particularly noteworthy, that in both systems the reactive intermediates are 1,2-dioxetanones (**7**) (*iv*) [23]. If these compounds decompose thermally (*v*) with the elimination of CO_2, the resulting molecule is in both cases electronically excited. In the case of the 2-coumaranones, the emitter of luminescence is an acylated salicylamide derivative,[24] which represents an efficient fluorophore due to a postulated proton transfer in the excited state (ESIPT); while in the firefly system the emitter is suspected to be a deprotonated form of the firefly oxyluciferin [33-36]. In the historical context, the addition of biradical oxygen (out of the air) represented a difficult step in the understanding of the chemical mechanism of firefly bioluminescence. To explain this, various processes have been postulated in the literature, some of which involve radical as well as ionic intermediates. Moreover, the necessary presence of an additional molecule acting as a single-electron acceptor was postulated, which should allow the spin-forbidden reaction. However, recent studies by Branchini et al. [37] show, that this single-electron acceptor is oxygen itself. It reacts in the course of a single electron transfer (SET) reaction with the phosphorylated firefly luciferin (**3**). This leads to the formation of superoxide radical anions, which in turn react with the intermediately present radical molecule forming the peroxy species (**5**). Using modern electron spin resonance spectroscopy (ESR) techniques and spin trapping reagents, the same reaction was also observed for the 2-coumaranones (*iii*) [25]. However, when singlet oxygen (*viii*) is used for the chemiluminescent reaction, the blocking of unwanted side reaction pathways (*xi*) also makes it possible to increase the chemical yield of the emitting species. Thus, the chemiluminescence quantum yields of 2-coumaranone reach values that exceed those of established systems such as luminol by almost one magnitude. This enhances the suitability of the 2-coumaranone for the use as an analytical detection reagent e.g., in ELISAs. A higher quantum yield contributes to a much better detection limit in such essays.

Recent studies show that the bioluminescence reactions of many different organisms, such as the above-mentioned Siberian earthworm *Fridericia heliota* or of fungi [38] might be oxidative decarboxylations, which might also proceed via a 1,2-dioxetanone as a reactive intermediate. In other living organisms, such as bioluminescent jellyfish of the species *Aequorea victoria*, this has been demonstrated earlier [7] and this phenomenon is also suspected in the New Zealand glowworm (*Arachnocampa luminosa*) [39] Thus, it appears to be a general mechanism of light generation in nature. The use of the chemiluminescent 2-coumaranone makes an important contribution to its investigation.

4. THE CRYSTALLINE FIREFLIES

2-Coumaranones offer a convenient way of producing *in situ* 1,2-dioxetanones, enabling the studies of these highly reactive compounds in solution without the need of bulk purification. This reaction is of interest since 1,2-dioxetanones are known to be explosive in pure solid state and should, therefore, be handled with extreme care. As a matter of fact, only a handful of 1,2-dioxetanones in their pure form have been prepared in the past [40]. In some reports the accidental explosive decomposition of similar peroxy compounds during their synthesis, indicated by a characteristic hissing sound was described [41]. It should serve as an indicator for giving the immediate signal of evacuating the laboratory in proximity to the reaction mixture. Stabilizing substituents such as 1,2-dioxetanes (41), [42] hydroperoxides (45), [43] endo-peroxides (44) [44] and other peroxy-species that are common HEIs in many chemiluminescence and bioluminescence reactions are mostly less labile than 1,2-dioxetanones, but they are known to be fairly unstable over prolonged periods of time at room temperature. The HEI of the peroxy-oxalate chemiluminescence, 1,2-dioxetandione (43) [45-50] has not be isolated in pure form so far.

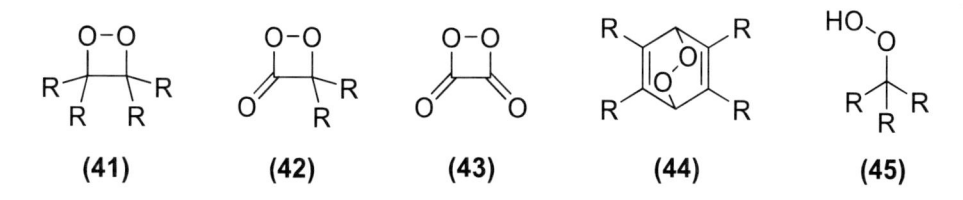

(41) (42) (43) (44) (45)

Figure 10. The formulas of some common high energy intermediates (HEI) of chemiluminescent and bioluminescent reaction. 1,2-dioxetanes (41), 1,2-dioxetanones (42), 1,2-dioxetandione (43), endo-peroxides (44), hydro-peroxides (45).

Therefore, chemiluminescence studies of these molecules focused mostly to the solution state and chemiluminescence reactions in the pure crystalline solid state were not reported until recently.

By incorporating 1,2-dioxetane moieties into polymers, Sijbesma and co-workers [51, 52] were able to produce an organic material that emits light when mechanically stressed. It is argued, that when mechanical force is applied to this polymeric material, the 1,2-dioxetane is cleaved, resulting in exited state polymer molecules that release their energy in form of light. (Figure 11) The color of light emission can be changed thought the visible spectrum by doping the polymer with sensitizer dyes like diphenyl-anthracene or perylene-bisimides.

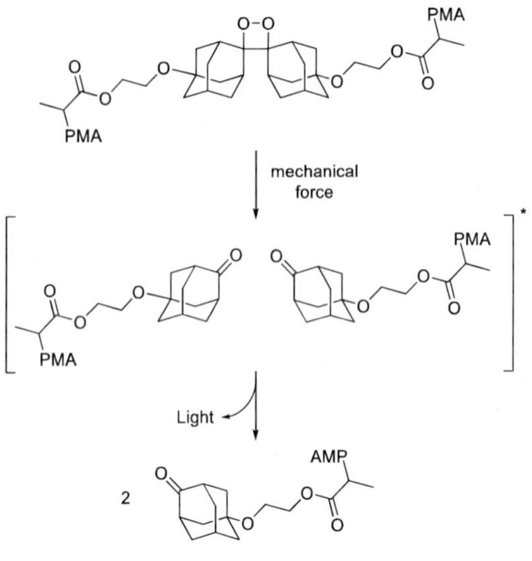

Figure 11. The mechanically induced chemiluminescence of polymers by incorporating a 1,2-dioxetane unit in their main chain.

Another way of triggering a chemiluminescence reaction in the solid state is its induction by heat. A HEI that is stable at room temperature is heated until its decomposition point. Its thermolysis results in the generation of excited state molecules and thus can lead to the release of light. Such reactions are commonly referred to as thermochemilumine-scence reactions.

Roda and co-workers applied this principle to develop thermochemiluminescence-based probes, assays and analytical methods. Such methods are especially attractive due to their easy way of reagent less reaction induction and the ultra-sensitive light detection capabilities of simple analytical instruments. In fact, a camera module of modern smartphones offers sufficient sensitivity for detecting small analytical traces [53, 54]. It has been reported that astronauts in space have applied successfully these assays and methods to monitor their health and to detect life markers in extraterrestrial environments [55-57]. Roda et al. found that acridinium-1,2-dioxetanes suit well the applicational need for themochemiluminescent probes [58]. By binding these 1,2-dioxetanes to doped silica nanoparticles or polymer dots the emission characteristics can be tuned to the desired criteria [59, 60].

By applying the principles of thermochemiluminescence, Schramm et al. were recently able to show that the concepts that have been preciously used to study solution chemiluminescence can also be transferred to the pure solid state of macroscopic crystals from organic peroxides [61]. For this, the authors explored the solid-state decomposition behavior of lophine-hydroperoxide (47) crystals. Lophine-hydroperoxide is a derivative of Lophine (46) (*2,4,5-triphenyl-1H-imidazole*), the first ever reported chemiluminescent organic molecule. Found in 1877 by the Polish-German scientist Radziszewski [62], lophine can undergo base induced decomposition in a variety of solvent accompanied by the release of light. When lophine is reacted with singlet oxygen it forms within a Schenk-Ene reaction lophine-hydroperoxide (Figure 12).

Lophine-hydroperoxide can be crystallized from a variety of solvents like ethanol or carbondisulfide in large block shaped crystals up to at least 15 mm in size. While purified solid samples of lophine-hydroperoxide appear to be stable at room temperature over several months, and thermo-gravimetric analysis indicate no long-term decomposition below 50°C, a storage below room temperature is advised as measure of precaution, a common practice for the handling of peroxides.

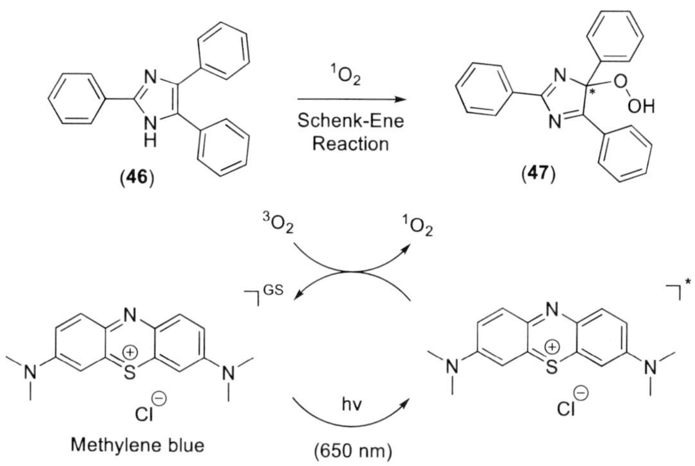

Figure 12. Synthesis of lophine-hydroperoxide (47) via the reaction of singlet oxygen with lophine (46). The singlet oxygen can be conveniently generated in situ by a photoreaction of methylene blue with triplet oxygen that is dissolved in the reaction solvent.

Figure 13. Macroscopic crystal of lophine-hydroperoxide about 15 mm in size and 700 mg in weight.

It was noticed early in the literature that lophine-hydroperoxide is thermochemilumine-scent [63, 64]. However, this reaction was not studied in macroscopic crystalline samples. When a crystal of lophine-hydroperoxide is heated above its decomposition, temperature (ca. 115°C) it emits light with an emission maximum around 530 nm. The solid-state kinetics of this reaction can be followed by applying powder x-ray diffraction and synchrotron µIR spectroscopy. These techniques indicate that lophine-hydroperoxide thermally decomposes in the solid state almost exclusively to lophine. This agrees with the observation that heating a crystal of lophine-hydroperoxide in oil results in the formation oxygen gas, and the bubbles indicate the cleavage of the hydroperoxide moiety of the molecule. Lophine was also chemically isolated and characterized as the main reaction product of this decomposition process.

Since crystals are intrinsically anisotropic objects it is possible to follow the progression of light through the crystal during the thermochemiluminescence reaction by applying low light microscopic techniques (Figure 14). [65] For this experiment, a crystal of lophine-hydroperoxide is placed on a heating stage below a microscopic setup capable of detecting even small amounts of light. After the initiation of the thermochemilumine-scence reaction by heating the crystal to about 120°C light can be observed, starting from the side of crystal, which is faced towards the heating plate. The front of light emission is progressing in the curse of the reaction thought the crystal to its top.

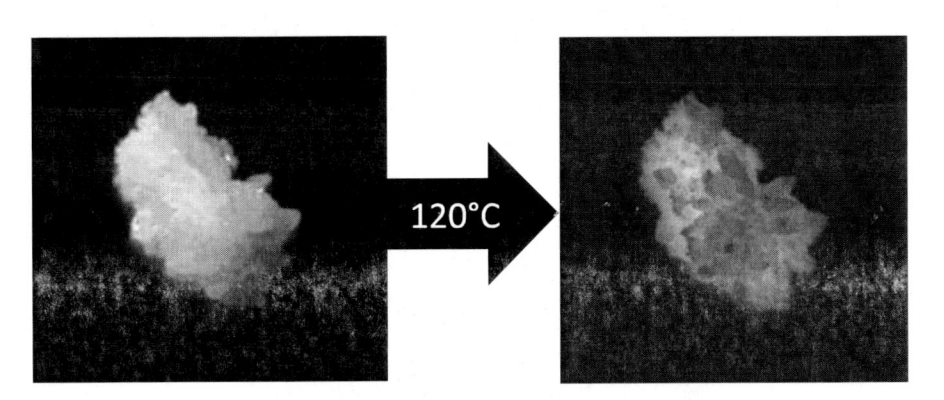

Figure 14. Thermochemiluminescence of a crystal of lophine-hydroperoxide during its thermal decomposition at 120°C, Left: A crystal at daylight, Right: Thermochemiluminescence image of the same crystal heated to 120°C in the dark.

The solid-state thermochemiluminescence reaction decays kinetically within the orders of several minutes at 115°C to few seconds at 160°C. Activation parameters derived from isothermal thermochemiluminescence kinetics at different temperatures within this interval are all higher than the corresponding values for the thermochemiluminescence reaction in polar and nonpolar solvents. This observation can be rationalized considering the lattice energy of the crystal that has to be overcome in order to initiate its decomposition. Additionally, the highly negative activation entropy points towards a disorganized transition state in case of the solid-state reaction. The absolute thermochemiluminescence quantum yield lies with $(2.1 \pm 0.1) \times 10^{-7}$ E mol^{-1} ($n = 10$) close to the literature reported value for the solution reaction.

The photo-physics of the main decomposition product lophine can be exploited in order to explain the observed solid-state thermochemiluminescence emission spectrum and to propose a candidate structure for the emitter of this reaction. While the fluorescence and phosphorescence spectra of a pure solid lophine sample do not fit the experimentally observed solid-state chemiluminescence spectrum very well, a close similarity of the phosphorescence spectrum of deprotonated lophine to the solid-state thermochemilumine-scence spectrum is striking. Hence, a deprotonated form of lophine is proposed to be the emitter of the thermochemiluminescence reaction.

From the point of understanding the solid-state thermochemiluminescence behavior of lophine-hydroperoxide, one might ask the question if this solid-state phenomenon is only limited to lophine-hydroperoxide and its derivatives or if it can also be found in a broad variety of different structures. In order find an answer to this, the author prepared crystalline samples of bis-adamantyl-1,2-dioxatane, rubrene-endoperoxide and dibenzoyl-peroxide. All these peroxides were found to be rather stable at room temperature. No explosive decomposition was encountered. But when crystals of these peroxides where heated to elevated temperature (110°C–160°C) they all gave rise to thermochemiluminescence. The color of light emission was characteristic for the used peroxide and ranged from blue for bis-adamantyl-1,2-dioxatane, to green for dibenzoyl-peroxide and red for rubrene-endoperoxide.

These preliminary results may serve as an indication that solid state chemilumine-scence is not a phenomenon only limited to a very small number of molecules, but it may be common for a wide variety of peroxides. Some of the tested peroxides, like rubrene endoperoxide, are commonly known as photo-oxidation products that can also be found in aged dye-sensitized solar cells. By heating these aged devices, one might find a way of restoring their energy harvesting capabilities and at the same time to generate energy in the form of light. Hence, the solid state thermochemiluminescence might also be used as a form of solar energy storage. Moreover, these sensitive luminometers could be used to quantify the content of dibenzoyl-peroxide in a given solid sample by simple heating. Given the fact that dibenzoyl-peroxide is used each year on a multi-hundred-ton scale in polymer industry as a polymerization initiator, its easy quantification by the application of solid-state thermochemiluminescence methods might introduce an enormous potential for financial savings and a reduction of chemical waste that could help transform chemical industry into a business model harmonizing with the aim of a sustainable economy. Finally, one might see thermochemiluminescent peroxide crystals also as a philosophical metaphor, learning from nature and transforming one of its most splendid creations into a technological product – the crystalline fireflies.

REFERENCES

[1]   Mitani, Y., Futahashi, R., Liu, Z., Liang, X. and Ohmiya, Y. (2017). Tibetan Firefly Luciferase with Low Temperature Adaptation. *Photochem. Photobiol.* 93: 466-472.

[2]   Vacher, M., Galván, I. F., Ding, B. W., Schramm, S., Berraud-Pache, R., Naumov, P., Ferré, N., Liu, Y. J., Navizet, I., Roca-Sanjuán, D., Baader, W. J., and Lindh, R. (2018). Chemi- and Bioluminescence of Cyclic Peroxides. *Chem. Rev.* 118: 6927-6974.

[3]   Haddock, S. H. D., Moline, M. A., and Case, J. A. (2010). Bioluminescence in the Sea. *Annu. Rev. Mar. Sic.* 2: 443-493.

[4]   Nair, R., and Maseeh, A. (2012). Vitamin D: The "sunshine" vitamin. *J. Pharmacol. Pharmacother.* 3: 118-126.

[5]   Seliger, H. H., and McElroy, W. D. (1959). Quantum yield in the oxidation of firefly luciferin. *Biochem. Biophys. Res. Commun.* 1: 21-24.

[6]   White, E. H., McCapra, F., and Field, G. F. (1963). The Structure and Synthesis of Firefly Luciferin. *J. Am. Chem. Soc.* 85: 337-343.

[7]   Kaskova, Z. M., Tsarkova, A. S., and Yampolsky, I. V. (2016). 1001 lights: luciferins, luciferases, their mechanisms of action and applications in chemical analysis, biology and medicine. *Chem. Soc. Rev.* 45: 6048-6077.

[8]   Ouyang, L., Huang, Y., Zhao, Y., He, G., Xie, Y., Liu, J., He, J., Liu, B., and Wei, Y. (2012). Preparation, antibacterial evaluation and preliminary structure-activity relationship (SAR) study of benzothiazol- and benzoxazol-2-amine derivatives. *Bioorg. Med. Chem. Lett.* 22: 3044-3049

[9]   Yoshiaki, T., Masaharu, T., Hisao, N., Nobutaka, S., Minoru, I., and Toshio. G. (1992). A Convenient Synthetic Method of 2-Cyano-6-methoxybenzothiazole, –A Key Intermediate for the Synthesis of Firefly Luciferin. *Bull. Chem. Soc. Jpn.* 65: 392-395.

[10] Würfel, H., Weiß, D., and Beckert, R. (2012). The mild thiolation: a useful tool for the synthesis of activated building blocks. *J. Sulfur. Chem.* 33: 619-638.

[11] Würfel, H., Weiss, D., Beckert, R., and Güther, A. (2012). A new application of the "mild thiolation" concept for an efficient three-step synthesis of 2-cyanobenzothiazoles: a new approach to Firefly-luciferin precursors. *J. Sulfur Chem.* 33: 9-16.

[12] Meroni, G., Ciana, P., Maggi, A., and Santaniello, E. (2009). A New Synthesis of 2-Cyano-6-hydroxybenzothiazole, the Key Intermediate of d-Luciferin, Starting from 1,4-Benzoquinone. *Synlett* 2009: 2682-2684.

[13] Suzuki, N., Nomoto, T., Toya, Y., Kanamori, N., Yoda, B., and Saeki, A. (1993). Synthetic Reactions in PEG: PEG-Assisted Synthesis of 2-Cyano-6-methoxybenzothiazole, A Key Intermediate for the Synthesis of Firefly Luciferin. *Biosci. Biotechnol. Biochem.* 57: 1561-1562.

[14] Hickson, J., Ackler, S., Klaubert, D., Bouska, J., Ellis, P., Foster, K., Oleksijew, A., Rodriguez, L., Schlessinger, S., Wang, B., and Frost. D. (2010). Noninvasive molecular imaging of apoptosis in vivo using a modified firefly luciferase substrate, Z-DEVD-aminoluciferin. *Cell Death Differ.* 17: 1003.

[15] Wu, W., Su, J., Tang, C., Bai, H., Ma, Z., Zhang, T., Yuan, Z., Li, Z., Zhou, W., Zhang, H., Liu, Z., Wang, Y., Zhou, Y., Du, L., Gu, L., and Li, M. (2017). cybLuc: An Effective Aminoluciferin Derivative for Deep Bioluminescence Imaging. *Anal. Chem.* 89: 4808-4816.

[16] Sun, Y. Q., Liu, J., Wang, P., Zhang, J., and Guo, W. (2012). D-luciferin analogues: a multicolor toolbox for bioluminescence imaging. *Angew. Chem. Int. Ed.* 51: 8428-8430.

[17] Schramm, S., Karothu, D. P., Raj, G., Laptenok, S. P., Solntsev, K. and Naumov, P. (2018). Turning on Solid-State Fluorescence with Light. *Angew. Chem. Int. Ed.* 57: 9538-9542.

[18] Skoko, Ž., Zamir, S., Naumov, P., and Bernstein, J. (2010). The Thermosalient Phenomenon. "Jumping Crystals" and Crystal Chemistry of the Anticholinergic Agent Oxitropium Bromide. *J. Am. Chem. Soc.* 132: 14191-14202.

[19] Medishetty, R., Sahoo, S. C., Mulijanto, C. E., Naumov, P., and Vittal, J. J. (2015). Photosalient Behavior of Photoreactive Crystals. *Chem. Mater.* 27: 1821-1829.

[20] Naumov, P., Chizhik, S., Panda, M. K., Nath, N. K., and Boldyreva, E. (2015). Mechanically Responsive Molecular Crystals. *Chem. Rev.* 115: 12440-12490.

[21] Schramm, S., Weiss, D., Navizet, I., Roca-Sanjuan, D., Beckert, R., and Görls, H. (2013). Investigations on the synthesis and chemiluminescence of novel 2-coumaranones. *ARKIVOC* 3: 174-188.

[22] Schramm, S., Ciscato, L. F. M. L., Oesau, P., Krieg, R., Richter, J. F., Navizet, I., Roca-Sanjuan, D., Weiss, D., and Beckert, R. (2015). Investigations on the synthesis and chemiluminescence of novel 2-coumaranones - II. *ARKIVOC* 5: 44-59.

[23] Ciscato, L. F. M. L., Bartoloni, F. H., Colavite, A. S., Weiss, D., Beckert, R., and Schramm, S. (2014). Evidence supporting a 1,2-dioxetanone as an intermediate in the benzofuran-2(3H)-one chemiluminescence. *Photochem. Photobiol. Sci.* 13: 32-37.

[24] Schramm, S., Navizet, I., Naumov, P., Nath, N. K., Berraud-Pache, R., Oesau, P., Weiss, D., and Beckert, R. (2016). The Light Emitter of the 2-Coumaranone Chemiluminescence: Theoretical and Experimental Elucidation of a Possible Model for Bioluminescent Systems. *Eur. J. Org. Chem.* 2016: 678-681.

[25] Schramm, S., Navizet, I., Karothu, D. P., Oesau, P., Bensmann, V., Weiss, D., Beckert, R., and Naumov, P. (2017). Mechanistic investigations of the 2-coumaranone chemiluminescence. *Phys. Chem. Chem. Phys.* 19: 22852-22859.

[26] Schramm, S. 2016. *Die Chemilumineszenz der 2-Coumaranone. Synthese, Lumineszenzmechanismus und Applikation.* [Chemiluminescent 3-coumaranones – synthesis, luminescence mechanism and applications] Berlin: Logos.

[27] Lofthouse, G. J., Suschitzky, H., Wakefield, B. J., Whittaker, R. A., and Tuck, B. (1979). Synthesis and chemiluminescent reactions of some 3-alkoxycarbamoyl benzo[b]furan-2(3H)-ones. *J. Chem. Soc. Perkin Trans. 1*: 1634-1639.

[28] Schramm, S., Weiss, D., and Beckert, R. (2018). Biolumineszenz - Das Leuchten der Natur verstehen. [Bioluminescence – Understanding the glowing of nature] *GIT Labor-Fachzeitschrift* 4: 27-29.

[29] Schramm, S., Weiss, D., and Beckert, R. (2017). Biophysikalische Chemie: Leuchten nach dem Vorbild der Natur. [Biophysical chemistry – Glowing with nature as a model] *Nachrichten aus der Chemie* 65: 132-134.

[30] Matuszczak, B. (1998). Linear and cyclic N-acyl-α-arylglycines. IV. Novel 3-Substituted 3-Acylaminobenzo[b]furan-2(3H)-ones: Synthesis and chemilumine-scence studies. *Journal für Praktische Chemie/Chemiker-Zeitung* 340: 20-25.

[31] Matuszczak, B. (1996). Linear and cyclic N-acetyl-α-aryl-glycines: Synthesis and chemiluminescence studies. *Monatshefte für Chemie / Chemical Monthly* 127: 1291-1303.

[32] Schramm, S., Navizet, I., Karothu, D. P., Oesau, P., Bensmann, V., Weiss, D., Beckert, R., and Naumov, P. (2017). Mechanistic investigations of the 2-coumaranone chemiluminescence. *Phys. Chem. Chem. Phys.* 19: 22852-22859.

[33] Lui, N. M., Schramm, S., and Naumov, P. (2018). pH-dependent Fluorescence from Firefly Oxyluciferin in Agarose Thin Films. *New J. Chem.* 43: 1122-1126.

[34] Solntsev, K. M., Laptenok, S. P., and Naumov, P. (2012). Photoinduced Dynamics of Oxyluciferin Analogues: Unusual Enol "Super"photoacidity and Evidence for Keto–Enol Isomerization. *J. Am. Chem. Soc.* 134: 16452-16455.

[35] Maltsev, O. V., Yue, L., Rebarz, M., Hintermann, L., Sliwa, M., Ruckebusch, C., Pejov, L., Liu, Y. J., and Naumov, P. (2014). Vibrational Spectra of Chemical and Isotopic Variants of Oxyluciferin, the Light Emitter of Firefly Bioluminescence. *Chem. Eur. J.* 20: 10782-10790.

[36] Rebarz, M., Kukovec, B. M., Maltsev, O. V., Ruckebusch, C., Hintermann, L., Naumov, P., and Sliwa, M. (2013). Deciphering the protonation and tautomeric equilibria of firefly oxyluciferin by molecular engineering and multivariate curve resolution. *Chem. Sci.* 4: 3803-3809.

[37] Branchini, B. R., Behney, C., Southworth, T. L., Fontaine, D. M., Gulick, A. M., Vinyard, D. J., and Brudvig, G. W. (2015). Experimental Support for a Single Electron-Transfer Oxidation Mechanism in Firefly Bioluminescence. *J. Am. Chem. Soc.* 137: 7592-7595.

[38] Purtov, K. V., Petushkov, V. N., Baranov, M., Mineev, K. S., Rodionova, N. S., Kaskova, Z. M., Tsarkova, A., Petunin, A. I., Bondar, V. S., Rodicheva, E. K.,

Medvedeva, S. E., Oba, Y., Oba, Y., Arseniev, A. S., Lukyanov, S., Gitelson, J. I., and Yampolsky, I. V. (2015). The Chemical Basis of Fungal Bioluminescence. *Angew. Chem. Int. Ed.* 54: 8124-8128.

[39] Watkins, O. C., Sharpe, M. L., Perry, N. B., and Krause, K. L. (2018). New Zealand glowworm (Arachnocampa luminosa) bioluminescence is produced by a firefly-like luciferase but an entirely new luciferin. *Sci. Rep.* 8: 3278.

[40] Bartoloni, F. H., Oliveira, M. A., Augusto, F. A., Ciscato, L. F. M. L., Bastos, E. L., and Baader, W.J. (2012). Synthesis of unstable cyclic peroxides for chemiluminescence studies. *J. Braz. Chem. Soc.* 23: 2093-2103.

[41] Kopecky, K. R. (1982). 3 - Synthesis of 1,2-Dioxetanes. In *Chemical and Biological Generation of Excited States*, edited by Waldemar Adam and Giuseppe Cilento, 85-114. San Diego: Academic Press.

[42] Meijer, E. W., and Wynberg, H. (1982). The synthesis and chemiluminescence of stable 1,2-dioxetane: An organic chemistry laboratory experiment. *J. Chem. Educ.* 59: 1071.

[43] Kimura, M., Lu, G. H., Nishigawa, H., Zhang, Z. Q., and Hu, Z. Z. (2007). Singlet oxygen generation from lophine hydroperoxides. *Luminescence* 22: 72-76.

[44] Wilson, T.. 1969. CHEMILUMINESCENCE FROM THE ENDOPEROXIDE OF 1,4-DIMETHOXY-9,10-DIPHENYLANTHRACENE. *Photochem. Photobiol.* 10: 441-444.

[45] Bos, R., Barnett, N. W., Dyson, G. A., Lim, K. F., Russell, R. A., and Watson, S. P. (2004). Studies on the mechanism of the peroxyoxalate chemiluminescence reaction: Part 1. Confirmation of 1,2-dioxetanedione as an intermediate using 13C nuclear magnetic resonance spectroscopy. *Anal. Chim. Acta* 502: 141-147.

[46] Stevani, C. V., Lima, D. F., Toscano, V., and Baader, W. J. (1996). Kinetic studies on the peroxyoxalate chemiluminescent reaction: imidazole as a nucleophilic catalyst. *J. Chem. Soc. Perkin Trans.* 2: 989-995.

[47] Silva, S. M., Wagner, K., Weiss, D., Beckert, R., Stevani, C. V., and Baader, W .J. (2002). Studies on the chemiexcitation step in peroxyoxalate chemiluminescence using steroid-substituted activators. *Luminescence* 17: 362-369.

[48] Stevani, C. V., and Baader, W. J. (1997). Kinetic studies on the chemiluminescent decomposition of an isolated intermediate in the peroxyoxalate reaction. *J. Phys. Org. Chem.* 10: 593-599.

[49] Stevani, C. V., Silva, S. M., and Baader, W. J. (2000). Studies on the Mechanism of the Excitation Step in Peroxyoxalate Chemiluminescence. *Eur. J. Org. Chem.* 2000: 4037-4046.

[50] Ciscato, L. F. M. L., Bartoloni, F. H., Bastos, E. L., and Baader, W. J. (2009). Direct Kinetic Observation of the Chemiexcitation Step in Peroxyoxalate Chemilumine-scence. *J. Org. Chem.* 74: 8974-8979.

[51] Chen, Y., and Sijbesma, R. P. (2014). Dioxetanes as Mechanoluminescent Probes in Thermoplastic Elastomers. *Macromolecules* 47: 3797-3805.

[52] Chen, Y., Spiering, A. J. H., KarthikeyanS, Peters, Gerrit W. M., Meijer, E. W., and Sijbesma, R. P. (2012). Mechanically induced chemiluminescence from polymers incorporating a 1,2-dioxetane unit in the main chain. *Nat. Chem.* 4: 559-562.

[53] Michelini, E., Calabretta, M. D., Cevenini, L., Lopreside, A., Southworth, T., Fontaine, D. M., Simoni, P., Branchini, B. R., and Roda, A. (2019). Smartphone-based multicolor bioluminescent 3D spheroid biosensors for monitoring inflammatory activity. *Biosens. Bioelectron.* 123: 269-277.

[54] Roda, A., Zangheri, M., Calabria, D., Mirasoli, M., Caliceti, C., Quintavalla, A., Lombardo, M., Trombini, C., and Simoni, P. (2019). A simple smartphone-based thermochemiluminescent immunosensor for valproic acid detection using 1,2-dioxetane analogue-doped nanoparticles as a label. *Sens. Actuator B-Chem.* 279: 327-333.

[55] Nascetti, A., Mirasoli, M., Marchegiani, E., Zangheri, M., Costantini, F., Porchetta, A., Iannascoli, L., Lovecchio, N., Caputo, D., de Cesare, G., Pirrotta, S., and Roda, A. (2019). Integrated chemiluminescence-based lab-on-chip for detection of life markers in extraterrestrial environments. *Biosens. Bioelectron.* 123: 195-203.

[56] Zangheri, M., Mirasoli, M., Guardigli, M., Di Nardo, F., Anfossi, L., Baggiani, C., Simoni, P., Benassai, M., and Roda, A. (2019). Chemiluminescence-based biosensor for monitoring astronauts' health status during space missions: Results from the International Space Station. *Biosens. Bioelectron.* 129: 160-268.

[57] Roda, A., Mirasoli, M., Guardigli, M., Zangheri, M., Caliceti, C., Calabria, D., and Simoni, P. (2018). Advanced biosensors for monitoring astronauts' health during long-duration space missions. *Biosens. Bioelectron.* 111: 18-26.

[58] Andronico, L. A., Quintavalla, A., Lombardo, M., Mirasoli, M., Guardigli, M., Trombini, C., and Roda, A. (2016). Synthesis of 1,2-Dioxetanes as Thermochemiluminescent Labels for Ultrasensitive Bioassays: Rational Prediction of Olefin Photooxygenation Outcome by Using a Chemometric Approach. *Chem. Eur. J.* 22: 18156-18168.

[59] Andronico, L. A., Chen, L., Mirasoli, M., Guardigli, M., Quintavalla, A., Lombardo, M., Trombini, C., Chiu, D., and Roda, A. (2018). Thermochemiluminescent semiconducting polymer dots as sensitive nanoprobes for reagentless immunoassay. *Nanoscale* 10: 14012-14021.

[60] Di Fusco, M., Quintavalla, A., Lombardo, M., Guardigli, M., Mirasoli, M., Trombini, C., and Roda, A. (2015). Organically modified silica nanoparticles doped with new acridine-1,2-dioxetane analogues as thermochemiluminescence reagentless labels for ultrasensitive immunoassays. *Anal. Bioanal. Chem.* 407: 1567-1576.

[61] Schramm, S., Karothu, D. P., Lui, N. M., Commins, P., Ahmed, E.; Catalano, L.; Weston, J., Moriwaki, T., Solnstev, K. M., Naumov, P. (2019). Thermochemiluminescent Peroxide Crystals. *Nat. Commun.* 10: 997.

[62] Radziszewski, B. (1877.) Untersuchungen über Hydrobenzamid, Amarin und Lophin. *Berichte der deutschen chemischen Gesellschaft* 10: 70-75.

[63] Sonnenberg, J., and White, D. (1964). Chemiluminescent and Thermochemiluminescent Lophine Hydroperoxide. *J. Am. Chem. Soc.* 86:5685-5686.

[64] McCapra, F., Richardson, D. G., and Chang, Y. C. (1965). Chemiluminescence Involving Peroxide Decompositions. *Photochem. Photobiol.* 4: 1111-1121.

[65] Kim, T. J., Turkcan, S., and Pratx, G. (2017). Modular low-light microscope for imaging cellular bioluminescence and radioluminescence. *Nat. Protocols* 12: 1055-1076.

In: A Comprehensive Guide to Chemiluminescence
Editor: Luís Pinto da Silva
ISBN: 978-1-53616-170-0
© 2019 Nova Science Publishers, Inc.

Chapter 10

THERMOSTABILIZATION OF FIREFLY LUCIFERASE

Roghaye Hamidi and Saman Hosseinkhani[*]
Department of Biochemistry, Faculty of Biological Science,
Tarbiat Modares University, Tehran, Iran

ABSTRACT

Low thermal resistance of firefly luciferase and rapid loss of activity at room temperature have hindered its functional application. Various attempts have been carried out to enhance the luciferase thermostability. Since proteins are dynamic objects, their functions are wholly dependent on their intrinsic structural flexibility. Structural flexibility of firefly luciferase is responsible for conformational fluctuations in response to changes of environmental factors, e.g., presence of other molecules or elevated temperature. In this approach, experimental and theoretical methods based on crystal structure of firefly luciferase were applied including site-directed mutagenesis, random mutagenesis and rational design, introduction of disulfide bonds, rational substitution of histidine by salt-bridge forming residues and the structural and sequence comparative studies. Introduction of disulfide bridges followed by a series of kinetic and thermodynamic analysis which showed the highest resistance against urea denaturation was observed for a mutant (L306C-L309C mutant) which was accompanied with increase of kinetic stability and further confirmed by accessibility of hydrophobic patches of the protein to ANS upon unfolding.

---

[*] Corresponding Author's Email: saman_h@modares.ac.ir.

1. INTRODUCTION

Bioluminescence (BL) is the emission of living light through a chemical reaction in a diverse group of organisms including bacteria, insects, shrimps, squid, fish, algae and fungi [1-4]. In this process, visible light is produced through an enzymatic reaction of a luciferase (enzyme) and a luciferin (substrate) [5, 6]. The most common luciferase in basic and applied research is the firefly luciferase (EC 1.13.12.7) due to its high quantum yield, high catalytic efficiency and substrate specificity [6-10]. Firefly luciferase uses ATP and O_2 to convert luciferin to oxyluciferin in the presence of magnesium ions with emission of light and production of CO_2 and AMP [5]. Firefly luciferase is a 62 KDa protein with 550 amino acids. The crystal structure of this enzyme has been solved from different species in which it is shown two separated globular domains including a large N-terminal domain (residues 1-436) and a small C-terminal domain (residues 440-550) with a flexible hinge (residues 437-439) and a wide cleft between them [11, 12]. Capability of bioluminescence for labeling and monitoring in modern biotechnology made firefly luciferase one of the most sensitive tools for monitoring of biological processes, such as gene expression [13], ATP detection for measuring microbial contamination [14, 15], detection of phosphatase activity [16], monitoring of protein-protein interaction [17], cell death, apoptosis and imaging of tumor growth and metastasis in whole animals [18-20].

2. PROTEIN THERMOSTABILITY ENGINEERING

Industrial and biomedical applications of proteins like enzymes and drugs are limited to their low resistance against elevated temperature and their rapid loss of activity at ambient temperatures [21]. So, protein engineering have been implemented for proteins of low thermostability to improve their thermal stability and expand their applications [21, 22]. Subtilisin is a well-known example of protein stability engineering [22, 23]. Strategies in the field of protein thermostability engineering can be divided in two categories: (1) Experimental methods, which are based on the structure and sequence comparison of various natural thermostable proteins. Nevertheless, they are time-consuming and costing. (2) Computational methods, i.e., rational structure-based design by using of molecular dynamics (MD) simulations and biochemical data like residue conservation and structural, sequence and dynamical features [21, 24]. Thermodynamic effect of certain types of interactions on protein stability has led to the implementation of several procedures to enhance protein stability such as entropic stabilization by proline and disulfide bridge introduction [25-28], introduction of salt bridges or increase the number of hydrogen bonds [29-35], and introducing clusters of aromatic-aromatic interactions [35-37]. One of the most important approaches in protein stability engineering is the comparative analysis of thermophilic and mesophilic homologous proteins to find the thermostability mechanisms through structural and sequence differences, e.g., Arg saturation in exposed state of thermophilic protein structures [38-40].

3. FIREFLY LUCIFERASE THERMOSTABILITY ENGINEERING

Since the low stability of luciferase and rapid loss of activity at room and physiological temperatures have hampered its application, many attempts have been carried out to enhance the luciferase thermostability. In this way, some experimental methods using crystal structure of enzyme as a model (PDB: 1lci) were applied such as site-directed mutagenesis [41, 42], random mutagenesis [43, 44] and rational design of mutations on the basis of the 3D structure, e.g., the mutagenesis of solvent exposed residues [42], introduction of disulfide bonds [45], rational substitution of histidine by salt-bridge forming residues [46] and the structural and sequence comparative studies [38].

4. INTRODUCTION OF DISULFIDE BONDS

Disulfide Bond is one of the most fundamental physical forces stabilizes the tertiary structure of proteins [47]. So, it is proposed that introduction of disulfide bridges can increase the protein stability by decreasing the configurational entropy of unfolded state [22]. Luciferases from different sources have been determined to possess 4-13 free reactive SH-groups; the larger the number of these groups, the smaller the stability of the enzyme [48-50]. For instance, *P.pyralis* luciferase with possession of four reactive SH-groups is more stable than the *Luciola* luciferases with seven or eight SH-groups [49]. However, no reports have been made of the presence of disulfide bridges in luciferases of luminous families in the order of Coleoptera including the Elateridae (click-beetles), the Phengodidae (railroad-worms) and the Lampyridae (fireflies) with the exception of the Metridia luciferase from marine copepods as the only case of cysteine-rich enzyme which uses coelenterazine to produce blue light and contains disulfide bridge(s) [51-53]. We designed five separate mutants of *P. pyralis* luciferase to introduce single disulfide bridges including A103C-S121C, A296C-A326C, C81-A105C, L306C-L309C and P451C-V469C and another two mutants to introduce double disulfide bridges including C81-A105C/P451C-V469C and A296C-A326C/P451C-V469C using site-specific mutagenesis to develop a thermostable enzymes in order to investigate effect of those bridges on structure and function of *P. pyralis* luciferase [45, 50]. The crystal structure of *P. pyralis* luciferase (PDB: 1lci) was used as a model for selecting the suitable location of cysteine residues to introduce disulfide bridges. The three-dimensional structure of the wild type and engineered enzymes were examined by Swiss-PDB Viewer. Comparison of time-dependent thermal stability and thermal inactivation of mutant and wild type luciferases showed that they are more stable than the wild type. Our results suggest that introduction of disulfide bonds enhances thermal resistance of mutant proteins in comparison with the wild type especially at higher temperatures and longer times [45, 50, 54]. As shown in Table 1, the effects of disulfide bond introduction on the kinetic properties of the *P. pyralis* luciferase K_m for both substrates (ATP and D-Luciferin), relative specific activity, optimum temperature and optimum pH were investigated. Measurement of relative specific activity showed almost 7-fold and 2-fold increase for A296C-A326C and C81-A105C respectively; whereas A103C-S121C, L306C-L309C and P451C-V469C mutants showed only 80%, 4.3% and 51% of wild type specific activity respectively. On the other hand, part of the loss of specific activity upon introduction of disulfide bonds may stand up from the increase of decay rates due to product inhibition. The optimum temperature of enzymatic activity for mutants exhibited a 5 to 10°C increase with exception of A103C-S121C and P451C-V469C, when compared to wild type luciferase (Table 1).

Table 1. Kinetic properties of wild type and mutants of *P.pyralis* luciferases

	Specificactivity ($\times 10^{10}$ RLU s^{-1} mg^{-1})	Relative activity (%)	Km(µM) ATP	Km(µM) LH$_2$	Optimum Temperature	pH	References
Wild type	2.7	100	105	10.2	25	8.00	[58]
A103C-S121C	2.2	81	78	2.5	20	8.00	[58]
A296C-A326C	19.6	725	62	10.3	35	8.50	
C81-A105C	5.66	208.12	91	10.0	35	9.00	
L306C-L309C	0.118	4.36	200	20.8	30	7.88	
P451C-V469C	1.39	50.97	120	41.3	25	7.95	
A296C-A326C/P451C-V469C	1.185	43	180	45.0	30	9.00	
C81-A105C/P451C-V469C	0.65	32	200	11.0	30	8.50	
D474K	5.2	55	150	16.0	25	8.50	[59]
D476N	17	181	100	8.0	25	7.00	
H461D	-	57.6	100	5.0	10	8.50	[46]
H489K	-	115	60	2.0	30	8.50	
H489D	-	112	90	2.5	25	8.50	
H489M	-	103	40	2.3	20	8.50	

A set of spectroscopic techniques were used to evaluate the effect of disulfide bridges on firefly luciferase structure and stability. Structural studies on mutants in comparison with wild type revealed that increase of helicity index, protein integrity and active site rigidity upon introduction of disulfide bonds could be regarded as reasons for increase of thermostability. The intensity of the intrinsic fluorescence spectrum of the L306C-L309C mutant decreased compared with the wild type protein. Also, the intensity of its far-UV CD spectrum is significantly decreased not only in the range of 190-200 nm with the positive CD signal but also at around 200-250 nm where it is negative. The observed changes in CD and fluorescence spectra implicate changes of secondary and tertiary structures of the mutant enzymes. Moreover, ANS as a hydrophilic reporter probe bound more efficiently to mutant luciferases, indicating exposure of the more hydrophobic groups to the solvent upon the mutation. Unlike the L306C-L309C mutant, the hydrophobic groups of other mutants become more hidden. In general, the spectroscopic results demonstrate dependence on the location of the introduced disulfide bond, both increasing and decreasing in observed Trp emissions, which show changes in the protein structure. As expected, decreases and increases in compactness of the structure are accompanied by exposure and hiding of the hydrophobic groups (for more information please refer to [45, 50, 54]).

Following these results, the conformational stability of all mutants and wild type was assessed *via* equilibrium unfolding studies, using urea as a protein denaturant and fluorescence spectroscopy analysis [54]. Denaturation curves were obtained by plotting the fluorescence intensities at 330 nm, the emission maximum for the protein, *versus* the denaturant molarity. Data presented in the urea denaturation curve was analyzed using the linear extrapolation method as described by Pace [55], Santoro and Bolen [56], and Pace et al. [57] to calculate ΔG_U as the free energy of unfolding, m as a measure of the dependence of ΔG_U on the denaturant concentration, and $D_{1/2}$ as the denaturant concentration at which the protein is half-unfolded.

To determine the spectroscopic contributions of the *P. pyralis* Trp residues, Trp417 and Trp426, which are in N-terminal domain near the protein surface, the conformational changes in mutants and wild type were evaluated by measuring the intrinsic fluorescence intensity upon treatment with different concentrations of urea. Intrinsic fluorescence intensity was reduced in wild type and all mutants except for L306C-L309C with a continuous increase in urea concentration. An increase in the fluorescence emission of the mutant L306C-L309C was detected between 0.0-1.5 M urea. Continuous drop in the intensity of the spectra indicates that denaturant increase induces changes in the tertiary structure of the protein and exposure of excitable amino acids to a more polar environment. The data obtained from fluorescence spectra was utilized to draw the denaturation curves, which were then analyzed to obtain ΔG_U. ΔG_U was plotted as a function of concentration of urea to obtain $D_{1/2}$ and ΔG (H_2O) as the free energy of unfolding extrapolated to zero denaturant concentration. As is clear in Table 2, introduction of disulfide bonds in all mutant forms results in an increase in ΔG (H_2O) and $D_{1/2}$. The highest values of $D_{1/2}$ and ΔG (H_2O) were obtained for L306C-L309C. The large Δ (ΔG) of 7.16 kJ mol $^{-1}$ for the L306C-L309C mutant indicates that the introduction of the disulfide bridge has a major stabilizing effect on the protein structure. Moreover, the highest resistance against urea denaturation was observed for the L306C-L309C mutant according to the accessibility of hydrophobic patches of the protein to ANS upon unfolding [54].

Table 2. Chemical denaturation parameters of wild type and mutant luciferases

Luciferase	$\Delta G(H2O)$ (kJ mol^{-1})	$\Delta (\Delta G)$ (kJ mol^{-1})	$D_{1/2}$	m value
P. pyralis	2.54+	-	1.53	1.69
P451C-V469C	4.36+	1.82	2.53	1.73
L306C-L309C	9.70+	7.16	4.25	2.3
C81-A105C	3.11+	0.57	2.15	1.44
C81-A105C/P451C-V469C	4.8+	2.26	2.98	1.61
A269C-A326C/P451C- V469C	3.27+	0.73	1.75	1.87

Reference: Nazari et al., 2013 [54].

5. THE MUTAGENESIS OF SOLVENT EXPOSED RESIDUES

One of the most important and basic strategy in protein engineering is finding the correlation between amino acids composition of proteins and their thermostability through the sequence and structural comparison of mesophilic and thermophilic proteins to found the way which can be used to improve the thermal resistance of proteins [38]. High occurrence of Arginine (Arg), especially in exposed state of thermophilic proteins, is reported as a factor, which would thermally stabilize the protein structure [39, 40, 60]. It is supposed that the high ratio of surface charged residues increases the interactions with solvent molecules, which would enhance the protein resistance against higher temperatures [61-63]. Based upon this, we performed insertion of Arg356 and substitution of Glutamate at position 354 (E354) with positively charged residues in an important flexible surface loop and substitution of multiple uncharged polar and hydrophobic solvent-exposed residues with Arg in the form of single, double and triple mutation on some flexible surface loops of E354Q/Arg356 mutant of *Lampyris turkestanicus* (*L.tu*) luciferase [64-66]. Uncharged polar residue of Gln at position 35 and hydrophobic residues of Ile at positions 182 and 232 and Leu at position 300 was chosen by homology modeling and visual examination of the structure of the green emitter mutant of *L.tu* luciferase (E354Q/Arg356) [66]. Insertion of an additional residue (Arg356) using site-directed mutagenesis not only led to color change from green to red, but also enhanced the optimum temperature of activity [64]. The homology modeling of Arg356 mutant and wild type luciferases revealed a significant conformational change of the flexible loop with a new ionic interaction and change in polarity of the emitter site, thereby leading to red emission and thermal resistance. The optimum temperature for activity of Arg356 mutant was improved to 34°C, which is 10°C higher than wild type. In addition, the optimum pH of Arg356 mutant also showed a slightly increase (Table 3). On the other hand, structural studies by CD spectra showed an increase in helical structure which is accompanied by a little decrease in unordered structures upon insertion of Arg356. So, both molecular modeling and structural studies revealed that insertion of this residue has disrupted some interactions and made new ionic and hydrogen bonds which led to higher thermostability. In another study, a Glutamate at position 354 in an important flexible loop of firefly *L.tu* luciferase was substituted with positively charged residues in form of single mutant including E354K and E354R and double mutant including E354Q/Arg356, E354K/Arg356 and E354R/Arg356 [66]. All mutant forms showed a good improvement in thermostability in comparison with wild type enzyme (Table 3). It should be noted that upon mutagenesis of firefly *L.tu* luciferase at position 354 and insertion of Arg356, the decay rate of light emission was decreased in comparison with wild type. In the next experiment our results indicated that arginine substitution on exposed flexible loops of firefly *L.tu* luciferase at position 35, 182, 232 and 300 leads to enhancement of thermostability and optimum temperature [66].

Table 3. Kinetic properties of wild type and mutants of *L.tu* luciferases

L.tu Luciferase wild/mutant types	Specific activity ($\times 10^{10}$ RLU s^{-1} mg^{-1})	Relative activity (%)	Km(μM) ATP	LH$_2$	Optimum Temperature	pH	References
Wild type	1.5±0.18	100	71	23	25	8.50	[64-66]
Arg356	-	81.6	142	24	34	8.50	[64]
E354K	1.9±0.02	126	63	13	30	8.00	[65, 66]
E354R	2±0.05	133	115	16	35	7.50	
E354Q-Arg356	1.2±0.03	80	50	12	28	8.50	
E354K-Arg356	1.8±0.07	120	40	11	32	8.00	
E354R-Arg356	2.3±0.14	153	95	14	32	8.00	
E354Q-Arg356-Q35R	1.27	86	62	7	28	8.00	
E354Q-Arg356-I232R	1.35	91	57	8	40	8.00	
E354Q-Arg356-I182R	1.21	82	60	10	28	8.00	
E354Q-Arg356-L300R	0.0003	0.02	75	30	28	8.50	
E354Q-Arg356-Q35R-I232R	1.67	113	42	10	40	8.50	
E354Q-Arg356-Q35R-L182R-I232R	1.70	115	44	10	40	8.50	

As shown in Table 3 some mutations are more effective on thermostability of luciferase. By introducing of I232R, Q35R/I232R and Q35R/I232R/I182R mutations, optimum temperature of activity was increased to 40°C which are 12 and 15°C higher than E354Q/Arg356 and wild type luciferases. Moreover, structural analysis of mutants and wild type luciferases revealed that surface Arginine saturation induces redirection of exposed hydrophobic residues toward the interior and more protein rigidity. Thereby, improving thermal stability of firefly luciferases with exception of L300R. Meanwhile, L300R confirms that ultra-rigidity of luciferase mutant opposed to appropriate activity [66]. Therefore, it should be noted that total increase of thermostability does not result of the additional Arginine charge [67].

6. EFFECTS OF SALT BRIDGES AND ION PAIRS ON LUCIFERASE THERMOSTABILITY

There are strong evidences in which salt bridges contribute to the protein thermal stability at high temperatures especially in the thermophilic proteins [31]. On the other hand, thermal resistance of thermophilic proteins seems to be the result of smaller fluctuations and reduction in conformational entropy of them [68]. Hence, tendency of thermophilic proteins for having more salt bridges was considered as a good factor for enhancing the thermal tolerance in protein engineering [29, 34, 61, 62]. Based on this concept, we designed some *p. pyralis* luciferase mutants and evaluated their thermal stability and structural flexibility in order to investigate the relationship between flexibility and stability. In this way, ranking of *p. pyralis* luciferase amino acids based on the atomic displacement parameter (B-factor, which is determined by high resolution x-ray crystallography) revealed some region in the structure of luciferase which have high density of amino acids with highest B-factor (Table 4). A comparison between N-terminal and C-terminal domains of *p. pyralis* luciferase based on B-factor analysis showed that C-terminal domain with high density of highest B-factor amino acids is more flexible than N-terminal domain. These analysis determined residues 473-477 as the most flexible region of *P.pyralis* luciferase [59]. Existence of three successive negatively charged amino acids (Asp474, Asp475 and Asp476) and repulsion between them in the most flexible region of firefly luciferase make this region suitable for mutagenesis. So, Asp474 and Asp476 were selected for mutagenesis as they are more conserved than Asp475 and two mutations (D474K and D476N) were constructed and evaluated for thermal stability and structural flexibility [59]. Structural flexibility analysis through dynamic quenching, limited proteolysis and circular dichroism spectroscopy on both mutants and wild type luciferase indicated that D474K is more flexible than wild type while no difference was observed in D476N. The specific contributions of both mutants to stability was also determined by analysis of kinetic stability in D474K and D476N mutants. The results showed that introduce of D474K mutation led to destabilization of luciferase, but no significant difference was observed in D476N relative to wild type. Based upon these observations, it was proposed that two possible reasons would destabilize the D474K mutant. According to bioinformatics analysis, D474 interacts with nonadjacent residue K445 electrostatically, which is possibly critical in linking two β-strands and stabilizing the tertiary structure of luciferase. Hence, removal of this ionic interaction by substitution of D474 with positively charged amino acid Lysine (K) is followed by perturbation of the local rigidity due to unzipping of the beta-sheet and by exposing a hydrophobic region of protein. The second possible reason for the destabilizing effect of D474 mutation is the breaking of hydrogen bond between side chain carbonyl group of Asp474 and the main chain nitrogen of Ala477, which may increase its flexibility and so destabilize the luciferase.

8. Rational Substitution of Histidines by Salt-Bridge Forming Residues Using Molecular Dynamics Simulations

Previous studies on thermophilic and mesophilic proteins have revealed an inverse relation between the number of histidines and thermal stability of proteins and higher portion of charged amino acids (Lys, Arg, Asp, Glu) in thermophilic proteins in nparison with the mesophilic ones [75, 76]. Nevertheless, it has been shown that there 14 histidine residues in firefly luciferase, in which two of them are located in C-terminal nain [11]. At the first step the root mean square fluctuation (RMSF) of *P. pyralis* firefly iferase backbone atoms was calculated by using a 50 ns long MD simulation in a listic environment at two temperatures (300 and 340 K) to find the thermal sensitive exible) regions of luciferase in which there are histidine residues. Since RMSF value is neasure of atomic fluctuation along MD trajectories, the most flexible residues (like face ones) exhibit higher RMSF values [74]. RMSF evaluation of *P. pyralis* firefly iferase revealed higher flexibility of C-terminal domain (residues 440-550) than the N-minal domain (residues 1-436) (Figure 1). Two histidine residues (His461 and His489), ich are located in flexible surface loops with high RMSF values, were substituted with negatively charged residue, Asp, or a positively charged residue, Lys, to assess how otein thermostability is conducted by the new salt bridges formation and altered drogen bonding propensities of these regions [46].

As shown in Table 5 the possibility of salt bridge interactions between residues in the inity of mutation sites was searched during the 50 ns simulation. The data shows a crease in the number of neighboring salt bridges from five in the wild type luciferase to ir upon introduction of H461D mutation, due to the absence of the Glu497-Lys491 salt dge (Table 5). Contrary to H461D mutation, introduction of the H489D mutation led to crease in the number of neighboring salt bridges from one in the wild type luciferase (at 0 K) to five. Among those salt bridges, the alternate pairing of Glu488 and Asp489 with g112 decreases the average distance between the N- and C-terminal domains of the 89D mutant to 3.60 Å (Figure 2) through a stable contact between the two domains. It ould be considered that these salt bridges were highly sensitive to temperature and will lost at the increased temperature of 340 K. A similar pattern of the increased salt bridges is observed for H489K mutation, while unlike the H489D mutant form, the increased salt idges were highly resistant to incremental change of temperature. Substitution of His489 th methionine, also led to increase in salt bridges of luciferase but with different pattern H489D and H489K mutations neighboring salt bridges.

Table 4. The amino acids in P. pyralis luciferase with the highest B-fa

Residue	Sequence number	B-value	Rank
Asp	475	73.23	1
Asp	476	72.76	2
Ala	477	66.88	3
Asp	474	65.24	4
Lys	544	62.95	5
Gly	478	62.50	6
Gln	448	60.98	7
Leu	204	60.83	8
Pro	359	60.45	9
Tyr	447	59.90	10
Asp	377	59.85	11

Reference: Amini-Bayat et al., 2012 [59].

7. MOLECULAR DYNAMICS SIMULATIONS

As mentioned above, most of former works aimed to engineer thermally luciferase were based on experimental approaches using its crystal structure. Howev dynamic nature of molecular atoms and its effect on functional properties whic commonly targeted in experimental methods of protein engineering are disregarded Since proteins are dynamic entities, their functions are completely dependent on intrinsic structural flexibility. Structural flexibility of proteins is responsible conformational fluctuations in response to changes of environmental factors, e.g., pres of other molecules or elevated temperature. Most of biochemical processes like lig receptor binding in signal transduction, protein transport, antigen recognition and enz catalysis depends on flexibility [70, 71]. On the other hand, the most crucial step in s directed mutagenesis (SDM) of proteins is the selection of residues at specific regi which are probably responsible for improved thermostability [72]. So computatio methods like molecular dynamics (MD) simulations, which simulates the motion proteins, have been employed to protein engineering, alongside with experimen techniques and rational design strategies [73]. MD simulations presents realistic models protein dynamics by providing atomistic details of the dynamic molecular interactio which underlie the protein stability and function, while also clarifying the relationshi between sequence, structure, dynamics, and function [73, 74]. MD simulation method wa used to survey the factors influencing the *in vitro* thermodynamic stability of firefl luciferase in order to propose a number of candidate mutations for improving protei performance at elevated temperature [46, 69].

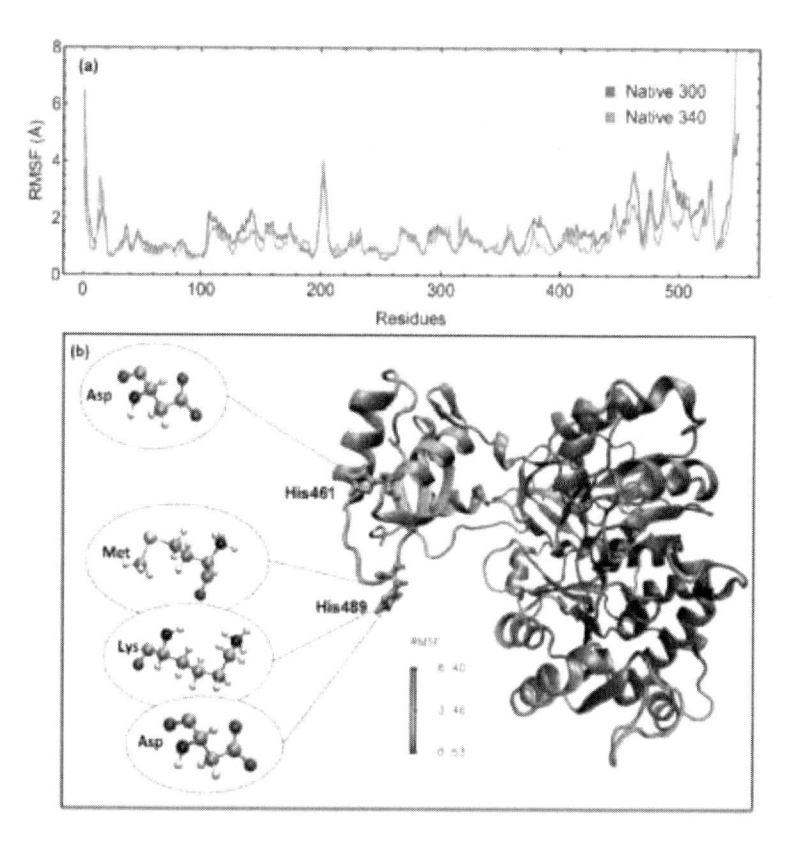

Figure 1. (a) Evaluation of RMSF for *P.pyralis* firefly luciferase in 50 ns MD simulation at two temperatures 300 and 340 K. (b) Schematic representation of *P.pyralis* luciferase colored by variable RMSF value from blue (low RMSF) to red (high RMSF). His461 and His489, which are located in C-terminal domain of *P.pyralis* firefly luciferase, are shown. His461 was substituted by Asp and His489 was substituted by Asp, Met and Lys. (Reprinted from [46])

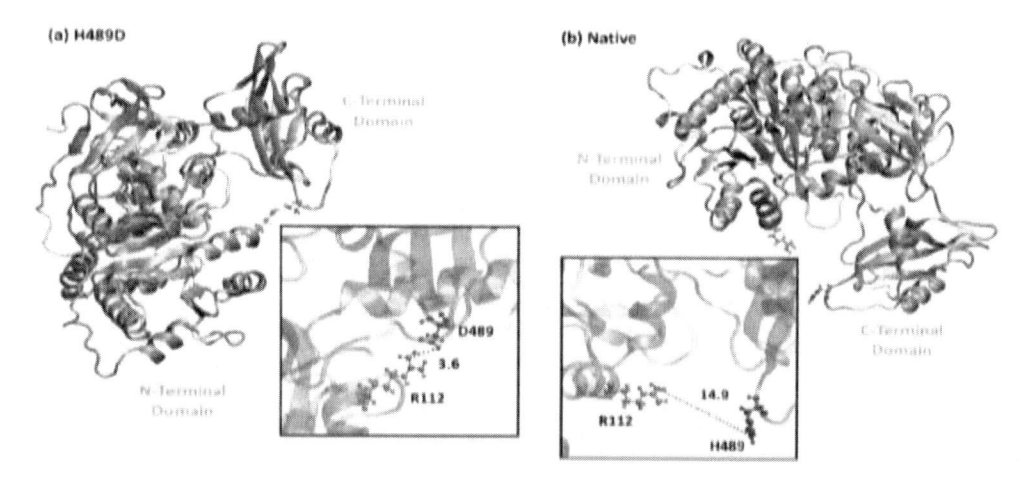

Figure 2. Schematic representation of wild type and mutant of luciferase (H489D). Substitution of His 489 by Asp in C-terminal led to formation of a new salt bridge between Asp489 and Arg112 at the C-terminal and N-terminal domains respectively. (a) For mutant (H489D) the distance between Asp489 and Arg112 is shorter (3.6 Å) (b) And for wild type, the distance is 14.99 Å. (Reprinted from [46])

Table 5. The most possibility of salt bridge interactions between residues during the 50 ns simulation (wild type and mutants)

	300 K	340 K
P.pyralis (H461)	ASP500-LYS496 GLU488- LYS491 GLU497-LYS496 GLU495-LYS496 GLU497-LYS491	ASP500-LYS496 GLU488- LYS491 GLU497-LYS496 GLU495-LYS496 GLU497-LYS491
H461D	ASP500-LYS496 GLU488- LYS491 GLU495-LYS496 GLU497-LYS496	ASP500-LYS496 GLU488- LYS491 GLU495-LYS496 GLU497-LYS496
H489K	ASP520-LYS489 ASP520-LYS549 GLU488-LYS489 GLU488-LYS491	ASP520-LYS489 ASP520-LYS549 GLU488-LYS489 GLU488-LYS491
H489D	ASP489-ARG112 ASP489-LYS491 ASP520-LYS549 GLU488-ARG112 GLU488-LYS491	ASP489-ARG112 GLU488-LYS491
H489M	GLU495-LYS547 GLU495-LYS549 GLU497-LYS491 GLU497-LYS496 GLU521-LYS524	ASP466- LYS524 GLU113-ARG112 GLU488-LYS491 GLU521-LYS524
P.pyralis (H489)	GLU488-LYS491	ASP520-LYS549 GLU488-LYS491

Reference: Rahban et al., 2017 [46].

In addition to salt bridges, hydrogen bonding networks affect the thermostability of proteins [77]. Hence, alterations of hydrogen bond networks around the mutations site was surveyed by spanning of hydrogen bonds in 6 Å proximity of residues 461 and 489 (Table 6). Despite a decrease of the total number of hydrogen bonds in proximity to His461 upon mutation to Aspartate, the number of hydrogen bonds correlated to residue 461 itself increased from three in the wild type luciferase to six, which were temperature-resistant without getting lost at elevated temperature of 340 K. Furthermore, it was found that H489D mutation led to disruption of some hydrogen bonds between Phe465 and Val486, while more new hydrogen bonds were formed between Glu488 and Lys491. In a similar manner, H489K and H489M mutations changed the hydrogen bonding networks surrounding the mutation sites, in which Glu488 with irregular hydrogen bonding tendency, could interact not only with Lys491 but also Asp463 and Lys489 (Table 6).

**Table 6. The most possibility of hydrogen bond interactions
between residues during the 50 ns simulation (wild type and mutants)**

	300 K	340 K
P.pyralis (**H461**)	ASN463 ... His461	ASN463 ... HIS461
	VAL502 ... ILE498	TYR501 ... GLU497
	TYR501 ... GLU497	TYR501 ... ILE457
	TYR501 ... ILE457	HIS461 ... LEU458
	GLN505 ... VAL502	ILE464 ... HIS461
	HIS461 ... LEU458	
	GLN505 ... TYR501	
	GLN505 ... TYR501	
	ILE464 ... HIS461	
	GLN460 ... SER456	
	GLN460 ... TYR501	
	GLN460 ... SER456	
H461D	TYR501 ... ILE457	TYR501 ... ILE457
	ASN463 ... ASP461(4)[a]	ASN463 ... ASP461(4)[a]
	ASP 461 ... LEU458	GLN505 ... TYR501
	ILE464 ... ASP461	ILE464 ... ASP461
		ASP461 ... LEU458
H489K	GLU488 ... ASN463	GLN505 ... TYR501
	LYS491 ... GLU488	PHE465 ... VAL486
	LYS489 ... GLU488(2)[a]	LYS491 ... GLU488(3)[a]
H489D	LYS491 ... GLU488 (3)[a]	LYS491 ... GLU488(7)[a]
H489M	VAL519 ... VAL485	GLU488 ... ASN463
	GLU488 ... ASN 463	PHE 465 ... VAL486
	ASN463 ... GLU488 (2)[a]	LYS491 ... GLU488 (5)[a]
P.pyralis (**H489**)	PHE465 ... VAL486	LYS491 ... GLU488(4)[a]
	LYS49 ... GLU488	

[a] The number of interactions.
Source: Rahban et al., 2017 [46].

The RMSF of Cα atoms of all mutants were calculated and compared with various luciferase mutants to address how the altered interactions due to Histidine substitution affect protein conformational dynamics. The overall results showed that both H461D and H489K mutants and wild type luciferase had higher general mobility than H489M and H489D. In spite of high general flexibility of mutant luciferases, the local mobility of residues 455-470 was limited upon H461D mutation, which is due to strengthening of hydrogen bond network [46]. Tryptophans are natural probes that are highly sensitive to their local microenvironment. Using such probes, we are able to predict their micro environmental changes, protein dynamics or conformational transitions [78]. *P.pyralis* firefly luciferase has two tryptophans, Trp417 and Trp46, which are located in the N-terminal domain near the hinge region [79, 80]. Flexibility of Trp residues were investigated by RMSF calculation of the two Trp in mutants and wild type: average RMSF of both Trp residues are 1.6, 1.4, 1.1, 0.9 and ~0.8 Å for the H489D, H461D, H489M, H489K mutants and the wild type respectively [46]. To investigate these computational results experimentally, single mutants of H461D, H489K, H489D and H489M were constructed using the Quick-change PCR. As shown in table 1 to survey the effect of these mutations on enzymatic activity, various activity parameters were determined. As indicated in table 1 relative specific activity of all mutants was like wild type with exception of H461D with 40% of wild type. An intriguing shift was observed in optimum temperature of luciferase activity by His461 and His489 substitutions: The H489K mutant indicated a 5°C increase in the optimum temperature, while a surprising decrease was revealed for H489M and particularly H461D with a cold adapted behavior. The optimum temperature of luciferase remained the same as wild type by H489D.

Assessment of thermal inactivation and thermal stability of these mutants in comparison with wild type luciferase showed that H461D had a higher rate of inactivation against incremental change of temperature, while H489D was more temperature tolerant. To better understanding of the mutated enzymes stability, thermal denaturation midpoint (T_m) were determined by DSC and intrinsic fluorescence experiments (for detailed please refer to [46]).The H461D luciferase with the melting point of 26.2°C had the lowest stability and the H489D with the melting point of 42°C was the most stable, which could be caused by the increasing number of salt bridges compared to wild type. The lowest stability of H461D may possibly be due to the higher global flexibility and loss of Glu497-Lys491 salt bridge and a decrease in the number of hydrogen bonds. Furthermore, intrinsic tryptophan fluorescence experiment showed an increase in fluorescence intensity of H489D and H489M and a decrease in H489K and H461D mutants compared to the wild type.

ACKNOWLEDGMENT

Research Council of Tarbiat Modares University is acknowleged in financial support of this project.

REFERENCES

[1]   Campbell, A.K, Campbell, A.K. *Chemiluminescence: principles and applications in biology and medicine.* 1988.

[2]   Herring, P.J. *Bioluminescence in action*: Academic Press; 1978.

[3]   Johnson, F.H., Haneda, Y. *Bioluminescence in Progress.* Princeton Univ NJ, 1966.

[4]   Shimomura, O. *Bioluminescence: chemical principles and methods: World Scientific;* 2012.

[5]   Hastings, J.W., McElroy, W.D., Coulombre, J. (1953). The effect of oxygen upon the immobilization reaction in firefly luminescence. *J. Cell. Comp. Physiol.* 42: 137-50.

[6]   Deluca, M. (1976). Firefly luciferase. *Adv. Enzymol. Relat. Areas Mol. Biol.* 44: 37-68.

[7]   De Wet, J.R., Wood, K., DeLuca, M., Helinski, D.R., Subramani, S. (1987). Firefly luciferase gene: structure and expression in mammalian cells. *Mol. Cell. Biol.* 7: 725-737.

[8]   De Wet, J.R., Wood, K.V., Helinski, D.R., DeLuca, M. (1985). Cloning of firefly luciferase cDNA and the expression of active luciferase in Escherichia coli. *Proc. Natl. Acad. Sci. U.S.A.* 82: 7870-7873.

[9]   Niwa, K., Ichino, Y., Kumata, S., Nakajima, Y., Hiraishi, Y., Kato, D., et al. (2010). Quantum yields and kinetics of the firefly bioluminescence reaction of beetle luciferases. *Photochem. Photobiol.* 86: 1046-1049.

[10] Hosseinkhani, S. (2011). Molecular enigma of multicolor bioluminescence of firefly luciferase. *Cell. Mol. Life. Sci.* 68: 1167-1182.

[11] Conti, E., Franks, N.P., Brick, P. (1996). Crystal structure of firefly luciferase throws light on a superfamily of adenylate-forming enzymes. *Structure.* 4: 287-298.

[12] Kheirabadi, M., Sharafian, Z., Naderi-Manesh, H., Heineman, U., Gohlke, U., Hosseinkhani, S. (2013). Crystal structure of native and a mutant of Lampyris turkestanicus luciferase implicate in bioluminescence color shift. *BBA - Proteins Proteom.* 1834: 2729-2735.

[13] Contag, C.H., Bachmann, M.H. (2002). Advances in in vivo bioluminescence imaging of gene expression. *Annu. Rev. Biomed. Eng.* 4: 235-260.

[14] Lundin, A., Deluca, M. *Bioluminescence and chemiluminescence.* Academic Press New York; 1981.

[15] Lundin, A. (2000). Use of firefly luciferase in ATP-related assays of biomass, enzymes, and metabolites. *Methods Enzymol.* 305: 346-370.

[16] Miska, W., Geiger, R. (1988). A new type of ultrasensitive bioluminogenic enzyme substrates. I. Enzyme substrates with D-luciferin as leaving group. *Biol. Chem. Hoppe Seyler* 369: 407-412.

[17] Azad, T., Tashakor, A., Hosseinkhani, S. (2014). Split-luciferase complementary assay: applications, recent developments, and future perspectives. *Anal. Bioanal. Chem.* 406: 5541-5560.

[18] Dickson, P.V., Hamner, B., Ng, C.Y., Hall, M.M., Zhou, J., Hargrove, P.W., et al. (2007). In vivo bioluminescence imaging for early detection and monitoring of disease progression in a murine model of neuroblastoma. *J. Pediatr. Surg.* 42: 1172-1179.

[19] Torkzadeh-Mahani, M., Ataei, F., Nikkhah, M., Hosseinkhani, S. (2012). Design and development of a whole-cell luminescent biosensor for detection of early-stage of apoptosis. *Biosens. Bioelectron.* 38: 362-368.

[20] Azad, T., Tashakor, A., Rahmati, F., Hemmati, R., Hosseinkhani, S. (2015). Oscillation of apoptosome formation through assembly of truncated Apaf-1. *Eur. J. Pharmacol.* 760: 64-71.

[21] Modarres, H.P., Mofrad, M., Sanati-Nezhad, A. (2016). Protein thermostability engineering. *RSC Adv.* 6: 115252-115270.

[22] Eijsink, V.G., Bjørk, A., Gåseidnes, S., Sirevåg, R., Synstad, B., van den Burg, B-, et al. (2004). Rational engineering of enzyme stability. *J. Biotechnol.* 113: 105-120.

[23] Bryan, P.N. (2000). Protein engineering of subtilisin. *Biochim. Biophys. Acta Protein Struct. Molec. Enzym.* 1543: 203-22.

[24] Pucci, F., Bourgeas, R., Rooman, M. (2016). Predicting protein thermal stability changes upon point mutations using statistical potentials: Introducing HoTMuSiC. *Sci. Rep..* 6: 23257.

[25] Matthews, B.W. (1987). Genetic and structural analysis of the protein stability problem. *Biochemistry.* 26: 6885-6888.

[26] Matsumura, M., Signor, G., Matthews, B.W. (1989). Substantial increase of protein stability by multiple disulphide bonds. *Nature.* 342: 291.

[27] Mansfeld, J., Vriend, G., Dijkstra, B.W., Veltman, O.R., Van den Burg, B., Venema, G., et al. (1997). Extreme stabilization of a thermolysin-like protease by an engineered disulfide bond. *J. Biol. Chem.* 272: 11152-11156.

[28] Clarke, J., Fersht, A.R. (1993). Engineered disulfide bonds as probes of the folding pathway of barnase: increasing the stability of proteins against the rate of denaturation. *Biochemistry.* 32: 4322-4329.

[29] Pace, N.C., Alston, R.W., Shaw, K.L. (2000). Charge–charge interactions influence the denatured state ensemble and contribute to protein stability. *Protein Sci.* 9: 1395-1398.

[30] Schwehm, J.M., Fitch, C.A., Dang, B.N., García-Moreno, E.B., Stites, W.E. (2003). Changes in stability upon charge reversal and neutralization substitution in staphylococcal nuclease are dominated by favorable electrostatic effects. *Biochemistry.* 42: 1118-1128.

[31] Sun, D.P., Sauer, U., Nicholson, H., Matthews, B.W. (1991). Contributions of engineered surface salt bridges to the stability of T4 lysozyme determined by directed mutagenesis. *Biochemistry.* 30: 7142-7153.

[32] Strop, P., Mayo, S.L. (2000). Contribution of surface salt bridges to protein stability. *Biochemistry.* 39: 1251-1255.

[33] Waldburger, C.D., Schildbach, J.F., Sauer, R.T. (1995). Are buried salt bridges important for protein stability and conformational specificity? *Nature Struct. Biol.* 2: 122.

[34] Makhatadze, G.I., Loladze, V.V., Ermolenko, D.N., Chen, X., Thomas, S.T. (2003). Contribution of surface salt bridges to protein stability: guidelines for protein engineering. *J. Mol. Biol.* 327: 1135-1148.

[35] Serrano, L., Bycroft, M., Fersht, A.R. (1991). Aromatic-aromatic interactions and protein stability: investigation by double-mutant cycles. *J. Mol. Biol.* 218: 465-75.

[36] Burley, S., Petsko, G.A. (1985). Aromatic-aromatic interaction: a mechanism of protein structure stabilization. *Science*. 229: 23-28.

[37] Puchkaev, A.V., Koo, L.S., de Montellano, P.R.O. (2003). Aromatic stacking as a determinant of the thermal stability of CYP119 from Sulfolobus solfataricus. *Arch. Biochem. Biophys*. 409: 52-58.

[38] Zhou, X.X., Wang, Y.B., Pan, Y.J., Li, W.F. (2008). Differences in amino acids composition and coupling patterns between mesophilic and thermophilic proteins. *Amino acids*. 34: 25-33.

[39] Chakravarty, S., Varadarajan, R. (2000). Elucidation of determinants of protein stability through genome sequence analysis. *FEBS Lett*. 470: 65-69.

[40] Kumar, S., Tsai, C.J., Nussinov, R. (2000). Factors enhancing protein thermostability. *Protein Eng*. 13: 179-91.

[41] Branchini, B.R., Ablamsky, D.M., Murtiashaw, M.H., Uzasci, L., Fraga, H-, Southworth, T.L. (2007). Thermostable red and green light-producing firefly luciferase mutants for bioluminescent reporter applications. *Anal. Biochem*. 361: 253-62.

[42] Law, G.E., Gandelman, O.A., Tisi, L.C., Lowe, C.R., Murray, J.A. (2006). Mutagenesis of solvent-exposed amino acids in Photinus pyralis luciferase improves thermostability and pH-tolerance. *Biochem. J*. 397: 305-312.

[43] White, P.J., Squirell, D.J., Arnaud, P., Christopher, R.L., Murray, J.A. (1996). Improved thermostability of the North American firefly luciferase: saturation mutagenesis at position 354. *Biochem. J*. 319: 343-350.

[44] Kajiyama, N., Nakano, E. (1993). Thermostabilization of firefly luciferase by a single amino acid substitution at position 217. *Biochemistry*. 32: 13795-13799.

[45] Imani, M., Hosseinkhani, S., Ahmadian, S., Nazari, M. (2010). Design and introduction of a disulfide bridge in firefly luciferase: increase of thermostability and decrease of pH sensitivity. *Photochem. Photobiol. Sci*. 9: 1167-1177.

[46] Rahban, M., Salehi, N., Saboury, A.A., Hosseinkhani, S., Karimi-Jafari, M.H., Firouzi, R., et al. (2017). Histidine substitution in the most flexible fragments of firefly luciferase modifies its thermal stability. *Arch. Biochem. Biophys*. 629: 8-18.

[47] Creighton, T.E. (1988). Disulphide bonds and protein stability. *BioEssays*. 8: 57-63.

[48] Branchini, B.R., Southworth, T.L., Murtiashaw, M.H., Magyar, R.A., Gonzalez, S.A., Ruggiero, M.C., et al. (2004). An alternative mechanism of bioluminescence color determination in firefly luciferase. *Biochemistry*. 43: 7255-7262.

[49] Devine, J.H., Kutuzova, G.D., Green, V.A., Ugarova, N.N., Baldwin, T.O. (1993). Luciferase from the east European firefly Luciola mingrelica: cloning and nucleotide sequence of the cDNA, overexpression in Escherichia coli and purification of the enzyme. *Biochim. Biophys. Gene Struct. Expression* 1173: 121-132.

[50] Nazari, M., Hosseinkhani, S. (2011). Design of disulfide bridge as an alternative mechanism for color shift in firefly luciferase and development of secreted luciferase. *Photochem. Photobiol. Sci.* 10: 1203-1215.

[51] Viviani, V., Silva, A., Perez, G., Santelli, R., Bechara, E., Reinach, F. (1999). Cloning and molecular characterization of the cDNA for the Brazilian larval click-beetle Pyrearinus termitilluminans luciferase. *Photochem. Photobiol.* 70: 254-260.

[52] Viviani, V., Arnoldi, F., Venkatesh, B., Neto, A.S., Ogawa, F., Oehlmeyer, A., et al. (2006). Active-site properties of Phrixotrix railroad worm green and red bioluminescence-eliciting luciferases. *J. Biochem.* 140: 467-474.

[53] Stepanyuk, G.A., Xu, H., Wu, C.K., Markova, S.V., Lee, J., Vysotski, E.S., et al. (2008). Expression, purification and characterization of the secreted luciferase of the copepod Metridia longa from Sf9 insect cells. *Protein Expression and Purif.* 61: 142-148.

[54] Nazari, M., Hosseinkhani, S., Hassani, L. (2013). Step-wise addition of disulfide bridge in firefly luciferase controls color shift through a flexible loop: a thermodynamic perspective. *Photochem. Photobiol. Sci.* 12: 298-308.

[55] Pace, C. Determination and analysis of urea and guanidine hydrochloride denaturation curves Methods Enzymol. Academic Press; 1986.

[56] Santoro, M.M., Bolen, D. (1988). Unfolding free energy changes determined by the linear extrapolation method. 1. Unfolding of phenylmethanesulfonyl. alpha.-chymotrypsin using different denaturants. *Biochemistry.* 27: 8063-8068.

[57] Pace, C.N., Scholtz, J.M. (1997). Measuring the conformational stability of a protein. *Protein Struct.* 2: 299-321.

[58] Naderi, M., Moosavi-Movahedi, A.A., Hosseinkhani, S., Nazari, M., Bohlooli, M., Hong, J., Hadi-Alijanvand, H., Sheibani, N. (2015). Implication of disulfide bridge induced thermal reversibility, structural and functional stability for luciferase. *Protein Pept. Lett.* 22: 23-30.

[59] Amini-Bayat, Z., Hosseinkhani, S., Jafari, R., Khajeh, K. (2012). Relationship between stability and flexibility in the most flexible region of Photinus pyralis luciferase. *Biochim. Biophys. Acta Proteins Proteomics.* 1824: 350-358.

[60] Das, S., Paul, S., Bag, S.K., Dutta, C. (2006). Analysis of Nanoarchaeum equitans genome and proteome composition: indications for hyperthermophilic and parasitic adaptation. *BMC genomics.* 7: 186.

[61] Xiao, L., Honig. B. (1999). Electrostatic contributions to the stability of hyperthermophilic proteins. *J. Mol. Biol.* 289: 1435-1444.

[62] Szilágyi, A., Závodszky, P. (2000). Structural differences between mesophilic, moderately thermophilic and extremely thermophilic protein subunits: results of a comprehensive survey. *Structure.* 8: 493-504.

[63] Suhre, K., Claverie, J.M. (2003). Genomic correlates of hyperthermostability, an update. *J. Biol. Chem.* 278: 17198-17202.

[64] Tafreshi, N.K., Hosseinkhani, S., Sadeghizadeh, M., Sadeghi, M., Ranjbar, B., Naderi-Manesh, H. (2007). The influence of insertion of a critical residue (Arg356) in structure and bioluminescence spectra of firefly luciferase. *J. Biol. Chem.* 282: 8641-8647.

[65] Alipour, B.S., Hosseinkhani, S., Ardestani, S.K., Moradi, A. (2009). The effective role of positive charge saturation in bioluminescence color and thermostability of firefly luciferase. *Photochem. Photobiol. Sci.* 8: 847-855.

[66] Mortazavi, M., Hosseinkhani, S. (2011). Design of thermostable luciferases through arginine saturation in solvent-exposed loops. *Protein Eng. Des. Sel.* 24: 893-903.

[67] Mortazavi, M., Hosseinkhani, S. (2017). Surface charge modification increases firefly luciferase rigidity without alteration in bioluminescence spectra. *Enzyme Microb. Technol.* 96: 47-59.

[68] Karshikoff, A., Nilsson, L., Ladenstein, R. (2015). Rigidity versus flexibility: the dilemma of understanding protein thermal stability. *FEBS Lett.* 282: 3899-3917.

[69] Jazayeri, F.S., Amininasab, M., Hosseinkhani, S. (2017). Structural and dynamical insight into thermally induced functional inactivation of firefly luciferase. *PloS one.* 12: e0180667.

[70] Teilum, K., Olsen, J.G., Kragelund, B.B. (2009). Functional aspects of protein flexibility. *Cell. Mol. Life. Sci.* 66: 2231.

[71] McGeagh, J.D., Ranaghan, K.E., Mulholland, A.J. (2011). Protein dynamics and enzyme catalysis: insights from simulations. *Biochim. Biophys. Acta Proteins Proteomics* 1814: 1077-1092.

[72] Yang, H., Liu, L., Shin, H.D., Chen, R.R., Li, J., Du, G., et al. (2013). Structure-based engineering of histidine residues in the catalytic domain of α-amylase from Bacillus subtilis for improved protein stability and catalytic efficiency under acidic conditions. *J. Biotechnol.* 164: 59-66.

[73] Childers, M.C., Daggett, V. (2017). Insights from molecular dynamics simulations for computational protein design. *Mol. Syst. Des. Eng.* 2: 9-33.

[74] Vijayakumar, S., Vishveshwara, S., Ravishanker, G., Beveridge, D. (1994). Analysis of hydrogen bonding and stability of protein secondary structures in molecular dynamics simulation. *ACS Symposium Series* 569(11): 175-193.

[75] Chakravarty, S., Varadarajan, R. (2002). Elucidation of factors responsible for enhanced thermal stability of proteins: a structural genomics based study. *Biochemistry.* 41: 8152-8161.

[76] Tompa, D.R., Gromiha, M.M., Saraboji, K. (2016). Contribution of main chain and side chain atoms and their locations to the stability of thermophilic proteins. *J. Mol. Graphics Modell.* 64: 85-93.

[77] Paul, M., Hazra, M., Barman, A., Hazra, S. (2014). Comparative molecular dynamics simulation studies for determining factors contributing to the thermostability of chemotaxis protein "CheY". *J. Biomol. Struct. Dyn.* 32: 928-949.

[78] Lackowicz, J. Principles of Fluorescence spectroscopy. 1999. New York: Kluwer Academic/Plenum Publishers.

[79] Hosseinkhani, S., Szittner, R., Meighen, E.A. (2005). Random mutagenesis of bacterial luciferase: critical role of Glu175 in the control of luminescence decay. *Biochem. J.* 385: 575-580.

[80] Matthews, B.W. (1993). Structural and genetic analysis of protein folding and stability: Current Opinion in Sturctural Biology 1993, 3: 589–593. *Curr. Opin. Struct. Biol.* 3: 589-593.

In: A Comprehensive Guide to Chemiluminescence ISBN: 978-1-53616-170-0
Editor: Luís Pinto da Silva © 2019 Nova Science Publishers, Inc.

Chapter 11

THEORETICAL STUDIES OF CHEMILUMINESCENCE

Cristina García-Iriepa[1,*], Romain Berraud-Pache[2,†] and Isabelle Navizet[1,‡]

[1]Laboratoire Modélisation et Simulation Multi Echelle, Université Paris-Est, MSME, UPEM, Marne-la-Vallée, France
[2]Max-Planck-Institut für Kohlenforschung, Mülheim an der Ruhr, Germany

ABSTRACT

Chemiluminescent reactions are characterized by producing a species in the excited state responsible of the light emission, so-called light emitter. Theoretical chemistry's studies have gain power to give insights in the different steps of the reaction scheme leading to the light emission. The present chapter presents some of the theoretical studies performed for a better understanding of the chemiluminescent reactions.

1. INTRODUCTION

Chemiluminescent reactions are characterized by producing a species in the excited state responsible of the light emission, so-called light emitter. The observed light comes from the decay of the light emitter to its ground state. This fascinating phenomenom can be studied both experimentally and computationally. In this context, computational chemistry is a valuable tool as it can afford data inaccessible from an experimental point of view.

---

* Corresponding Author's Email: cristina.garciairiepa@u-pem.fr.
† Corresponding Email: berraudpache@kofo.mpg.de.
‡ Corresponding Email: isabelle.navizet@u-pem.fr.

Regarding chemiluminescence, we can discern three parts in the mechanism: first the generation of a high-energy intermediate (HEI) second, the decomposition of the HEI leading to the production of the excited light emitter and third, the physical electronic phenomena of light emission. In the following, we discuss the contribution of computational chemistry in a better understanding of each of these three parts. In particular, the chemiluminescent systems discussed in this chapter are shown in Figure 1 and the acronyms used are detailed in Table 1.

Table 1. Acronyms used in this chapter

Acronym	Meaning
ACT	Activator moiety (either inter- or intra-molecular to the HEI)
AMP	Adenosine mono-phosphate
m-AMPD	3-(2'-spiroadamantyl)-4-methoxy-4-(3"- hydroxyphenyl)-1,2-dioxetane
p-AMPD	3-(2'-spiroadamantyl)-4-methoxy-4-(4"- hydroxyphenyl)-1,2-dioxetane
ATP	Adenosine tri-phosphate
BCT	Back charge transfer
BS	Bioluminescent state
CI	Conical intersection
CIEEL	Chemically Initiated Electron-Exchange Luminescence
CS	Chemiexcited state
CT	Charge transfer
CTIL	Charge-Transfer Induced Luminescence
ESPT	Excited state proton transfer
FS	Fluorescent state
GS	Ground state
HEI	High-energy intermediate
IRC	Intrinsic reaction coordinate
ISC	Intersystem crossing
MEP	Minimum energy path
PES	Potential energy surface
RC	Reaction coordinate
S_1	First singlet excited state
TS	Transition state

The first part of the reaction concerns the study of the formation of the HEI, the precursor of the light emitter. This intermediate is not very stable and usually shows a peroxy bond (often included a dioxetanone cycle) which dissociates leading to a compound in its excited state, the light emitter. Theoretical studies on the formation of the HEI in some chemiluminescent systems are presented. The difficulty of the theoretical study of this reaction step is to exhaustively explore all the possible paths leading (or not) to the HEI. However, as the formation of HEI takes place only in the ground state (GS), so called thermal reactions, the study of this first reaction step is less challenging than the investigation of the HEI decomposition (second step) which involves also excited states.

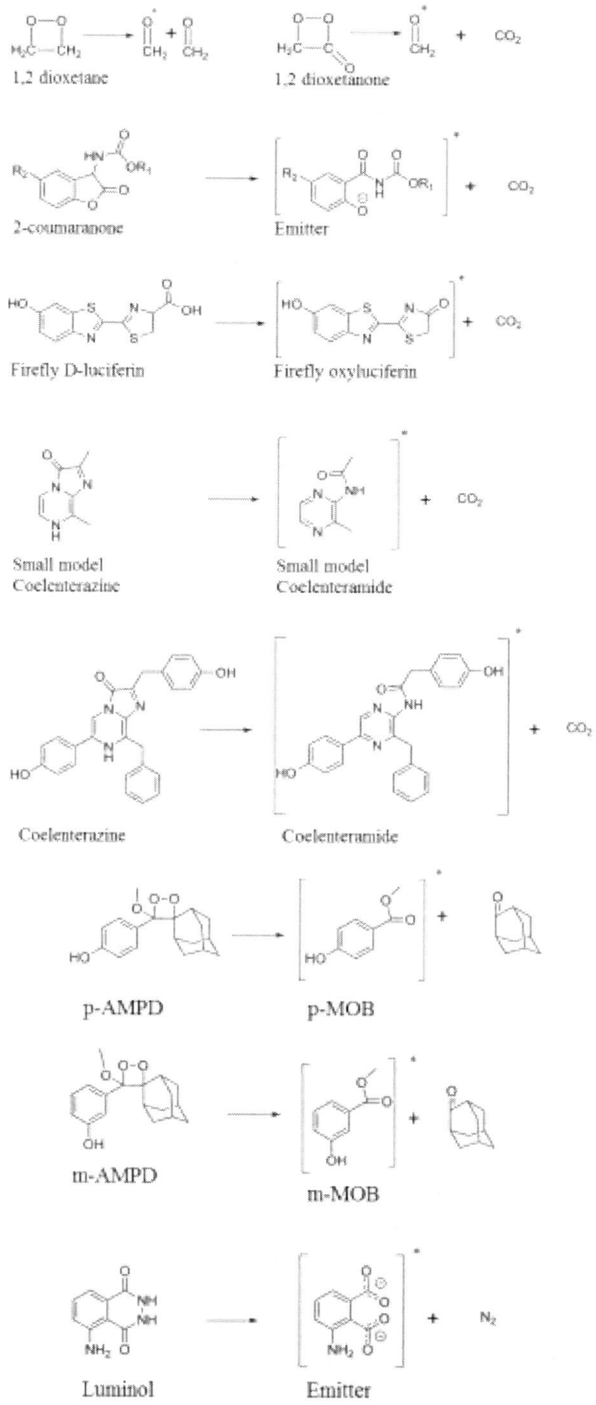

Figure 1. Graphical representation of the chemiluminescent substrates and emitters described in this chapter regardless their protonation state. The star symbol corresponds to the excited molecule.

The second part of the chemiluminescent reaction is the decomposition of the HEI. This part is a big challenge for the modelling as it involves electronic state crossings and so, accurate methods capable of describing more than one state should be used. Very time-consuming multiconfigurational wavefunction methods are the most reliable (see chapter 12 from Roca-Sanjuan et al.). In addition, the complexity of the theoretical study is further increased as efficient chemiluminescent systems show an activator (part of the molecule or external catalyser) that favours the formation of the light emitter over the decomposition of the HEI without light emission.

The last part of a chemiluminescent reaction is of crucial importance as the structure and properties of the light emitter govern the emission color. In this respect, computational chemistry could aid to understand the origin of the experimental emission spectra and the color modulation. For instance, thanks to the computation of the emission energy, the chemical form of the light emitter can be assigned. Moreover, once discerned the light emitter structure, different properties such as the electronic nature of the transition and charge transfer character can be analysed from a computational point of view. In addition, computational simulations can explain the experimentally observed emission color modulation by the environment from the solvent nature to the protein surrounding (in the case of bioluminescence). Finally, the emission spectra of artificial light emitters can be computed to explain the color modulation or to predict structural modifications, which could shift the emission to the blue or to the red, as desired.

2. REACTIVITY IN CHEMILUMINESCENT REACTIONS

Even if the global mechanism of chemiluminescence has been explored for several centuries, the structures and properties of the different chemical intermediates are known for less than 50 years. In general, the elucidation of the structure of the high-energy intermediate (HEI) is the most challenging task, being in some cases still unknown, and limiting the comprehension of the chemiluminescent reaction. For example, the chemical mechanism for the light emission of luminol is still under debate nowadays while being the most used chemiluminescent reactant [1–3].

To get further insights into the mechanism of chemiluminescent reactions, the use of theoretical chemistry is helpful, as it can not only identify the main intermediates of the reaction, but also compute the energetic barriers, the stability of the products and the reaction rates.

In this section we start by describing how to model a generic chemical reaction using theoretical chemistry. Then, we describe the key parts of a chemiluminescent reaction and which type of calculations can be done to get insight into its mechanism.

2.1. Theoretical Study of a Chemical Reaction

Chemical reactions can be modelled by using computational chemistry. In most chemical reactions, one or more reactants react to give one or several products, by the breaking and/or formation of bonds. The coordinate corresponding to this process (e.g., bond distance) is generally considered as the so-called reaction coordinate (RC) (Figure 2). Starting from the reactant, the system evolves along the RC by increasing its energy, upon reaching a maximum so-called transition state (TS). From the TS, the system can continue along the RC decreasing the energy, leading to the product (Figure 2).

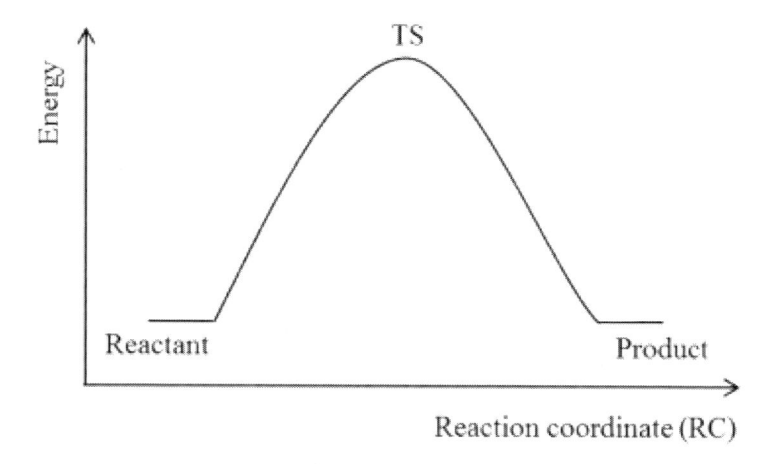

Figure 2. Schematic representation of a chemical reaction. Starting from the reactant, the product is formed by overcoming the energy barrier of the transition state (TS).

Within this reaction scheme, the analysis of the TS is fundamental. The TS corresponds to a structure with only one imaginary frequency, corresponding to the vibrational mode linking the reactants and the products. Contrary, local energy minima are characterized by having all frequencies real positive. The optimization of a TS is a challenging task because it requires information of the reactant and product structures and in some cases, of an initial guess geometry of the TS. To evaluate if the reaction is chemically feasible, the energy difference between the TS and the reactant (i.e., energy barrier) is usually computed. Moreover, the TS geometry is analysed in detail as it gives information about the stereoselectivity (e.g., in the nucleophilic substitution S_N2) and highlights the parts of the molecule involved in the effective vibrational mode driving the reaction to the product. In addition, the computation of the minimum energy path (MEP) starting from the TS allows determining the product structure. The path from the TS to the reactant and to the product is named the intrinsic reaction coordinate (IRC) approach.

2.2. Theoretical Study of a Chemiluminescent Reaction

The brief description of a general chemical reaction given in the previous section is valid when considering only one electronic state, usually the ground state (GS). By definition, in a chemiluminescent reaction, the product responsible of the light emission is obtained in its excited state. To model this type of reaction, we need therefore to take into account more than one electronic state.

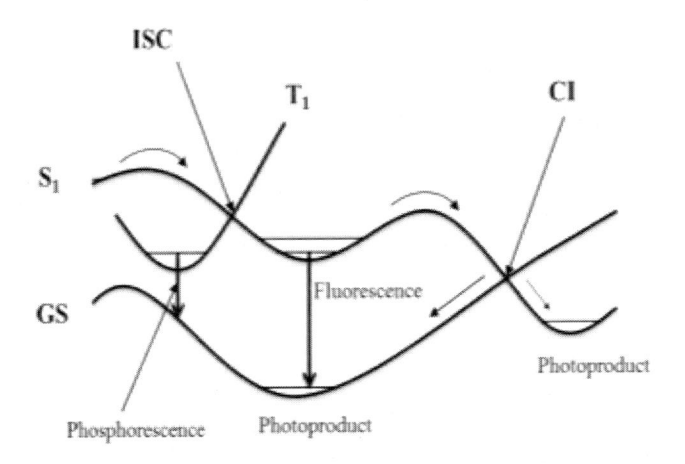

Figure 3. Schematic representation of photochemical processes after population of the first singlet excited state (S_1). The first triplet state (T_1) and the ground state (GS) are also depicted.

Thus, theoretical studies on chemiluminescent reactions take more effort than studying processes in which the system stays in the GS (i.e., thermal reactions). In particular, the computational time and cost increase, as at least two electronic states have to be considered for the calculations. Moreover, computing reactions in the excited state is less trivial than in the GS. One needs to use precise methods to get accurate energies. Finally, crossing points can exist between these different electronic states when their energies are degenerated (Figure 3). Determining the geometry and energy of these intersections is key to model correctly chemiluminescent processes. Among the different crossing points, conical intersections (CI) connect two states with the same spin multiplicity, for example the ground state (GS) and the first singlet excited state (S_1). Conical intersections are often found in chemiluminescent reactions as they allow populating the excited state responsible to the observed light emission. Other types of intersections are the intersystem crossings (ISC), which connect two states with different spin multiplicity, i.e., singlet and triplet states. After population of a triplet state, the system can then emit light by phosphorescence. Figure 3 shows the photochemical processes after population of S_1. From this state to the generation of the photoproduct, the system can: *i)* decay to the ground state by photon emission that is, fluorescence; *ii)* populate a triplet state through an ISC with the subsequent photon emission that is, phosphorescence and *iii)* decay to the GS through a CI (non-radiative decay).

In case of a chemiluminescent reaction, all the processes previously described are possible, being the ISC and CI competitive pathways of the light emission. In detail, the chemiluminescent reaction can be decomposed in three main steps (Figure 4). In the first one, a HEI is formed, which structure is characterized by one peroxide O-O bond, generally included in a 1,2-dioxetanone moiety. In the second step, and contrary to most of thermal reactions in which the system stays in the GS, one product is formed in the excited state after the decomposition of the HEI (driven by the breaking of the peroxide bond). Finally, the third step consists on the relaxation of the product in the exited state and its radiative decay to the GS by photon emission that is, light emission.

In the following sections, we describe in detail the use of computational chemistry to get insight into these three steps of a chemiluminescent reaction.

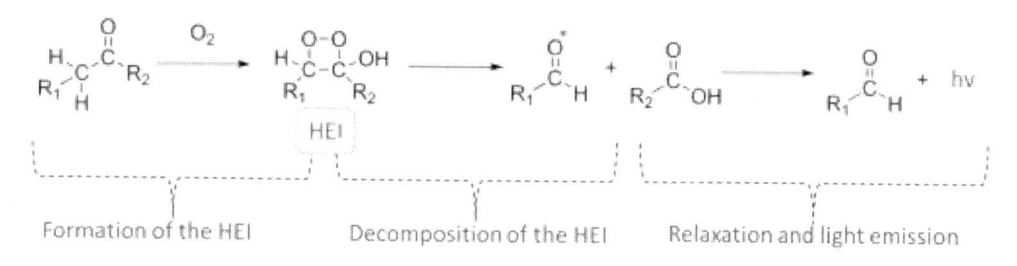

Formation of the HEI   Decomposition of the HEI   Relaxation and light emission

Figure 4. Graphical representation of the three steps of chemiluminescent reactions. The hv symbol corresponds to the energy of the emitted photon (i.e., the light emission).

3. FORMATION OF THE HEI

The first step of a chemiluminescent reaction consists in the formation of the HEI, the precursor of the light emitter. Information (both theoretical and experimental) of the formation of the HEI is sparser than those about its dissociation, as the key intermediates are not stable, complicating their isolation. Moreover, in most cases this step of the reaction is catalysed, making more difficult the study. Some examples are the case of bioluminescent reactions, where the role of the protein and some cofactors in the catalysis of the reaction are still not well understood, or the chemiluminescence of luminol, triggered by the presence of a metal (usually FeIII) [4]. Nevertheless, some computational studies have tried to model the formation of the HEI as is presented in the following.

In order to form the HEI dioxetanone moiety (Figure 4), a dioxygen molecule reacts with a carbonyl group of the chemiluminescent reactant. During the reaction the carbonyl reactant has a singlet GS while the dioxygen is in its stable triplet state. To form the singlet dioxetanone product, the spin rules states that the reaction has to undergo a spin-forbidden process (from the triplet reactant system, carbonyl and dioxygen, to the singlet dioxetanone). Spin-forbidden reactions are difficult to study due to the presence (and therefore the determination) of ISC events. The environment (either the solvent, the protein or other catalysers) has an impact on the energetics and mechanism of this step and makes the study even more difficult.

We first illustrate theoretical studies on the formation of HEI, taking as examples the formation of the dioxetanone moiety of 2-coumaranones and the one of the fireflies' bioluminescent reaction.

3.1. Formation of HEI from the 2-Coumaranone Compounds

2-coumaranones (Figure 1 and 5) are organic compounds which chemiluminescent mechanism is similar to the one observed in several bioluminescent systems, like fireflies. The absence of protein makes this system easier to study and may serve as a case of study to get insight into the behaviour of bioluminescent reactions.

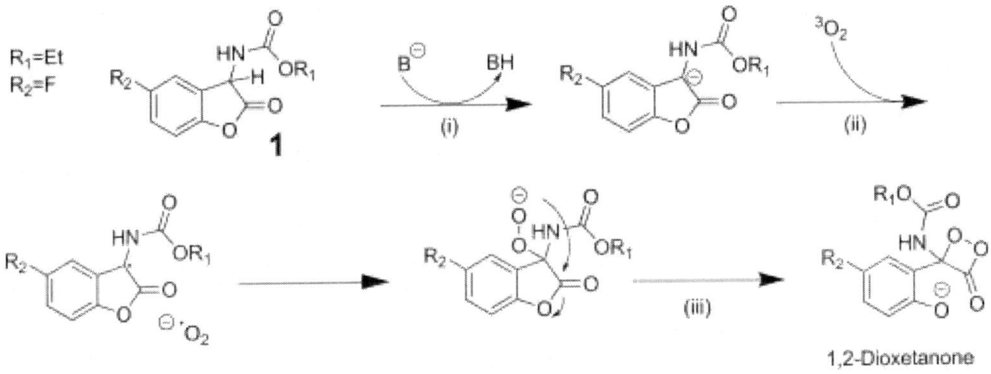

Figure 5. Proposed reaction mechanism of the formation of the HEI in 2-coumaranone. (i) Initial deprotonation, (ii) single electron transfer with molecular dioxygen, (iii) formation of the 1,2-dioxetanone HEI. " SCHRAMM, Stefan et al. Phys. Chem. Chem. Phys., 2017, vol. 19, p. 22852- Reproduced by permission of the PCCP Owner Societies."

A recent joint experimental/theoretical study has explored the global mechanism of 2-coumaranone [5]. Starting from the reactant (**1** in Figure 5), the first step consists in the deprotonation at α-position of the carbonyl group. The calculations estimate that the deprotonation goes through an energy barrier of 11 kcal/mol when ammonia is modelled in the neighbourhood of the proton, making the reaction chemically feasible.

To study the second step of the reaction, corresponding to the approach of the triplet dioxygen towards the deprotonated carbon atom, both the singlet and triplet states have been computed. The calculations show that when the distance between the two reactants is smaller than 2.2 Å, the singlet state is lower in energy than the triplet state. This means that the coordination of dioxygen, which happens around 1.4 Å, is feasible in its singlet state.

Indeed, an ISC is found around 2.2 Å allowing the population of the singlet ground state from the triplet state.

A hypothesis to explain how the system containing the substrate and the dioxygen can reach a singlet ground state claims the presence of one chemical species to abstract one electron from the deprotonated substrate before the coordination of the dioxygen. In the present study, dioxygen plays the role of electron acceptor thus becoming a superoxide radical anion O_2^- (Figure 5). The same behaviour has also been proposed in fireflies' bioluminescent mechanism [6] (see next section).

Finally, the peroxy group of the newly obtained peroxyanion reacts with the nearby lactone resulting in a ring opening and the formation of the dioxetanone moiety, the HEI. The computed energy barriers are relatively low (less than 5 kcal/mol) making the two reaction steps almost immediate.

3.2. Formation of HEI in Fireflies' Bioluminescence

In fireflies, the bioluminescent reaction is catalysed by the protein (luciferase) but also by other co-factors such as the adenosine tri-phosphate (ATP) and magnesium ions. ATP decomposes in the presence of magnesium ions into adenosine mono-phosphate (AMP) and binds to the substrate (luciferin) at the beginning of the bioluminescent reaction. The role of magnesium ions as catalysers of the decomposition of ATP substrate is known but not their role in the rest of the bioluminescent reaction. As a result, magnesium ions are usually not included in the models used for studying the bioluminescent reaction. However, even if the impact of the protein is complex to model and understand, the active site of the protein, where the substrate is present, has to be somehow modelled due to its clear effect on the bioluminescent reaction.

We describe here the HEI formation in fireflies, where the hypothesis of the deprotonated benzophenol of the luciferin is made [6]. Starting with the D-luciferyl adenylate substrate (Figure 6) inside the protein, the first step of the formation of the dioxetanone corresponds to the deprotonation at the α-position of the carbonyl group. Here, contrary to the 2-coumaranones' HEI formation, no additional base is necessary. Indeed, in the protein cavity, a close amino acid or the dioxygen itself, can abstract the proton. Depending on the molecule acting as the base, two hypotheses can be explored resulting in three proposed mechanisms (Figure 6). In mechanism (a), similar to the one of 2-coumaranone, a basic amino acid abstracts a proton from the luciferin, leading to a negatively charged carbon. A single electron is transferred from the negative charged carbon to dioxygen, resulting in the creation of a radical superoxide anion and a radical substrate. Both radicals react to form a C-O bond. Proposed mechanism (b) starts with the same first step as mechanism (a) but the negatively charged carbon attacks directly the triplet dioxygen. Mechanism (c) starts with the abstraction of the proton by dioxygen, giving rise to a hydroperoxide molecule, which has a singlet ground state and can react easily with the substrate.

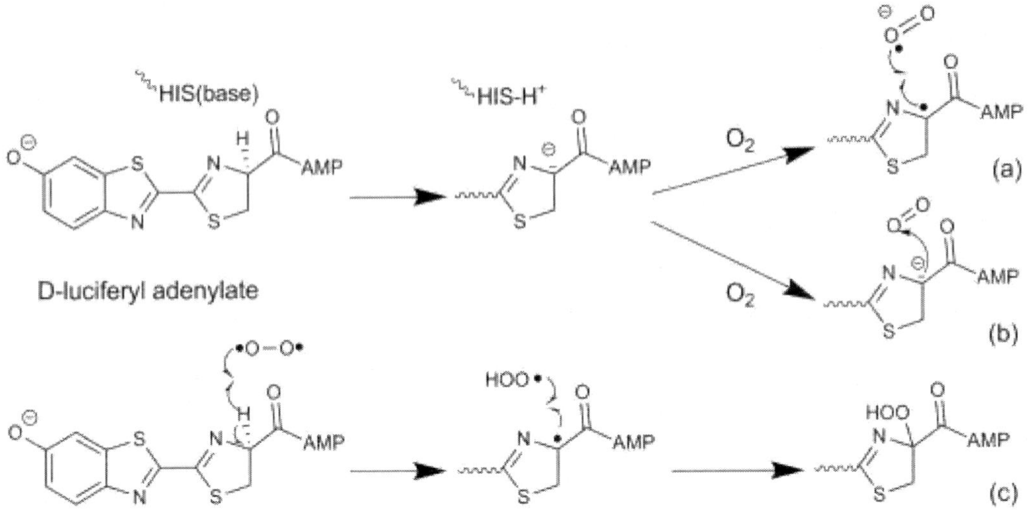

Figure 6. Suggested pathways for the reaction of triplet O_2 and the deprotonated D-luciferyl adenylate leading to the singlet 1,2-dioxetanone. (a) Deprotonation mechanism followed by a single electron transfer to form a superoxide anion; (b) deprotonation mechanism followed by direct attack of dioxygen; (c) hydrogen abstraction mechanism to form a hydroperoxide followed by homolytic formation of C-O bond. "Reprinted (adapted) with permission from (VACHER, Morgane et al. Chemical Reviews, 2018, vol. 118, no 15, p. 6924-6974.). Copyright (2018) American Chemical Society.".

While hypothesis (c) has only been experimentally studied [7], the first and the second ones have been also computationally analysed [6, 8]. The calculations show that the single electron transfer is instantaneous as soon as a dioxygen molecule is inside the cavity near the deprotonated D-luciferyl adenylate. During the approach of the dioxygen to the substrate, the two first singlet and triplet states (GS, S_1, T_1 and T_2) are degenerated until the formation of the first C-O bond. This favours the system initially in its triplet ground state to undergo the ISC leading to a singlet dioxetanone intermediate. The theoretical study also shows that the formation of the dioxetanone ring is only possible when the bond between the AMP moiety and the D-luciferyl moiety is broken. The protein also plays an important role as the dioxygen molecule is well stabilized inside the protein through interactions with both water molecules and the amino acid that has deprotonated the luciferin. As a result, the transferred proton might move from the amino acid to the superoxide anion leading to the hydroperoxide molecule. The reaction would then finish by following mechanism (c). Therefore, the mechanisms (a) and (c) may coexist or be part of only one global mechanism.

As we have seen, a lot of unanswered questions remain unsolved on the formation of the HEI and theoretical studies can help to decipher the correct mechanisms. In the following section, we present theoretical studies of the HEI decomposition and the information that can be drawn from them.

4. DECOMPOSITION OF CYCLIC PEROXIDE

To understand the mechanism of the decomposition of the HEI within complex systems, we first consider the smallest possible substrates, i.e., 1,2-dioxetane and 1,2-dioxetanone (Figure 1). The first one has been extensively studied both experimentally and theoretically, while the second is less studied although is a better model of chemiluminescent systems [9]. It should be pointed out that these two substrates are far from the exact description of the chemiluminescent and bioluminescent HEI. However, they are great systems to explain the principle of chemiluminescent reactions.

4.1. Decomposition of 1,2-Dioxetane

The decomposition of 1,2-dioxetane starts with the breaking of the peroxide O-O bond (Figure 7). Calculations have shown that this first step is induced by the torsion of the O-C-C-O dihedral angle increasing from 19° to 44° at a first TS (TS$_1$). The computed activation barrier is around 23 kcal/mol and corresponds to the rate-limiting step of the mechanism [9–12].

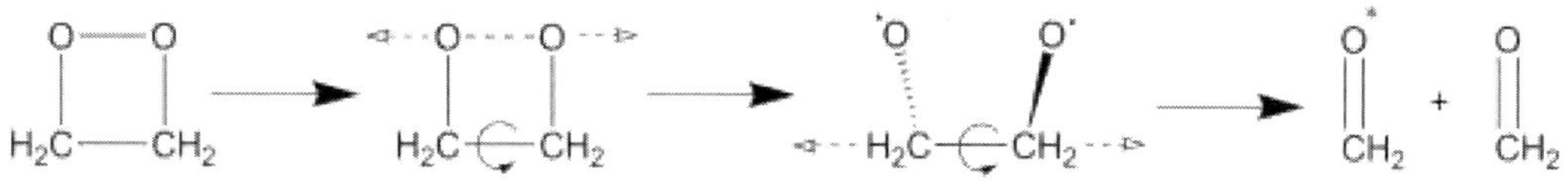

Figure 7. Simplified reaction scheme of the decomposition of 1,2-dioxetane.

During this step, the breaking of the peroxide bond corresponds to a homolytic dissociation, leading to a biradical singlet species where each oxygen atom has 3 electrons distributed in two lone pairs. At the TS$_1$, the GS becomes degenerated with S$_1$ and T$_1$, giving rise to some possible crossings, either CI or ISC. After TS$_1$, the main reaction coordinate corresponds to the breaking of the C-C bond. As soon as the C-C bond starts to elongate, the degeneracy between the GS, S$_1$ and T$_1$ is lifted. Then it has been shown that two formaldehyde molecules are formed in T$_1$. [13]. Indeed, the triplet quantum yield for this reaction is high, around 50% resulting in very low chemiluminescent efficiency. The computed energy barrier leading to a second TS (carbon-carbon dissociation) is lower along the triplet state pathway (around 5 kcal/mol less than in S$_1$) [14]. Recently, quantum molecular dynamic simulations have been performed starting from the geometry corresponding to TS$_1$ (after breaking of O-O bond) [15]. In order to have a successful dissociation, specific geometrical conditions have to be achieved, with the O-C-C-O dihedral angle larger than 55° and the O-C-C angle smaller than 117°. Otherwise the system remains trapped in the state-degenerated region. Moreover, the dissociation of the dioxetane has only been achieved with the system in the T$_1$ state, supporting the formation of triplet products.

4.2. Decomposition of 1,2-Dioxetanone

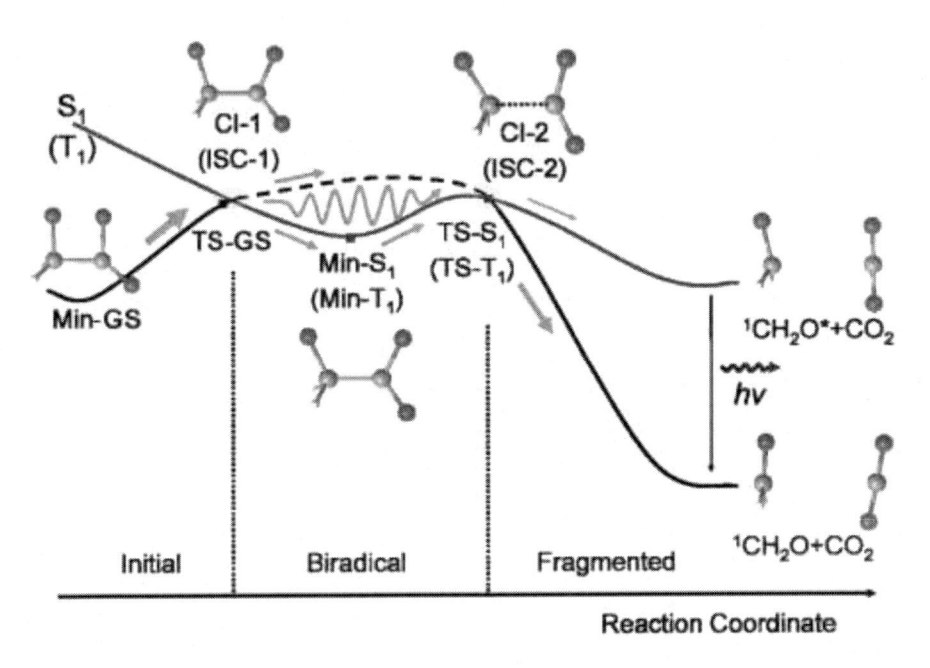

Figure 8. Potential energy profile of the dissociation of 1,2-dioxetanone. "Reprinted (adapted) with permission from (LIU, Fengyi et al. Journal of the American Chemical Society, 2009, vol. 131, no 17, p. 6181-6188.). Copyright (2009) American Chemical Society."

The 1,2-dioxetanone moiety is often found in the HEI of many chemiluminescent systems. The dissociation mechanism of 1,2-dioxetanone starts with an O-O bond elongation as for 1,2-dioxetane. The computed energy barrier is similar for these two systems, around 26 kcal/mol. At the TS (TS-GS in Figure 8), the system reaches a CI and an ISC due to degeneracy of GS, S$_1$ and T$_1$. Unlike for 1,2-dioxetane, the degeneracy of these three states is kept during the elongation of the C-C bond (Figure 8). The three states gather close to the second TS (TS-S$_1$ in Figure 8), corresponding to the C-C bond breaking. Dissociation leads to one excited formaldehyde fragment (S$_1$ or T$_1$) and one carbon dioxide in its GS. The resulting chemiluminescent efficiency is however still weak.

Despite the fundamental information obtained during the study of 1,2-dioxetane and 1,2-dioxetanone, the low efficiency of the formation of the S_1 products as well as the high activation barrier for the dissociation reaction do not reproduce correctly the behaviour of chemiluminescence observed in living species.

4.3. Dissociation of the Cyclic Peroxide in Chemiluminescent Systems

4.3.1. Description of the Mechanisms

The previously described mechanisms cannot be used to model the chemi- and bio-luminescent reactions because of two main issues. First, the activation barriers for 1,2-dioxetane and 1,2-dioxetanone dissociation are too high to be reached. The second reason is the presence of a highly populated triplet state after the decomposition, resulting in a low efficiency of light emission.

Usually, in chemiluminescent reactions an activator (ACT), either inter- or intra-molecular to the HEI, facilitates the overall reaction by decreasing the energy barriers found along the path. This ACT presents an electron reservoir and a charge transfer controlling group. Two different mechanisms have been described to explicit the effect of these activators (Figure 9 and 10) [9, 16–20].

The Chemically Initiated Electron-Exchange Luminescence (CIEEL) corresponds to a stepwise mechanism. Here we consider the ACT as an external catalytic fluorescent activator, i.e., responsible of the light emission. The activator interacts first with the HEI forming a charge transfer complex. Then, the ACT transfers one electron on the peroxide bond (Figure 9, step 2). The O-O bond breaks, followed by the carbon-carbon bond opening (Figure 9, step 3), like in the uncatalysed mechanism. A neutral species is eliminated (CO_2 in dioxetanone) and the remaining radical anion carbonyl transfers back an electron to the ACT (Figure 9, step 4). A neutral carbonyl compound is achieved as well as the activator in its electronically singlet excited state which decays to the GS by emitting light (Figure 9, step 5).

The second mechanism proposed is the Charge-Transfer Induced Luminescence (CTIL). Contrary to the stepwise CIEEL, the CTIL is a concerted mechanism where only a partial electron is transferred to the peroxide bond. Thus, no radical anion species are involved. Thanks to the presence of activators, the two main drawbacks to efficient light emission found for 1,2-dioxetane and 1,2-dioxetanone can be avoided. The activation barrier is lowered because the O-O bond is destabilized by the (full or partial) transfer of an electron from the ACT. The population of S_1 is preponderant compared to the population of T_1, increasing the chemiluminescent efficiency. In the following, the bioluminescence of fireflies is presented as it is a clear example of the role of an ACT on the overall mechanism.

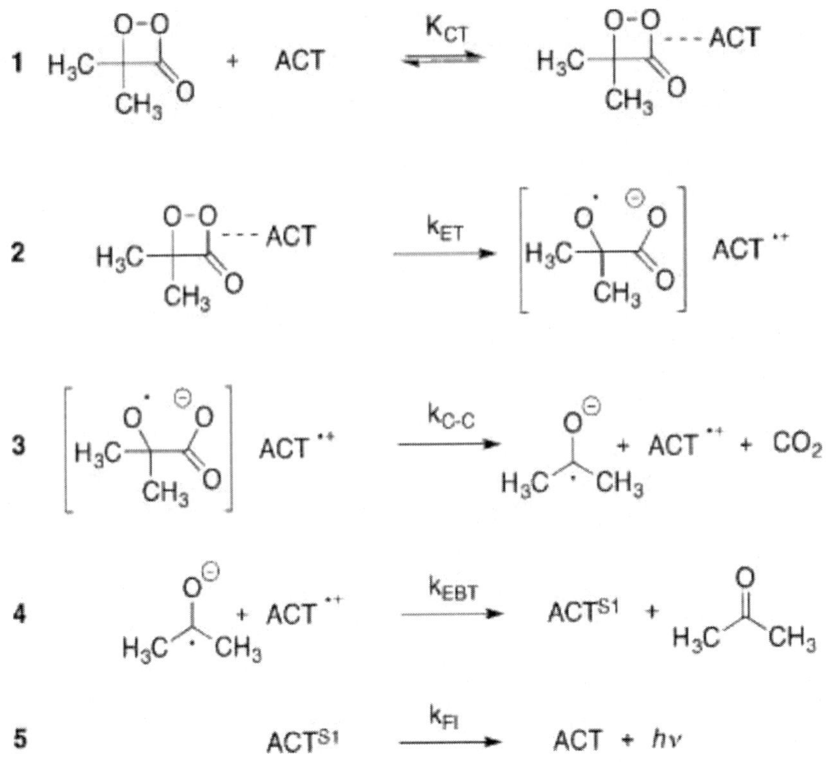

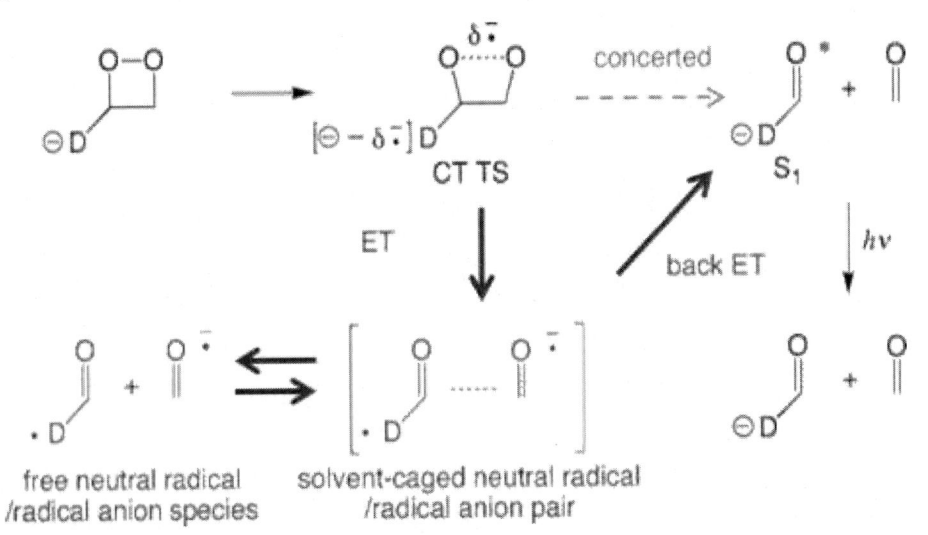

Figure 9. Individual chemical reaction steps of the chemically initiated electron-exchange luminescence (CIEEL) mechanism for the catalysed decomposition of dimethyl-1,2-dioxetanone with a catalytic fluorescent activator (ACT). "Reprinted (adapted) with permission from (VACHER, Morgane et al. Chemical Reviews, 2018, vol. 118, no 15, p. 6924-6974.). Copyright (2018) American Chemical Society.".

Figure 10. Chemical representation of the Charge Transfer Induced Luminescence (CTIL) mechanism (dash lines) compared with the Chemically Initiated Electron Exchange Luminescence (CIEEL) mechanism (bold lines). In this figure, the ACT (described in Figure 9) is the "D" moiety of the molecule. "Reprinted (adapted) with permission from (ISOBE, Hiroshi et al. Journal of the American Chemical Society, 2005, vol. 137, no 24, p. 8667-8679.). Copyright (2005) American Chemical Society.".

4.3.2. Decomposition of Firefly's Dioxetanone

In fireflies, the HEI corresponds to the firefly's dioxetanone and decomposes into carbon dioxide and oxyluciferin, the emissive substrate. The bioluminescent efficiency is around 40%, which makes firefly one of the most efficient systems. To achieve this, the dioxetanone moiety is bonded to an extended π-conjugated system (Figure 11). The extended π-conjugated system plays the role of activator for the reaction, as it is rich in electrons. However, the protonation state of the substrate, i.e., neutral or anionic firefly's dioxetanone inside the protein environment, is still under debate. Computational calculations were performed in solvent to decipher the protonation state of the activator and the CIEEL or CTIL mechanism taking place in firefly's dioxetanone dissociation [21].

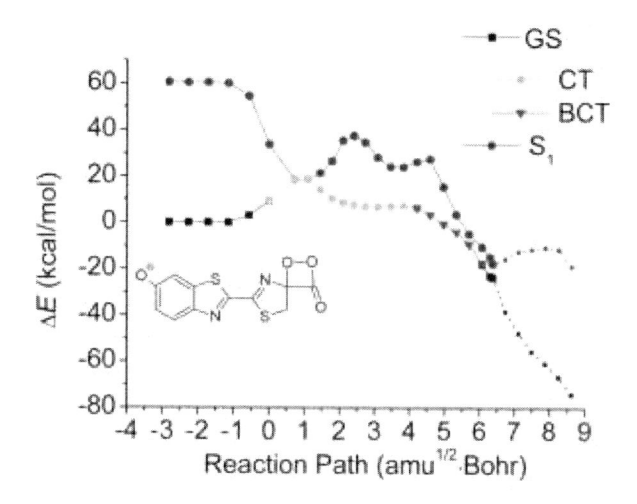

Figure 11. Calculated potential energy plot of GS and S_1 along the dissociation path of the deprotonated firefly's dioxetanone in benzene solvent. CT, charge transfer; BCT, back-charge transfer. "Reprinted (adapted) with permission from (YUE, Ling et al. Journal of the American Chemical Society, 2012, vol. 134, no 28, p. 11632-11639.). Copyright (2012) American Chemical Society."

The pathway of the decomposition of the neutral form is similar to the one of 1,2-dioxetanone. After the elongation of the O-O bond, the system reaches a biradical region where the GS, S_1 and T_1 states are degenerated. This degeneration is kept along the dissociation path, which favours the population of the triplet state by ISC, leading to a very low chemiluminescent efficiency. Moreover, the activation barrier for reaching this region is high, around 26 kcal/mol. Therefore, the neutral form of the substrate is not probably the HEI in fireflies. Moreover, as no electron transfer is observed along the path, neither CIEEL nor CTIL mechanisms can explain the decomposition of this neutral form.

However, the pathway of the anionic firefly's dioxetanone decomposition is different (Figure 11). The reaction starts with the elongation of the O-O bond until reaching the TS. Then the nature of the GS changes to a biradical state. Indeed, a charge transfer from the phenolate to the peroxide bond occurs, destabilizing the O-O bond. As a result, the activation barrier for this step is decreased to 15 kcal/mol, facilitating the chemiluminescent reaction [11, 22, 23]. Then, during the dissociation of the C-C bond a CI is reached, from where the system can decay to the GS (non-radiative deactivation) or continue in S_1 giving light emission. The decomposition of anionic firefly's dioxetanone occurs via a charge transfer asynchronous-concerted process as described in CIEEL or CTIL mechanisms. To decipher between the two mechanisms, the charge population on the oxyluciferin and on the carbon dioxide along the dissociation pathway are analysed. Two different processes can be identified (Figure 11), one corresponding to a charge transfer (CT) from oxyluciferin to carbon dioxide and one to a back-charge transfer (BCT), where the electron goes back from carbon dioxide to the carbonic substrate. In the CT the absolute total charge of oxyluciferin gradually decreases (charge from -0.96e to -0.28e) while the one of CO_2 gradually increases (charge from -0.04e to -0.72e). During the BCT process, the absolute total charge gradually increases on the oxyluciferin and gradually decreases on CO_2. No full electron is transferred during the dissociation. Besides, the gradually charge transfer suggests the presence of a CTIL mechanism. In conclusion, in the decomposition of firefly's dioxetanone, the deprotonation of the activator is necessary to explain the efficiency of the chemiluminescence through a CTIL mechanism.

The dissociation of the HEI can be explained if an ACT activates the reaction. This leads to the formation of a chemiexcited state (CS) that is responsible for the chemiluminescence emission as described in the next section.

5. LIGHT EMISSION IN CHEMILUMINESCENCE

The previous sections deal with the steps leading to a chemical species in its first singlet excited state, the light emitter. Herein, we focus on the last step of the chemiluminescent reaction that is, light emission. In a first part, we describe how computational chemistry can figure out if the chemiluminescence and fluorescence correspond to the de-excitation of the same species. In a second part, we discuss the possibility of having different chemical forms for the light emitter and the necessity of clarifying which one is responsible of the emission. Indeed, for some light emitters, different chemical forms could be possible due to inter-exchange reactions (e.g., protonation, tautomerization). Thanks to a better understanding of the nature of the light emitter, numerous experimental and theoretical studies were performed to change the color of the emitted light by modification of its structure, leading to artificial light emitters. The structural modifications of the light emitter designed to tune the emitted color can be explained and predicted by computational studies as presented in the third part of this section. Finally, the effect of the environment, from the solvent nature to the protein surrounding (in case of bioluminescence), is presented.

From a theoretical point of view, the study of the light emitter implies to find the structure of the most stable local minimum in S_1 via an optimization algorithm. When the system is in the excited state optimized structure, the difference of energy between the GS and S_1 gives an estimation of the energy of the emitted light (so the emitted wavelength and color). Moreover, an estimation of the relative expected intensity of the light is given from the calculations by the oscillator strength (f) of the transition.

5.1. Light Emission through Chemiexcited or Fluorescent State

Because of the complexity of carrying out a chemiluminescent reaction, the study of the concomitant light emission is often difficult to perform. To solve this issue, recording the fluorescent properties of the product obtained after the chemiluminescent reaction is a commonly used strategy. Doing so, it is assumed that the chemiluminescent minimum on S_1 (or chemiexcited state, CS) is the same as the fluorescent minimum (or fluorescent state, FS) obtained by photoexcitation (Figure 12). However, depending on the studied system, this assumption may not always be true. Indeed, if the energy barrier in S_1 is significantly high, then the CS and FS correspond to two different chemical species (Figure 12A), leading the photoexcitation and chemiluminescent experiments to different colors of the emitted light. If the energy barrier connecting the CS and FS is small, then the chemiluminescent emitted light could derive from both CS and FS (Figure 12B). Finally, if there is no energy barrier connecting the CS and FS, then CS and FS correspond to the same species (Figure 12C). Computational studies can unveil the geometries and electronic structures of both CS and FS, explain the relation between them and discern between the three previously discussed scenarios.

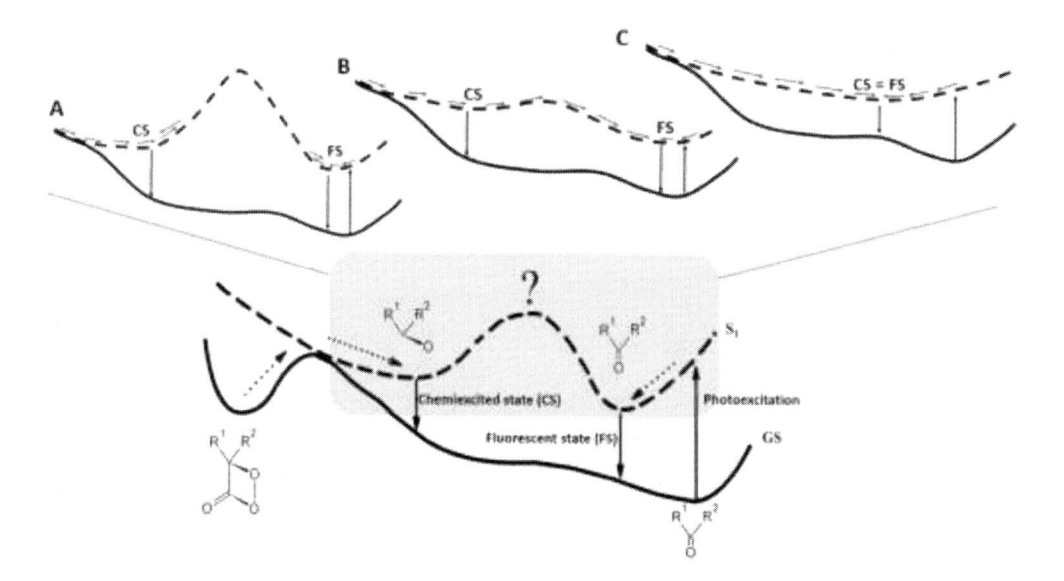

Figure 12. Schematic potential energy surface highlighting the difference between chemiexcited and fluorescent states. "Reprinted (adapted) with permission from (NAVIZET, Isabelle et al. Photochemistry and Photobiology, 2013, vol. 89, issue 2, p. 319-325.). Copyright (2012) Wiley Periodicals, Inc. Photochemistry and Photobiology and (2012) The American Society of Photobiology."

The case of different FS and CS due to a high energy barrier connecting them was first raised in a theoretical study on a small model of the coelenterazine and cypridinid luciferins [24]. To obtain the CS, the reaction is studied from the decomposition of the HEI and departure of CO_2 (strategy 2 in Figure 13). The obtained chemiexcited state has a C=O group which shows a sp^3 hybridization (Figure 13). It corresponds to a charge transfer from the π orbital of the pyrazine ring to the π^* of the C=O group. On the other hand, the FS corresponds to a minimum in S_1 after excitation of the light emitter from the GS (strategy 1 in Figure 13). In this structure, the carbonyl C=O group exhibits an sp^2 hybridization and is a result of excitation delocalized over the π-conjugated system (Figure 13). Hence, the electronic nature of the excited species responsible for the chemiluminescence and the fluorescence is different, as well as the corresponding geometrical structures. The color of the light emitted during these two experiments is therefore not necessary the same. As the easier way to (theoretically or experimentally) study a chemiluminescent process is to perform fluorescent experiments, these results warn the scientific community that both states can be completely different.

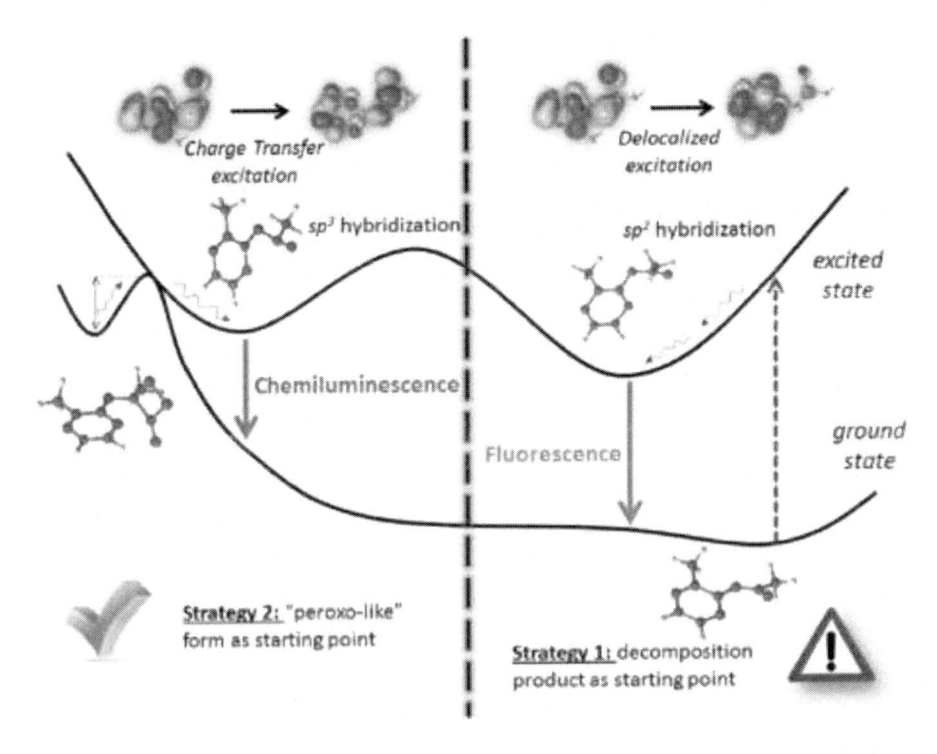

Figure 13. Scheme of the comparison between chemiluminescent pathway (left) and photoexcited pathway (right). The geometries and electronic transitions of chemiexcited and fluorescent states are highlighted. "Reprinted (adapted) with permission from (ROCA-SANJUAN, Daniel et al. Luminescence, 2012, vol. 27, p. 155-156.). Copyright (2012) John Wiley & Sons, Ltd.".

Following this work, several other studies were published to tackle the problem of CS against FS [23, 25, 26]. In that respect, another study has been done on the obelin system but using this time the whole substrate instead of a minimal model [25]. Here the CS is computed as a dark state, i.e., not light emission because of a small intensity (negligible f value). The hypothesis proposed is that the substrate has accumulated enough energy at the end of the chemiluminescent reaction to overcome the energy barrier (TS) between the CS and FS. Hence, the CS proceeds to the FS leading to light emission. The main differences between CS and FS are that, in CS, the amide moiety is not planar and charge transfer character is larger.

Another example is the decomposition of the meta and the para isomers of the 3-(2'-spiroadamantyl)-4-methoxy-4-(3"(4")-hydroxyphenyl)-1,2-dioxetane, m-AMPD and p-AMPD [26]. As illustrated in Figure 1, the only difference between these two chemiluminescent substrates is the position of the phenol group either in meta (m-AMPD) or para (p-AMPD) position. However, the computed excited state path between the CS and the FS is dependent of the position of the phenol group. In both cases, the chemiexcited states are stable, but for m-AMPD the energy barrier between the two emissive states is so small, around 3 kcal/mol that the CS is rapidly converted into the FS. The theoretical study concludes that the emitted light from either chemiluminescence or fluorescence should be the same, which is confirmed by experiments. However, for p-AMPD, the computed energy barrier between the CS and FS is higher, around 16 kcal/mol and the FS cannot be reached. The low value of the oscillator strength for the chemiexcited state also favours a non-radiative decay, i.e., a low chemiluminescence yield (as experimentally observed).

Finally, some other systems do not show any difference between CS and FS. For example, in the case of fireflies' bioluminescence, the calculations of different optimized structures in the first excited state of the light emitter in its phenolate-keto form shows that the calculated CS and FS structures are the same. No energetic barrier has been computed between these two states, in gas phase, in solvent or inside the protein (Figure 14) [23]. However, several factors discussed in the next paragraph can affect the color of the light, e.g the chemical form of the emitter.

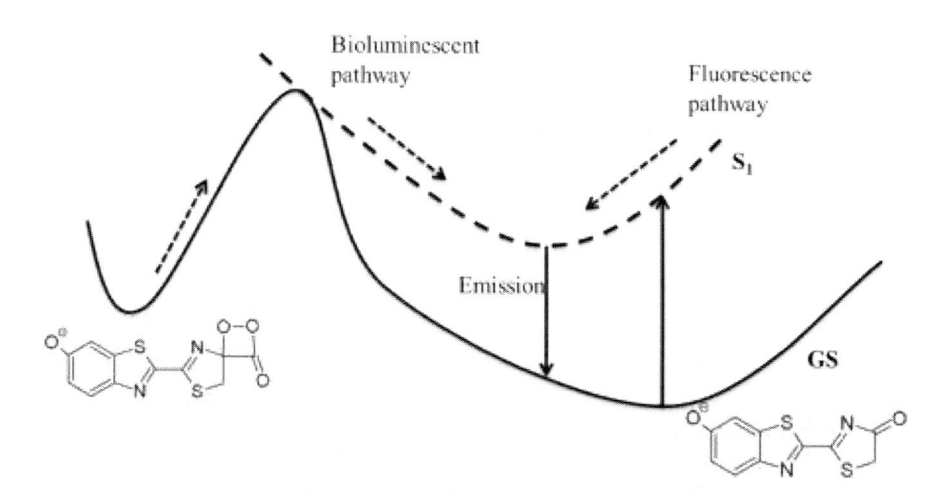

Figure 14. Schematic potential energy surface between chemiexcited and fluorescent states in fireflies.

5.2. Chemical Nature of the Light Emitter

In some chemiluminescent systems, the structure of the light emitter can lead to diverse inter-exchange reactions such as protonation/deprotonation of hydroxyl or amine groups or tautomerization reactions. When deprotonation occurs in the excited state, the process is known as an excited state proton transfer (ESPT). This way, an equilibrium of the different chemical species can take place both in the ground and in the excited state. As a consequence, two possibilities arise: *i)* an equilibrium in the excited state between two or more forms leading to light emission or *ii)* one of these forms is the most stable one in the excited state and hence, it is the only responsible of the light emission, which is the general case. From an experimental point of view, it is not straightforward to assign which form(s) is(are) responsible of the emission. For this aim, computational studies have been performed to check if the emission wavelengths and intensities of these forms are different enough to attribute the light emission to one chemical form of the emitter. Clear examples are the chemiluminescence of 2-coumaranone, the light emitter of fireflies (oxyluciferin), and the one of diverse marine organisms, known as coelenteramide (Figure 1).

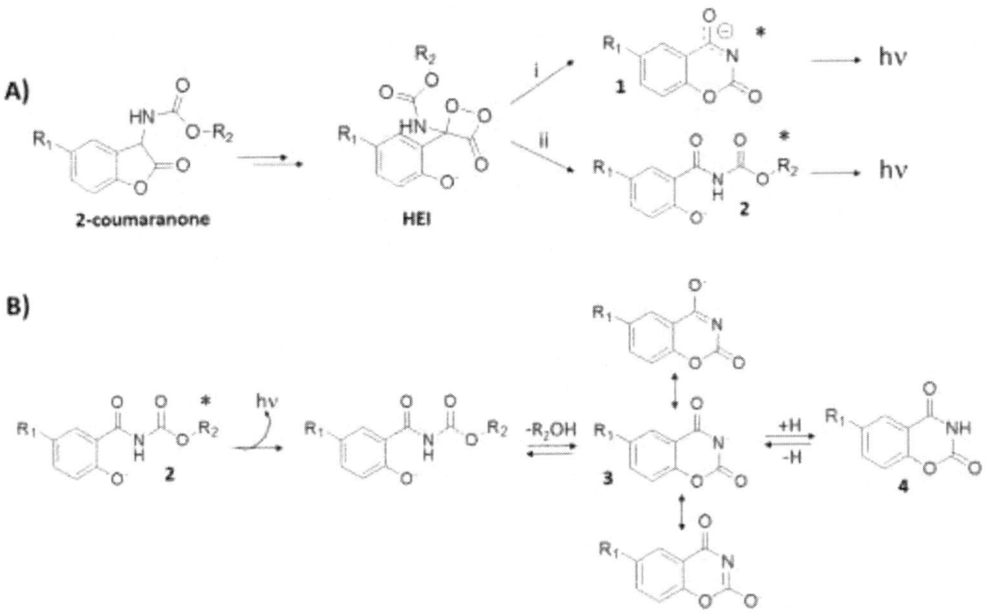

Figure 15. A) Chemiluminescent reaction of 2-coumaranone showing the two possible light emitters 1 and 2. B) Thermal reactions after light emission of 1, leading to the experimentally isolated compound 4.

Starting with the chemiluminescence of 2-coumaranone, two different light emitters could be possible resulting from two different paths of the HEI decomposition (1 and 2 in Figure 15A). To get insight into this finding the emission energies and intensities (i.e., oscillator strength) of 1 and 2 have been computed [27]. The results show that the emission computed for 1 is blue-shifted compared to the chemiluminescence and its computed intensity (i.e., oscillator strength) is quite low. However, the calculated emission energy for 2 matches with the chemiluminescent emission and its computed intensity (i.e., oscillator strength) is significantly high. For this reason, compound 2 has been assigned as the light emitter of 2-coumaranones. In fact, 2 comes directly from the decomposition of the dioxetanone HEI and corresponds to the CS [5]. Nevertheless, compound 4 has been experimentally isolated from the chemiluminescent reaction mixture. To explain this finding, it has been proposed that 2 could be transformed to the cyclic compound 3 (three resonance structures in equilibrium), which finally leads to 4 in more acidic media (Figure 15B).

Regarding firefly's oxyluciferin, the light emitter could coexist in six different chemical forms due to protonation/deprotonation reactions of the phenol and enol groups together with a keto-enol tautomerization (Figure 16). Despite numerous studies, it is still difficult to know which form is the main emitter. To get insight into this enigma, the emission energies have been computed [28–31], and the emission spectra have been simulated in water solution [32] for the six chemical forms. By analysing the computed data, it has been observed that the emission from the protonated neutral forms (phenol-keto and phenol-enol) are blue-shifted compared to the deprotonated ones. Specially, for phenol-enol the emission energy is clearly blue-shifted compared to the experimental emission spectra. For this reason, the neutral forms were excluded. Moreover, the enolate-phenol form was excluded for two reasons: first, the oscillator strength of the emission (i.e., emission intensity) is lower than the ones computed for the other deprotonated forms and second, guided by the reaction leading to the excited oxyluciferin species, the oxygen atom of the phenol group should be negatively charged [33–36]. In fact, by comparing the simulated and experimental emission spectra of the different chemical forms of oxyluciferin and some synthetic analogues it has been demonstrated that an efficient ESPT takes place and so, only the emission of the deprotonated forms is observed [32, 37].

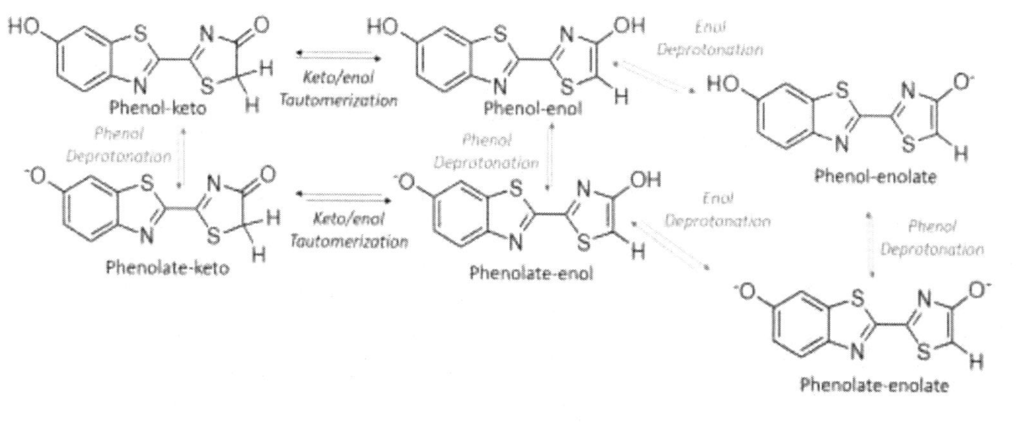

Figure 16. Chemical structure of the six possible forms of firefly's oxyluciferin and the inter-exchange reactions connecting them.

However, the origin of the color modulation from green-yellow to red observed by changing the pH is still under debate. Different hypotheses have been proposed but more computational studies are needed to clarify this experimental finding [28, 38, 39].

Similarly to firefly's oxyluciferin, the light emitter of a wide range of bioluminescent processes of marine organisms, so-called coelenteramide, could coexist in different chemical forms. In fact, for coelenteramide the general picture is even more complicated as the computed emission intensity of the chemiexcited state (CS) is quite low and so not detectable. Hence, the CS is not the species responsible of the light emission. In fact, the system continues to evolve in the excited state to a structure corresponding to the FS, as the energy barrier connecting CS and FS is not so high (Figure 12B). The calculated FS structure shows a large emission intensity [24, 25, 40, 41]. Although it has been clarified that the FS is responsible of the light emission, different chemical forms are possible due to deprotonation of the hydroxyl and amino groups together with phenolate-amide tautomerization (Figure 17). Several computational studies have been published focused on the calculation of the emission energies and intensities of the possible chemical forms of coelenteramide. For instance, the equilibrium constants in the excited state of the processes connecting the different forms have been calculated in order to estimate their relative stability [40]. It has been experimentally demonstrated that the neutral form (Figure 17A) is the primary excited state product formed after decarboxylation, i.e., the CS [42, 43]. However, the computation of the emission energies of both the neutral and phenolate forms demonstrates that the value found for the neutral derivative is blue-shifted compared to the experimental emission [25]. However, the one calculated for the phenolate form (Figure 17B) matches the experimental emission. In fact, thanks to the computational study of coelenteramide inside the protein, it has been shown that an efficient excited state proton transfer (ESPT) can take place favoured by the interaction with the side chain of close amino acids such as His16, Trp86 and Tyr82 [41]. This way, the neutral form is quickly converted into the phenolate one in the excited state, the last being the blue-light emitter.

Computational chemistry has played in this study an essential role to get insight into the chemical nature of the light-emitter. This point is of crucial importance as the modulation of the emission properties (e.g., color emission, intensity) are directly related to the chemical structure. Only the calculations were able to determine the structure of the emitter and rationalize the emission properties and their modulation.

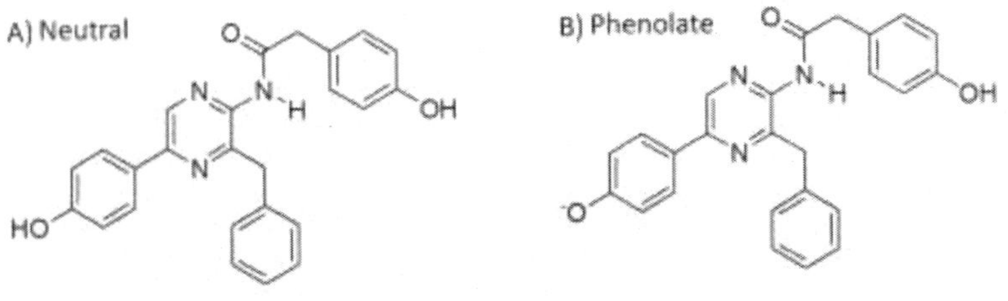

Figure 17. Chemical structure of A) the neutral and B) phenolate forms of coelenteramide.

5.3. Modifying/Tuning the Light Emitter

Once the chemical form of the chemiluminescent reactant or the light emitter is known and its emission color origin and modulation analysed, usually synthetic derivatives are proposed. The aim of the design of these derivatives is mainly the enhancement of the chemiluminescent properties, that is, the emission intensity, the chemiluminescent quantum yield or the tuning of the emission color. For instance, for biological applications green or red emitted light is needed, as low in energy light is required to ensure a deep tissue penetration. In this regard, computational studies can be applied to: *i)* rationalize the experimental chemiluminescent properties of the synthetic derivatives compared to the parent compound and *ii)* compute the emission properties of novel derivatives prior to the synthesis, as a screening tool to select the ones with the desired properties.

Computational studies have been performed to analyze the change of the emission color or intensity when modifying the chemical structure of a light emitter. For instance, different derivatives of luminol have been synthetized, containing either a quinoxaline or a o-hydroxyphenyl benzimidazole unit (Figure 18A and B, respectively) [44,45]. Experimentally, it has been observed a red-shift of the emission spectra for many of the proposed derivatives compared to luminol. To get insight into this finding, the emission energies of these derivatives were computed, getting the same trend, which could be probably due to the higher conjugation extent of the chromophore and to more energy losses in the excited state due to dissipation. The molecular orbitals involved in the transition were analyzed, giving an idea of the charge transfer character of the emission, directly related to the emission properties of the derivative.

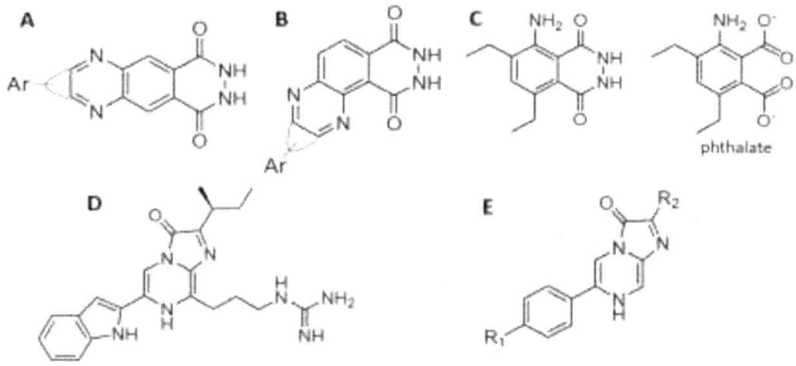

Figure 18. Schematic representation of A) quinoxaline, B) o-hydroxyphenyl benzimidazole and C) dialkyl luminol (and phthalate form) synthetic derivatives of luminol. D) Natural cypridina luciferin and E) its computationally studied derivatives.

Similarly, the broad application of fireflies' bioluminescence has driven the design of numerous derivatives, mainly motivated by the idea of achieving a red-shift of the emitted light, crucial for biological applications as aforementioned [46]. Computational studies have been performed to both analyze the spectroscopical properties of already synthetized derivatives and to predict the emission modulation by structural modifications. This way, a wide set of luciferin substrates has been proposed and synthetized (from amino-substituted, cyclic amino, derivatives with increasing conjugated chain or derivatives to block the keto or enol form) covering a great range of emission properties (Figure 19) [32, 46–49].

Furthermore, computational studies could be useful for understanding the modification of the chemiluminescent properties (i.e., efficiency, emission color and intensity) of synthetic derivatives. An example is dialkyl luminol derivatives (Figure 18C) for which a 20-fold increase in chemiluminescence efficiency compared to luminol was achieved [4]. To rationalize this finding, the chemiluminescent mechanism has been computed for both luminol and dialkyl luminol derivatives. The increase of the chemiluminescence efficiency in dialkyl luminol derivatives is due to steric gearing, which eases the conversion of the HEI to the electronically excited phthalate (Figure 18C). A similar computational study has been performed for cypridina luciferin (substrate for the bioluminescence of the luminous ostracod Cypridina) and its synthetic analogues (Figure 18D and E, respectively) [51]. The aim of this study was to analyze the origin of the different chemiluminescent efficiency found for the diverse derivatives in order to propose a derivative whose chemiluminescent efficiency is close to the bioluminescent one. To do that, the mechanism of representative analogues has been computed and compared. This way, structural substitutions decreasing or increasing the efficiency were identified, facilitating the design of novel cypridina luciferin derivatives with enhanced chemiluminescent efficiency.

The examples below illustrate how calculations can explain the color, brightness and efficiency of synthetic derivatives of known chemiluminescent systems. Hence, the same computational procedures can be used to design new derivatives with target structurally related properties, like for example with even more red-shifted emission.

5.4. Influence of the Environment

One of the experimental factors that could influence the emission derived from the chemiluminescent reaction is the nature of the selected solvent. That is, the emission color could be different when using apolar, polar or protic solvents. In this regard, computational studies give information of the origin of this color emission modulation, allowing the rationalization of the experimental findings or predicting the possible tuning by computing the emission considering solvents of diverse nature. In fact, the modulation of the emission could be due mainly to: *i)* stabilization or destabilization of the ground or/and the excited states, especially when the excited state presents a large charge transfer (CT) character; *ii)* interaction between the solvent molecules and the light emitter (e.g., hydrogen-bond interactions) and *iii)* stabilization of different chemical forms of the light emitter depending on the solvent. In this section we are going to present some particular cases for which computational chemistry has get insight into the solvatochromism of the emission spectra.

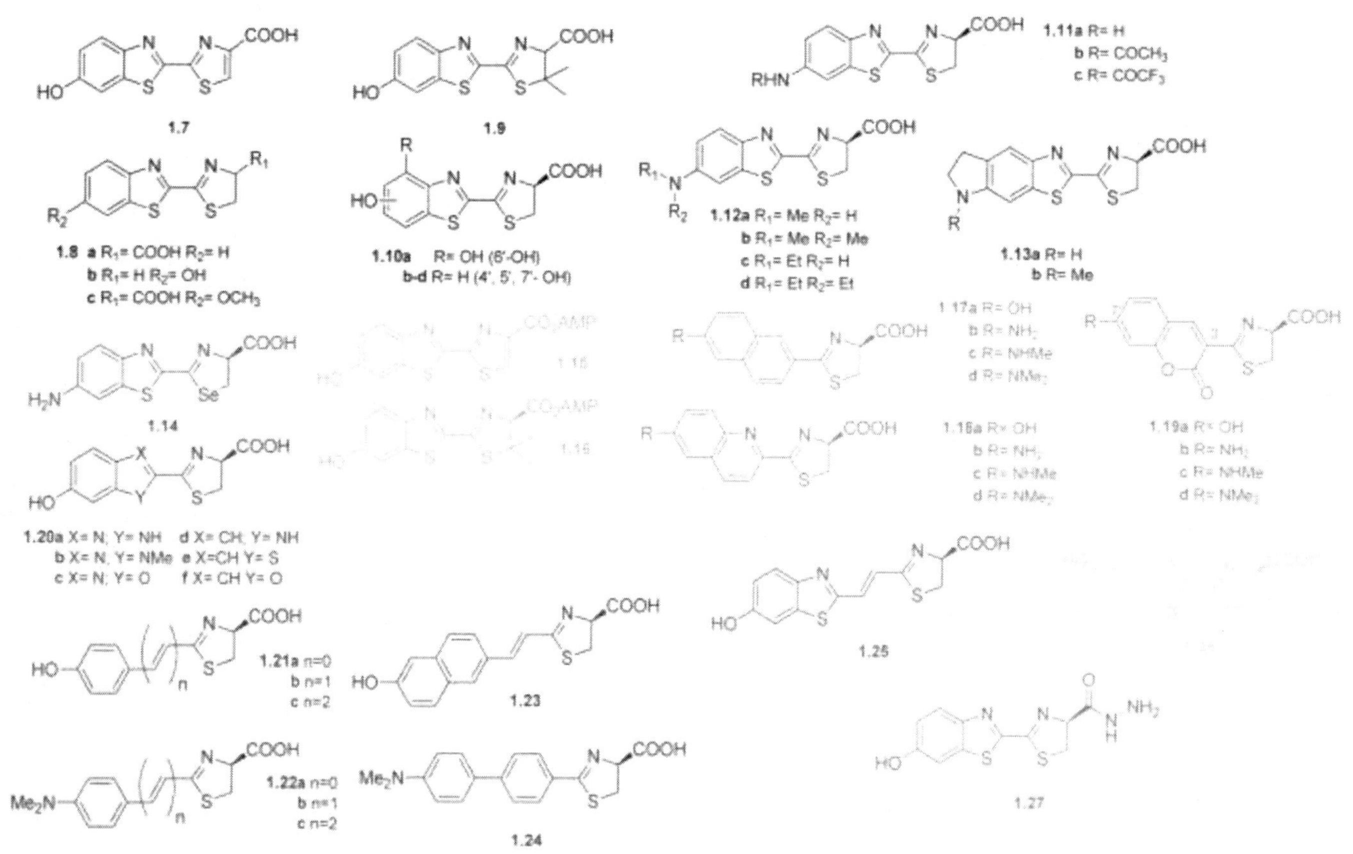

Figure 19. Graphical representation of fireflies' luciferin and synthetic analogues. Contrary to the luciferins drawn in the figure, both nitrogen and sulfur atoms are not face to face in the most stable form of luciferin [50]. "Reprinted (adapted) with permission from (KASKOVA, Zinaida M. et al. Chemical Reviews, 2016, vol 45, p. 6048-6077.). Copyright (2018) American Chemical Society."

It has been shown that for luminol, the emission is significantly shifted from blue to longer wavelengths in hydrogen-bonding solvents [52]. Experimentally, it has been proposed that this red-shift could be due to the formation in the excited state of a complex between luminol and the protic solvent. To elucidate this finding, computational studies have been performed to analyse the effect of hydrogen-bonding on the emission of luminol [53, 54] where water has been selected as the protic solvent. The first reported work covered the relative stability of two possible tautomeric forms of luminol hydrated with two water molecules and the computation of their spectroscopic properties [53]. It should be pointed out that only the first hydration layer, that is the layer of water that interacts directly with the substrate by H-bond, was considered. Including further hydration layers should drive to insignificant differences, increasing the computational time. The results shown that the iminol (form II in Figure 20) was more stable than the amide (form I in Figure 20) by 1.8 kcal/mol.

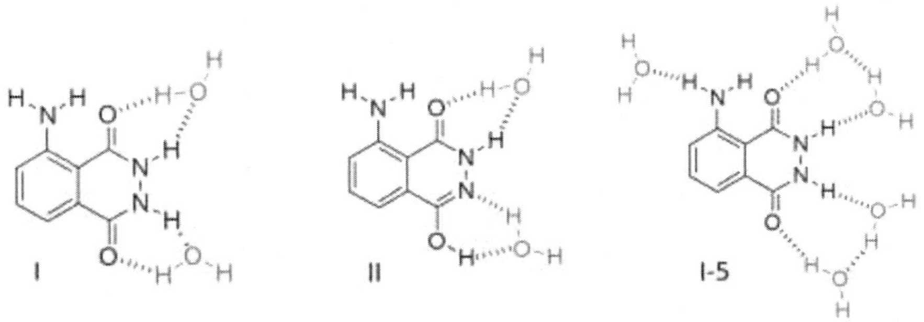

Figure 20. Structures of the tautomers I and II (with two hydrogen-bonded water molecules) and tautomer I (with five bonded water molecules) (I-5).

However, in the excited state the computed energy difference of these two tautomers is smaller (0.8 kcal/mol), being possible the population of both. In fact, the experimental emission spectrum recorded in water solution is quite broad, as a consequence of the averaged emission of the tautomers and hydrogen-bonded clusters in the excited states. It should be noted that for both tautomers, I and II, this interaction takes place at the localized electron-rich centres, which are the imine nitrogen and the carbonyl oxygen at the positions 2 and 4 of the phthalhydrazide ring.

A following computational study has considered an increasing number of water molecules, up to 5 (I-5 in Figure 20) on the structural parameters and spectroscopic properties of tautomer I [54]. This way, it was ensured that each hydrogen-bond site could be saturated with water. It was found that the complex with five water molecules was quite stable and that its structure was more planar compared to isolated luminol (in gas phase). The increase of luminol planarity by hydrogen-bond interactions, involves a rise of the conjugation extent, decreasing the excitation and emission energies and so, red-shifting the absorption and emission spectra as experimentally observed. Hence, the interaction between the water molecules and luminol is the reason of the experimentally observed red-shift for the emission spectrum when using protic solvents compared to the emission of luminol in non-protic ones.

Other example of solvatochromic effect on the emission of a chemiluminescent reactant is coelenterazine. In this case, the solvent effect is slighter than the one observed for luminol. In particular, for the neutral form of coelenterazine (Figure 17A), it has been experimentally found a 10 nm red-shift of the emission going from benzene to acetonitrile solutions [55]. The emission energies of diverse conformers of the neutral form have been computed in gas phase, in benzene and in acetonitrile [25] using the polarizable continuum model (PCM) [56]. It was found that for the most stable conformation in the excited state, the computed emission energy in acetonitrile is red-shifted compared to the ones calculated in benzene, in line with the experimental finding. As both are aprotic solvents, their different polarity should be responsible of the emission solvatochromism, modifying most probably the charge transfer character of the vertical transition.

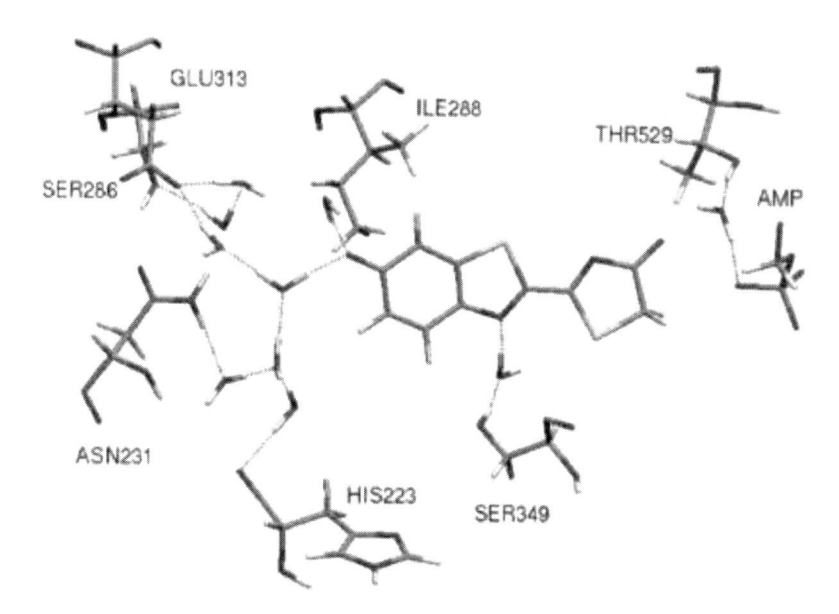

Figure 21. Representation of the amino acids involved in the hydrogen-bond network formed between phenolate-keto, water molecules and side chains inside the protein environment. Amino acids are depicted with the three-letters code and the number of the amino acid corresponding to the 2D1S pdb. "Reprinted (adapted) with permission from (NAVIZET, Isabelle et al. Journal of the American Chemical Society, 2009, vol. 132, no 2, p. 706-712.). Copyright (2009) American Chemical Society."

So, we have shown that the solvent nature could have a clear influence on the emission wavelength of given chemiluminescent reactants. But, the influence of the environment not only covers the solvent but other scenarios such as interaction with other molecules, being the protein surrounding an important example. For instance, different emission colors of fireflies' bioluminescent systems have been found when studying them in vivo or in vitro (with or without the protein). Regarding chemiluminescence in dimethyl sulfoxide

(DMSO), which is a solvent that possess more or less the same polarity of the active site of the protein, the phenolate-keto and the phenolate-enol forms emit red and green light respectively [57]. However, bioluminescence of fireflies leads to green light. By analogy, it was proposed that the phenolate-enol form was the responsible of the green emission inside the protein. To check this hypothesis a synthetic analogue of the phenolate-keto form, designed to block the keto moiety, was synthetized. Then, its bioluminescent process was studied, obtaining a green emission [58, 59]. Hence, the emission color derived from the chemiluminescent and bioluminescent processes for phenolate-keto are quite different, ranging from the red to the green. This fact can be explained by the influence of the protein surrounding (protein active site). To get insight into this experimental finding, computational calculations have been performed to evaluate the effect of the protein environment. Thanks to computational studies together with experimental findings, it has been concluded that the polarity and charged nature of some of the amino acids placed on the protein active site close to oxyluciferin (Figure 21) are the main factors modulating the emission color [60–62]. In particular, the hydrogen-bond network formed between oxyluciferin and the side chains of these amino acids governs the blue-shift of the emission found for phenolate-keto in the bioluminescent process compared to the chemiluminescent one.

Moreover, the phenolate-enol form could be formed only if the tautomerization reaction is energetically feasible in the excited state. A recent study demonstrates that this reaction is quite energy demanding inside the protein and so, most probably the phenolate-keto form is the light emitter [63].

The examples presented in this section evidence the possible effect of the environment, either solvent or protein surrounding, on the emission color.

CONCLUSION

In this chapter, we have highlighted with some examples how computational studies were able to give a better understanding on the chemiluminescent phenomena. Chemiluminescent reactions have the specificity to involve more than one electronic state during the reaction. It leads to difficulties in their theoretical study that explain why the research field is rather recent.

However, the above presented examples have shown that the computational studies could help rationalising the reaction scheme: from the formation of the HEI to its decomposition and the following formation of the light emitter in its excited state. The calculations are also useful to discern the chemical form of the light emitter and to understand the possible equilibriums in the excited state. In this regard, the different conformations and chemical forms possible in the excited state can be identified along with the influence on these equilibria of chemical modifications or changes of external factors such as the environment (solvation, protein or pH).

Moreover, computational studies could aid not only understanding the experimental findings, but also predicting qualitative trends which could assist the design of novel synthetic analogues.

The reader should not forget that combination of experimental and computational studies leads to a complete picture of the chemiluminescent process.

REFERENCES

[1] Barni, F., Lewis, S. W., Berti, A., Miskelly, G. M., Lago, G. (2007). Forensic application of the luminol reaction as a presumptive test for latent blood detection. *Talanta* 72: 896–913.

[2] Khan, P., Idrees, D., Moxley, M. A., Corbett, J. A., Ahmad, F., Von Figura, G., Sly, W. S., Waheed, A., Hassan, M. I. (2014). Luminol-based chemiluminescent signals: Clinical and non-clinical application and future uses. *Appl. Biochem. Biotechnol.* 173: 333–355.

[3] Marquette, C. A., Blum, L. J. (2006). Applications of the luminol chemiluminescent reaction in analytical chemistry. *Anal. Bioanal. Chem.* 385: 546–554.

[4] Griesbeck, A. G., Díaz-Miara, Y., Fichtler, R., Jacobi Von Wangelin, A., Pérez-Ruiz, R., Sampedro, D. (2015). Steric Enhancement of the Chemiluminescence of Luminols. *Chem. - A Eur. J.* 21: 9975–9979.

[5] Schramm, S., Navizet, I., Prasad Karothu, D., Oesau, P., Bensmann, V., Weiss, D., Beckert, R., Naumov, P. (2017). Mechanistic investigations of the 2-coumaranone chemiluminescence. *Phys. Chem. Chem. Phys.* 19: 22852–22859.

[6] Berraud-Pache, R., Lindh, R., Navizet, I. (2018). QM/MM Study of the Formation of the Dioxetanone Ring in Fireflies through a Superoxide Ion. *J. Phys. Chem. B* 122: 5173–5182.

[7] Sundlov, J. A., Fontaine, D. M., Southworth, T. L., Branchini, B. R., Gulick, A. M. (2012). Crystal structure of firefly luciferase in a second catalytic conformation supports a domain alternation mechanism. *Biochemistry* 51: 6493–6495.

[8] Min, C. G., Ren, A. M., Li, X. N., Guo, J. F., Zou, L. Y., Sun, Y., Goddard, J. D., Sun, C. C. (2011). The formation and decomposition of firefly dioxetanone. *Chem. Phys. Lett.* 506: 269–275.

[9] Vacher, M., Fdez Galván, I., Ding, B. W., Schramm, S., Berraud-Pache, R., Naumov, P., Ferré, N., Liu, Y. J., Navizet, I., Roca-Sanjuán, D., Baader, W. J., Lindh, R. (2018). Chemi- and Bioluminescence of Cyclic Peroxides. *Chem. Rev.* 118: 6927–6974.

[10] De Vico, L., Liu, Y. J., Krogh, J. W., Lindh, R. (2007). Chemiluminescence of 1,2-dioxetane. Reaction mechanism uncovered. *J. Phys. Chem. A* 111: 8013–8019.

[11] Navizet, I., Liu, Y.-J., Ferré, N., Roca-Sanjuán, D., Lindh, R. (2011). The Chemistry of Bioluminescence: An Analysis of Chemical Functionalities. *ChemPhysChem* 12: 3064–3076.

[12] Pinto da Silva, L., Esteves da Silva, J. C. G. (2013). Chemiluminescence of 1,2-dioxetanone studied by a closed-shell DFT approach. *Int. J. Quantum Chem.* 113: 1709–1716.

[13] Baader, W. J., Stevani, C. V., Bastos, E. L. *In the chemistry of peroxides*; Rappoport, Z., Ed., John Wiley & Sons, Ltd., 2006;

[14] Farahani, P., Roca-Sanjuán, D., Zapata, F., Lindh, R. (2013). Revisiting the nonadiabatic process in 1,2-dioxetane. *J. Chem. Theory Comput.* 9: 5404–5411.

[15] Vacher, M., Brakestad, A., Karlsson, H. O., Galván, I. F., Lindh, R. (2017). Dynamical Insights into the Decomposition of 1,2-Dioxetane. *J. Chem. Theory Comput.* 13: 2448–2457.

[16] Isobe, H., Takano, Y., Okumura, M., Kuramitsu, S., Yamaguchi, K. (2005). Mechanistic insights in charge-transfer-induced luminescence of 1,2-dioxetanones with a substituent of low oxidation potential. *J. Am. Chem. Soc.* 127: 8667–8679.

[17] Koo, J. young; Schuster, G. B. (1977). Chemically Initiated Electron Exchange Luminescence. A New Chemiluminescent Reaction Path for Organic Peroxides. *J. Am. Chem. Soc.* 99: 6107–6109.

[18] Koo, J. A., Schmidt, S. P., Schuster, G. B. (1978). Bioluminescence of the firefly: key steps in the formation of the electronically excited state for model systems. *Proc. Natl. Acad. Sci.* 75: 30–33.

[19] Koo, J. young; Schuster, G. B. (1978). Chemiluminescence of Diphenoyl Peroxide. Chemically Initiated Electron Exchange Luminescence. A New General Mechanism for Chemical Production of Electronically Excited States. *J. Am. Chem. Soc.* 100: 4496–4503.

[20] Takano, Y., Tsunesada, T., Isobe, H., Yoshioka, Y., Yamaguchi, K., Saito, I. (1999) Theoretical studies of decomposition reactions of dioxetane, dioxetanone, and related species. CT induced luminescence mechanism revisited. *Bull. Chem. Soc. Jpn.* 72: 213–225.

[21] Guo, X., Cao, Z. (2012). Low-lying electronic states and their nonradiative deactivation of thieno[3,4-b]pyrazine: An ab initio study. *J. Chem. Phys.* 137: 224313.

[22] Yue, L., Liu, Y. J., Fang, W. H. (2012). Mechanistic insight into the chemiluminescent decomposition of firefly dioxetanone. *J. Am. Chem. Soc.* 134: 11632–11639.

[23] Navizet, I., Roca-Sanjuán, D., Yue, L., Liu, Y. J., Ferré, N., Lindh, R. (2013). Are the bio- and chemiluminescence states of the firefly oxyluciferin the same as the fluorescence state? *Photochem. Photobiol.* 89: 319–325.

[24] Roca-Sanjuán, D., Delcey, M. G., Navizet, I., Ferré, N., Liu, Y. J., Lindh, R. (2011). Chemiluminescence and fluorescence states of a small model for coelenteramide and Cypridina oxyluciferin: A CASSCF/CASPT2 study. *J. Chem. Theory Comput.* 7: 4060–4069.

[25] Chen, S. F., Navizet, I., Roca-Sanjuán, D., Lindh, R., Liu, Y. J., Ferré, N. (2012). Chemiluminescence of coelenterazine and fluorescence of coelenteramide: A systematic theoretical study. *J. Chem. Theory Comput.* 8: 2796–2807.

[26] Yue, L., Liu, Y. J. (2013). Mechanism of AMPPD chemiluminescence in a different voice. *J. Chem. Theory Comput.* 9: 2300–2312.

[27] Schramm, S., Navizet, I., Naumov, P., Nath, N. K., Berraud-Pache, R., Oesau, P., Weiss, D., Beckert, R. (2016). The Light Emitter of the 2-Coumar-anone Chemiluminescence: Theoretical and Experimental Elucidation of a Possible Model for Bioluminescent Systems. *European J. Org. Chem.* 2016: 678–681.

[28] Min, C.-G., Ren, A.-M., Guo, J.-F., Li, Z.-W., Zou, L.-Y., Goddard, J. D., Feng, J.-K. (2010). A Time-Dependent Density Functional Theory Investigation on the Origin of Red Chemiluminescence. *ChemPhysChem* 11: 251–259.

[29] Chen, S.-F., Liu, Y.-J., Navizet, I., Ferré, N., Fang, W.-H., Lindh, R. (2011). Systematic Theoretical Investigation on the Light Emitter of Firefly. *J. Chem. Theory Comput.* 7: 798–803.

[30] Da Silva, L. P., Esteves Da Silva, J. C. G. (2011). Computational studies of the luciferase light-emitting product: Oxyluciferin. *J. Chem. Theory Comput.* 7: 809–817.

[31] Garcia Iriepa, C., Zemmouche, M., Ponce-Vargas, M., Navizet, I. (2019). The role of solvation models on the computed absorption and emission spectra: The case of fireflies oxyluciferin. *Phys. Chem. Chem. Phys.* 21: 4613-4623.

[32] García-Iriepa, C., Gosset, P., Berraud-Pache, R., Zemmouche, M., Taupier, G., Dorkenoo, K. D., Didier, P., Léonard, J., Ferré, N., Navizet, I. (2018). Simulation and Analysis of the Spectroscopic Properties of Oxyluciferin and Its Analogues in Water. *J. Chem. Theory Comput.* 14: 2117–2126.

[33] Liu, F., Liu, Y., Vico, L. De; Lindh, R. (2009). A CASSCF/CASPT2 approach to the decomposition of thiazole-substituted dioxetanone: Substitution effects and charge-transfer induced electron excitation. *Chem. Phys. Lett.* 484: 69–75.

[34] Liu, F., Liu, Y., De Vico, L., Lindh, R. (2009). Theoretical study of the chemiluminescent decomposition of dioxetanone. *J. Am. Chem. Soc.* 131: 6181–6188.

[35] Lung, W. C., Hayashi, S., Lundberg, M., Nakatsu, T., Kato, H., Morokuma, K. (2008). Mechanism of efficient firefly bioluminescence via adiabatic transition state and seam of sloped conical intersection. *J. Am. Chem. Soc.* 130: 12880–12881.

[36] Naumov, P., Kochunnoonny, M. (2010). Spectral–Structural Effects of the Keto–Enol–Enolate and Phenol–Phenolate Equilibria of Oxyluciferin. *J. Am. Chem. Soc.* 132: 11566–11579.

[37] Pinto Da Silva, L., Simkovitch, R., Huppert, D., Da Silva, J. C. G. E. (2013). Oxyluciferin photoacidity: The missing element for solving the keto-enol mystery? *ChemPhysChem* 14: 3441–3446.

[38] Pinto Da Silva, L., Esteves Da Silva, J. C. G. (2015). Chemiexcitation induced proton transfer: Enolate oxyluciferin as the firefly bioluminophore. *J. Phys. Chem. B* 119: 2140–2184.

[39] Pinto Da Silva, L., Esteves Da Silva, J. C. G. (2011). Computational investigation of the effect of pH on the color of firefly bioluminescence by DFT. *ChemPhysChem* 12: 951–960.

[40] Min, C. G., Pinto da Silva, L., Esteves da Silva, J. C. G., Yang, X. K., Huang, S. J., Ren, A. M., Zhu, Y. Q. (2017). A Computational Investigation of the Equilibrium Constants for the Fluorescent and Chemiluminescent States of Coelenteramide. *ChemPhysChem* 18: 117–123.

[41] Chen, S. F., Ferré, N., Liu, Y. J. (2013). QM/MM study on the light emitters of aequorin chemiluminescence, bioluminescence, and fluorescence: A general understanding of the bioluminescence of several marine organisms. *Chem. - A Eur. J.* 19: 8466–8472.

[42] Van Oort, B., Eremeeva, E. V., Koehorst, R. B. M., Laptenok, S. P., Van Amerongen, H., Van Berkel, W. J. H., Malikova, N. P., Markova, S. V., Vysotski, E. S., Visser, A. J. W. G., Lee, J. (2009). Picosecond fluorescence relaxation spectroscopy of the calcium-discharged photoproteins aequorin and obelin. *Biochemistry* 48: 10486–10491.

[43] Liu, Z.-J., Stepanyuk, G. A., Vysotski, E. S., Lee, J., Markova, S. V., Malikova, N. P., Wang, B.-C. (2006). Crystal structure of obelin after Ca2+-triggered bioluminescence suggests neutral coelenteramide as the primary excited state. *Proc. Natl. Acad. Sci.* 103: 2570–2575.

[44] Deshmukh, M. S., Sekar, N. (2015). Chemiluminescence properties of luminol related o-hydroxybenzimidazole analogues: Experimental and DFT based approach to photophysical properties. *Dye. Pigment.* 113: 189–199.

[45] Deshmukh, M. S., Sekar, N. (2015). Chemiluminescence properties of luminol related quinoxaline analogs: Experimental and DFT based approach to photophysical properties. *Dye. Pigment.* 117: 49–60.

[46] Kaskova, Z. M., Tsarkova, A. S., Yampolsky, I. V. (2016). 1001 lights: luciferins, luciferases, their mechanisms of action and applications in chemical analysis, biology and medicine. *Chem. Soc. Rev.* 45: 6048–6077.

[47] Ran, X. Q., Zhou, X., Goddard, J. D. (2015). The spectral-structural relationship of a series of oxyluciferin derivatives. *ChemPhysChem* 16: 396–402.

[48] Min, C.-G., Yan Leng; Yan-QinZhu; Xi-Kun Yang; Shao-Jun Huang; Ai-MinRen (2017). Modification of firefly cyclic amino oxyluciferin analogues emitting multicolor light for OLED and near-Infrared biological window light for bioluminescence imaging: A theoretical study. *J. Photochem. Photobiol. A Chem.* 336: 115–122.

[49] Kakiuchi, M., Ito, S., Yamaji, M., Viviani, V. R., Maki, S., Hirano, T. (2017). Spectroscopic Properties of Amine-substituted Analogues of Firefly Luciferin and Oxyluciferin. *Photochem. Photobiol.* 93: 486–494.

[50] Yang, T., Goddard, J. D. (2007). Predictions of the Geometries and Fluorescence Emission Energies of Oxyluciferins. *J. Phys. Chem. A* 111: 4489–4497.

[51] Ishii, Y., Hayashi, C., Suzuki, Y., Hirano, T. (2014). Chemiluminescent 2,6-diphenylimidazo[1,2-a]pyrazin-3(7H)-ones: A new entry to Cypridina luciferin analogues. *Photochem. Photobiol. Sci.* 13: 182–189.

[52] Mitra, S., Das, R., Mukherjee, S. (1995). Complex formation and photophysical properties of luminol: solvent effects. *J. Photochem. Photobiol. A Chem.* 87: 225–230.

[53] Moyon, N. S., Chandra, A. K., Mitra, S. (2010). Effect of solvent hydrogen bonding on excited-state properties of luminol: A combined fluorescence and DFT study. *J. Phys. Chem. A* 114: 60–67.

[54] Xue, B., Zhang, C., Liu, C., Liu, E. (2014). Luminol: Extended hydrogen bond network in water solution. *Comput. Theor. Chem.* 1028: 81–86.

[55] Shimomura, O., Teranishi, K. (2000). Light-emitters involved in the luminescence of coelenterazine. *Luminescence* 15: 51–58.

[56] Tomasi, J., Mennucci, B., Cammi, R. (2005). Quantum mechanical continuum solvation models. *Chem. Rev.* 105: 2999–3093.

[57] White, E. H., Rapaport, E., Hopkins, T. A., Seliger, H. H. (1969). Chemi- and Bioluminescence of Firefly Luciferin. *J. Am. Chem. Soc.* 91: 2178–2180.

[58] Branchini, B. R., Martha H. Murtiashaw, Rachelle A. Magyar, Nathan C. Portier, M. C. R., Stroh, J. G. (2002). Yellow-Green and Red Firefly Bioluminescence from 5,5-Dimethyloxyluciferin. *J. Am. Chem. Soc.* 124: 2112–2113.

[59] Branchini, B. R., Southworth, T. L., Murtiashaw, M. H., Magyar, R. A., Gonzalez, S. A., Ruggiero, M. C., Stroh, J. G. (2004). An alternative mechanism of bioluminescence color determination in firefly luciferase. *Biochemistry* 43: 7255–7262.

[60] Navizet, I., Liu, Y.-J., Ferré, N., Xiao, H.-Y., Fang, W.-H., Lindh, R. (2010). Color-Tuning Mechanism of Firefly Investigated by Multi-Configurational Perturbation Method. *J. Am. Chem. Soc.* 132: 706–712.

[61] Nakatani, N., Hasegawa, J., Nakatsuji, H. (2007). Red Light in Chemiluminescence and Yellow-Green Light in Bioluminescence: Color-Tuning Mechanism of Firefly, Photinus pyralis, Studied by the Symmetry-Adapted Cluster–Configuration Interaction Method. *J. Am. Chem. Soc.* 129: 8756–8765.

[62] Hirano, T., Hasumi, Y., Ohtsuka, K., Maki, S., Niwa, H., Yamaji, M., Hashizume, D. (2009). Spectroscopic Studies of the Light-Color Modulation Mechanism of Firefly (Beetle) Bioluminescence. *J. Am. Chem. Soc.* 131: 2385–2396.

[63] Berraud-Pache, R., Garcia-Iriepa, C., Navizet, I. (2018). Modelling chemical reactions by QM/MM calculations: the case of the tautomerization in fireflies bioluminescent systems. *Front. Chem.* 6: 116.

In: A Comprehensive Guide to Chemiluminescence
Editor: Luís Pinto da Silva

ISBN: 978-1-53616-170-0
© 2019 Nova Science Publishers, Inc.

Chapter 12

MULTICONFIGURATIONAL QUANTUM CHEMILUMINESCENCE

Daniel Roca-Sanjuán[1,*], Javier Carmona-García[1],
Miriam Navarrete-Miguel[1], Antonio Francés-Monerris[1,2]
and Angelo Giussani[1]

[1]Institut de Ciència Molecular, Universitat de València, València, Spain
[2]Laboratoire de Physique et Chimie Théoriques,
Universitè de Lorraine and CNRS, Nancy, France

ABSTRACT

Chemiluminescence is a phenomenon of non-adiabatic chemistry, involving crossings (conical intersections and singlet-triplet crossings) between distinct electronic states. Such crossings mediate the population transfer from the ground to the excited state and require multiconfigurational methodologies for an accurate description. In this chapter, we shall briefly describe the fundamental theoretical aspects of the most common methods based on multiconfigurational wave functions and show illustrative examples of application of those methodologies, mainly using small models to clearly identify the key electronic-structure features that characterize chemiluminescence.

1. INTRODUCTION

Chemiluminescence is essentially a non-adiabatic quantum-chemistry phenomenon involving ground and excited electronic states [1, 2]. Non-adiabatic chemistry (NAC)

---

* Corresponding Author's Email: daniel.roca@uv.es.

refers to chemical phenomena taking place via internal conversions or intersystem crossings [3, 4]. Within the framework of the Born-Oppenheimer approximation to solve the time-independent Schrödinger equation, NAC corresponds to electronic transitions between potential energy surfaces (PESs) without the absorption or emission of light. Conical intersections (CIs) and singlet-triplet crossings (STCs) are in this context the key structures. Such crossings are characterized by the interaction of at least two electronic configurations and therefore they have multiconfigurational nature.

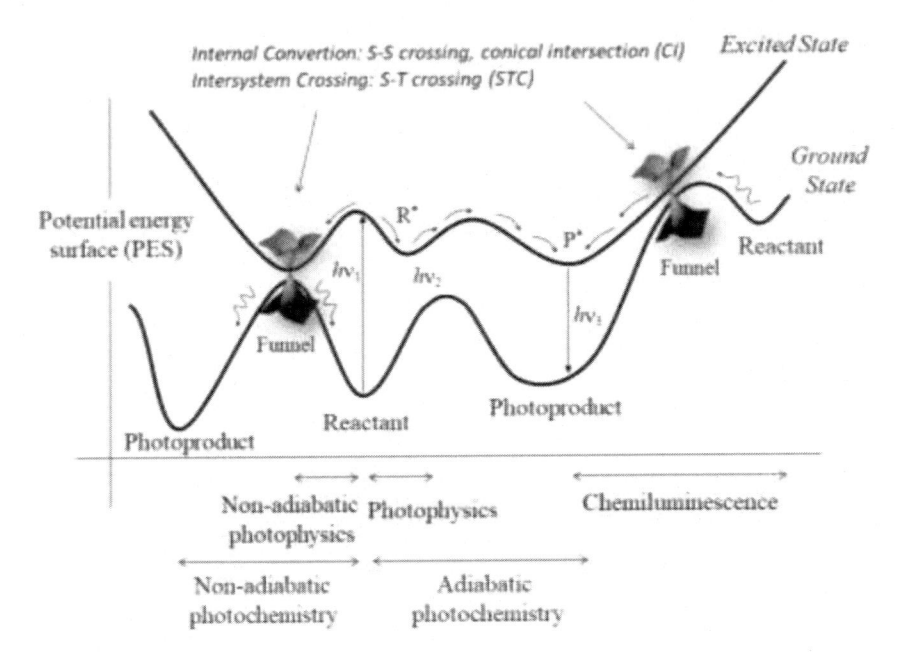

Figure 1. Scheme based on potential energy surfaces (PESs) comparing photophysical, photochemical and chemiluminescence.

From a mechanistic viewpoint, chemiluminescence can be better understood by comparing it with photophysics and photochemistry as illustrated in Figure 1 [3, 4]. Photophysical and photochemical processes begin with the promotion of the reactant from the ground to the excited electronic state by means of light irradiation. Once there, the molecule will try to return to the ground state by using the most energetically favorable decay paths which are distinct depending on the intrinsic electronic and nuclear structure of the molecules. Decay to the original ground-state equilibrium geometry by means of a radiative process is often defined as photophysics (no reaction took place). Light emission can also occur after a chemical transformation along the excited-state PES, via an adiabatic process, and decay then to a different ground-state structure, giving rise to an adiabatic photochemical process. Decay to the ground state might also occur involving crossings between the PES of the excited and ground states (CIs or STCs) rather than by radiative decay. In this case we refer to non-adiabatic photochemistry or photophysics depending on the occurrence or not, respectively, of a reaction. On the other hand, in chemiluminescence, the process begins with the activation of a reactant by means of thermal energy from the ground-state equilibrium geometry to the transition state (TS) of a decomposition reaction. Near the TS, there are CIs and/or STCs between the PES of the ground and excited state and the molecule can access the excited state. Once there, the molecule vibrationally relaxes to an excited-state equilibrium geometry from where light emission takes place.

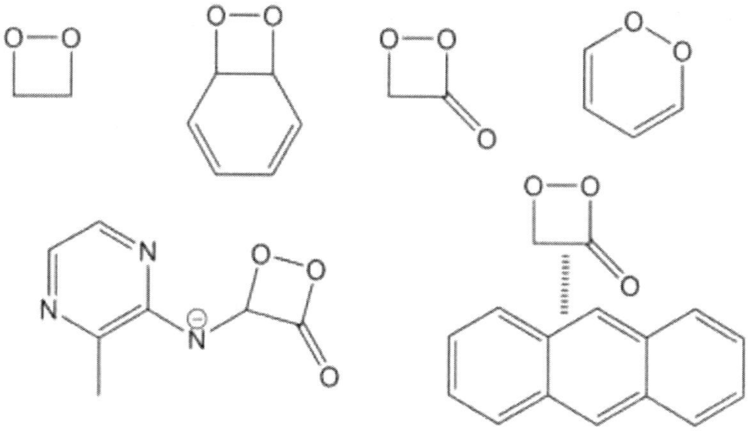

Figure 2. Chemical structure of the peroxide models revised in this work. Top, from left to right: 1,2-dioxetane, Dewar dioxetane, 1,2-dioxetanone, 1,2-dioxin. Bottom, from left to right: deprotonated 2-acetamido-3-methylpyrazine dioxetanone, 1,2-dioxetanone – anthracene complex.

Multiconfigurational quantum chemistry provides the user with high-level *ab initio* methods based on the principles of quantum mechanics applied to chemistry able to accurately describe all the aforementioned NAC phenomena and to determine the most relevant reaction paths over the PESs and the geometrical and electronic structure of the key intermediates (minima), CI, and/or STC which characterizes any chemiluminescence reaction of interest. During the last decades, it has been applied to the study of several chemi- and bioluminescence systems which has allowed to establish the molecular basis of the chemically-induced light emission phenomenon in the framework of the excited state chemistry [1,2]. Herein, we shall firstly describe the main concepts of probably the most practical and popular multiconfigurational methods, the complete-active-space self-consistent field (CASSCF) and complete-active-space second-order perturbation theory (CASPT2) methods [5]. Next, we will present the main findings obtained in the studies of small-size molecular models of chemi-/bioluminescent systems in which the relatively high cost of multiconfigurational quantum chemistry is affordable and deep analyses have been performed to get the molecular basis of chemiluminescence. Those systems correspond to 1,2-dioxetane [6], Dewar dioxetane [7], 1,2-dioxetanone [8–10], deprotonated 2-acetamido-3-methylpyrazine dioxetanone [11], 1,2-dioxetanone – anthracene complex [12], and 1,2-dioxin [13] (see Figure 2). Findings related to the presence or absence of intermediates, the concerted or stepwise character of the mechanisms, the nature of the electronic states involved, and relevant details to comprehend uncatalyzed and catalyzed chemiluminescence will be provided. Finally, the contents are summarized, and the conclusions are clearly listed.

2. MULTICONFIGURATIONAL METHODS

To correctly describe CIs and STCs and consequently to accurately determine the mechanisms of chemiluminescence, methods able to deal with the multiconfigurational nature of the crossings are needed [14]. As two illustrative examples to demonstrate the need of using methodologies based on several electronic configurations, we show here the electronic structure of the CIs and STCs occurring at the intermediate region in the decomposition reaction of 1,2-dioxetane and the CIs taking place in the excited-state isomerization of a double bond (see Figure 3). In the first case, the intermediate in which the -O-O- bond is broken corresponds to a diradical with the unpaired electrons in the $2p$ atomic-like orbitals of the oxygen atoms perpendicular to the direction of the -C-O bonds. In total, there are 4 possible occupations of the orbitals as shown in Figure 3 top. Moreover, the unpaired electrons can have parallel or opposite spin. This gives rise to a total amount of 4 singlet and 4 triplet electronic states energetically degenerated. In the case of the E/Z isomerization of a double bond, the CI which characterizes the process corresponds to a twisted and pyramidalized structure. The electronic configurations of the two electronic states that cross is a biradicaloid one with unpaired electrons at the carbon atoms and a zwitterionc one with the two electrons in the pyramidalized carbon (carbanion-like) and zero in the other carbon atom (carbocation-like) (see Figure 3 bottom).

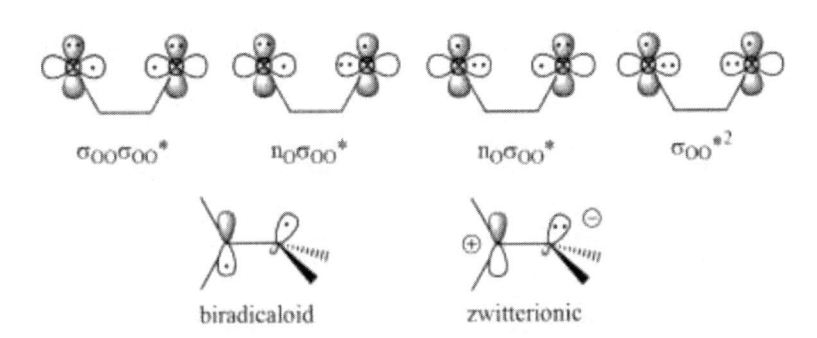

Figure 3. Electronic configurations of the CIs and STCs occurring at the intermediate in the decomposition reaction of 1,2-dioxetane (top) and in the excited-state isomerization of a double bond in ethylene-based molecules (bottom).

Correct wavefunctions for the examples described above are those which include all the distributions of the "chemically active" electrons in the "chemically active" orbitals. Moreover, orbitals must not be optimal only for one configuration, like in single-reference methods based on only one Hartree-Fock (HF) configuration such as the well-known density functional theory (DFT) method. Orbitals must be optimal for the combination of electronic configurations. These conditions are satisfied by the multiconfigurational self-consistent field methods, from which one of the most practical is the CASSCF [15]. Here, the wavefunction is built by considering all possible distributions of the active electrons among the active orbitals (full configuration interaction approach) and the wavefunction optimization procedure is based on changing the contributions of each configuration (or coefficients of the configuration interaction linear expansion) and at the same time allowing to change the way in which the atomic orbitals are combined to produce the molecular orbitals (coefficients of the linear combination of atomic orbitals) until reaching a minimum of the associated energy. Unrestricted or open-shell single-reference approaches have been demonstrated [8, 9] to still provide a reasonable and approximated description although benchmarking with multiconfigurational methods is recommended.

By means of the CASSCF method, the interaction or correlation between different electronic configurations is computed (so-called long-range or static correlation). Due to the large increase of the computational cost by considering more active orbitals, practical applications in chemi- and bioluminescence are limited to no more than around 14 chemically-active orbitals. Many orbitals are therefore not correlated in the CASSCF method; electron-electron repulsion at short distances in the space of the remaining orbitals is not accurately computed. Consequently, the CASSCF wavefunction can be accurate enough but not the energies. To solve this problem, a two-step approach can be used, in which first the CASSCF wavefunction is generated and next it is used as reference in a perturbation theory treatment to improve the energy-adding corrections up to second order in a Taylor expansion. In such manner, the dynamical correlations attributed to the electron repulsion at short distances (short-range correlation) are computed. This is done by the popular CASPT2 method [16–20] which can decrease the energy errors up to 0.1-0.3 eV and therefore achieve reasonably good quantitative determinations in chemi-/bioluminescence when appropriate reference functions are used [5].

3. DIOXETANE MODELS

3.1. 1,2-Dioxetane

Probably one of the most popular and smallest models of chemi- and bioluminescence is 1,2-dioxetane. It constitutes the smallest chemical functionality with chemiluminescence properties formed by a cyclic peroxide with a 4-membered ring. It can be generated by a [1+2] cycloaddition of $^1O_2$ to ethylene. The chemiluminophore properties of 1,2-dioxetane are well-understood and have been recently reviewed in detail by some of the authors of this work in collaboration by the theoretical groups of R. Lindh, N. Ferré, Y.-J. Liu and I. Navizet, and the experimental groups of W. J. Baader and P. Naumov, led by R. Lindh [1].

The first and most simplistic theoretical approach to comprehend the electronic structure of 1,2-dioxetane chemiluminescence mechanism corresponds to the drawing of Walsh correlation diagrams of orbitals and electronic states along the ring-opening reaction to produce two formaldehyde molecules, and the symmetry analysis of the molecular orbitals considering the Woodward-Hoffmann rules [21,22], analogously as done in Ref. [1]. The process preserves a C_2 symmetry axis perpendicular to the peroxide bond (see Figure 4 top). Therefore, the orbitals can be classified as symmetric or antisymmetric with respect to the C_2 axis (hereafter indicated as a capital S and A as a subscript, respectively). In the -O-O- bond breaking process leading from the cyclic peroxide (reactant) to the dialdehyde structure (product), the bonding σ_S and antibonding σ^*_A orbitals of the reactant located on the -O-O- bond correlate with the antibonding π^*_S and bonding π_A orbitals of the product, while the oxygen lone pair orbital, n_S, is maintained as the highest occupied molecular orbital along the transformation. These correlations consequently describe a change of the bonding/antibonding nature, which in turn translates into the presence of crossings or avoided crossings between the corresponding electronic states along the reaction coordinate describing the chemical passage from the reactant to the product. As can be seen in the Walsh correlation diagrams of states (Figure 4 bottom), in the reactant, the ground state describes a closed-shell structure in which the π_S and n_S orbitals are double occupied, $(\sigma_S)^2(n_S)^2$ state; the lowest-lying (first) excited state corresponds to a single electronic promotion from n_S to σ^*_A, $(\sigma_S)^2(n_S\sigma^*_A)$ state; and at much higher energies, an excited state appears which is the result of a double excitation from σ_S to σ^*_A, $(\sigma_S)^2(\sigma^*_A)^2$ state. In the product, the ground state is a closed-shell structure in which the n_S and π_A are doubly occupied, $(n_S)^2(\pi_A)^2$ state; the first excited state corresponds to a single electronic promotion from n_S to π^*_S, $(n_S)^2(n_S\pi^*_S)$ state; and a high-energy excited state can be found resulting from a double n_S to π^*_S excitation, $(n_S)^2(\pi^*_S)^2$ state. The ground and doubly excited states of the reactant correlate with the doubly excited and ground states of the product, respectively, and the reaction consequently describes, having the mentioned states the same symmetry, an avoided crossing between them located in the region of the transition state (TS) separating the two S_0 minima. The singly excited states of the reactant and product correlates, and being instead of A symmetry, can cross with both the ground and doubly excited states. The presence of crossings, which connect the original ground state to the excited manifold, allows the chemiexcitation process.

The simplistic below analysis provides the basis for a qualitative description of the chemiexcitation in 1,2-dioxetane. However, higher-level quantum-chemistry computations were needed to more accurately characterize the reaction mechanism and to interpret the experimental observations. In this context, it was reported in experimental studies that the activation energy for the chemiluminescence in 1,2-dioxetane is 22.7 ± 0.8 kcal/mol and the yield for triplet chemiexcitation is 1000 times higher than that for singlet chemical excitation [23].

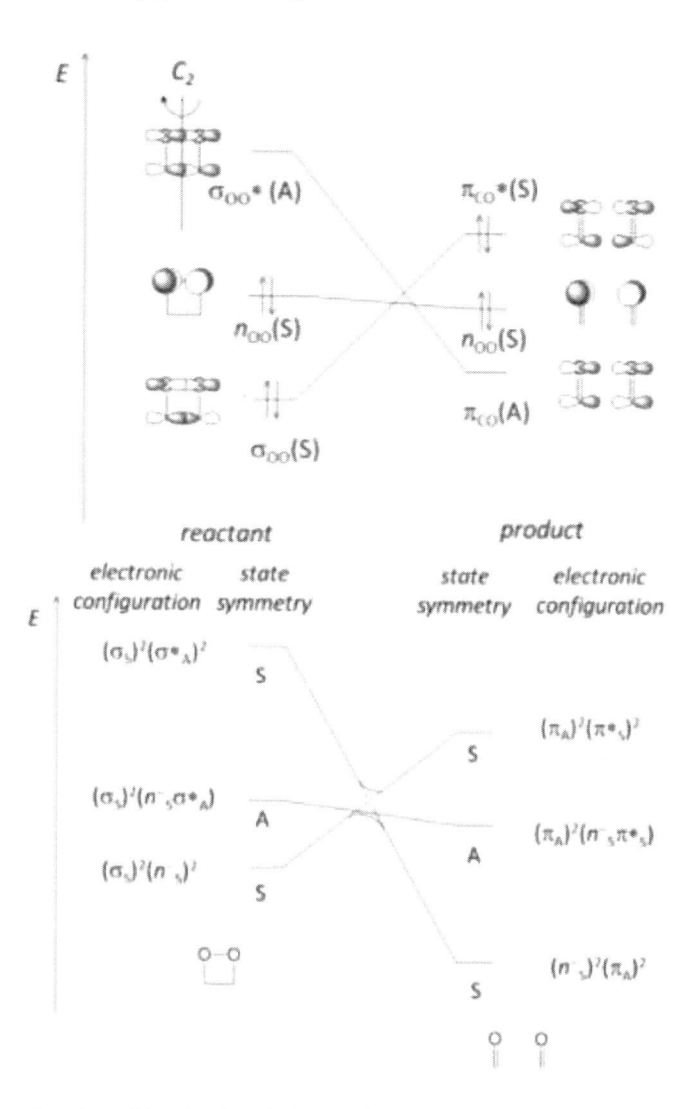

Figure 4. Walsh molecular orbital (top) and electronic state (bottom) correlation diagrams for the ring opening or 1,2-dioxetane. The notation (A) or (S) stands for antisymmetric or symmetric, respectively, and denotes the symmetry property of each orbital with respect to the C_2 axis. Dash and solid lines (bottom) denote diabatic and adiabatic representations, respectively. State symmetry labels are obtained by multiplying the symmetry of the orbitals with unpaired electrons according to the following rules: S×S=A×A=S, S×A=A×S=A; doubly occupied orbitals contribute with S.

Multiconfigurational quantum-chemistry studies with the CASPT2//CASSCF protocol [24] showed that the reaction corresponds to a stepwise process in which first the -O-O-bond is broken via the TS_{OO} giving rise to a diradical intermediate in which the unpaired electrons are localized in $2p$ atomic-like orbitals of the oxygen atoms (see Figure 3). This intermediate is characterized by a 8-folded crossing point involving 4 singlet and 4 triplet states, each one having a distinct distribution of the unpaired electrons ($\sigma_{OO}\sigma_{OO}^*$, $n_O\sigma_{OO}^*$, or σ_{OO}^{*2}) as shown in Figure 3. Due to the energetic degeneracy and the significantly high spin-orbit coupling (SOC) between the singlet and triplet states (49 cm^{-1}) [6], the molecules can populate all the electronic states. The next step corresponds to the dissociation of the -

C-C- bond via the TS_{CC} which can take place in three different decomposition routes, (i) thermal decomposition (on the S_0 state, (ii) triplet excited state decomposition (on the $^3n_O\sigma_{OO}*$ state) and (iii) singlet excited state decomposition (on the $^1n_O\sigma_{OO}*$ state). Highly-accurate CASPT2//CASPT2 computations and analyses based on free energies rather than electronic energies produced activation energy barriers of 23.0, 23.9, and 29.2 kcal/mol for path (i), (ii), and (iii), respectively [6]. Thus, it was shown that path (i) requires the least activation energy and path (ii) is energetically more favourable than path (iii) with an energy difference (ΔE) of 5.3 kcal/mol, which allowed to interpret the low chemiexcitation and chemiluminescence yields and the higher triplet chemiexcitation yield as compared to the singlet one observed in the experiments [23]. It is worth noting that to reach such conclusions, energies had to be corrected taking into account the zero-point vibrational energy and entropic factors, which decrease the barriers of TS_{OO} and TS_{CC} by 2.1 and 5.9 kcal/mol, respectively, as compared to the profiles based on electronic energies [6]. By this means, previous theoretical estimations of the chemiluminescence activation energy [24] were improved reaching a higher agreement with the experimental data.

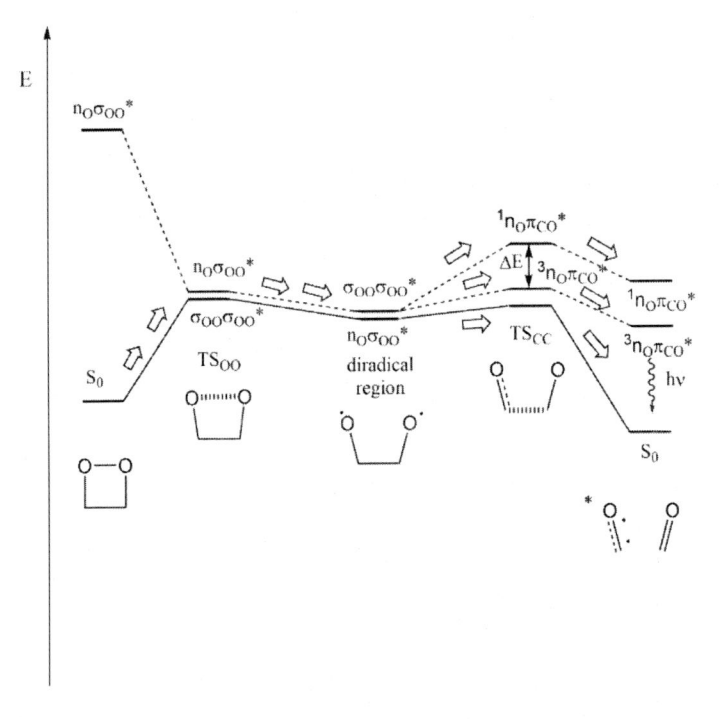

Figure 5. Chemiluminescence mechanism of 1,2-dioxetane. Empty arrows represent evolution from the ground-state (S_0) structure of the reactant to the three distinct decomposition paths corresponding to the thermal process on the S_0 PES and the lowest-lying singlet ($^1n_O\pi_{CO}*$) and triplet ($^3n_O\pi_{CO}*$) excited states. The energy difference (ΔE) between the transition state related to the CC bond breaking (TS_{CC}) on the $^1n_O\pi_{CO}*$ and $^3n_O\pi_{CO}*$ states is highlighted (see text).

3.2. Dewar Dioxetane

Dewar dioxetane can be considered as the result of adding $^1O_2$ to adjacent positions of benzene giving rise to two chemical structures, 1,2-dioxetane and *E*-1,3-butadiene (see Figure 2). The study of the thermal decomposition of the dioxetane ring in this molecule [7] allowed to better understand the excited-state chemistry of dioxetane-based molecules and especially to know how the initial excitation localized in the dioxetane ring is transferred to the rest of the molecule. Furthermore, in this system a new concept arises, the so-called photochemistry in the dark. It corresponds to the production of a photochemical reaction (in this case, E/Z isomerization) by means of the chemical -O-O- and -C-C- bonds dissociation, therefore using chemical energy rather than light [25].

The initial steps of the decomposition reaction of Dewar dioxetane are almost identical to those taking place in the isolated 1,2-dioxetane. Thus, the stepwise mechanism takes place in which firstly the -O-O- bond is broken; secondly, the diradical intermediate is formed in the chemiexcitation region with 4 singlet and 4 triplet electronic states energetically degenerated and thirdly the -C-C- bond rupture occurs (see Figure 6). As in 1,2-dioxetane (Figure 5), three distinct decomposition paths are here possible, on the ground state (S_0), on the lowest-lying singlet state ($^1n_O\pi_{CO}^*$), or on the lowest-lying triplet state ($^3n_O\pi_{CO}^*$).

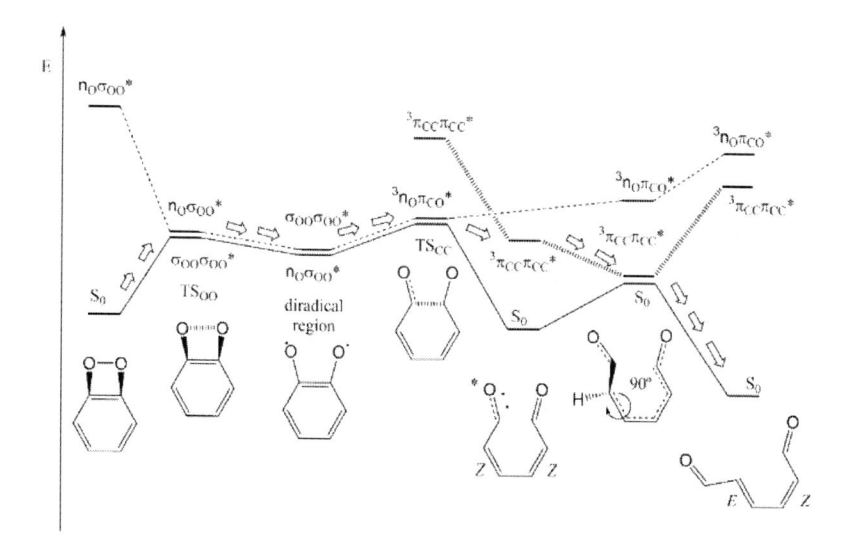

Figure 6. Dark photochemistry mechanism of Dewar dioxetane. Empty arrows represent evolution from the ground-state (S_0) structure of the reactant along the path that gives rise to the excited-state E/Z isomerization on the triplet $^3\pi_{CC}\pi_{CC}^*$ state (see text).

Decomposition on the singlet excited state manifold has the same features as in 1,2-dioxetane. Here, the excited state accessed along the diradical region and the TS$_{CC}$ corresponds to an excitation localized in one of the carbonyl groups related to the electron promotion from the lone pair of the oxygen atom (n_O) to the antibonding π-type orbital of the carbonyl group ($\pi_{CO}{}^*$). Such excitation remains localized along the vibrational relaxation on the excited state PES towards the equilibrium structure responsible for light emission.

Triplet excited state evolution along the decomposition shows a different behavior. Immediately after TS_{CC}, the $^3n_O\pi_{CO}{}^*$ state strongly interact with a close-lying triplet state corresponding to a delocalized excitation over the whole π-conjugated double bonds ($^3\pi_{CC}\pi_{CC}{}^*$). This state becomes at some point more stabilized than the $^3n_O\pi_{CO}{}^*$ state and a transformation from $^3n_O\pi_{CO}{}^*$ to $^3\pi_{CC}\pi_{CC}{}^*$ occurs along the triplet excited state vibrational relaxation. Once in the new type of excited state, the photochemical behavior of such state will be displayed. In this case, the photochemical behaviour of the excited $^3\pi_{CC}\pi_{CC}{}^*$ state of the π-conjugated chain is the E/Z-isomerization. This situation occurring in the triplet manifold and not on the singlet one is due to the fact that while the $^3\pi_{CC}\pi_{CC}{}^*$ state is much lower in energy than $^1\pi_{CC}\pi_{CC}{}^*$, the $^1n_O\pi_{CO}{}^*$ and $^3n_O\pi_{CO}{}^*$ states have basically the same energy.

4. DIOXETANONE MODELS

4.1. 1,2-Dioxetanone

In Firefly luciferin and many other bioluminescent molecular systems, the 1,2-dioxetane ring appears together with a carbonile group, forming therefore 1,2-dioxetanone (see Figure 2). In the Firefly luciferase-luciferin system, this chemical functionality arises from the activation and addition of $^3O_2$ to the carbon atom and the α-position of an ester with a good leaving group (AMP) [1,26]. Due to its role in bioluminescence, 1,2-dioxetanone has been profusely studied in the literature (see [1] and references therein). Like in 1,2-dioxetane, experimental reports indicate an inefficient chemiluminescent process for the 1,2-dioxetanone molecule, giving rise to larger excited-state triplet quantum yields as compared to those for the excited singlet-state population (for instance, singlet and triplet chemiexcitation yields for dimethyldioxetanone are reported to be 0.05 and 0.15, respectively [27,28]; yields for 1,2-dioxetanone are expected to be lower because methylation increases chemiexcitation efficiency [29]). Quantum chemistry was also used in this system to explain this striking difference and to provide a mechanistic description of the chemiluminescence process. In such studies, comparison of the performance of distinct methodologies allowed to decipher whether the chemiluminescence mechanism of 1,2-dioxetanone is stepwise or concerted (Figure 7) and to obtain methodological hints for an accurate determination, which shall be briefly revised in the following lines.

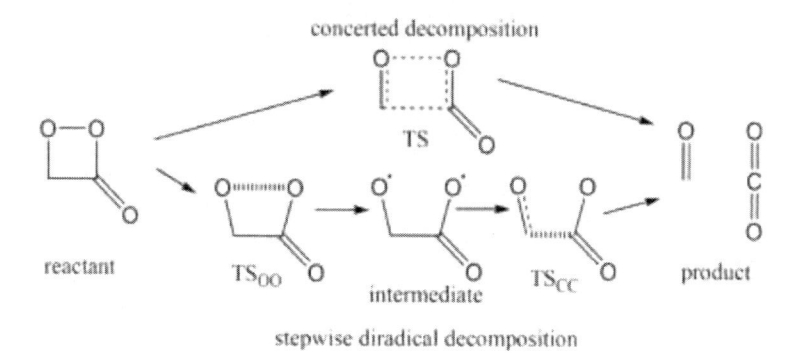

Figure 7. Concerted and stepwise diradical decomposition mechanisms of 1,2-dioxetanone.

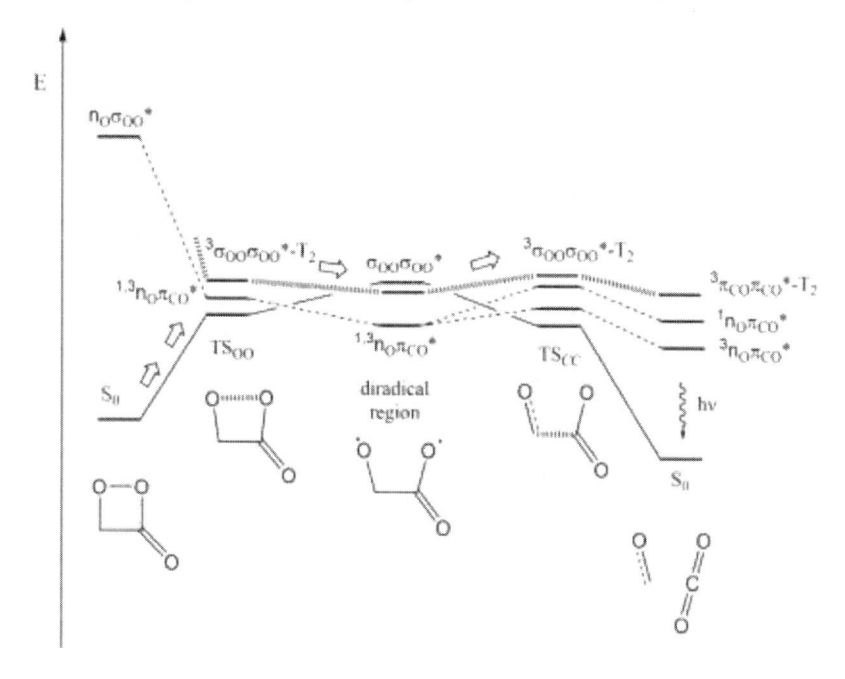

Figure 8. Chemiluminescence mechanism of 1,2-dioxetanone. Empty arrows highlight the decomposition path involving the second triplet excited state, $^3n_o\pi_{CO}*$-T_2 (see text).

The chemiluminescent pathways of 1,2-dioxetanone were initially studied by using highly accurate *ab initio* CASPT2 and two-electron reduced density matrix (2RDM) methods, suggesting a two-step diradical mechanism although with a less-stable intermediate as compared to that of 1,2-dioxetane, as displayed in Figure 8 [30,31]. However, energetics related to the singlet and triplet decomposition paths were not able to explain the distinct triplet and singlet chemiexcitation yields. Later, the use of restricted DFT and time dependent (TD)DFT gave rise to a closed-shell electronic configuration of the molecule along all the decomposition path, pointing to a concerted mechanism for the decomposition of 1,2-dioxetanone [32] which allowed to provide a plaussible interpretation of the yields. Thus, the larger triplet population was ascribed to the much higher energy of the excited S_1 state as compared to that of T_1, which was greater than 14 kcal/mol along the entire reaction path. Further TDDFT studies assisted by CASSCF analyses [8] revealed an open-shell singlet electronic configuration lower in energy than the closed-shell description already reported, supporting the stepwise mechanism as the most favourable one.

Additional studies were also carried out to evaluate the performance of closed-shell DFT making use of hybrid functionals with distinct amounts of exact HF exchange to correctly describe the thermolysis of 1,2-dioxetanone [9]. Five different percentages of exact HF exchange in the B3LYP functional were assessed, extending also the study to other functionals such as PBE1KCIS, PBE1PBE, MPW1B95, MPW1K and MPWB1K, all of them having different HF and DFT exchange admixtures. In this work, the authors tested the stability of the Kohn-Sham wavefunction along the entire S_0 and T_1 PESs. Yue et al. reported that the larger the amount of HF exchange, the greater the reaction barrier and the instability of the closed-shell S_0 and S_1 wavefunction [9]. The authors also studied the performance of unrestricted DFT. These open-shell calculations predicted a stepwise diradical mechanism in agreement with previous multiconfigurational *ab initio* results. Interestingly, all the unrestricted Kohn-Sham wavefunctions are stable along the reaction coordinate regardless of the amount of HF exchange. Using unrestricted DFT, Yue et al. showed that the larger the amount of HF exchange, the lower the reaction barrier, varying between 17.5 and 10.1 kcal/mol. Considering the experimental activation barriers determined in the range of 16-24 kcal/mol and the descriptions provided by previous *ab initio* results, it was concluded that unrestricted DFT methods can describe the decomposition of 1,2-dioxetanone correctly, whereas restricted closed-shell DFT fails due to instabilities in the Kohn-Sham wavefunctions [9].

Since previous attempts to rationalize the notably larger triplet population as compared to the singlet one during the chemiexcitation step in 1,2-dioxetanone were not successful, further CASPT2 computations were performed considering the suggestion made by Da Silva and Da Silva [32] pointing to a relevant role of the T_2 state in the decomposition mechanism [10]. Results showed that the T_2 state corresponds to a $^3\sigma_{OO}\sigma_{OO}*$ electronic configuration and it is indeed near-degenerated energetically with the S_0 - $^1\sigma_{OO}\sigma_{OO}*$, S_1 - $^1n_O\pi_{CO}*$, and T_1 - $^1n_O\pi_{CO}*$ states at regions close to the transition state that mediates the -O-O- bond scission, being thus accessible during the chemiexcitation process (see Figure 8). Reaction-path calculations on the T_2 state following the -O-O- and -C-C- cleavages evidenced a quite planar energy profile, almost barrierless, toward the generation of photoexcited H_2CO* and CO_2 in the T_2 manifold which moves from $^3\sigma_{OO}\sigma_{OO}*$ to $^3\pi_{CO}\pi_{CO}*$ nature. Therefore, the presence of two triplet decomposition pathways (T_1 and T_2) and only one singlet channel (S_1) was proposed to explain the controversial larger quantum yield of triplet population as compared to the singlet one [10].

4.2. 2-Acetamido-3-Methylpyrazine Dioxetanone

Apart from Firefly luciferin/oxyluciferin, the dioxetanone ring described in the previous section (1,2-dioxetanone) is also present in the coelenterazine/coelenteramide and *Cypridina* luciferin/oxyluciferin chemiluminescence systems, which are responsible of bioluminescence processes in certain marine organisms [33,34]. In these cases, the cyclic peroxide is formed by activation and addition of $^3O_2$ at the central imidazopyrazinone moiety that have these luciferins. Multiconfigurational quantum-chemistry studies of the dioxetanone structure of such imidazopyrazinone unit (see Figure 2) and especially the ketone formed after releasing CO_2 were carried out by Roca-Sanjuán and co-workers [11] finding relevant electronic-structure properties which allow to find fine differences between the chemi-induced emission and photo-induced emission.

Photo-induced and chemically-induced luminescence are two light emitting phenomena that imply the emission of light after electronic excitation promoted from distinct sources (photon energy and chemical energy, respectively). It has been always common in the studies of chemi- and bioluminescence to consider that the excited-state species emitting in chemiluminescence (chemiluminescence state, CS) is the same as that emitting upon exciting the product of the chemiluminescence decomposition reaction (fluorescence state, FS). However, both excited-state species are not necessarily the same in both cases. In the former, the excited state is populated from the geometry of the TS related to the decomposition of the dioxetanone ring (see CS in Figure 9). Meanwhile, in the latter, excited-state population takes place from chemiluminescence decomposition product (see FS in Figure 9). Excited-state evolution from these distinct points monitored by geometry optimizations and reaction-path computations with the CASSCF/CASPT2 method was different arriving to distinct excited-state minima, CS and FS, respectively (see Figure 9). While the former is characterized by a CO group with a single-bond nature and *sp3* hybridization of the carbon atom, the latter has a CO double bond and *sp2* hybridization. Thus, both structures maintained the CO character of the starting geometries. Another significant difference is that CS corresponds to a charge transfer (CT) state from the π-conjugated ring or donor (don) to the π^* orbital of the CO group ($\pi_{don}\pi_{CO}^*$). In contrast, the excitation is delocalized over the whole molecule in FS, mainly in the ACT part ($\pi_{don}\pi_{don}^*$).

The findings obtained by Lindh and co-workers [11] indicate that experimental determination of the fluorescence spectrum of the chemiluminescent decomposition product might be misleading to predict chemiluminescent properties of the reactant. Similarly, in theoretical studies, it might be not always accurate to try to find the chemiluminescence emitter (CS) by starting from the chemiluminescence reaction product. If not done, it must be verified that CS and FS are the same. Otherwise, the whole chemiluminescence reaction must be carried out to obtain correct analyses.

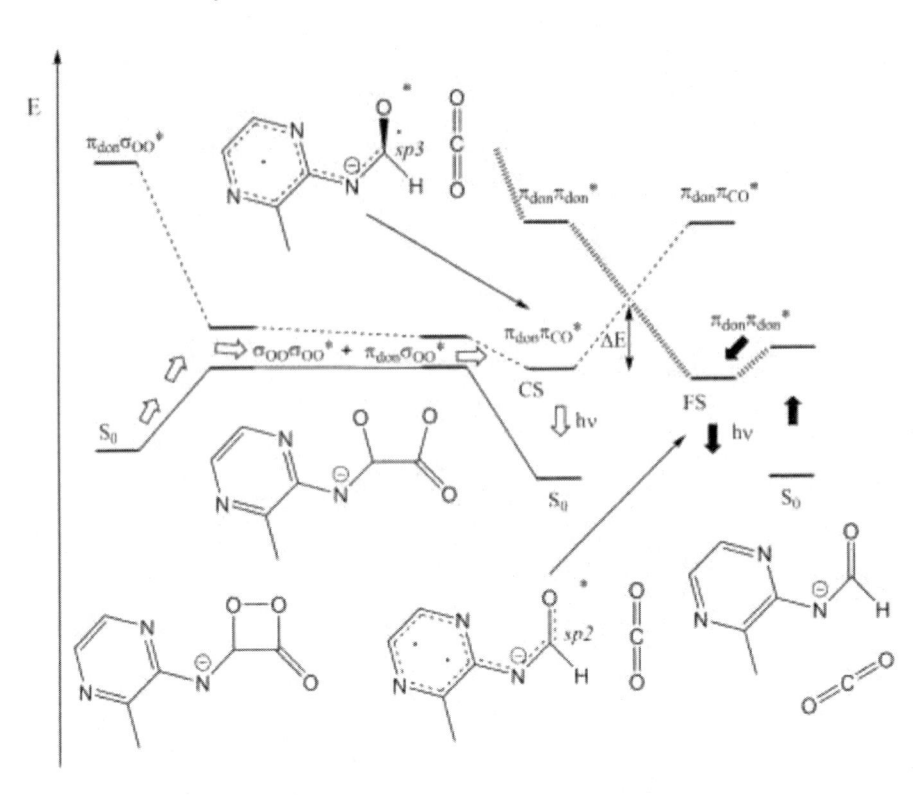

Figure 9. Chemiluminescence mechanism in the imidazopyrazinone-related dioxetanone. Empty arrows represent the decomposition path from the ground-state (S_0) structure of the reactant to the charge transfer emitting species so-called chemiluminescence state (CS) with $\pi_{don}\pi_{CO}$* nature. Filled arrows represent the photo-induced path from the ground-state (S_0) structure of the product to emitting species (fluorescence state, FS, with $\pi_{don}\pi_{don}$* nature). Evolution from CS to FS is hindered by an energy barrier ΔE (see text).

The mentioned study on 2-acetamido-3-methylpyrazine [11] prompted subsequent studies in other chemi- and bioluminescence systems to verify whether the CS and FS were the same or different (see Figure 10). CS and FS were found to be the same species in Firefly oxyluciferin [35] and also in the whole coelenteramide molecule [36]. On the other hand, the dioxetanone of a coumaranone derivative shows different CS and FS [37]. By comparing the chemical structure and electronic-structure properties of all the systems, it can be suggested that different CS and FS emitters are favored by small π-conjugated systems and flexible CO groups. In contrast, when π-conjugation increases as in the whole coelenteramide molecule, the $\pi_{don}\pi_{don}$* state becomes much more stable than $\pi_{don}\pi_{CO}$* eliminating the energy barrier between them (ΔE in Figure 9). This is also happening in Firefly luciferin, where the CO group is part of the 5-member ring and therefore more constrained to become planar (*sp2* hybridization) than in 2-acetamido-3-methylpyrazine (Figure 10).

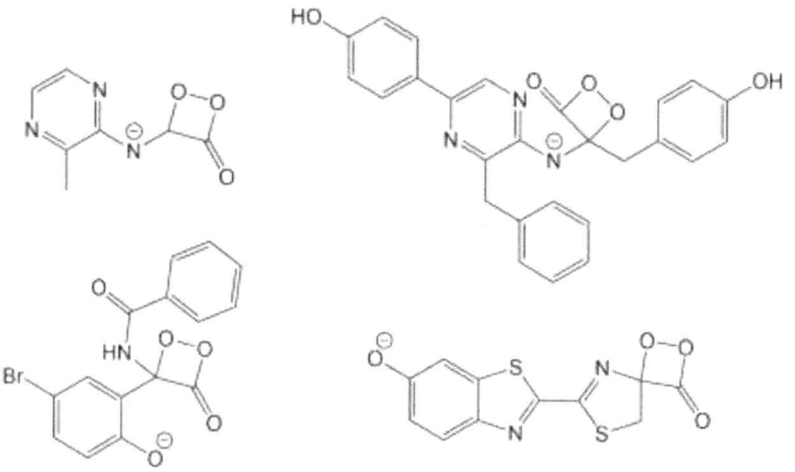

Figure 10. Chemiluminescence molecules in which CS and FS are the same structure, 2-acetamido-3-methylpyrazine dioxetanone (left top) and coumaranone-derivative dioxetanone (left bottom) and those for which are different, coelenterazine (right top) and oxyluciferin (right bottom) (see text).

Roca-Sanjuán et al. [11] also compared the electronic-structure properties of the anionic (deprotonated) and neutral (protonated) forms of 2-acetamido-3-methylpyrazine, which brought about relevant aspects on the differences between the catalyzed and uncatalyzed intra-molecular decomposition of dioxetanones. In the anion, as mentioned above, the lowest-lying excited state at the peroxo-like geometry corresponds to a CT state ($\pi_{don}\pi_{CO}*$), while the lowest excited states of the neutral form with the peroxo-like correspond to $n_O\pi_{CO}*$ at lower energies or $\pi_{don}\pi_{don}*$ at much higher energies. This can be attributed to the fact that the charged nitrogen in the anion facilitates the electron donation from the π system by lowering the ionization potential of the π_{don} orbital. This electron or charge transfer is the first step of the so-called chemically initiated electron exchange luminescence (CIEEL) and charge transfer induced luminescence (CTIL) mechanisms which were proposed by Schuster [38] and Takano et al.[39], respectively, to describe the catalyzed chemiluminescence processes as those taking place in bioluminescence (see Figure 11 left). In these mechanisms, the ionization potential of the electron- or charge-donating group is low enough to become an activator (ACT) of the reaction, producing a significant stabilization of the CT state and consequently a decrease of the energy barrier catalyzing the chemiluminescence process. The CIEEL/CTIL mechanisms also involve a back electron/charge transfer which gives rise to the excitation of ACT. According to the CASSCF/CASPT2 results [11], in the dioxetanone related to 2-acetamido-3-methylpyrazine, the back electron/charge transfer step is not favorable due to the fact that the $\pi_{don}\pi_{don}*$ state is not enough stabilized as compared to the CT ($\pi_{don}\pi_{CO}*$) stabilization. Such $\pi_{don}\pi_{CO}*$ to $\pi_{don}\pi_{don}*$ evolution occurs for instance in the Firefly luciferin and coelenterazine molecules of Figure 10 [2,35,36].

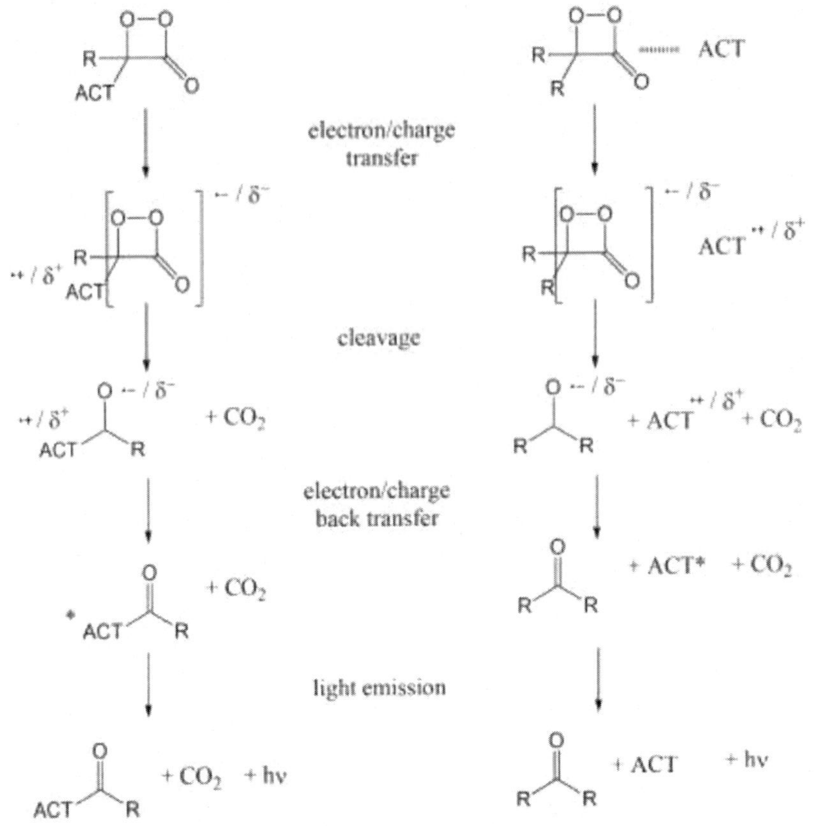

Figure 11. Reaction mechanisms for the intramolecular (left) and intermolecular (right) chemically iniciated electron exchange luminescence (CIEEL) and charge transfer induced luminescence (CTIL).

4.3. 1,2-Dioxetanone – Anthracene

As discussed in the previous section, the presence of a closeby dye activator, ACT, attached covalently to the chemiluminophore facilitates the -O-O- decomposition due to the charge transfer phenomena from the π orbital of the aromatic system to the σ_{OO}^* orbital of the -O-O- bond. Additionally, the chemiluminescence of peroxides can be also activated when the ACT is not linked covalently (intermolecular chemiluminescence), suggesting the formation of a CT complex adhered by non-covalent interactions, mainly of electrostatic and π-stacking nature [40, 41]. The CIEEL has also been proposed to rationalize the enhanced chemiluminescence of the intermolecular system, where the excitation energy released by the high-energy -O-O- bond is employed to excite the ACT, which ultimately emits fluorescence during the decay process to the ground state (see Figure 11 right). The mechanism was originally proposed mostly on the grounds of experimental observations, and in spite of all the advances conducted in the theoretical understanding of chemiluminescence, the intermolecular mechanism has only been studied from its first principles very recently, which shall be described in the following lines.

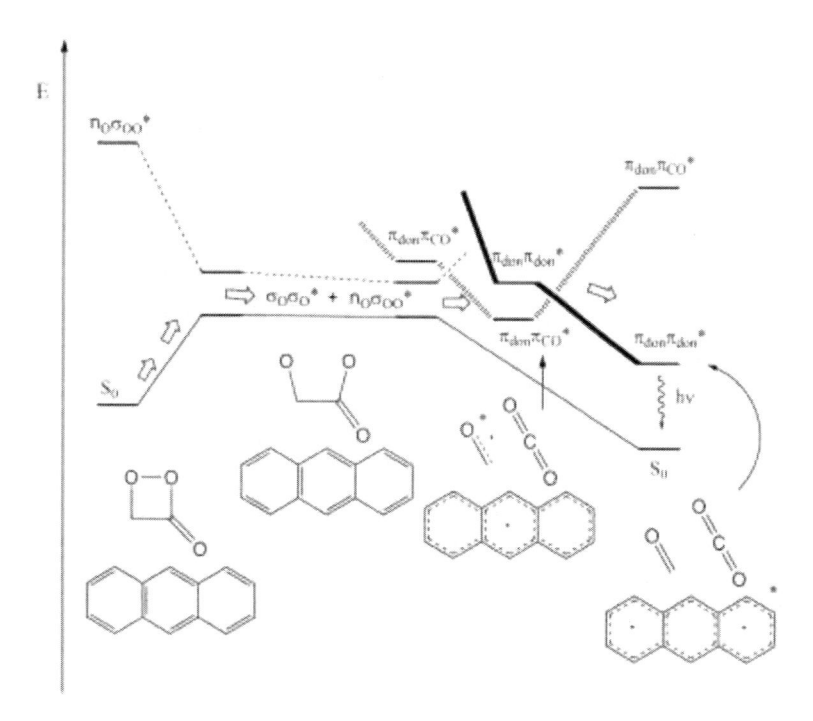

Figure 12. Intermolecular chemiluminescence mechanism in the 1,2-dioxetanone – anthracene complex. Empty arrows represent the decomposition path from the ground-state (S_0) structure of the reactant to the final $\pi_{don}\pi_{don}*$ excited state localized on anthracene passing by the charge transfer $\pi_{don}\pi_{CO}*$ state.

In 2017, Augusto *et al.* reported a study on a model system composed by 1,2-dioxetanone and anthracene to track the nature and energetics of the most relevant excited states upon the peroxide decomposition by using the CASPT2 method (see Figure 12) [12]. Interestingly, the results allowed the authors to capture important aspects of the intermolecular CIEEL/CTIL mechanism and consequently determine the molecular basis of the phenomena. Like in the intramolecular CIEEL/CTIL, the reaction path involves the population of a CT state $\pi_{don}\pi_{CO}*$, which corresponds to the electronic excitation from a π orbital from the donor group (anthracene) to the $\pi_{CO}*$ or 1,2-dioxetanone. Such state ultimately crosses with the $\pi_{don}\pi_{don}*$ state localized over anthracene. This transformation corresponds to the back electron/charge transfer from the dioxetanone moiety to ACT (Figure 11). In the 1,2-dioxetanone-anthracene system, it was observed that the presence of the anthracene does not decrease the thermal energy barrier needed to break the -O-O-bond. As for the deprotonated 2-acetamido-3-methylpyrazine dioxetanone system analyzed in the previous section, this is due to the fact that the electron/charge donor group has not low enough ionization potential to become an ACT. This finding in 1,2-dioxetanone-anthracene agrees with the low yield of chemiluminescence registered experimentally for similar systems [12, 41]. Catalyzed chemiluminescence in the framework of the CIEEL/CTIL mechanism requires better electron/charge-donating ACT able to stabilize much more the $\pi_{ACT}\pi_{ACT}*$ state and make it the S_0 state rather than $\sigma_{OO}\sigma_{OO}*$ or $n_O\sigma_{OO}*$ at early stages of the peroxide bond-breaking.

5. DIOXIN MODELS

5.1. 1,2-Dioxin

This molecule corresponds to the smallest chemical functionality of another type of chemiluminescence systems which slightly differ from those based on the 1,2-dioxetane cycle. 1,2-dioxin constitutes the basic molecular structure responsible for luminol chemiluminescence. In DMSO solution and in the presense of a strong base, luminol reacts with $^3O_2$ incorporating a -O-O- bond and releasing $^1N_2$. This gives rise to 1,2-dioxane-3,6-dione cyclic peroxide dianion, CP^{-2}, as intermediate, which has 1,2-dioxin as central unit. Thermally-activated ring opening in CP^{-2} produces 3-aminophtalate dianion, $3AP^{-2}$, in the excited state, which decays by light emission [13].

The chemiexcitation process originated by 1,2-dioxin passes through the rupture of the peroxide bond and, as we did for 1,2-dioxetane, it can be qualitatively described by correlating the frontier orbitals and resulting electronic states before and after the -O-O- breaking, according to Walsh correlation diagrams. The process also preserves a C_2 symmetry axis (see Figure 13) therefore having symmetric (S) and antisymmetric (A) orbitals and states. From reactant to product, the bonding σ_S and antibonding σ^*_A orbitals located on the -O-O- bond correlate with the two oxygen lone pairs (n_S and n_A, respectively), while the π_S orbital describing the two CC double bonds in the reactant corresponds to a π^*_S orbital in the product. For the state correlations, the $(\sigma_S)^2(\pi_S)^2$ ground state of the reactant becomes an upper-lying $(n_S)^2(\pi^*_S)^2$ excited state of the product, the upper-lying $(\sigma_S)^2(\sigma^*_S)^2$ of the reactant becomes the $(n_S)^2(n_A)^2$ ground state of the product, and the lowest-lying $(\sigma_S)^2(\pi_S\sigma^*_A)$ is transformed to $(n_S)^2(n_A\pi^*_S)$.

To obtain further details in the ring-opening mechanism and determine the differences with the mechanism in the 1,2-dioxetane and 1,2-dioxetanone systems, Giussani et al. have recently used the CASSCF/CASPT2 method and reaction-path computations giving rise to findings which are schematized in Figure 14 [13]. In contrast to 1,2-dioxetane and 1,2-dioxetanone, only one energy barrier characterizes the decomposition and it has a much lower $\Delta E^{\#}$ value of 11.1 kcal/mol (~24 kcal/mol for 1,2-dioxetane [6, 24] and ~27 kcal/mol for 1,2-dioxetanone [30]). At this TS, the energy degeneracy between the ground and excited singlet and triplet states occurs. Therefore, it represents the chemiexcitation structure in 1,2-dioxin. Regarding the thermodynamics, the reaction releases ΔE = -59.7 kcal/mol, which means a higher exergonicity as compared to 1,2-dioxetane (-27.1 kcal/mol [6, 24]) and lower than that of 1,2-dioxetanone (-78.6 kcal/mol [30]).

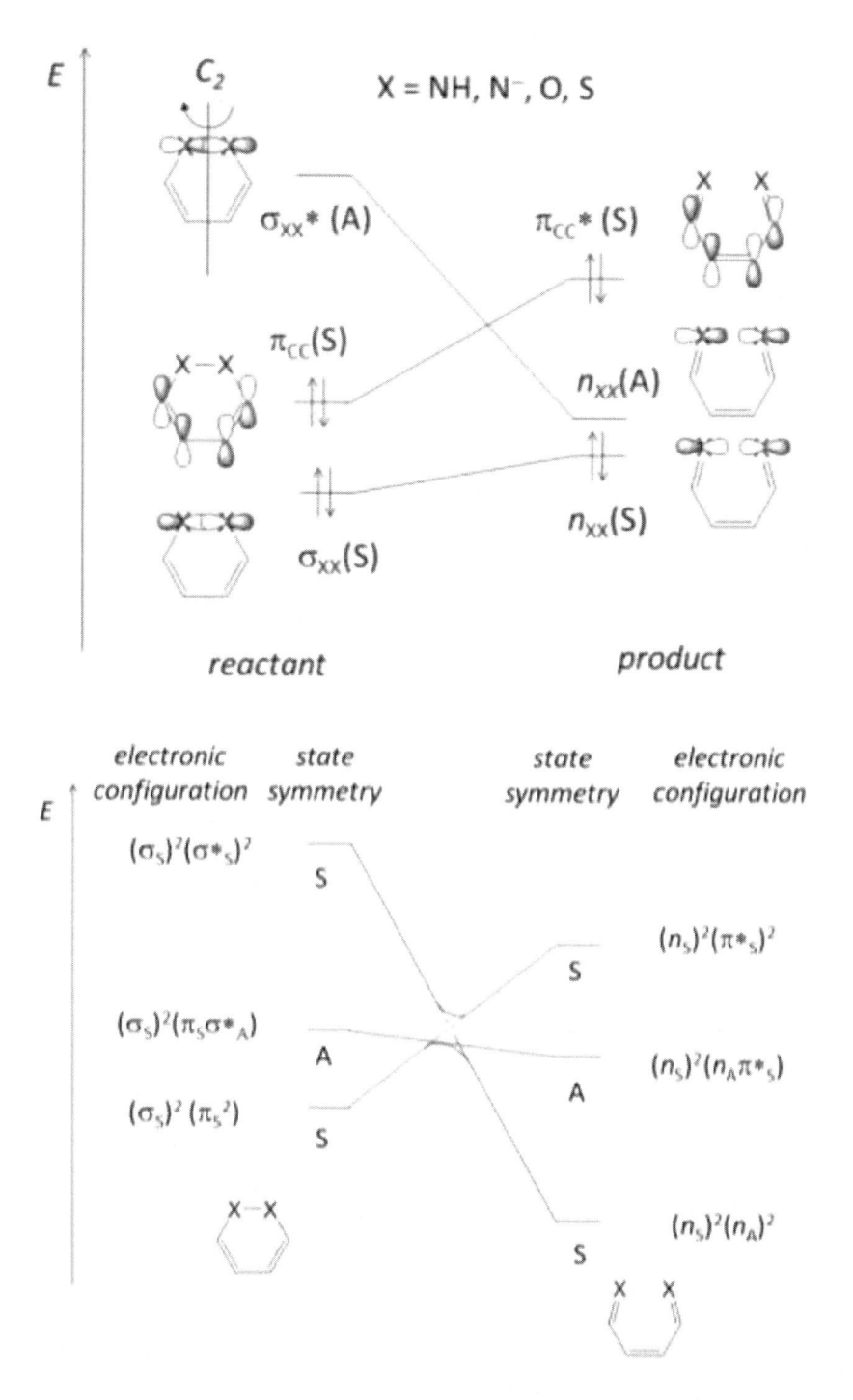

Figure 13. Walsh molecular orbital (top) and electronic state (bottom) correlation diagrams for the ring opening or 1,2-dioxin. See Figure 4 for the definition of the notation. Adapted with permission from ref. 13. Copyright 2019 John Wiley and Sons.

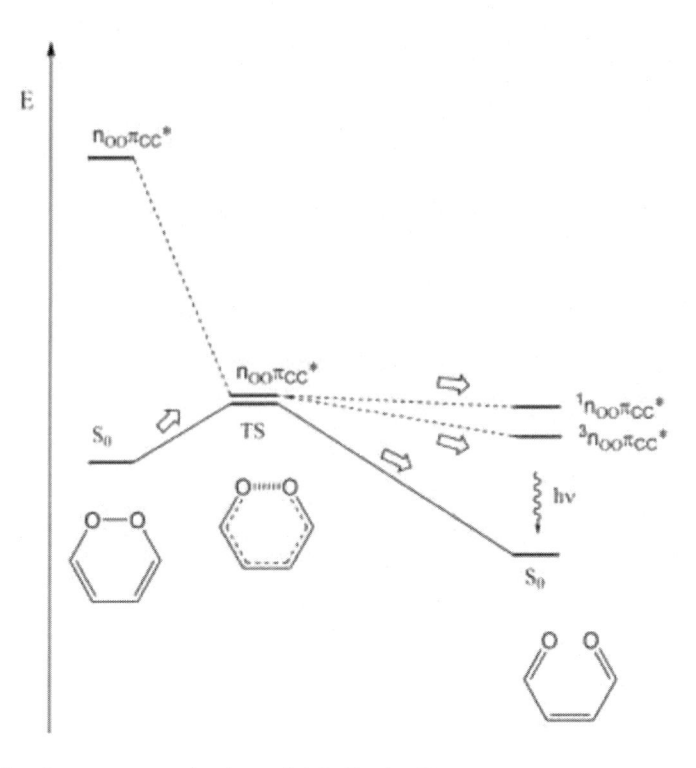

Figure 14. Chemiluminescence mechanism of 1,2-dioxin. Empty arrows represent evolution from the ground-state (S_0) structure of the reactant to the three distinct decomposition paths corresponding to the thermal process on the S_0 PES and the lowest-lying singlet ($^1n_{OO}\pi_{CC}^*$) and triplet ($^3n_{OO}\pi_{CC}^*$) excited states (see text).

The kinetic and thermodynamic properties of 1,2-dioxin have been also compared with those of isoelectronic systems substituting the -O-O- bond with the -S-S-, -NH-NH-, and -N⁻-N⁻- bonds. According to the Walsh diagrams (Figure 13), all the resulting rings have analogous abilities for chemiexcitation. Reaction-path CASSCF/CASPT2 computations show however significant differences [13]. -S-S- and -N⁻-N⁻- present a S_0-S_1-T_1 degeneracy at the TS, while only S_0 and T_1 are significantly close in -NH-NH-. Moreover, $\Delta E^{\#} = 34.5$, 36.7, and 27.0 kcal/mol; $\Delta E = 19.1$, -12.2, and -30.7 kcal/mol, for -S-S-, -NH-NH-, and -N⁻-N⁻-, respectively. Therefore, the -O-O- bond lead to the most favourable (efficient) situation for chemiexcitation kinetically (lower chemiexcitation barrier) and thermodynacally (larger product stabilization). This result is particularly interesting since it provides a molecular interpretation of why luminol, in which a -NH-NH- bond is present, requires the incorporation of a peroxide bond in order to produce efficient chemiluminescence. For the others, it is worth noting that chemiexcitation is more favorable in -N⁻-N⁻- than in -NH-NH- and the process in -S-S- is endothermic, which might be interesting for the search of reversible chemiluminescence processes.

CONCLUSION

Multiconfigurational quantum chemiluminescence refers to the study of chemically-induced luminescence phenomena by means of computational and theoretical approaches using *ab initio* quantum-chemistry methods like CASSCF/CASPT2. Such methods are required in chemiluminescence for an accurate determination of the radiationless change from the ground to the excited electronic state (chemiexcitation) taking place via conical intersections (CIs) and singlet-triplet crossings (STC). Based on CASSCF/CASPT2 applications to distinct peroxide small model molecules, relevant results have been obtained during the last decades, which have allowed our community to achieve a better comprehension of the chemi- and bioluminescence phenomena. The main findings are summarized in the following lines:

- Chemiluminescence involves a chemiexcitation step of population transfer from the ground to the excited state which is characterized by a multistate crossing between singlet and triplet states (for example, 4 singlet and 4 triplet states in 1,2-dioxetane and 2 singlets and 1 triplet in 1,2-dioxin).
- Chemiexcitation has biradical nature, which can be a well-defined intermediate or diradical (in 1,2-dioxetane), a less stable intermediate (in 1,2-dioxetanone), or only a biradicaloid TS (for instance, in 1,2-dioxin).
- Thermal and entropy corrections of the electronic energies are in some occasions relevant for a fine interpretation of the experimental data as in 1,2-dioxetane. On the contrary, in 1,2-dioxetanone, the highest triplet yield was found to be related to the easier accessibility of 2 triplet excited states as compared to the access to the manifold of the singlet excited state (only 1 accessible).
- The peroxide bond is much better chemiluminophore (much lower activation barriers) in the 1,2-dioxin chemical functionality than in 1,2-dioxetane.
- 1,2-dioxin related isoelectronic systems in which oxygen atoms are substituted by N-, NH, or S are still chemiluminophores but much less efficient due to a higher activation barrier, lower exothermicity, and/or lower probability for excited state access at the chemiexcitation region.
- Chemiluminescence can be catalyzed intra- and inter-molecularly in the framework of the CIEEL/CTIL mechanisms. In both intra- and inter-molecular processes, an electron/charge transfer (CT) from the activator group (with good donating properties) to the σ^* antibonding orbital of the peroxide bond is involved in the chemiluminescence. For an effective catalysis, this state must be populated at early stages of the -O-O- bond breaking. This occurs in Firefly luciferin and coelenterazine. When the donor group has not low enough ionization potential, the CT state appears and is populated at a late stage not decreasing significantly the

reaction barrier (as in 2-acetamido-3-pyrazine anion or the 1,2-dioxetanone-anthracene complex).

- In the conventional CIEEL or CTIL mechanisms, the CT state is transformed to a localized $\pi\pi^*$ state in the electron-/charge-donating group which has intense emission efficiency by a back electron/charge transfer step. This transformation might not take place if the $\pi\pi^*$ is not significantly stabilized as compared to the CT state (as it occurs in 2-acetamido-3-pyrazine anion).

ACKNOWLEDGMENTS

This work has received financial support from the Spanish MINECO/FEDER (projects CTQ2017-87054-C2-2-P; "Ramón y Cajal" grant RYC-2015-19234; the Generalitat Valenciana (PROMETEO/2016/135) and from the French ANR and 'Région Grand Est' government. J.C.-G. acknowledges the Universitat de València for his master scholarship.

REFERENCES

[1]   Vacher, M., Fdez Galván, I., Ding, B.W., Schramm, S., Berraud-Pache, R., Naumov, P., Ferré, N., Liu, Y. J., Navizet, I., Roca-Sanjuán, D., Baader, W. J., Lindh, R. (2018) Chemi- and Bioluminescence of Cyclic Peroxides. *Chem. Rev.*, 118: 6927–6974.

[2]   Navizet, I., Liu, Y. J., Ferré, N., Roca-Sanjuán, D., Lindh, R. (2011) The chemistry of bioluminescence: An analysis of chemical functionalities. *ChemPhysChem.*, 12: 3065–3076.

[3]   Klessinger, M., Michl, J. (1995) Excited states and photochemistry of organic molecules, VCH.

[4]   Turro, N. J. (1991) Modern molecular photochemistry, University Science Books.

[5]   Roca-Sanjuán, D., Aquilante, F., Lindh, R. (2012) Multiconfiguration second-order perturbation theory approach to strong electron correlation in chemistry and photochemistry. *Wiley Interdiscip. Rev. Comput. Mol. Sci.*, 2: 585–603.

[6]   Farahani, P., Roca-Sanjuán, D., Zapata, F., Lindh, R. (2013) Revisiting the Nonadiabatic Process in 1,2-Dioxetane. *J. Chem. Theory Comput.*, 9: 5404–5411.

[7]   Farahani, P., Lundberg, M., Lindh, R., Roca-Sanjuán, D. (2015) Theoretical study of the dark photochemistry of 1,3-butadiene via the chemiexcitation of Dewar dioxetane. *Phys. Chem. Chem. Phys.*, 17: 18653–18664.

[8]   Roca-Sanjuán, D., Lundberg, M., Mazziotti, D. A., Lindh, R. (2012) Comment on "Density functional theory study of 1,2-dioxetanone decomposition in condensed phase." *J. Comput. Chem.*, 33: 2124–2126.

[9]   Yue, L., Roca-Sanjuán, D., Lindh, R., Ferré, N., Liu, Y. J. (2012) Can the closed-shell DFT methods describe the thermolysis of 1,2-dioxetanone? *J. Chem. Theory Comput.*, 8: 4359–4363.

[10]   Francés-Monerris, A., Fdez. Galván, I., Lindh, R., Roca-Sanjuán, D. (2017) Triplet versus singlet chemiexcitation mechanism in dioxetanone: a CASSCF/CASPT2 study. *Theor. Chem. Acc.*, 136: 70.

[11]   Roca-Sanjuán, D., Delcey, M. G., Navizet, I., Ferré, N., Liu, Y. J., Lindh, R. (2011) Chemiluminescence and fluorescence states of a small model for coelenteramide and Cypridina oxyluciferin: A CASSCF/CASPT2 study. *J. Chem. Theory Comput.*, 7: 4060–4069.

[12]   Augusto, F. A., Francés-Monerris, A., Fdez Galván, I., Roca-Sanjuán, D., Bastos, E. L., Baader, W. J., Lindh, R., Fdez. Galván, I., Roca-Sanjuán, D., Bastos, E. L., Baader, W. J., Lindh, R. (2017) Mechanism of activated chemiluminescence of cyclic peroxides: 1,2-dioxetanes and 1,2-dioxetanones. *Phys. Chem. Chem. Phys.*, 19: 3955–3962.

[13]   Giussani, A., Farahani, P., Martínez-Muñoz, D., Lundberg, M., Lindh, R., Roca-Sanjuán, D. (2019) Molecular Basis of the Chemiluminescence Mechanism of Luminol. *Chem. – A Eur. J.*, 25: 5202-5213.

[14]   Roos, B. O., Andersson, K., Fulscher, M. P., Malmqvist, P. Å., Serrano-Andrés, L., Pierloot, K., Merchán, M. (1996) Multiconfigurational perturbation theory: Applications in electronic spectroscopy. *Adv. Chem. Phys.*, 93: 219–331.

[15]   Roos, B. O., Taylor, P. R., Sigbahn, P. E. M. (1980) A complete active space SCF method (CASSCF) using a density matrix formulated super-CI approach. *Chem. Phys.*, 48: 157–173.

[16]   Andersson, K., Malmqvist, P. Å., Roos, B. O., Sadlej, A. J., Wolinski, K. (1990) Second-order perturbation theory with a CASSCF reference function. *J. Phys. Chem.*, 94: 5483–5488.

[17]   Andersson, K., Malmqvist, P. Å., Roos, B. O. (1992) Second-order perturbation theory with a complete active space self-consistent field reference function. *J. Chem. Phys.*, 96: 1218–1226.

[18]   Finley, J., Malmqvist, P. Å., Roos, B. O., Serrano-Andrés, L. (1998) The multi-state CASPT2 method. *Chem. Phys. Lett.*, 288: 299–306.

[19]   Forsberg, N., Malmqvist, P. Å. (1997) Multiconfiguration perturbation theory with imaginary level shift. *Chem. Phys. Lett.*, 274: 196–204.

[20]   Ghigo, G., Roos, B. O., Malmqvist, P. Å. (2004) A modified definition of the zeroth-order Hamiltonian in multiconfigurational perturbation theory (CASPT2). *Chem. Phys. Lett.*, 396: 142–149.

[21]   Woodward, R. B., Hoffmann, R. (1969) The Conservation of Orbital Symmetry, *Angew. Chemie Int. Ed. Eng.*, 8: 781–853.

[22]   Woodward, R. B., Hoffmann, R. (1965) Stereochemistry of Electrocyclic Reactions.

J. Am. Chem. Soc., 87: 395–397.

[23] Adam, W., Baader, W. J. (1985) Effects of methylation on the thermal stability and chemiluminescence properties of 1,2-dioxetanes. *J. Am. Chem. Soc.*, 107 : 410–416.

[24] De Vico, L., Liu, Y. J., Krogh, J. W., Lindh, R. (2007) Chemiluminescence of 1,2-Dioxetane. Reaction Mechanism Uncovered. *J. Phys. Chem. A.*, 111: 8013–8019.

[25] Baader, W. J., Stevani, C. V., Bechara, E. J. H. (2015) "Photo" Chemistry Without Light? *J. Braz. Chem. Soc.*, 26: 2430–2447.

[26] Sundlov, J. A., Fontaine, D. M., Southworth, T. L., Branchini, B. R., Gulick, A. M. (2012) Crystal Structure of Firefly Luciferase in a Second Catalytic Conformation Supports a Domain Alternation Mechanism. *Biochemistry*, 51: 6493–6495.

[27] Schmidt, S. P., Schuster, G. B. (1978) Kinetics of unimolecular dioxetanone chemiluminescence. Competitive parallel reaction paths. *J. Am. Chem. Soc.*, 100: 5559–5561.

[28] Adam, W., Simpson, G. A., Yany, F. (1974) Mechanism of direct and rubrene enhanced chemiluminescence during .alpha.-peroxylactone decarboxylation. *J. Phys. Chem.*, 78: 2559–2569.

[29] Vacher, M., Farahani, P., Valentini, A., Frutos, L. M., Karlsson, H. O., Fdez. Galván, I., Lindh, R. (2017) How Do Methyl Groups Enhance the Triplet Chemiexcitation Yield of Dioxetane? *J. Phys. Chem. Lett.*, 8: 3790–3794.

[30] Liu, F., Liu, Y., De Vico, L., Lindh, R. (2009) Theoretical Study of the Chemiluminescent Decomposition of Dioxetanone. *J. Am. Chem. Soc.*, 131: 6181–6188.

[31] Greenman, L., Mazziotti, D. A. (2010) Strong electron correlation in the decomposition reaction of dioxetanone with implications for firefly bioluminescence. *J. Chem. Phys.*, 133: 164110.

[32] Da Silva, L. P., da Silva, J. C. G. (2012) Density functional theory study of 1,2-dioxetanone decomposition in condensed phase. *J. Comput. Chem.*, 33: 2118–2123.

[33] Shimomura, O., Johnson, F. H., Saiga, Y. (1962) Extraction, Purification and Properties of Aequorin, a Bioluminescent Protein from the Luminous Hydromedusan Aequorea. *J. Cell. Comp. Physiol.*, 59: 223–239.

[34] Morin, J. G., Hastings, J. W. (1971) Energy transfer in a bioluminescent system. *J. Cell. Physiol.*, 77: 313–318.

[35] Navizet, I., Roca-Sanjuán, D., Yue, L., Liu, Y. J., Ferré, N., Lindh, R. (2013) Are the bio- and chemiluminescence states of the firefly oxyluciferin the same as the fluorescence state? *Photochem. Photobiol.*, 89: 319–325.

[36] Chen, S. F., Navizet, I., Roca-Sanjuán, D., Lindh, R., Liu, Y. J., Ferré, N. (2012) Chemiluminescence of coelenterazine and fluorescence of coelenteramide: A systematic theoretical study. *J. Chem. Theory Comput.*, 8: 2796–2807.

[37] Schramm, S., Ciscato, L. F. M. L., Oesau, P., Krieg, R., Richter, J. F., Navizet, I., Roca-Sanjuán, D., Weiss, D., Beckert, R. (2015) Investigations on the synthesis and

chemiluminescence of novel 2-coumaranones - II. *Arkivoc.*, 44–59.

[38] Schuster, G. B. (1979) Chemiluminescence of organic peroxides. Conversion of ground-state reactants to excited-state products by the chemically initiated electron-exchange luminescence mechanism. *Acc. Chem. Res.*, 12: 366–373.

[39] Takano, Y., Tsunesada, T., Isobe, H., Yoshioka, Y., Yamaguchi, K., Saito, I. (1999) Theoretical studies of decomposition reactions of dioxetane, dioxetanone, and related species. CT induced luminescence mechanism revisited. *Bull. Chem. Soc. Jpn.*, 72: 213–225.

[40] De Oliveira, M. A., Bartoloni, F. H., Augusto, F. A., Ciscato, L., Bastos, E. L., Baader, W. J. (2012) Revision of Singlet Quantum Yields in the Catalyzed Decomposition of Cyclic Peroxides. *J. Org. Chem.*, 77: 10537–10544.

[41] Augusto, F. A., De Souza, G. A., De Souza, S. P., Khalid, M., Baader, W. J. (2013) Efficiency of Electron Transfer Initiated Chemiluminescence. *Photochem. Photobiol.*, 89: 1299–1317.

In: A Comprehensive Guide to Chemiluminescence
Editor: Luís Pinto da Silva

ISBN: 978-1-53616-170-0
© 2019 Nova Science Publishers, Inc.

Chapter 13

THE ART OF CHEMILUMINESCENCE EXPERIMENTS: EASY TO EXECUTE CHEMILUMINESCENCE DEMONSTRATIONS AS AN EDUCATIONAL TOOL FOR TEACHING FUNDAMENTAL ENERGY TRANSDUCTIONS AND ATTRACTING HIGH-SCHOOL STUDENTS TOWARDS CHEMISTRY

Dieter Weiß[1,] and Stefan Schramm[1,2,†]*
[1]Friedrich-Schiller University Jena, Germany
[2]New York University Abu Dhabi, UAE

Dedicated to the memory of Herbert Brandl (1947–2018)

ABSTRACT

Skillfully executed chemical experiments are among the highlights of any chemistry lectures. There are numerous publications on demonstration experiments which are suitable to impress pupils and students [1]. These experiments, however, are usually intended in a way that they must be performed by experienced experimenters [2]. This also applies to chemiluminescence demonstrations. Experiments for students or hobby chemists often fail since the chemicals required are expensive, toxic or difficult to obtain, that there are problems with waste disposal or that the preparation and execution of the experiments

---

*Corresponding Author's Email: Dieter.Weiss@uni-jena.de.
† Corresponding Email: Dr.Stefan.Schramm@gmail.com, Stefan.Schramm@uni-jena.de, Stefan.Schramm@nyu.edu.

takes too long. Although many of them are very impressive, they are hardly self-explanatory for all students in a class. The following chapter discusses some chemiluminescent experiments that are mostly harmless, inexpensive and can be performed for students. Although some of these experiments can be entirely done with household chemicals, sometimes special chemicals such a luminol or 2-coumaranones are needed. These chemicals are either commercially available or very easy to synthesize. Additional content is provided for selected experiments in the form of QR-Codes that link to illustrative youtube-videos.

1. GENERAL SAFETY CONSIDERATIONS

Experiments on chemiluminescence are usually very impressive and convincing because light as the result of a reaction is immediately visible and therefore no advanced instrumentation is needed to observe the course of the reaction. However, the brightness of the reactions is often overestimated. Therefore, almost all experiments, should at least be performed in a partial, or even better a fully darkened room. The described chemilumine-scence experiments require the appropriate chemicals and solvents. Many of the substances used are harmful to health, toxic or environmentally hazardous. For some experiments, a concentrated solution of hydrogen peroxide is needed which is highly corrosive. There has been no lack of effort to make the experiments more user-friendly and environmentally friendly, partly with good results. Experiments that might also be suitable for demonstrations for children (always performed of a skilled adult!) are marked separately.

The authors do not take any responsibility for the results that may arise as an unwanted outcome of the described experiments or improper handling of chemicals or instruments. All experiments are for educational purposes only and should exclusively performed by a qualified person.

2. PICTOGRAM EXPLANATION

Corrosive	Environmentally hazardous	Harmful	Children friendly
Mortar	Round bottom flask	Beaker	Erlenmeyer flask

3. QR-CODES INCLUDING LINKS TO ILLUSTRATIVE YOUTUBE VIDEOS OF SELECTED EXPERIMENTS

4. EXPERIMENTS WITH LUMINOL

Luminol is the classic chemiluminescent reaction. It has long been known, is relatively non-toxic and easy to handle. The reaction can be carried out in a slightly alkaline aqueous solution and does not require extra organic solvent. The oxidant used is usually hydrogen peroxide, but it is also possible to use percarbonates which are commonly found in many heavy-duty detergents and oxygen based bleaching agents [3]. The catalyst used is usually hemin (blood dye) or potassium hexacyanoferrate ($K_3[Fe(CN)_6]$), but also peroxidases (e.g., horseradish peroxidase, a slice a freshly cut horseradish) are well suited.

Figure 1: Chemiluminescence reaction of luminol

4.1. Standard Procedure

Dissolve 2 g of NaOH in 50 ml of water. Add 1 g of luminol and stir until everything has dissolved. This solution is diluted with 350 ml of water and 2-3 ml of hydrogen peroxide (30%) are added. The reaction starts immediately after addition of the catalyst. This can be among others a drop of blood or a solution of a crystal of potassium hexacyanoferrate. As an alternative a fresh slice of horseradish or seaweed can be used. The later ingredients contain peroxidases as an catalyst.

4.2. Alternative Oxidizing Agents

In a mortar, thoroughly rub 1 g of heavy-duty detergent, 10 mg of luminol and a catalytic amount of hemin using a pestle. After addition of 5-10 ml of water, the grinding is continued. A few seconds later, the chemiluminescent reaction starts and the characteristic intense blue glow of the luminol appears. By using various detergents and bleaches, a suitable peroxide source can be found. A dry mixture of detergent and luminol can be stored for a long time and is then available for testing at any time. We call this mixture *LumoInstant* because it allows to start a chemiluminescence reaction quickly and without much preparation just by adding water and a catalyst to it.

4.3. Detection Limit of Hemin

Mix 1g heavy-duty detergent with 10 mg luminol or 1 g *LumoInstant* and 5 ml water. Je 1 ml of this suspension is added to a test tube and diluted with 5 ml of water. 10 - 20 mg of hemin and a spatula tip of sodium carbonate are dissolved in 10 ml of water in a small volumetric flask. 0.05 ml, 0.1 ml, 0.2 ml, 0.4 ml, 0.8 ml and 1.6 ml of this solution are placed in the test tubes containing the stock solution in ascending order. The chemiluminescence can be observed in a dark room. Even with 0.05 ml of hemin solution a clear blue glow is observed.

4.4. Criminological Blood Test

Cut a piece of cloth into 4 parts. Brush the pieces one by one with ketchup, red ink, red paint and blood and let it dry. Then fill in a spray bottle with an aqueous solution of *LumoInstant* and spray the fabric with it. Only the blood spot gives a blue chemiluminescence.

5. PEROXYOXALATE CHEMILUMINESCENCE

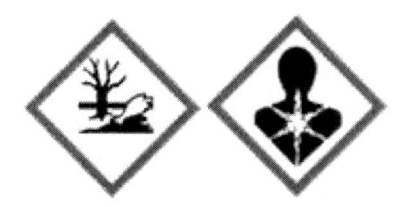

Peroxyoxalate chemiluminescence which is known from glow sticks is one of the brightest chemiluminescence reactions known. The chemicals and solvents commonly used are both harmful to health and the environment and are therefore safely trapped in the plastic tube of the glow stick. By selecting a suitable dye, any emission color can be produced.

Figure 2: Top: Chemiluminescence reaction of TCPO with hydrogen peroxide. Bottom: Further examples of oxalic acid esters that are able to undergo peroxyoxalate chemiluminescence

5.1. Standard Procedure

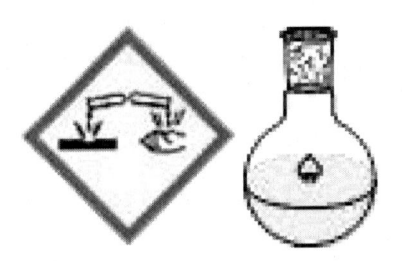

Dissolve a small amount of fluorescent dye (e.g., diphenyl anthracene, rubrene) in dibutyl phthalate and add as much TCPO or DNPO as is needed to form a saturated solution. Prepare in a second dish the oxidation mixture. For this add a spatula of sodium salicylate to 1 ml of hydrogen peroxide (30%) and add 1 ml of tert-butanol. The reaction starts when the oxidizer mixture is added to the solution of the oxalate. The emission color depends on the fluorescent dye.

5.2. Suitable Fluorescence Dyes and the Resulting Color

Diphenyl anthracene (blue), bis-phenylethynyl anthracene (green), coumarine-7 (green), eosine (orange), rubrene (violet), chlorophyll (red).

5.3. "Draculas Tea"

A transparent cup or a tea glass is filled with ethyl acetate. Add approximately 1 g TCPO and a tea bag with peppermint tea. After a few minutes the ethyl acetate has dissolved the chlorophyll from the peppermint tea and also the TCPO should have dissolved for the most part. Remove the tea bag, extinguish the light, add the oxidizer mixture and stir. The liquid in the cup shines in a red light. Do not drink!

5.4. Procedure with TCPO and a Heavy-Duty Detergent

Take a mortar and mix in it about 1g of heavy-duty detergent with a spatula tip of TCPO. Add a small amount of solvent (e.g., bio-diesel) to this mixture so that a viscous paste is formed. Extinguish the light and initiate the start the reaction by the addition a small amount of water. The result is a blue chemiluminescence in which the optical brightener from the detergent serves as the emissive fluorescence dye.

Less bright but also much less toxic as TCPO is BVO, an oxalic ester based on vanillin [4]. As solvent one can also use malonic acid diethyl ester which has no classification of hazardous substances.

5.5. Procedure with BVO and Malonic Acid Diethyl Ester

Mix approx. 200 mg BVO with 1.9 g sodium sulphate and 0.1 g potassium carbonate in a mortar. The resulting mixture is stable for months and can be used for experiments at any time. We have called it *LumiFlash*. Add a small amount of fluorescent dye to a few milliliters of malonic acid diethyl ester and add *LuminFlash*. The reaction starts after the addition of a few drops of hydrogen peroxide and vigorous stirring or shaking. A closed reaction vessel or a mortar with pestle are advantageous.

6. CHEMILUMINESCENCE OF SINGLET OXYGEN

Oxygen is one of the few molecules that are present as a triplet in the ground state. The energy difference between singlet and triplet oxygen is just big enough to emit visible light in the red spectral range.

It is key to chemically produce singlet oxygen from singlet molecules such as hydrogen peroxide and not oxygen from the air. A simple method to generate singlet oxygen in such a way is to introduce chlorine gas into hydrogen peroxide. The peroxide is oxidized to oxygen in the singlet state. During the transition to the thermodynamically more stable triplet state, energy is released in the form of red light. The handling of chlorine gas requires some practice but is easily possible if suitable equipment is used. Nevertheless, appropriate precautions and risk assessments should always be taken into account when working with chlorine in order to prevent accidents or at least limit their effects.

a) $\cdot$O-O$\cdot$ $\rightleftharpoons$ O=O ΔH_R = + 92 kJ/Mol
          Triplet oxygen   Singlet oxygen

b)   Cl-Cl + 2 OH$^{\ominus}$ $\longrightarrow$ Cl$^{\ominus}$ + OCl$^{\ominus}$ + H$_2$O

c)   HO-OH + OCl$^{\ominus}$ $\longrightarrow$ O=O + H$_2$O + Cl$^{\ominus}$

d)   O=O $\longrightarrow$ $\cdot$O-O$\cdot$ + hν

Figure 3: Reacation cascade, that can leads to the formation of singlet oxygen

6.1. Standard Procedure

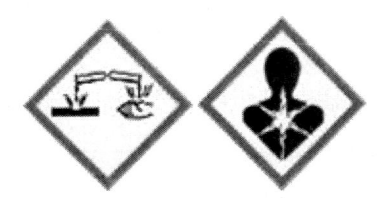

Place 5 g KMnO$_4$ in a 100 ml two-neck flask. Add a fermentation tube on the flask and fill it with a mixture of 18 ml H$_2$O, 2g NaOH and 4 ml H$_2$O$_2$. The absolute filling quantity depends on the size of the fermenting tube, nothing should splash out when gas bubbles are passed through. The second neck is closed with a septum. A syringe with a cannula is filled with 5 ml HCl (conc.) and the cannula is inserted through the septum. After this the light in the room is extinguished the hydrochloric acid is slowly injected to the permanganate. The reaction starts when the rising chlorine has displaced the trapped air. Each chlorine bubble produces a medium bright red flash of light.

If a fermentation tube is not available, one can also use an apparatus as shown in the illustration below. After completion of the reaction, the reaction vessels should be thoroughly ventilated to remove chlorine residues [5].

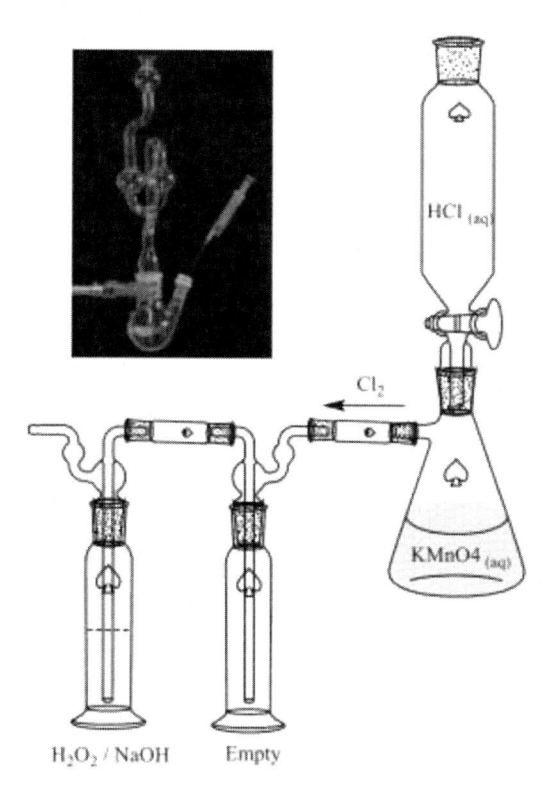

Figure 4: Instrumentation for generation of singlet oxygen

Hydrogen peroxide can also be oxidized with N-bromosuccinimide to singlet oxygen with significantly less effort and less toxicity.

N-bromosuccinimide (NBS)   succinimide

Figure 5: Singlet oxygen generation from N-bromosuccinimide

6.2. Procedure with N-Bromosuccinimide

Mix 20 ml of diluted hydrogen peroxide (10%) with 20 ml of a potassium hydroxide solution (10%) in a reaction vessel. Cool the solution with ice to about 0 - 3°C. Then extinguish the light in the room, add a spatula tip of N-bromosuccinimide and shake or stir. Immediately the red emission of the formed singlet oxygen can be seen.

7. CHEMILUMINESCENCE OF 2-COUMARANONES

This chemiluminescence reaction is very bright and particularly easy to perform [6]. Dissolve the 2-coumaranone in a polar aprotic solvent like acetone or DMF, add a base e.g., DBU, trimethylamine or some crystals of sodium carbonate and the light emission immediately starts. The limiting factor is the oxygen dissolved in the solvent. A large amount of 2-coumaranone does not produce linearity correlated more light because the oxygen is fastly depleted; the reaction extinguishes and only starts again when additional oxygen is supplied. By the addition of a suitable fluorophore (e.g., rubrene, coumarine-7 or BPEA) it is possible to shift the emission color from blue to other spectral regions. Detailed explanations on the synthesis of 2-coumaraones can be found in the chapter: "Crystalline Fireflies"

Figure 6: Chemiluminescence reaction of 2-coumaranones

7.1. General Procedure

Dissolve 10 mg of 2-coumaranone in 100 ml of acetone or DMF. The reaction starts after the addition of one drop of a base like DBU (1,8-Diazabicyclo[5.4.0]undec-7-ene). If some crystals of a suitable fluorescence dye like coumarine-7 are added to this solution, one can observe how the crystals dissolve and the pure blue emission of the 2-coumaranone changes into the green emission of the coumarine-7.

8. CHEMILUMINESCENCE OF PEROXIDES IN EDIBLE FATS AND OILS

Peroxides are our daily companions in life. They occur when fatty foods age and become rancid. What is needed is only the oxygen from air and some time. If these peroxides are heated to a temperature above 150°C, they begin to decompose, releasing energy. This energy can be transferred to an added dye, which leads to its luminescence. Magnesium phthalocyanine has proven its worth as such a dye. It is non-toxic and easy to produce. All fats and oils used in households are suitable as fats in this experiment. By this chemiluminescence one can estimate very well the age and the freshness of fats and oils. The oil from an intensively used frying machine, for example, is particularly suitable as it usually contains a particularly high concentration of peroxides. One can ask a local fast-food store for a sample of their oil (or just squeeze out their most fatty meal) [3].

Figure 7: Chemiluminescence reaction of linolenic acid with oxygen

8.1. General Procedure

Pour some cooking oil into a 100 ml beaker, leave to stand for several days and stir it occasionally. Depending on the quality of the oil, peroxides form more or less quickly. With "high-quality" oils, which contain many polyunsaturated fatty acids, the process occurs much faster than with "inferior" oils with little unsaturated fatty acids. After a few days pour a small amount of degraded oil into a test tube, add a few grains of magnesium phthalocyanine and heat the test tube with a Bunsen burner or a hot air gun. A red emission is easily visible even below the boiling point of the oil.

This experiment suits very well to illustrate the chemical difference between saturated and unsaturated fats. While lard or beef fat form practically no peroxides, linseed oil or olive oil show a clear reaction even shortly after purchase.

9. CHEMILUMINESCENCE OF 1,2-DIOXETANES

The chemiluminescence of 1,2-dioxetanes is particularly bright and impressive. Unfortunately, the starting materials can usually only be produced in well-equipped laboratories with the appropriate specialist personnel. 1,2-Dioxetanes are also not commonly stable for storage, depending on substitution they decompose quite quickly (and sometimes explosively) into two carbonyl fragments, whereby light is also emitted. Substitution with adamantane increases thermal stability to such an extent that the resulting compound can be stored and handled safely. However, in general there is no need to isolate 1,2-dioxetanes in order to see their luminescence. It is often enough to generate them *in situ* in order to be able to observe the emission of light during immediate decay.

9.1. Thermal Decomposition of Adamyntylidenadamantane-Dioxetane

This dioxetane is the most stable dioxetane known. It is not too difficult to produce and is stable in storage.

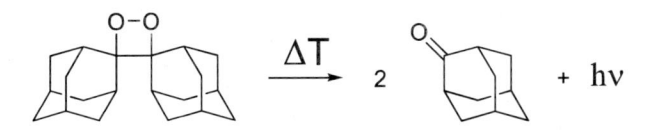

Figure 8: Chemiluminescence reaction of Adamyntylidenadamantane-Dioxetane

9.1.1. General Procedure

Heat in a beaker some cooking fat on a heating plate. Once the fat has been melted, add a spatula of adamantylidenadamantan-dioxetane, extinguish the light and continue heating. At about 150°C the contents of the beaker shine in a blue light.

9.2. *In Situ* Generation of 1,2-Dioxetanes

1,2-Dioxetanes can be produced by reacting electron-rich olefins with photochemically produced singlet oxygen. Some 1,2-dioxetanes are so unstable that they decay at room temperature and emit light. If an olefin containing a suitable emitter is used for this reaction, the light is so bright that it can be easily observed. In this case, red light is irradiated into the reaction mixture (e.g., by a red laser pointer) and blue light is emitted from the reaction mixture. We have called this an "uphill energy conversation" because the emitted light is more energetic than the incident light [7].

Figure 9: In-situ generation of 1,2-dioxetanes

9.2.1. *General Procedure*

A small amount of olefin is dissolved in acetonitrile. A spatula tip of methylene blue is added as a sensitizer dye. The red part of sunlight already triggers the reaction in ambient environments. It is better to work in the dark and to use a red-light source to start the reaction. A laser pointer is particularly suitable. The chemical reaction takes place along the light beam and the channel lights up blue when the laser is switched off.

10. THE TRAUTZ-SCHORGIN-REACTION

This reaction is one of the oldest known chemiluminescence reactions. Formaldehyde is oxidized with hydrogen peroxide in the presence of phenols. The mechanism is very complex, and glycerol or glutaraldehyde can be used instead of formaldehyde. Suitable phenol components are pyrogallol, tannin or natural polyphenols from tea or other plant material.

10.1. General Procedure

For this reaction 4 solutions are needed:

- Solution A: 1 g pyrogallol dissolved in 10 ml water
- Solution B: 5 g potassium carbonate in 10 ml water
- Solution C: 10 ml formaldehyde (35% in water)
- Solution D: 15 ml hydrogen peroxide (30%)

In an Erlenmeyer flask, the previously prepared solutions are combined and mixed. The light is extinguished. The content of the flask heats up strongly and after a short amount of time a beautiful orange emission can be seen. If a spatula tip of luminol is added at the beginning of the reaction, a blue emission after the decay of the orange emission can be seen. The temperature of the solution rises so much that luminol reacts with hydrogen peroxide even without a catalyst.

REFERENCES

[1] Brandl, H. (2010). *Trickkiste Chemie*, Aulis Verlag-

[2] Roesky, H., Möckel, K. (1996). *Chemische Kabinettstücke*, Wiley/VCH-

[3] Brandl, H., Täuscher, E., Weiß, D. (2016). Keine Angst vor Peroxiden. *Chem. Unserer Zeit 50:* 130–139.

[4] Jilani, O., Donahue, T.M., Mitchell, M.O. (2011). A Greener Chemiluminescence Demonstration. *J. Chem. Educ.88:* 786–787.

[5] Weiß, D., Brandl, H., Täuscher, E. (2018). Halogene Licht und Feuer. *CHEMKON* 25: 63–68.

[6]   Ciscato, L. F. M. L., Bartolini, F.H., Colavite, A.S., Weiss, D., Beckert, R., Schramm, S. (2014). Evidence supporting a 1,2-dioxetanone as an intermediate in the benzofuran-2(3H)-one chemiluminescence. *Photochem. Photobiol. Sci.* 13: 32–37.

[7]   Ciscato, L. F. M. L., Weiss, D., Beckert, R., Bastos, E.L., Bartoloni, F.H., Baader, W.J. (2011). Chemiluminescence-based uphill energy conversion. *New J. Chem. 35:* 773–775.

INDEX

U

W

T

Y

P

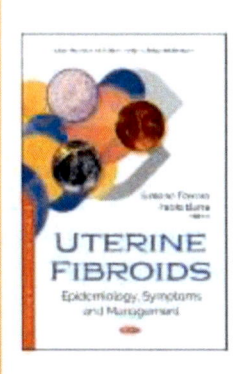